305 CIRCUITS

Elektor Electronics (Publishing)
P.O. Box 1414
Dorchester
England DT2 8YH

British Library Cataloguing in Publication Data
A catalogue record for this book is available from the British Library

ISBN 0 905705 36 X

First published in the United Kingdom 1994
Reprinted 1995

Printed in the Netherlands by Giethoorn/NND, Meppel

Contents

Radio & television

Test & measurement

General interest

WARNING: electricity is dangerous

The projects in this book are, to the best of the Publishers' knowledge and belief, both accurately described and safe. Nevertheless, great care must always be taken when assembling electronic circuits that carry mains voltages; the Publishers can not accept responsibility for any accidents that may occur.

Because electricity is dangerous, its use, application and transmission are subject to rules, regulations and guidance. These are laid down in numerous laws, Generating Board regulations, British Standards and recommendations of the Institution of Electrical Engineers (IEE). Some of these may be obtained from your local electricity showroom, but most, if not all, should be available for reference in your local library.

On/off delay for valve amplifiers

001 This circuit has been designed primarily for use with valve amplifiers. When the amplifier is switched on, filament voltage is supplied first and the anode potential a few minutes later. When there has been no input signal for a while, the anode voltage is switched off again automatically.

When the mains is switched on, a set pulse is supplied to bistable (flip-flop) IC_{3b} via R_{16-C6}. The Q output (pin 13) then goes high and the bistable resets itself via R_5-C_{15}, whereupon relay Re1 is energized via T_3 and the valves are provided with heater voltage. After a delay, dependent on time constant R_{13}-C_4, the potential at the clock input of IC_{3a} reaches a level that causes the bistable to toggle so that its Q output (pin 1) goes high. Relay Re_2 is then energized and switches the high voltage to the valves.

During operation of the amplifier, the inverting input of IC_1 and the non-inverting input of IC_2, both of which circuits are connected as comparator, are provided with a voltage of about 6 V by potential divider R_1-R_2-R_3-R_4. The audio signal from the preamplifier or output amplifier (one channel suffices) is fed to both ICs. The earth of this signal is connected to the potential divider, which means that the supply and earth lines of the amplifier and the delay circuit must be well isolated from one another. When the signal level is about 60 mV or greater, the output of either IC_1 or IC_2 will go high, depending on the polarity of the signal. Transistor T_1 is then switched on via R_6 or R_7, which results in C_3 discharging. When T_1 is off, C_3 is charged slowly via R_{18}. When there has been no signal input for a few minutes, the voltage across C_3 rises to a level where IC_{3b} gets a clock pulse. This results in IC_{3a} being reset, whereupon T_2 is switched off, so that relay Re_2 is deenergized and the high voltage is removed from the valves. Transistor T_1 is provided with base current via R_{11} and D_3, so that the clock input of IC_{3b} remains low. This bistable resets itself almost immediately, however, via R_{15}-C_5. The interval between clock pulse and reset is so short that Re_1 remains energized: heater voltage to the valves is, therefore, maintained.

There are two keys for user operation: S_1 resets IC_{3a} whereupon the high voltage is reapplied to the valves; S_2, when pressed, causes a clock pulse to be generated that switches the amplifier to standby.

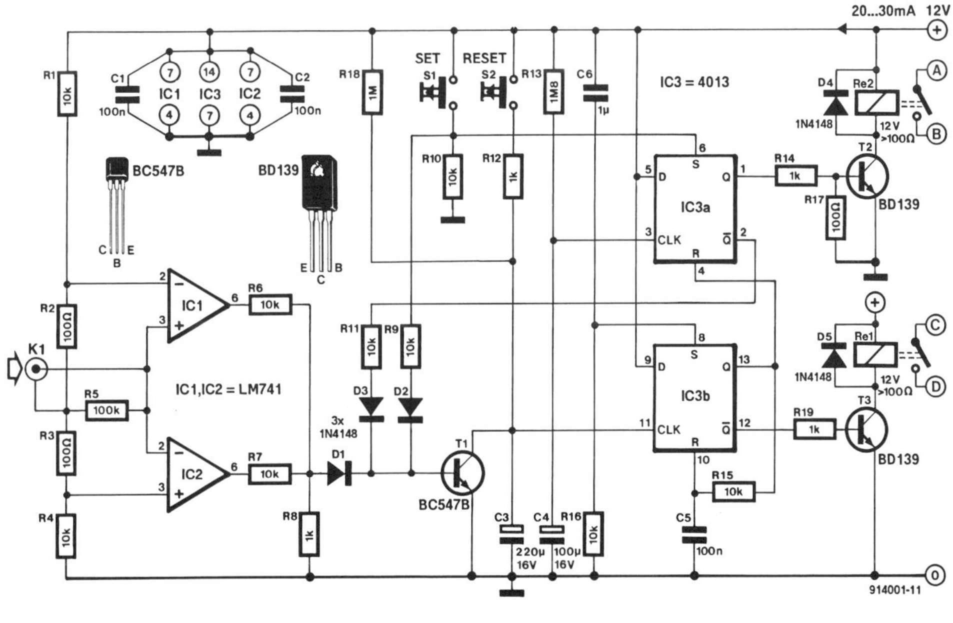

Auto power off for audio equipment

002 This little circuit switches off the equipment in your audio rack when this has not produced sound for some time.

The circuit is actuated by pressing S_1, which causes capacitor C_1 to be charged. Next, the output of opamp IC_{1B} goes high, and the audio equipment is powered from the mains via solid-state relay ISO_1.

The LINE OUT signal from the audio power amplifier is fed to the input of the circuit via connector K_1. Opamp IC_{1A} is set up to function as a signal detector with a trigger threshold of about 50 mV. Note that the ground potential of the audio amplifier is raised to about +4.5 V in the auto-power off circuit by means of R_1-R_2-R_3. When the audio signal is greater than 50 mV (i.e., 4.05 V with respect to the circuit ground), the output of IC_{1A} goes high, and transistor T_1 starts to conduct. Consequently, C_1 is charged rapidly, so that ISO_1 continues to conduct and power the equipment.

In the absence of an audio input signal, C_1 is discharged slowly via R_5 and R_6. Opamp IC_{1B} toggles, and the equipment is switched off via ISO_1 when the capacitor voltage drops below the voltage set with P_1 at the inverting input. It should be noted that the solid-state relay specified here has a maximum current rating of 1.5 A. When heavier loads are to be switched, it is recommended to use a conventional relay.

Since the relay outputs and the transformer primary are connected to the mains, great care should be taken to ensure the required electrical insulation. For reasons of safety, the circuit is best fitted in a mains adapter enclosure with a moulded mains plug. The two mains connections in the enclosure must be made with properly rated and secured screw terminal blocks. The output is connected to a mains cable with a 4- or 5-way distribution board.

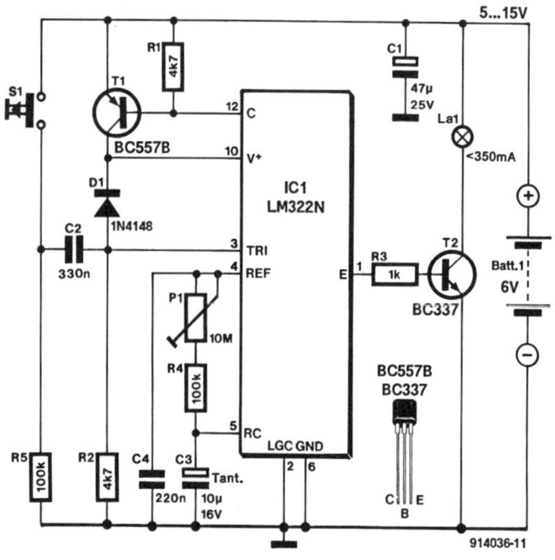

The delay before the equipment is switched off will depend on the time needed to rewind a tape, change a compact disc or record, etc. To adjust the delay, connect a 100 kΩ resistor across R_5. This reduces the actual delay by a factor of about 10. Turn P_1 fully in the direction of R_7, press the START button, and wait for the desired delay (divided by 10) to elapse, whereupon P_1 must be adjusted until the output of IC_{1B} goes high. Then remove the 100 kΩ resistor, press START again and time the actual delay. If necessary, re-adjust the preset.

Digital (tape) counter

003 Since there are still tape recorders about that have no mechanical tape counter, the circuit shown here offers an excellent, electronic add-on for these. It can, of course, also be used to replace a mechanical counter.Furthermore, it can be used for other applications, for instance, as hoist-height indicator for a model building crane, or for indicating the position of a chisel on a lathe.

The input of the circuit is formed by two optoisolators. The sequence of the pulse signals provided by these isolators depends on the direction into which a coding disc turns. The NAND gates following the isolators produce from those signals an up pulse or a down pulse, which enables up/down counter IC_3 to register the position irrespective of the direction of rotation. That position is made visible via a decoder on a seven-segment display. The number of digits the counter provides can be extended by adding more counter/decoder/display stages to the extension terminals, PC_1–PC_4, in the same way that IC_5, IC_6, LD_2 are connected to IC_3.

The optical input signal is provided by a coding disc that is sub-divided into a number of alternate reflecting and non-reflecting segments. The two optoisolators are positioned above the disc in such a way that when one is directly above a segment, the other is exactly above the line dividing two segments. It is, of course, possible to use a light barrier and a coding disc that has alternate transparent and opaque segments. Two LEDs and a pair of phototransistors can also be used.

The power supply is a 5 V regulated type that can deliver 250 mA. For each additional counter stage, 100 mA should be added to that figure.

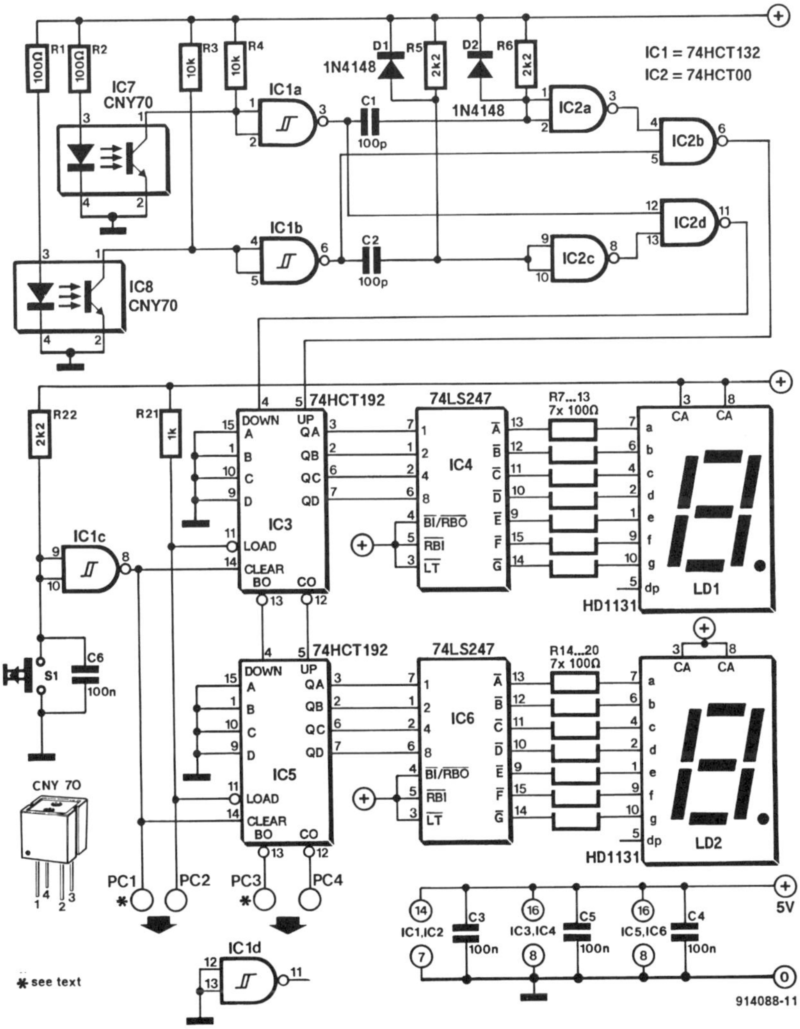

Amplification/attenuation selector

004 A Type TL081 opamp and some passive components are sufficient to construct a small amplifier whose amplification can be varied between +1 and −1 with a potentiometer – see the diagram.

The input signal is applied to both inputs of the opamp: to the inverting input via C_1 and R_1, and to the non-inverting input via C_1 and P_1. The amplification of the amplifier is R_2:R_1=1. However, the level of the signal at the +input is determined by the position of the wiper of P_1. When the wiper is at the centre of its travel, the two input signals cancel each other, so that there is no output. When the wiper is at the 'high' end of the potentiometer, the signal at the +input is larger than that at the inverting input and this is then available, amplified by 1, at the output. When the wiper is at earth potential, the opamp functions as a normal inverting amplifier with unity gain.

The input impedance of the circuit is about 50 kΩ. With a value of C_1 as shown, the amplifier can handle frequencies from 30 Hz upwards.

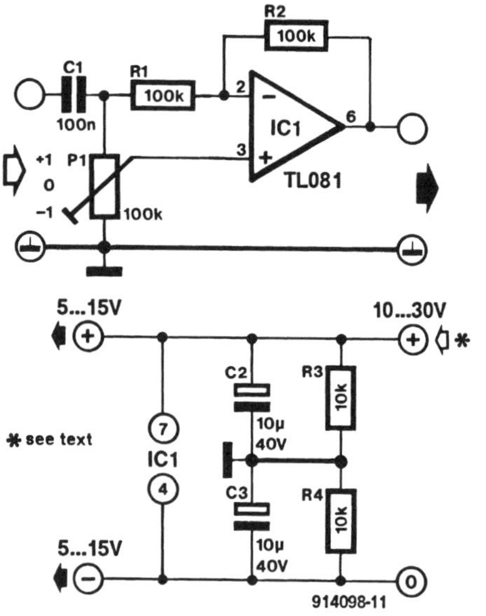

The circuit requires a power supply of ±5–15 V and draws a current of only a few mA. If such a supply is not available, it may be produced from a single 10–30 V supply as shown in the diagram.

Protection against direct voltage

005 Amplifiers that have no capacitor at their output may, in case of a defect, apply a direct voltage to the loudspeakers and this can destroy the drive units. The circuit shown can prevent such a catastrophe.

It is best to give the circuit a separate power supply: this minimizes any work on the amplifier(s). This supply must, however, be switched synchronously with that to the amplifier(s), since on power-on, T_1 ensures that the relay (which switches the loudspeaker inputs) is energized after some delay. The delay is determined by the time-constant R_3C_3.

Assuming that the amplifier operates correctly, its output signal at point b is linked to point a, and thence to the loudspeaker(s), via the relay contact. Owing to the

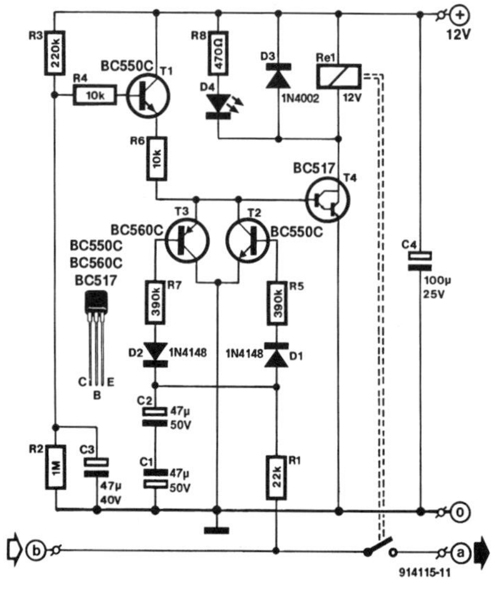

time constant R_3–C_3, bipolar capacitor C_1-C_2 cannot be charged by the a.c. signal at point b. If, however, a defect causes a direct voltage at that point, the capacitors will be charged via R_2. Depending on the polarity of the direct voltage, either T_2 or T_3 will be switched on, which removes the base current from T_4 and this results in the deenergizing of the relay: the amplifier output is then removed from the loudspeaker(s).

The supply to the protection circuit must be neither regulated nor smoothed. True, C_4 provides some smoothing, but the important thing is that after the amplifier has been switched off, this capacitor is discharged more rapidly than the smoothing capacitors in the amplifier power supply. This ensures that the relay is deenergized before the amplifier can produce a click in the loudspeaker(s). Depending on the relay, the current drawn by the circuit is about 50 mA.

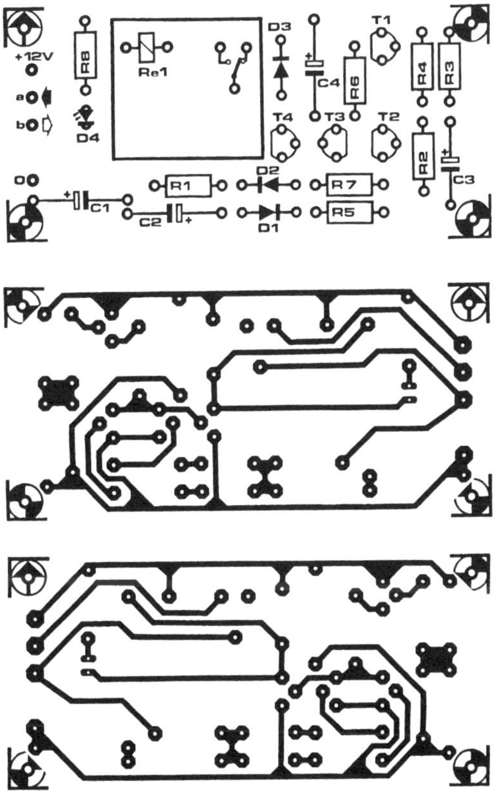

Parts list

Resistors:
R_1 = 22 kΩ
R_2 = 1 MΩ
R_3 = 220 kΩ
R_4, R6 = 10 kΩ
R_5, R7 = 390 kΩ
R_8 = 470 Ω

Capacitors:
C_1, C_2 = 47 µF, 50 V
C_3 = 47 µF, 40 V
C_4 = 100 µF, 25 V

Semiconductors:
D_1, D_2 = 1N4148
D_3 = 1N4001
D_4 = LED
T_1, T_2 = BC550C
T_3 = BC560C
T_4 = BC517

Miscellaneous:
Re_1 = 12 V relay,
e.g. Siemens V23217-A0002-A101

Overload indicator

006 The overload indicator consists of a window comparator that measures the magnitude of an a.f. signal. Two of the opamps contained in an TL072 are supplied with a reference voltage by potential divider R_1–R_2–R_3–P_1. The outputs of the opamps drive T_1 via diodes D_1 and D_2 (that function as half-wave rectifier), which in turn actuates D_3. Network R_5–R_6–C_2 ensures that the LED lights even during short signal peaks. Capacitor C_2 is charged fairly rapidly via D_1 (or D_2) and R_5,

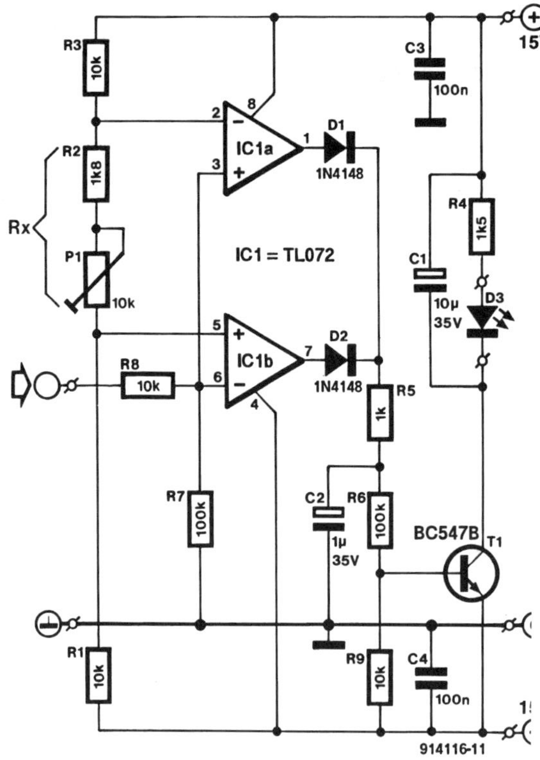

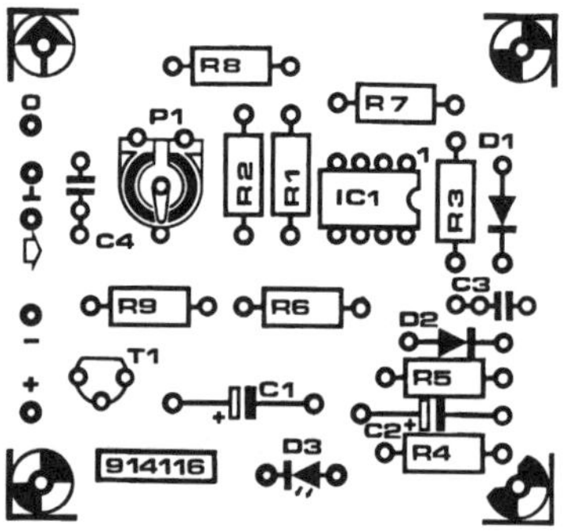

after which it discharges slowly via R_6, R_9 and the base-emitter junction of T_1. Capacitor C_1 also contributes to the longer lighting of the LED.

When the level of the signal at the input is high enough, IC_{1a} is toggled by the positive half periods of the signal and IC_{1b} by the negative halves. In this way, a peak above the maximum level will be indicated even when the signal is asymmetrical.

Because of the symmetrical power supply and design of the indicator, the reference voltage for both opamps can be set with one potentiometer.

The circuit draws a current of 5–6 mA when the LED is off. When an overload peak is indicated, the LED draws an additional 20 mA. With values as shown, the reference voltage can be set roughly between 0.9 V and 5.5 V.

The circuit can be connected to the output of a power amplifier, but potential divider R_7–R_8 then needs to be adapted and protected by diodes to the supply lines.

Parts list

Resistors:
R_1, R_3, R_8, R_9 = 10 kΩ
R_2 = 1.8 kΩ
R_4 = 1.5 kΩ
R_5 = 1 kΩ
R_6, R7 = 100 kΩ
P_1 = 10 kΩ preset

Capacitors:
C_1 = 10 µF, 35 V
C_2 = 1 µF, 35 V
C_3, C_4 = 100 nF

Semiconductors:
D_1, D_2 = 1N4148
D_3 = LED, red
T_1 = BC547B

Integrated circuits:
IC_1 = TL072 or TL082

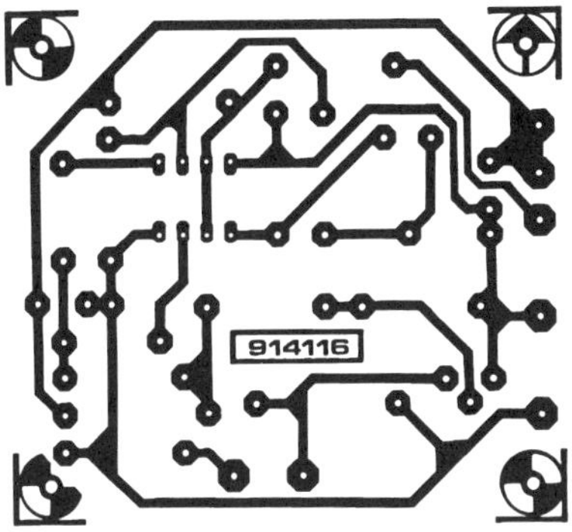

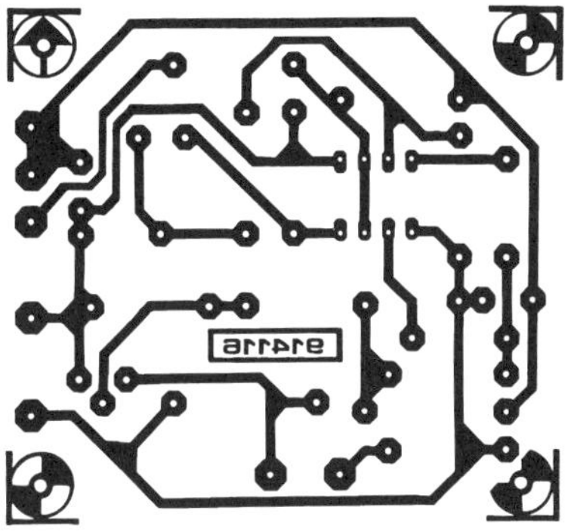

Power-on delay

007 The power supply for this analogue circuit, which affords delays of up to 330 seconds, is taken directly from the mains. The direct voltage at the output of the bridge rectifier is held at 22 V by zener diode D_5. Resistor R_6, which enables C_1 to discharge rapidly as soon as the mains is switched off, must be rated at 250Va.c. or 400Vd.c.

The delay is provided primarily by C_4, which is charged via C_3, whose impedance at 50 Hz is about 10 MΩ, and half-wave rectifier D_6–D_7. After a given period, the potential across C_4 will be 12 V higher than the source voltage of T_1, which is set with P_1. The gate of T_1 has the same potential as C_4. Network R_2–C_5 serves to suppress any spurious voltage peaks.

When the potential across C_4 becomes higher than the source voltage of T_1, the FET begins to conduct and this will result in T_2 being switched on. Moreover, the voltage across the relay is fed back to the gate of T_1 via D_8. This feedback ensures

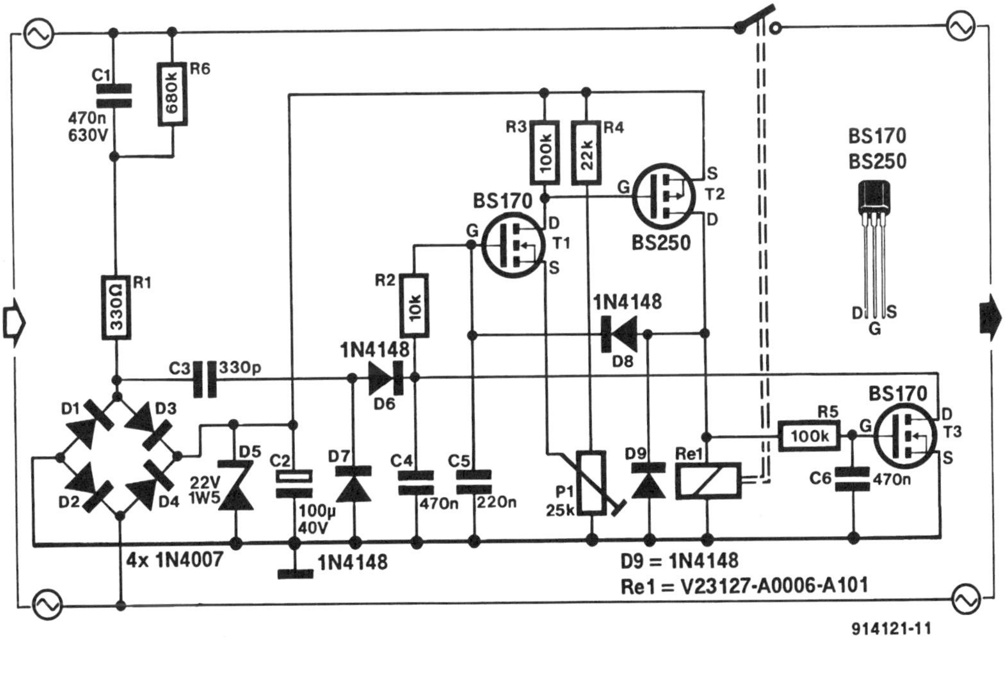

that T_1 and T_2 are quickly driven into saturation.
Once the relay has been energized, transistor T3 will be switched on via R_5–C_6. When this happens, C_4 will be discharged through the transistor, so that the circuit is quickly back in its initial state. The delay on power-up is, therefore, not shortened by the residual charge on C_4. In spite of C_4 being discharged, the relay remains actuated because the gate voltage of T_1 is held steady via D_8. Only when the supply voltage is switched off will the relay be deenergized.
Note that the circuit is connected electrically to the mains so that great caution should be observed during any testing and operation.

Electronic power-on delay

008 Power for the delay circuit is derived directly from the mains by a bridge rectifier, D_1–D_4. The rectifier must be able to handle a current of up to 1 A. For safety reasons, resistor R1 ensures that capacitor C_1 is discharged rapidly when the mains is switched off. The resistor should, therefore, be rated at 250 V a.c. or 400 V d.c.
The direct voltage output of the rectifier is held steady at 24 V by zener diodes D6 and D_7. The use of two zener diodes gives a stable voltage of 12 V for IC_1. The rectifier output is smoothed by C_2 and C_3.
Timing of the circuit is based on the mains frequency: the clock signal is rectified by D5 and then taken from junctionR_2–R_3 at a level of about 11 V from where it is applied to pin 10 of IC_1. After 2–10 clock pulses, output Q11 goes high. If contact 2 of DIP switch S_1 is closed, the high signal (12 V) is applied to the gate of T_1, an n-channel VMOS FET. This transistor then conducts, which causes the relay to be energized.
If, as shown in the diagram, all contacts of S_1 are open, the gate of T_1 is permanently connected to the positive supply rail. This means that as soon as the supply is switched on, the relay is energized.
When T_1 is on, the clock input of IC_1 is low via D_9, so that the counter is disabled.
The delays that can be selected with the

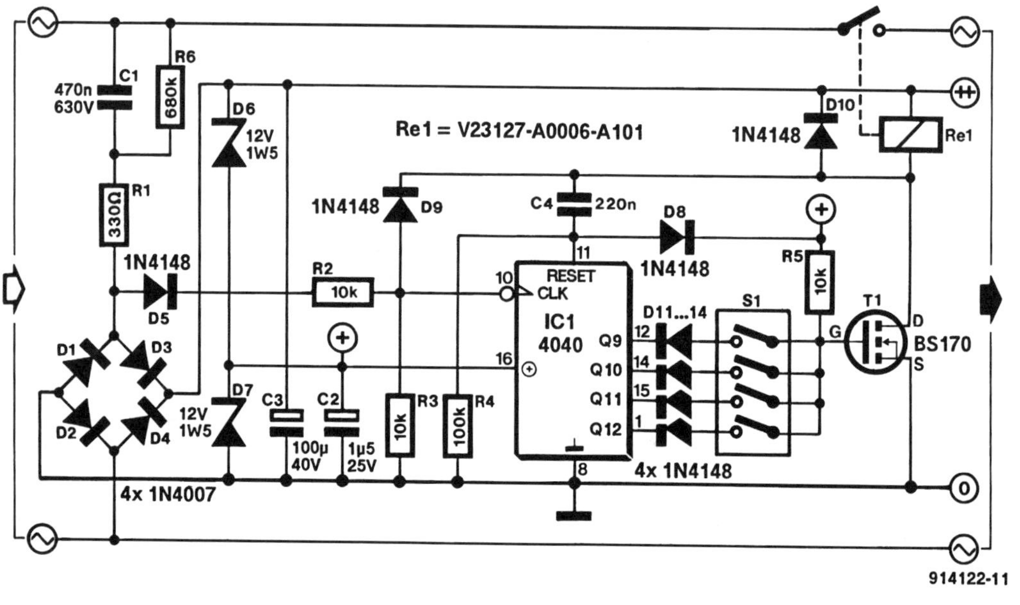

DIP switch are shown in the table. The DIP switch may be replaced by a binary-coded decimal (thumb-wheel) switch.
Immediately after the supply is switched on, assuming that T_1 does not begin to conduct, the IC is reset by a pulse generated by R_4-C_5. The level of the pulse is limited to a safe value by D_8. When a new count cycle is started, the IC is, therefore, always in its zero state.
Diode D_{10} is a free-wheeling device to suppress voltage peaks that are caused by the relay coil when the current through it is switched off.
The current drawn by the circuit is determined mainly by the value of C_1: in the diagram, it amounts to about 30 mA.
CAUTION must be exercised because the circuit is connected directly to the mains. The circuit must, therefore, be tested with the aid of an isolating transformer. Furthermore, it should be built into an enclosure that makes it impossible for mains-carrying parts to be touched.

Speech/sound memory

009 OKI's Series MSM6372–6375 enable the reproduction of speech or other sound stored in their internal ROM. The capacity of the ROM lies between 128 kbit and 1024 kbit, depending on the type. The type and the sampling frequency specified (4 kHz, 6.4 kHz or 8 kHz) determine the length of the stored speech, which is 4–64 seconds. This time may be divided into 111 words that can be addressed individually.
Each IC contains a 12-bit digital-to-analogue converter (DAC) and a fourth-order low-pass filter. The customer can specify the words that must be stored on the mask-programmable ROM. As an example, the Type MSM6374-007 is programmed to tell the time in English.
Based on this IC, the circuit shown in Fig. 1 is intended to be connected to the I^{2}C I/O card (see Ref. 1) of which the two converters have been omitted. The card then serves as interface between the I^{2}C bus and IC_2.
The wanted word is selected via inputs 10–16. Since the MSM6374-007 contains only words at addresses where 13 is zero, the corresponding input is strapped to earth. This arrangement leaves two of the eight available I/O bits for starting and timing of the words. These bits are available at st (start input) and NAR (next address request output). Briefly, the control is

- wait until NAR is high;
- key in the address;
- wait not less than 10 seconds;
- briefly render ST low (pulse duration 0.35–350 µs).

Normally the nar signal indicates that the next address may be keyed in before the entire word has been spoken. This arrangement provides a smooth transition between words or parts of words.
To simplify the control of the synthesizer, the start pulse is not generated by software, but by monoflop IC_{1a}. This stage is triggered by both the first and last transition of the start signal, which is applied to the trigger inputs via differentiating networks R_5-C_7 and R_7-C_8. The timing diagram in Fig. 2 shows what happens. Every time the computer writes data to IC_2, the software inverts bit P6. It then takes 30 µs before the onset of the start signal, which is 250 µs long. In this way, the control computer can start the 'utterance' of a word in one write operation. Without IC_{1a}, the computer would have to write the address first, then, after 10 µs, the start signal, and finally end the start signal.
Resistors R_2, R_3, and capacitor C_3, set the oscillator frequency to 64 kHz, resulting in a sampling frequency of 6.4 kHz.
Resistor R_4 and capacitor C_5 provide a power-up reset.
Capacitor C_4, which forms part of the output filter, provides an improvement of the signal-to-noise ratio.
A second channel enables the IC to give speech with echo, two tones, or a a tone with three different volumes. This facility cannot be used in this application owing to

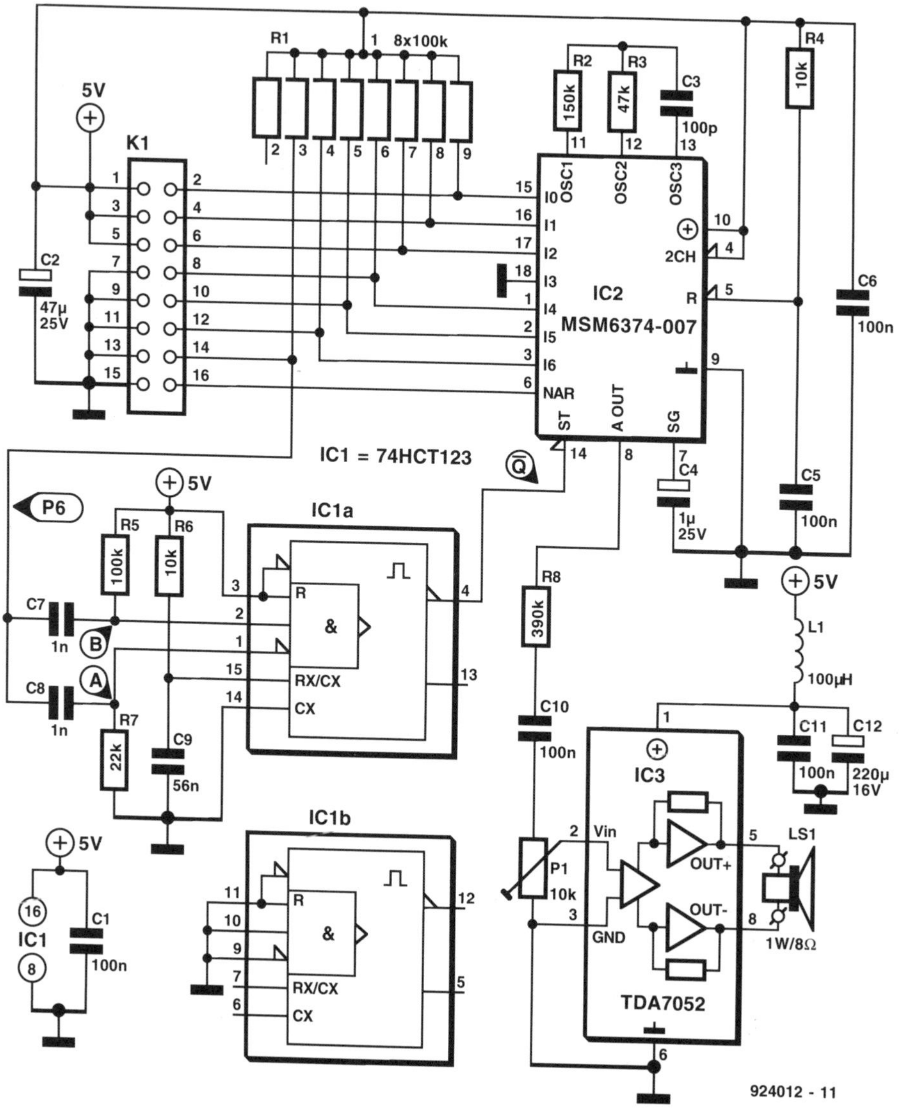

924012 - 11

lack of I/O bits. Input 2CH is, therefore, disabled by strapping it to the positive supply line.

The output signal is raised to about 1 W into 8 Ω by integrated bridge amplifier IC3. The circuit may be controlled via the computer-to-I²C interface by software that, once installed, may be enabled by a suitable key combination. It is possible for either the time to be spoken or an alarm time to be set. The use of this software presupposes that the I²C driver (Type 1671*) has been installed. When Program 1773* is run, it installs itself, after which it is enabled (even if another program is being

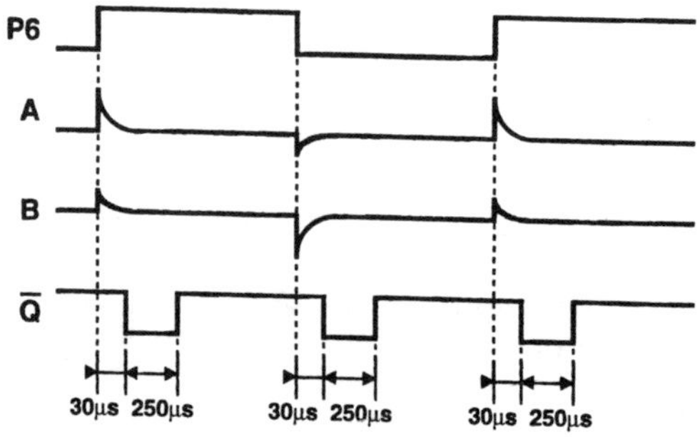

924012 - 12

run) by the simultaneous pressing of keys CTRL and F1. If that combination cannot be used, for instance, because it is used by another program, another combination may be chosen by loading the program with TALKTIME/H. You will then be asked to key in the alternative combination. Note that in the comination the program can recognize only the lefthand shift key.
When the combination has been keyed in, a self-evident menu appears after about a second, provided the screen is not in the graphics mode. In that case, a high tone is emitted, whereupon only functions 'tell time' and 'alarm on/off' are available.
The circuit draws a current of not more than 300 mA.

* Available from Elektor Electronics (Publishing)

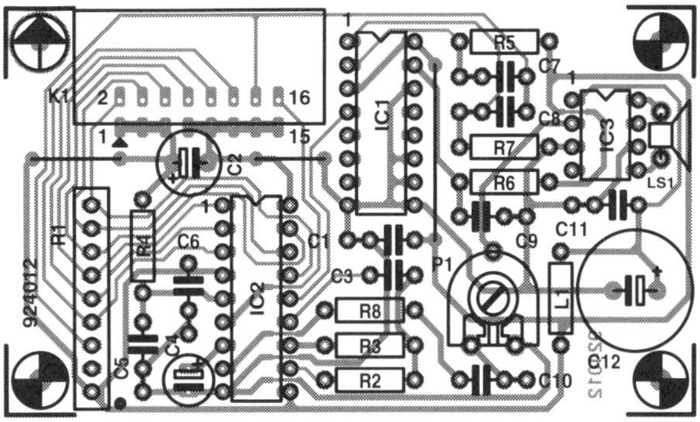

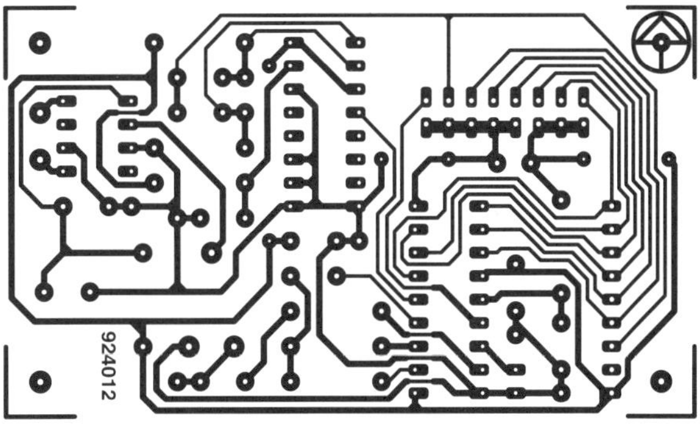

Stereo protector against d.c.

010 If a d.c. coupled output amplifier breaks down during operation, the loudspeakers, particularly the bass units, are at risk. The bass particularly so because it is not decoupled for d.c. by the capacitors in the cross-over network. If, for instance, the output transistor has given up the ghost, the bass units will get the full d.c. supply voltage at their terminals.
A suitable circuit to protect the loudspeakers in such an eventuality, and at the same time obviate the annoying 'plops' on switch-on, is shown in the diagram. Note that it operates from an unregulated, unsymmetrical power supply. Normally, it may be powered directly from the power supply of the output amplifier.
The a.c. components of the signals in the output stage is bypassed by R_1 and the two anti-series connected capacitors, C_2 and C_3. The signal at the junction R_1-R_2 is, therefore, the d.c. component of the loudspeaker signal. From there it is applied to potential divider R_2-R_3 and then to window comparator IC_{1a} and IC_{1b}. Since the supply voltage is fixed at 10 V by R_{13}-D_7, the window height is fixed at 2 V by R_5. In other words, u_2 = 6 V and u_3 = 4 V. In the absence of d.c. at the output of the power amplifier, u_1 = 5 V. In this situation, the outputs of 'OR gates' D_1 and D_2 is logic high.
When the d.c. component at the output of the power amplifier is greater than ±2 V, u_1 is greater or smaller than either u_2 or u_3. The output of one of the opamps will then be logic low.
When the power amplifier is switched on in step with the present circuit and u_1 lies within the window, C_4 is charged via R_8. After about 1.5 s, 'Schmitt trigger' IC_{1d} changes state and its output becomes logic high. The relay is then energized and connects the loudspeaker to the power amplifier: no 'plop'.

If a defect occurs, or if the direct voltage at the output of the power amplifier rises, C_4 is discharged via R_7 within 50 ms. The output of IC_{1d} then goes low, the relay is de-energized and the loudspeaker is disconnected from the output amplifier.

Resistor R_{13} and the operating voltage of the relay must be suitable for the supply voltage. If that voltage is 20–40 V, a good value and rating for R_{13} is 4.7 kΩ, 1 W; while for 12–20 V, 1 kΩ, $^1/_4$ W is right. If the supply voltage is, say, 36 V, the operating voltage of the relay should be 24 V. The difference of 12 V should be dropped across a suitable resistor. If, for instance, the relay draws 15 mA, R_{16} should have a value and rating of 820 Ω, $^1/_4$ W.

If there is likely to be a requirement for switching off the circuit, S_1 should be incorporated. When that switch is closed, the relay is energized.

For a circuit suitable for a stereo power amplifier, only components R_1–R_3, C_2, C_3, D_1, D_2, D_5, D_6, IC_{1a} and IC_{1b} need to be duplicated. The additional circuit is connected in parallel with S_1. Note that the relay should then have two working contacts or two relays with their contacts in series should be used.

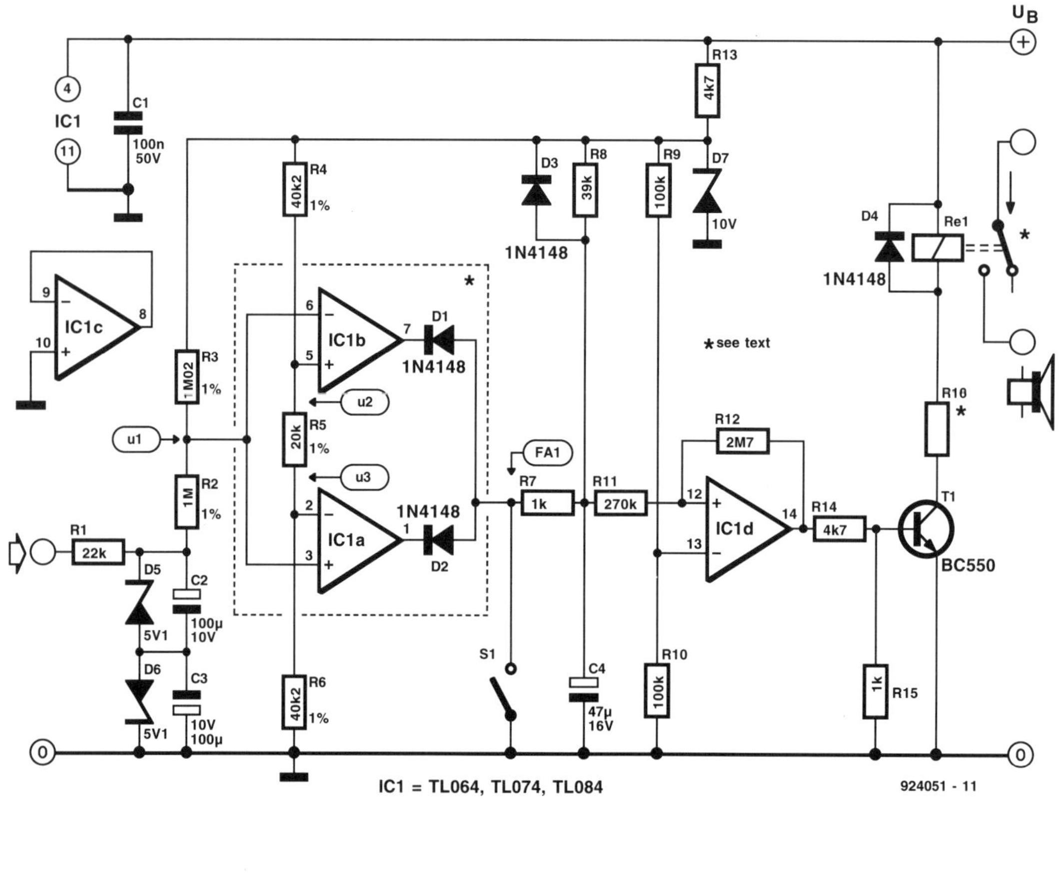

Scanner for preamplifier

011 The scanner is an extension for the 'All-solid-state preamplifier' published in *Elektor Electronics**. As its name implies, it scans all inputs of the preamplifier to ascertain where there is an audio signal present. That input remains selected. After there has been no signal for some time, scanning is resumed.
The scanning action is provided by rectangular-wave generator IC_{2c}.The output of this oscillator is applied to one of the input selector keys via buffer/inverter IC_{2d} and diode D_3. The diode prevents the key being disabled when the oscillator is off. The oscillator is switched on and off by IC_{2b}, which in turn is controlled by IC_{1b}. That opamp is configured as a comparator whose voltage threshold is preset with P_1.
The inputs of the scanner are linked to the audio inputs on the volume control board of the preamplifier. The relevant signal is amplified ×40 by IC_{1c} and IC_{1d}, after which summing amplifier IC_{1a} combines the left-hand and right-hand signals. As soon as there is music or speech on the input lines, C_1 will partly discharge rapidly. When the voltage across the capacitor drops below the level set by P_1, IC_{1b} disables the oscillator via IC_{2b} and the input selected at that instant remains actuated. As long as there is a signal coming in, part of the charge on C_1 will ebb away via IC_{1a}. If there is no signal for some time (presettable with P_2 between 3 s and 25 s), the capacitor will be charged almost completely via P_{23}, R_8 and D_1. Once the terminal voltage of C_1 rises above the comparator threshold, the oscillator is enabled again and the scanning action resumes.
A total scan of all inputs is completed in 3 s (determined by R_{11}-C_2).
The input sensitivity can be set between 10 mV and 4 V with P_1.
The current drawn by the scanner is not greater than 10 mA.

* December 1989/January 1990

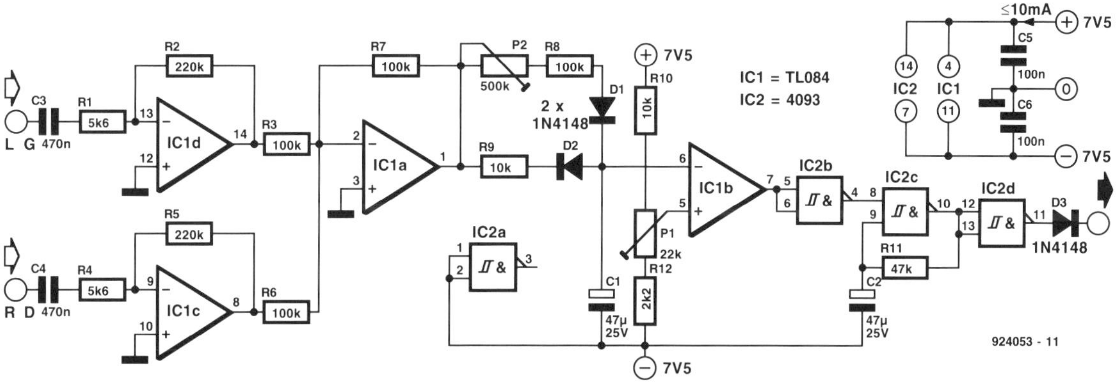

40 W output amplifier

012 Although there are a number of hybrid output modules on the market, very few of them combine compactness with reasonable price and good performance. One of these few is SGS's used in the present amplifier.
The design of the amplifier is straightforward: a power opamp followed by two output transistors.
The audio signal is applied to the non-inverting input of power opamp IC_1 via socket K_1 and capacitor C_1. The supply

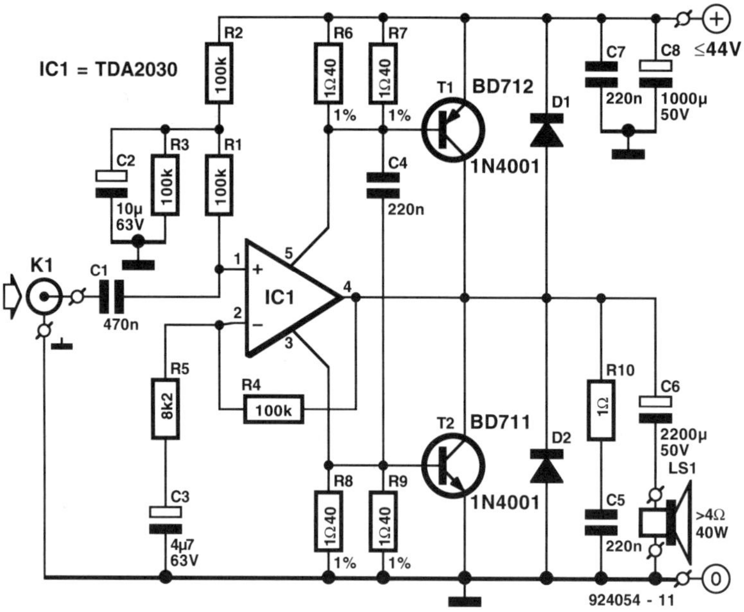

current to the IC varies in accordance with the input signal. Consequently, there will be a similarly varying voltage drop across resistors R_6, R_7, R_8, and R_9 since these are in the supply lines to the opamp.

As long as the current is lower than about 1 A, the voltage drop across the resistors will be insufficient to switch on transistors T_1 and T_2. This means that outputs up to 2 W into 4 Ω are provided entirely by the opamp. Once the output current exceeds a level of 1 A, the transistors are on and contribute to the power output.

When the input signal is small, there is no quiescent current through the transistor, but there is through the opamp. Crossover problems are thus obviated. Since the IC also provides thermal compensation, stability of the operating point is ensured.

The supply voltage may lie between 12 V and an absolute maximum of 44 V.

Construction of the amplifier on the printed-circuit board should be straightforward. The transistors as well as the IC must be fitted insulated on to a heat sink of about 2 k W^{-1}. Use plenty of heat conducting paste.

The supply line should be protected by a 3.15 A fuse.

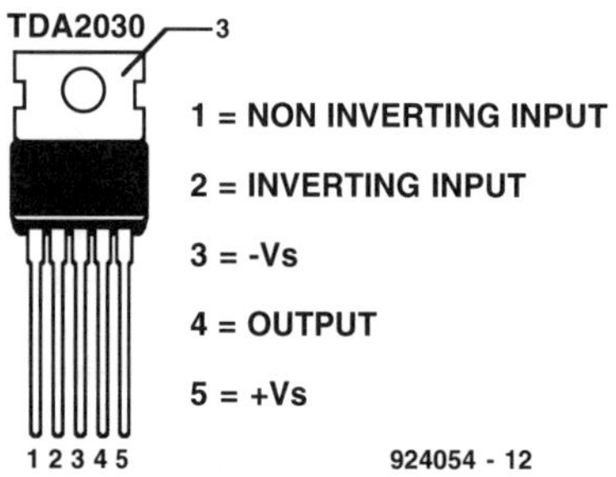

Technical data

Supply voltage	44 V
Maximum output	22 W into 8 Ω
(for THD=0.1%)	40 W into 4 Ω
Harmonic distortion	
1 kHz/8 Ω/11 W	0.012%
1 kHz/4 Ω/20 W	0.032%
20 kHz/8 Ω/11 W	0.074%
20 kHz/4 Ω/20 W	0.2%
1 kHz/8 Ω/1 W	0.038%
1 kHz/4 Ω/1 W	0.044%

Quiescent current	about 38 mA	
Efficiency	8 Ω	62.5%
(maximum load)	4 Ω	64%

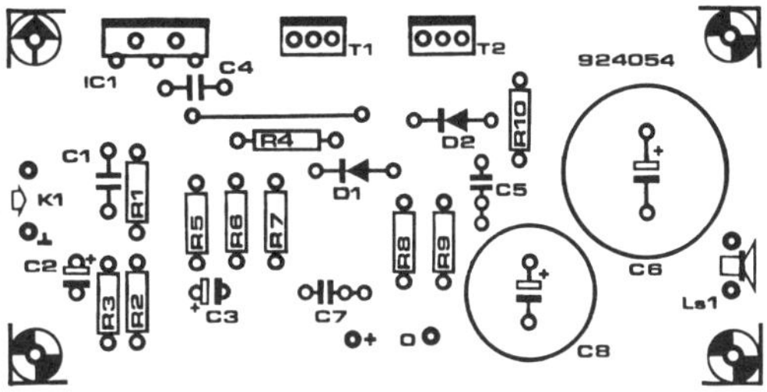

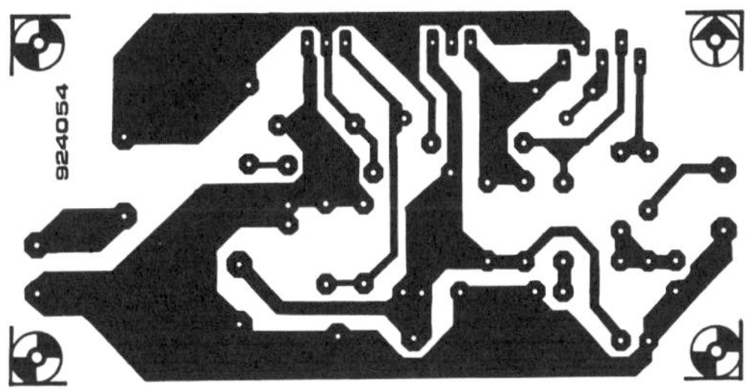

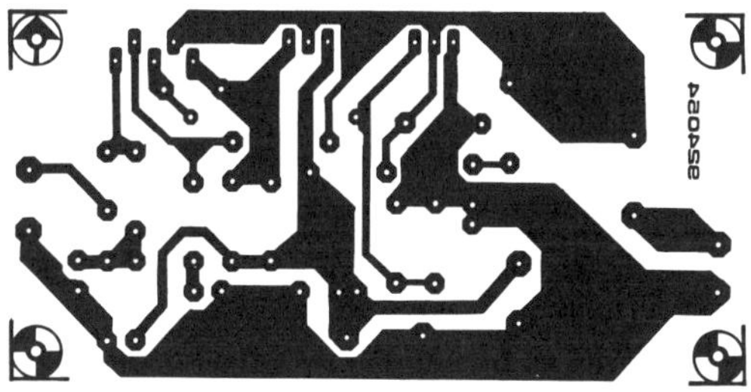

Parts list

Resistors:
R_1–R_4 = 100 kΩ
R_5 = 8.2 kΩ
R_6–R_9 = 1.4 Ω, 1%
R_{10} = 1 Ω

Capacitors:
C_1 = 470 nF
C_2 = 10 μF, 63 V, radial
C_3 = 4.7 μF, 63 V, radial
C_4, C_5, C_7 = 220 nF
C_6 = 2200 μF, 50 V, radial
C_8 = 100 μF, 50 V, radial

Semiconductors:
D_1, D_2 = 1N4001
T_1 = BD712
T_2 = BD711

Integrated circuits
IC_1 = TDA2030

Miscellaneous:
K_1 = audio socket
Heat sink 2 K W^{-1}
Insulating washers, etc, for IC_1, T_1, T_2

Pulse generator for AV recorders

013 AV recorders used in audio-video presentations contain an additional head to read and write the control pulses for the slide projector. These pulses are normally used to actuate a relay that operates the projector. The pulse data are usually written on one of the audio tracks at the non-used side of the tape. Since the extra head in modern recorders does not make use of the two standard audio tracks, it becomes possible to have stereo sound for the presentation. This does mean, however, that only one side of the cassette can be used.

The pulse generator is a complete circuit – see diagram – for utilizing the extra head (L_1) of an AV recorder. The pulses are written on the tape with the aid of S_1, while S_2 is used to select on/off, record, and play. When S_2 is set to its centre position (play), the pulse signal is applied to amplifier T_5-T_6 via C_5. At a certain signal strength the level at the collector of T_6 goes high, whereupon T_7 is switched on via C_{10} and S_{2d}. The relay is then actuated; C_{11} ensures that it does not clatter.
With S_2 in position 3, 6 (record), T_3 is connected to amplifier T_5-T_6. As long as S_1 is

not operated, T_2 is on and T_3 is off. However, the stage based on T_1 causes T_2 to switch off for a short time. During that time, T_3 is on and amplifier T_5-T_6 oscillates owing to the feedback via R_5. Since the emitter of T_3 is linked to L_1, the oscillatory signal is written on the tape. During the remainder of the time, T_2, and thus T_4, is on and the tape is erased.
When S_1 is pressed, T_1 conducts for a period determined by the time-constant R_4-C_1 (here 100 ms). Because of this, C_2 is discharged rapidly, and T_2 is switched off. As soon as T_1 is off again, C_2 is charged via R_5. After about one second, the potential across C_2 has risen to a level at which T_2 begins to conduct again. Thus, pressing S_1 causes a one-second pulse to be written on the tape. Because the collector of T_2 is connected to T_7 via R_{16} and D_3, the relay is actuated when S_1 is pressed, irrespective of whether S_2 is in position play or record.
If for a given projector the pulse duration is too long, it may be shortened by rducing the value of C_2.
The ITT AV recorder shown in the photograph contains a circuit similar to the one described here.

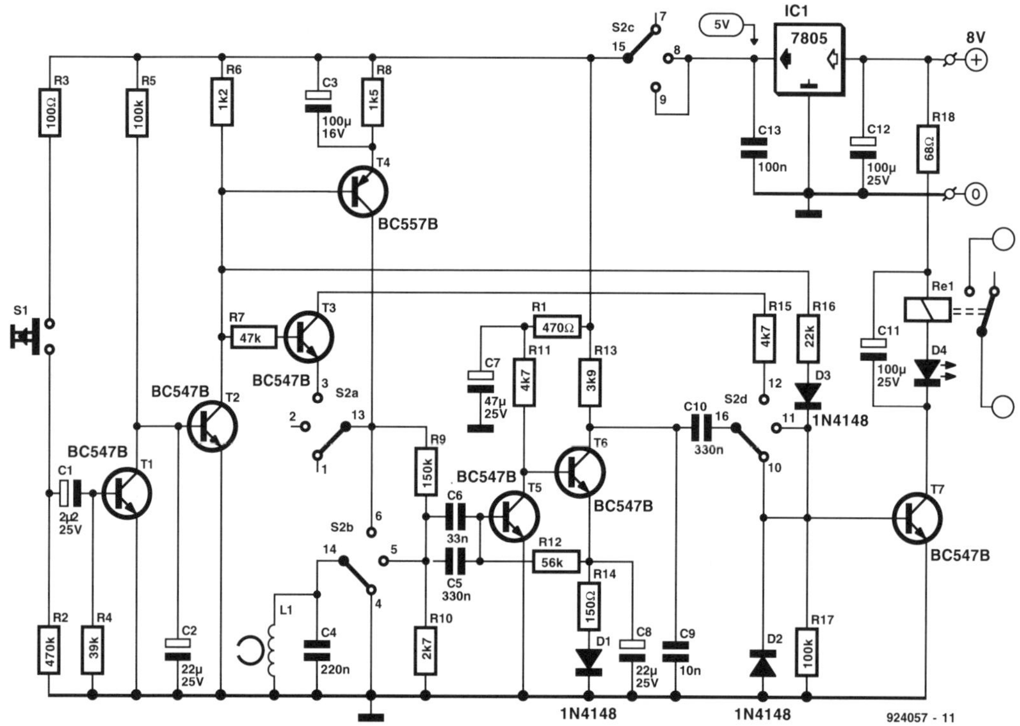

Decibel stepper

014 The stepper presented here makes use of nine positions of a 12-position rotary switch. It is, of course, possible to add the three steps. As usual, the amplification factors are arrived at by connecting a number of resistors in series in the feedback circuit. The resistors used here are from the E-96 series, which enable a fairly accurate approximation of the wanted amplification.

If greater precision is required, combinations of two resistors may be used. This has the drawback, however, that a make-before-break switch must be used to prevent the output constantly being switched to the supply line.

The bandwidth of the amplifier is determined by the set gain and the gain-bandwidth product of the opamp. If an LF351 is used as shown, the gain-bandwidth product is 4 MHz, while the slew rate is 13 V µs^{-1}.

The circuit draws a current not exceeding 2 mA.

Capacitor C_1 improves the stability when the amplification factor is large.

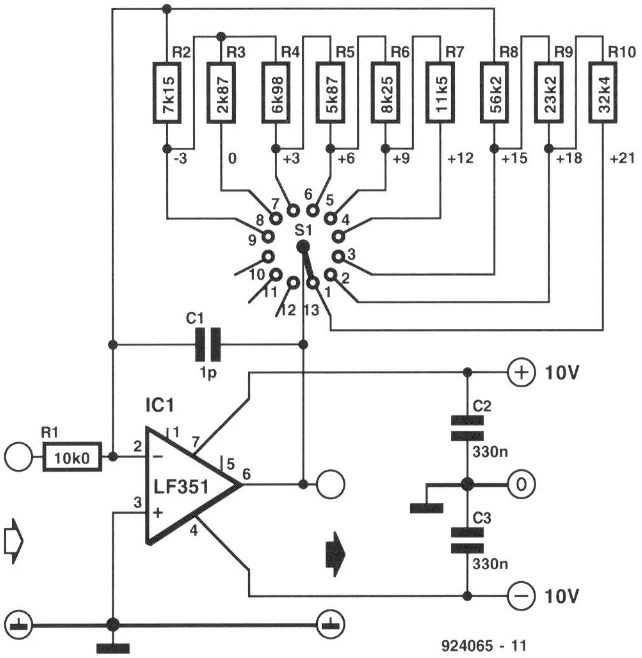

Digital volume control

015 Two ladder networks and a buffer each form a volume control with a range of 63 dB. Network R_3–R_{17} provides fine control in steps of 1 dB, whereas network R_{20}–R_{34} provides coarse control in steps of 8 dB. The desired attenuation is set with the aid of multiplexers IC_1 and IC_3, each of which is controlled via three digital inputs. The design is such that the binary code on the six-bit wide overall control input accords with the set attenuation. Resistor R_1 ensures that C_1 can discharge, even if K_1 is open-circuited. This resistor and network R_3–R_{17} form an input impedance of 46.3 kΩ. The resistor also determines the maximum permissible input voltage. That voltage depends, in the first instance, on the supply voltage to IC_1 and IC_3 (±8.2 V). Resistor R_1 plus R_3–R_{17} attenuate the input signal ×2.4 (7.6 dB). This means that the maximum input level must not exceed 20 V_{peak}, that is, 14 V r.m.s. That means also that IC_{2a} must not amplify to prevent too high an input to the second ladder network and buffer.

The amplification of the two opamps is determined by R_{18}-R_{19} and R_{35}-R_{36} respectively. As stated before, that of IC_{1a} should be unity, in which case R_{18}=0 Ω and R_{19} is omitted. If the amplification of IC_{2b} is also unity, the overall control range is –7.6 dB to –70.6 dB. To obtain a control range of 0–63 dB (when the binary code on the control inputs accords with the actual attenuation), IC_{2b} should provide an amplification of ×2.4.

The current drawn by the circuit is determined primarily by the dual opamp and amounts to about 10 mA.

The overall distortion is <0.003% over the range 20 Hz to 20 kHz and an input signal of 1 V.

The control has one small drawback: when the volume is set, weak clicks occur (which are typical of all normal CMOS switches). That makes it less suitable for super-de-luxe applications, although many listeners will not even notice the clicks. And, in any case, the volume is not varied constantly.

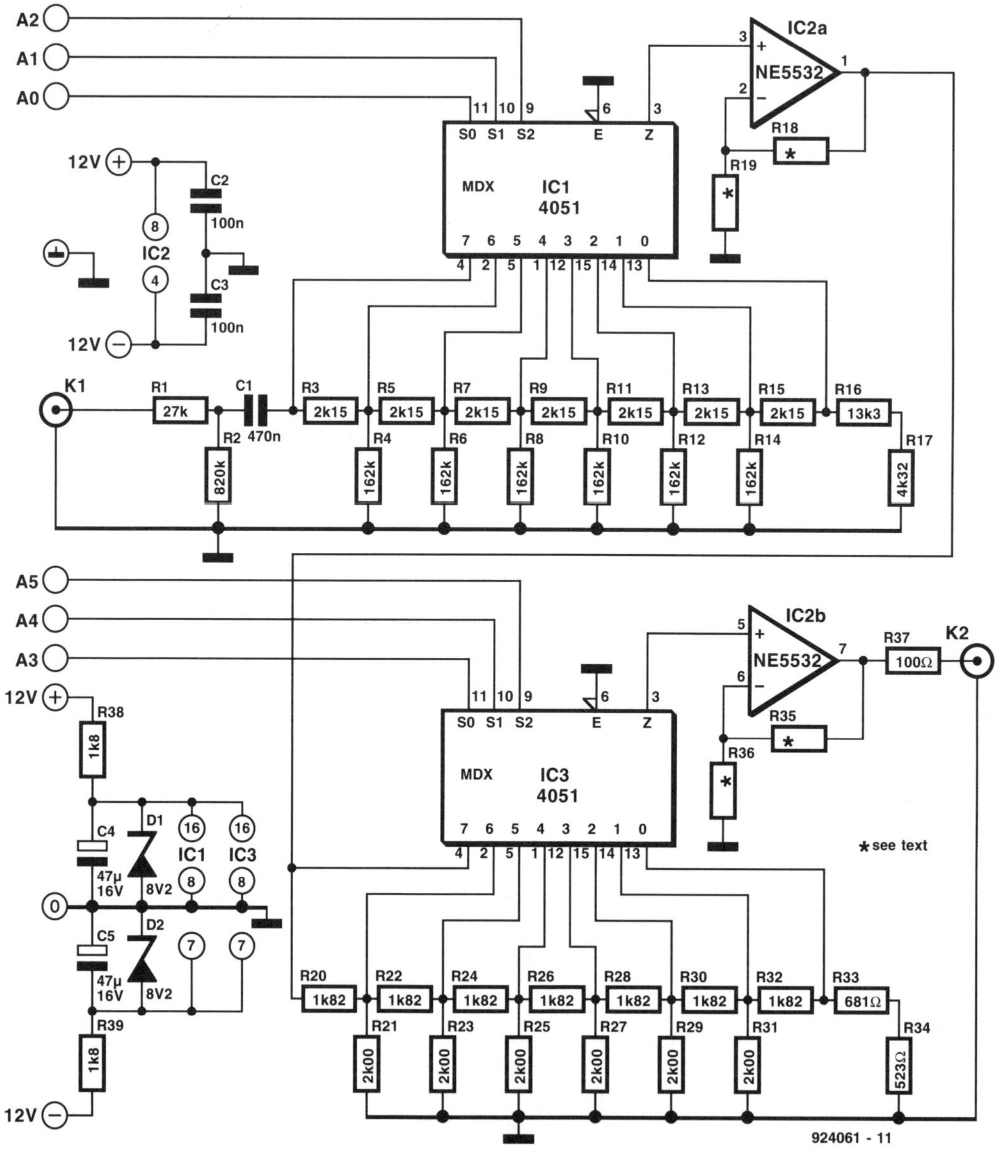

60 watt music amplifier

016 This is a robust, no-frills medium power amplifier that is particularly suited to use in 'combo' type portable amplifiers used by guitar players and jazz musicians. The amplifier is a straightforward combination of an integrated audio driver IC, the LM391-80, and a push-pull power output stage designed with bipolar transistors.
A few peculiarities of the design will be discussed. The NTC, which is in thermal contact with the power output transistors, enables the LM391 to switch off the power stage when this gets too hot. The onset point of this thermal protection lies at an NTC current of about 200 µA. The electrolytic capacitor that shunts the NTC serves to provide a 'soft start', which prevents a loud click or other disconcerting noises when the amplifier is switched on. It may happen that the protection is too sensitive, in which case some experimenting with the value of R_4, or that of the NTC, is in order.
It is possible to implement feedback in the amplifier by connecting R_{23} to series network C_5-R_7. The latter parts, together with R_{10}, determine the frequency response of the amplifier, which may need adjusting to meet individual requirements. The component values given here will, however, be all right for most applications.
The effect of different values of C_5 and R_7 is simple to measure (or hear) by shorting out R_{23} temporarily. For 4-Ω loudspeakers, R_{23} is lowered to 0.18 Ω. Unfortunately, the LM391-80 is prone to oscillation, which is suppressed by components R_x, C_6, C_8 and C_9 (in most cases, C_6 may be omitted). Resistor R_x in particular reduces the open-loop gain. If R_x is used, R_y must be fitted to compensate the resulting off-set voltage. Components R_{22} and C_{12} form a Boucherot network that serves to stabilize the amplifier at high frequencies.
The input of the amplifier should be driven by a low-impedance source capable of supplying 'line' level audio signals (0 dB). Network R_1-C_1 attenuates signals above 50 kHz or so.
The quiescent current of the amplifier is set by preset P_1. Set this control to 0 Ω ini-

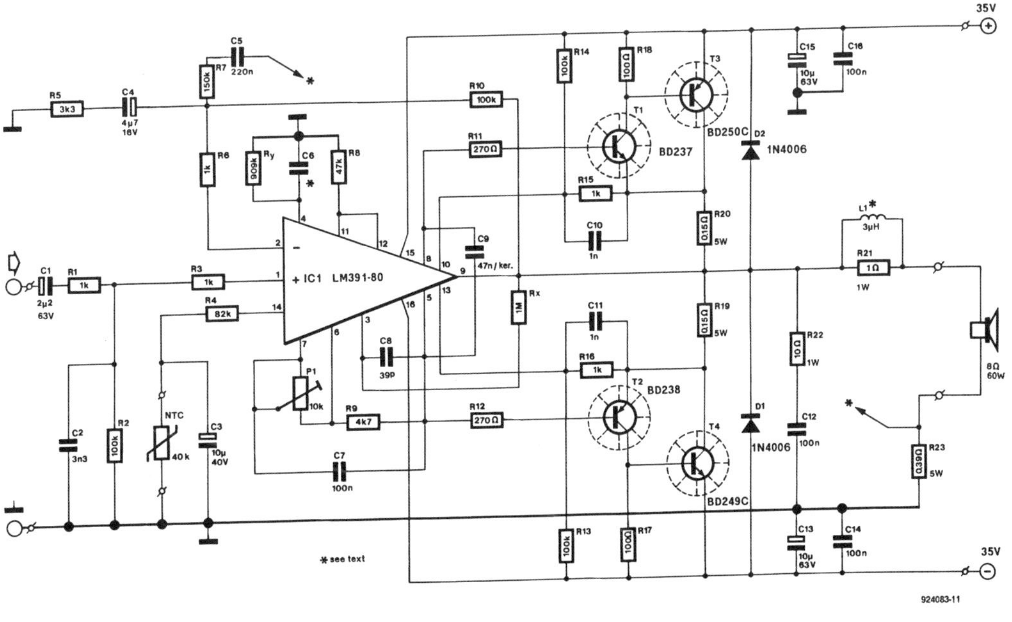

tially, and adjust it until a quiescent current of 50 mA flows. This may be increased to 400 mA if low distortion is desired.

The power transistors are all located at the same side of the PCB so that they can be bolted on to a common heatsink, together with the NTC. The heat sink should be fairly large and have a thermal resistance of 1 K/W or smaller.

Note that L_1 consists of 20 turns of 0.8 mm dia. enamelled copper wire wound around R_{21}. C_9 is a ceramic capacitor.

Finally, some measured data (supply voltage: ±35 V; R_{23} short-circuited):

- 3-dB bandwidth (8 Ω): approx. 11 Hz to 20 kHz
- THD (third harmonic distortion) at 1 kHz:
  1 W into 8 Ω: 0.006% ($I_q \approx$ 400mA)
  1 W into 8 Ω: 0.02% ($I_q \approx$ 50 mA)
  65 W into 8 Ω: 0.02% (U_{in} = 873 mV)
  80 W into 4 Ω: 0.2% ($U_{in} \approx$ 700 mV; onset level of current limit)

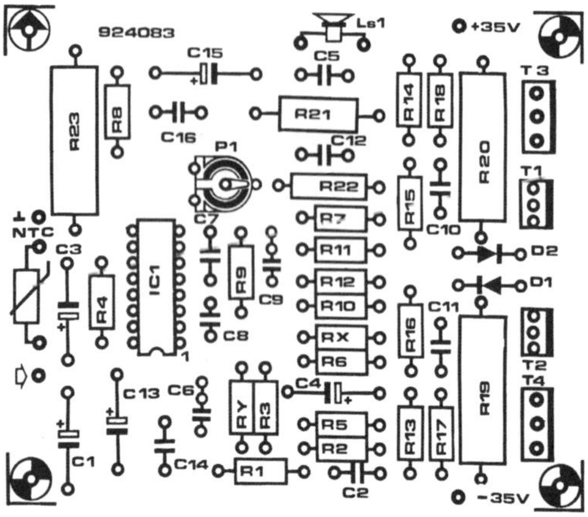

Parts list

Resistors:

R_1;R_3;R_6;R_{15};R_{16} = 1 kΩ
R_2;R_{10};R_{13};R_{14} = 100 kΩ
R_4 = 82 kΩ
R_5 = 3.3 kΩ
R_7 = 150 kΩ
R_8 = 47 kΩ
R_9 = 4.7 kΩ
R_{11};R_{12} = 270 Ω
R_{17};R_{18} = 100 Ω
R_{19};R_{20} = 0.15 Ω/5 W
R_{21} = 1 Ω/1 W
R_{22} = 10 Ω/1 W
R_{23} = 0.39 Ω/5 W
Rx = 1 MΩ (see text)
Ry = 909 kΩ (see text)
P_1 = 10 kΩ preset H

Capacitors:

C_1 = 2.2 μF, 63V
C_2 = 3.3 nF
C_3 = 10 μF, 40 V
C_4 = 4.7 μF, 16 V
C_5 = 220 nF
C_6 = not fitted (see text)
C_7;C_{12};C_{14};C_{16} = 100 nF
C_8 = 39 pF
C_9 = 47 nF, ceramic
C_{10};C_{11} = 1 nF
C_{13};C_{15} = 10 μF, 63 V

Semiconductors:

D_1;D_2 = 1N4006
T_1 = BD237
T_2 = BD238
T_3 = BD250C
T_4 = BD249C

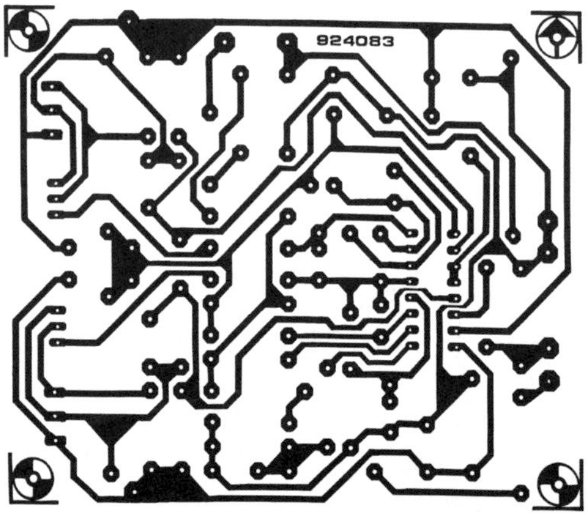

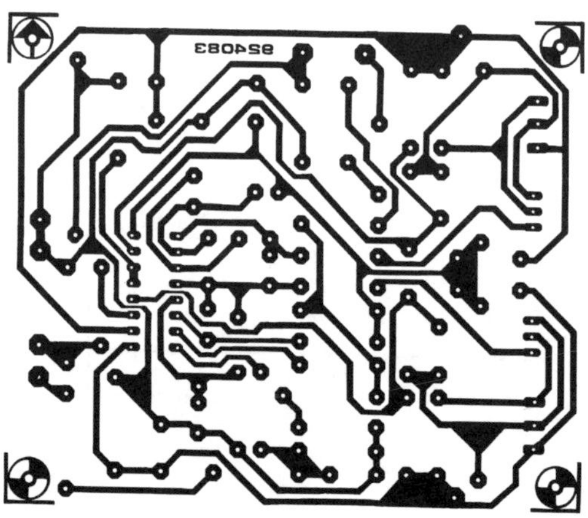

Integrated circuits:
IC_1 = LM391-80

Miscellaneous:
L_1 = see text
NTC = 40 kΩ stud type
Heatsink 1 K W^{-1}

Low-noise amplifier I

017 Opamp Type SSM2017 from PMI is eminently suitable for designing very-low-noise microphone pre-amplifier. According to the manufacturer's data sheet, the input noise is only 750 pV $Hz^{1/2}$. The amplification ($\alpha = 1 + 10^4/R_1$) may be set to a value of 1–1000 with a single resistor, R_1.

The internal compensation of the IC is such that stability is retained even at unity amplification, yet, when the amplification is ×1000, the bandwidth is 300 kHz.

The circuit in the diagram has a symmetrical input and a standard input impedance of 600 Ω. The amplification is ×1001, so that a microphone signal is converted directly into a line level. If the output of the amplifier is conveyed via a long cable (capacitive load), it is advisable to insert a 47–100 Ω resistor in series with it to obviate any likelihood of oscillation. Also, at high amplification the output may have an offset voltage, which, in certain applications may give rise to problems. If that is so, an output capacitor should be used.

The maximum output voltage of the IC is 9 V r.m.s. when the supply voltage is ±15 V. At a 1 kHz output at a level of 1 V r.m.s., the harmonic distortion does not exceed 0.005%. When the supply voltage is lower than ±9 V, the distortion rises to 0.01%. The noise factor of the prototype (including the noise contribution of R_1) was measured at 1 nV $Hz^{1/2}$. The current drawn by the amplifier from a ±15 V supply is about 11 mA.

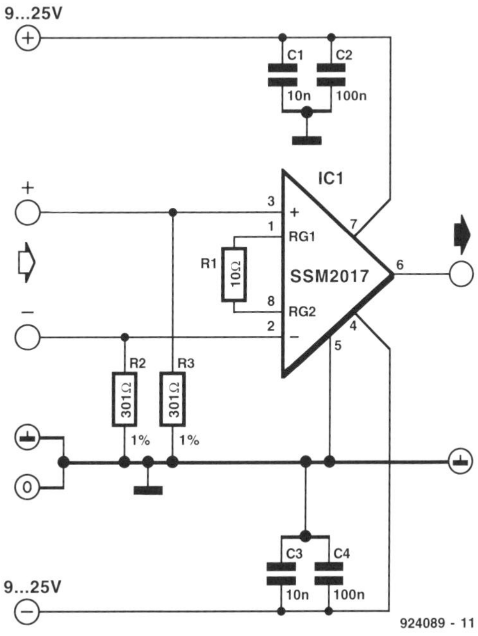

Low noise amplifier II

018 One of the available choices in the design of a low-noise amplifier is the shunting of several input stages. This reduces the overall noise, $u_n = u_i n^{-1/2}$, where u_n is the total noise potential of n stages; u_i is the noise voltage of one stage; and n is the number of stages. This is possible because noise is a randomly composed signal. Thus, the noise signals of a number of stages at any one moment are likely to have a diferent frequency and phase, so that they partly neutralize one another.

In the present amplifier, three low-noise opamps, IC_1–IC_3, are connected in parallel. According to the manufacturers' data sheet, the noise of a single LT1028 amounts to 0.9 nV Hz^{-1}. To this must be added the thermal noise generated by resistors R_2–R_{11}. Circuit IC_4 sums and amplifies the output signals of IC_1–IC_3.

Measurements on the prototype show a total noise of 0.67 nV Hz^{-1}. According to the earlier formula, the three opamps have an overall noise of 0.52 nV Hz^{-1}. The difference of 0.15 nV between this and the measured figure is caused by the resistors. This is a low figure bearing in mind that a 1 Ω resistor at room temperature generates

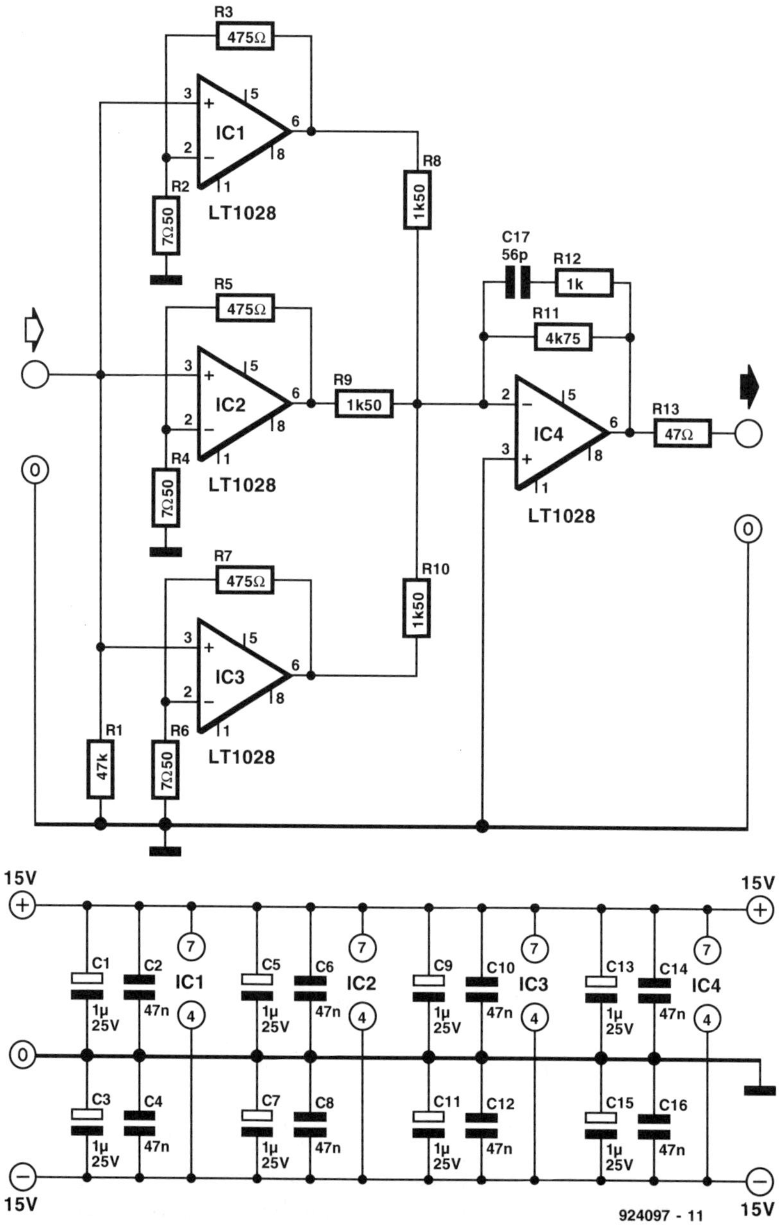

a thermal noise of 0.13 nV Hz^{-1}.

The amplification, α, of the circuit is computed from

$$\alpha = -n(1+R_3/R_2)\cdot R_{11}/R_8.$$

It is necessary that resistors R_3, R_5, and R_7; R_2, R_4, and R_6; and R_8, R_9, and R_{10}, have identical values. With values as shown, the circuit has an amplification of ×600.

Apart from having a low noise output, the Type LT1028 opamp is also fast: it has a slew rate of 15 V μs^{-1} and a bandwidth of 75 MHz for $\alpha = -1$. Even when the amplification is ×63, the bandwidth of the circuit as a whole, but without R_{12} and C_{17}, is 1.2 MHz. However, to avoid signal overshoot, the bandwidth is limited to 500 kHz by R_{12}–C_{17}, which is more than adequate for even the most demanding audio application. The THD+noise ratio at a 1 kHz output at a level of 1 V is only 0.008%.

Experimenters should bear in mind that the stability of the LT1028 is internally compensated for amplifications of >×2. Since resistors R_2, R_4, and R_6 contribute most to the overall noise, it is necessary that their value is kept as low as possible. Naturally, all resistor used should be metal film types.

With a supply voltage of ±15 V, each IC draws a current of about 7.5 mA.

Compressor/limiter

019 The compressor is based on two series-connected attenuator networks, whose attenuation is obtained by means of light-dependent resistors (LDRs) that are illuminated by light-emitting diodes (LEDs).

The input signal is applied to the non-inverting input of opamp IC_{1a} via R_1. In con-

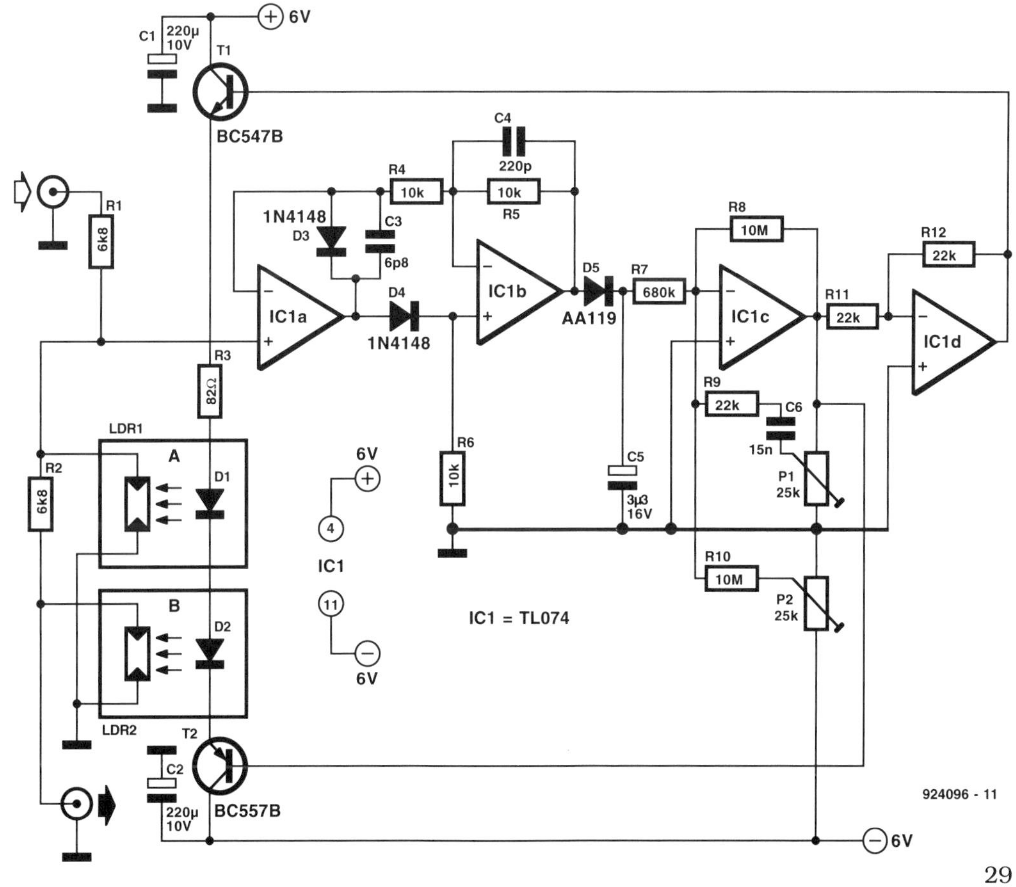

junction with D_3, D_4 and IC_{1b}, IC_{1a} provides full-wave rectification of the signal. The resulting direct voltage is used to charge C_5 via D_5. The diode allows fast charging of the capacitor, which can discharge only via R_7.
Compression proper is provided by IC_{1c}. Depending on the setting of P_1 and P_2, the output voltage of IC_{1c} drops when the potential across C_5 reaches a certain value. This causes T_2 and, via IC_{1d}, T_1 to conduct, whereupon the LEDs light and the input signal is attenuated.
The attack time of the circuit is determined by the speed of the LDRs and the setting of P_2.
The amplification of IC_{1c} and the setting of P_1 determine the point at which voltage limiting commences. The output voltage is held constant when the input signal is above a certain level until the current through the LEDs reaches a maximum (about 40 mA).
The circuit as shown acts as a limiter; if the +ve input of IC_{1a} is connected to the output of the circuit, a standard compressor is obtained.
The circuit is able to process signals between about 10 mV r.m.s. and 2 V r.m.s. This range can be extended by adding one or several attenuator sections or by increasing the value of R_1 and R_2.
The LDR/LED combinations must be housed in a light-tight enclosure.
The current draws by the circuit is determined largely by the LEDs and is 50 mA maximum.

Copying with AV recorders

020 Cassette recorders like ITT's SL537 AV or SL837 AV, used by tape-slide producers, have a four-track head. A pulse circuit is connected to one of these tracks. Pulses recorded on to the tape can be used to actuate a relay to control the slide projector.
The cassette that contains both the sound and the control pulses must be handled carefully and for that reason it is wise to make a copy of it for normal use. Making such a copy is not so easy, however, because the pulses need to be copied precisely. When two AV recorders are available, making a copy becomes appreciably easier, provided that a small modification can be made to one of them.
The ITT recorders have a remote control input socket that is in parallel with a 'to projector' socket. Both sockets are 6-way DIN types. If only the relay contacts in the recorder are used to send control pulses to the projector, the 'tact sw' can be linked to one of the unused pins of the remote control socket (pins 1, 4, 5 and 6).
When remote control is used, it needs to be ascertained whether there are free pins available. In the prototype, pins 4 and 5 are used—see Fig. 1a. If you cannot be certain, it is safer to bring out the two relay contacts via a 3.5 mm jack socket.

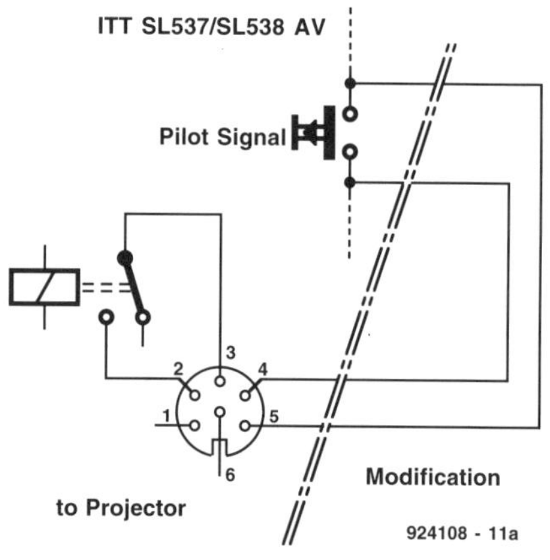

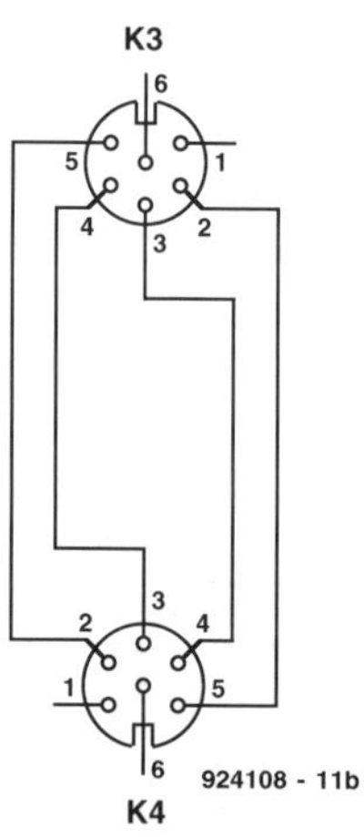

Once this work is done, pulses can be recorded on the tape from a second recorder with pulse output. But first a short cable has to be made to connect the contacts on the second recorder (pins 2 and 3 of the 'to projector' socket) to the modified socket on the first recorder In the case of two ITT recorders, this cable is shown schematically in Fig. 1b.
Furthermore, for copying a second short (record/playback) cable as shown in Fig. 1c is needed. This cable should be matched to the input levels of the recorders.
With the modified recorder set to record (audio as well as AV), the second recorder plays back the original cassette. The modified recorder will then make a perfect copy of the original tape.
The cable connections in Fig. 1 have been chosen so that two modified recorders can copy back and forth ad infinitum.

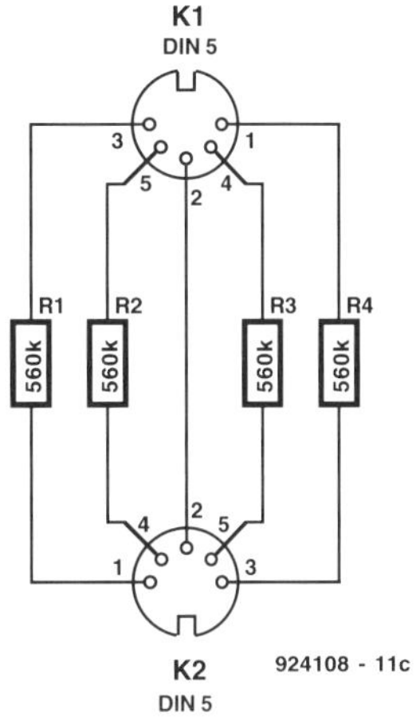

Bandpass filter with extra feedback

021 An important property of a bandpass filter is its shape factor, which is the ratio of its bandwidth at high attenuation to that at low attenuation. The smaller the shape factor, the better the filter.
Another important parameter is the group delay time, which determines how fast pulse-shaped and sinusoidal signals can traverse the filter. Within the pass band, the delay time should be constant to ensure the faithful transfer of the signal.
These parameters often do not show up very well in the usual active filter with local feedback for each second-order section (multiple feedback filter).

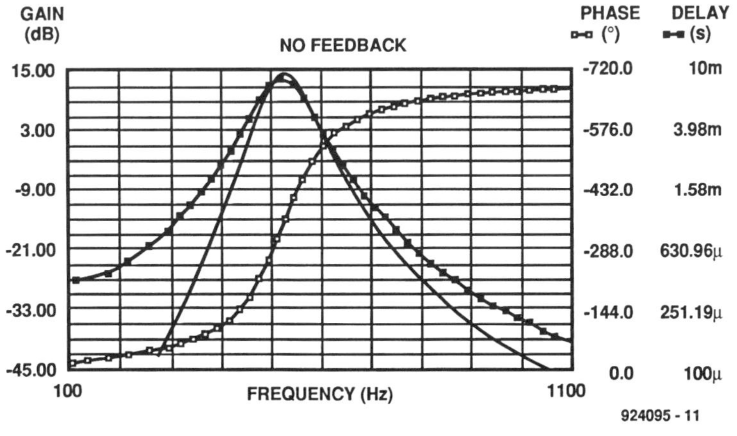

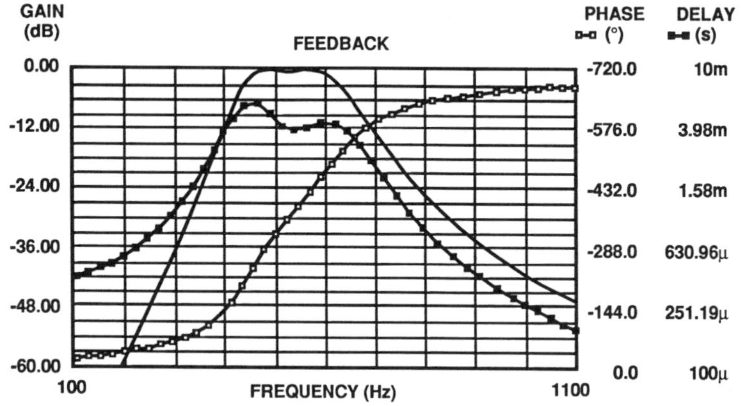

924095 - 12

The magic word in electronics is 'feedback'. The graphs in Fig. 1 and 2, pertaining to the circuit in Fig. 3, show the importance of feedback. Figure 1 gives the frequency vs gain response if R_{14} and R_{16} are omitted (no feedback). The bandwidth at –3 dB is 50 Hz and that at –40 dB is 450 Hz. That gives a shape factor of 9.
With the extra feedback provided by R_{14} and R_{16}, the filter has a bandwidth of 200 Hz at –3 dB and of 660 Hz at –40 dB. That gives a shape factor of 3.3.
Moreover, the group delay in the pass band is constant within acceptable limits.
This shows that with the extra feedback the filter gives a better performance than in its traditional set-up.

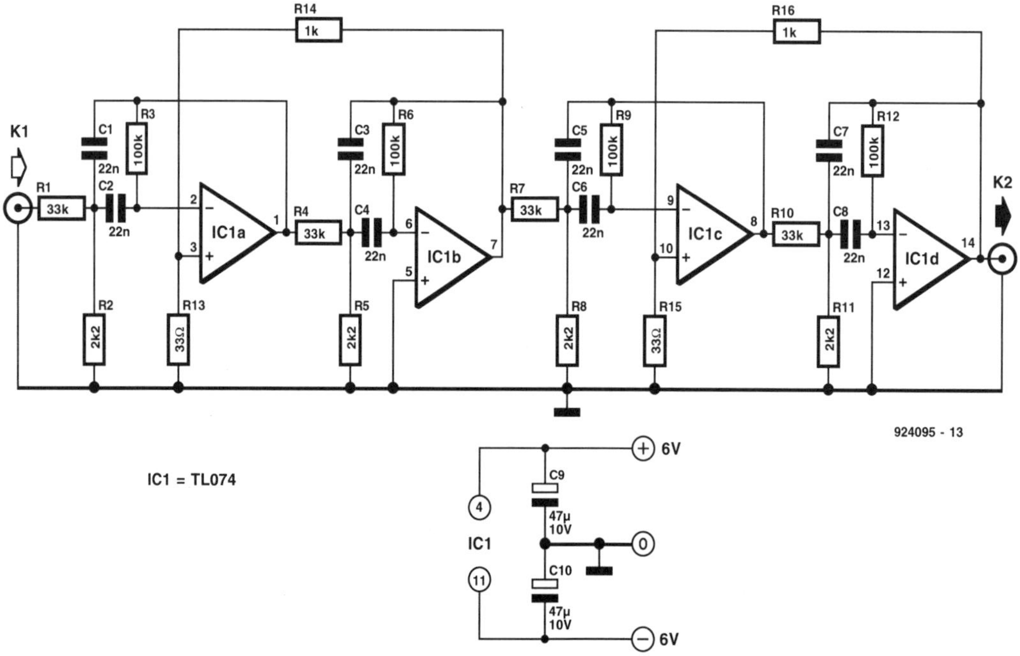

924095 - 13

Voice operated recording

022 Voice operated recording (VOR) is a feature found on most handheld memo-recorders. The function of the VOR control is to start the recorder's tape on detection of a speech signal. This relieves the user from pressing a 'record' button. Typical memo-recorder users include the too energetic manager and the detective on TV, unravelling a murder plot by talking to himself behind the wheel of his fast sportscar. The circuit shown here adds a VOR function to any standard cassette recorder, provided this has a 'remote' control input. Alternatively, it could be used as a VOX (voice operated switch) to control a transmitter, or a sensitive sound detector in an alarm system. Depending on the position of switch S_1, the output signal from the electret microphone (approx. 5 mV) is amplified ×520 or ×2,200 by opamp IC_1, which also functions as an active filter with a (speech) bandwidth from 160 Hz to about 9 kHz. Components D_1, C_5 and R_8 turn the amplified audio signal into a direct voltage, which is applied to the inverting input of comparator IC_{2a}. If the sound picked up by the microphone is sufficiently loud, the voltage at the inverting input falls below that at the non-inverting input, and the comparator output changes state. The resulting logic high at pin 7 of IC_{2a} causes relay Re_1 to be energized, so that its contact closes. If the contact is connected to the remote control input on the recorder, the tape will be started, and recording begins. This condition is indicated by the lighting of D_2. The supply voltage to the circuit and the microphone is switched on and off with S_3. Switch S_2 allows the VOR function only to

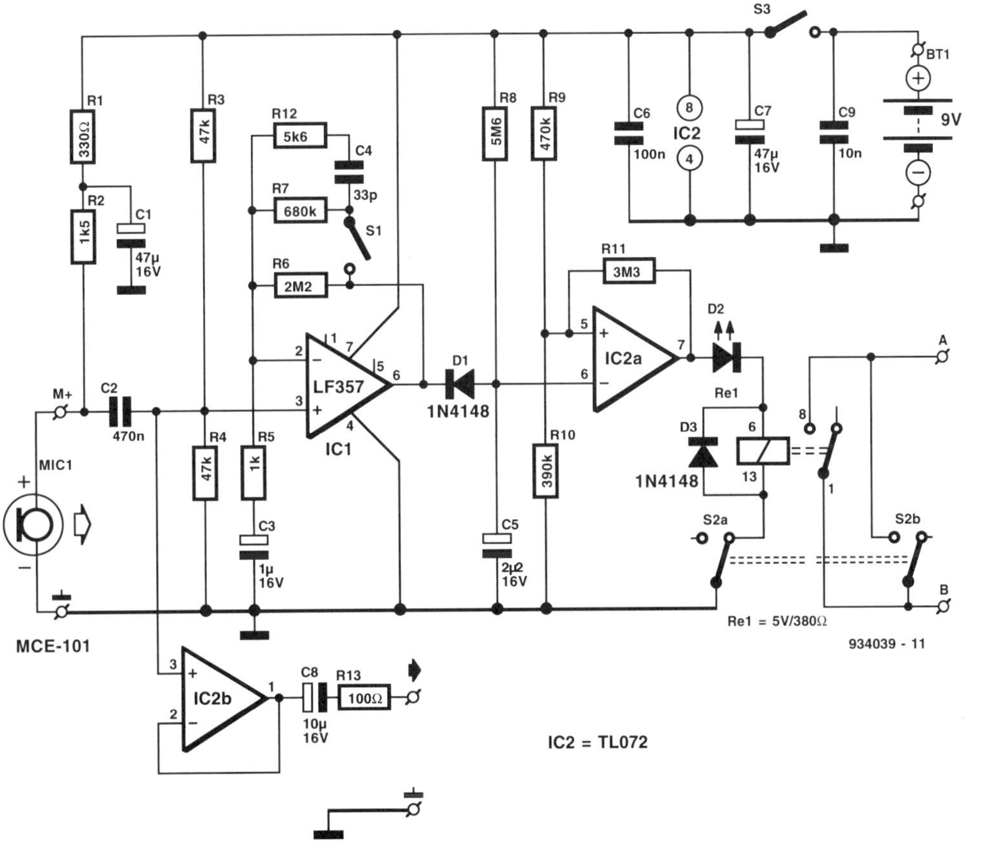

be switched off, while S_1 serves to select between 'low' and 'very high' sensitivity.

Since the gain-bandwidth product of the LF357 is about 20 MHz, the opamp will have a bandwidth of about 9 kHz if its amplification is ×2,200. With S_1 closed, the gain is reduced to ×520 times, which results in an effective bandwidth of 38.5 kHz. Since this is not practical or desirable for speech signals, capacitor C_4 is included in the feedback loop to give a bandwidth of 9 kHz at the lower gain also. Resistor R_{12} prevents the opamp oscillating by maintaining a minimum amplification of of ×5.

When the relay is energized, the circuit draws about 18 mA from the 9 V battery; if not, the current is roughly halved. The minimum battery voltage for reliable operation is about 6.8 V.

The VOR is an extremely sensitive circuit, and to ensure that it can not be upset by strong RF signals, it is best built on the printed circuit board shown here. The dimensions of the board allow it to be fitted into a ready-made plastic enclosure with an integral battery compartment. A suggestion for a nice looking front panel is also given. Notice that all inputs and outputs of the circuit are located at one side of the PCB – this is done to ensure that common-mode signals induced in the cables are not 'forced' to cross the PCB, where they could affect the operation of the electronics. All connections to ground are soldered to a wide copper pad that forms a low impedance. An exception has been made in the case of capacitor C_5: to prevent the discharge currents interfering with the microphone signal, they should not be allowed to travel through the ground pad.

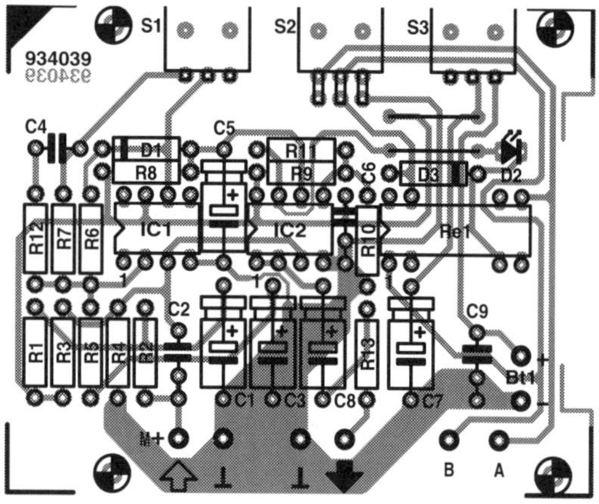

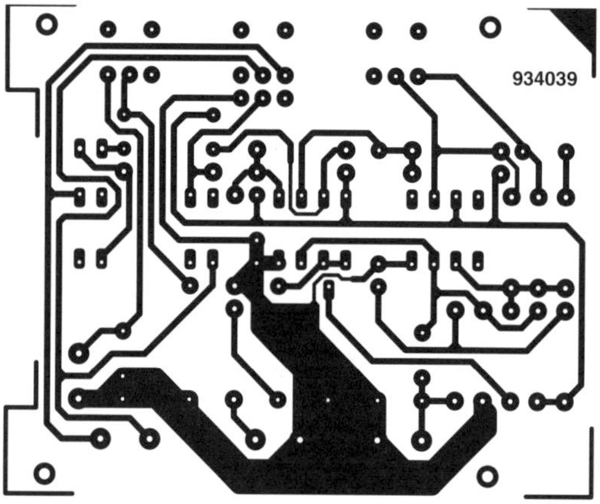

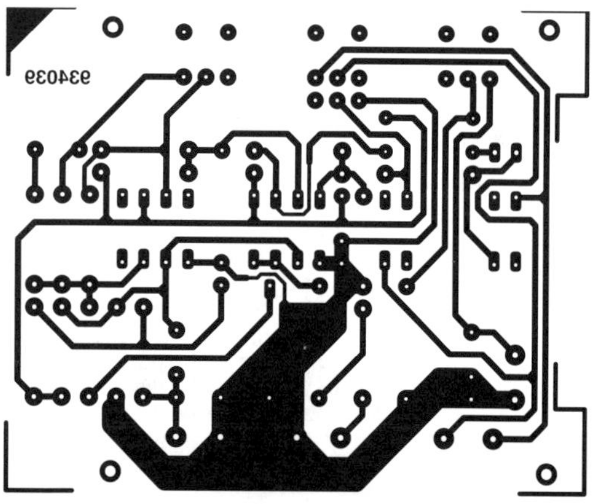

Parts list

Resistors:

R_1 = 330 Ω

R_2 = 1.5 kΩ

R_3, R_4 = 47 kΩ

R_5 = 1 kΩ

R_6 = 2.2 MΩ

R_7 = 680 kΩ

R_8 = 5.6 MΩ
R_9 = 470 kΩ
R_{10} = 390 kΩ
R_{11} = 3.3 MΩ
R_{12} = 5.6 kΩ
R_{13} = 100 Ω

Capacitors:
C_1, C_7 = 47 µF, 16 V
C_2 = 470 nF
C_3 = 1 µF, 16 V
C_4 = 33 pF
C_5 = 2.2 µF, 16 V
C_6 = 100 nF
C_8 = 10 µF, 16 V
C_9 = 10 nF

Semiconductors:
D_1, D_3 = 1N4148
D_2 = LED, red, 3 mm

Integrated circuits
IC_1 = LF357
IC_2 = TL072

Miscellaneous:
S_1, S_3 = AS1D-5M single-pole change-over slide switch,angled wires, PCB mount (Fujisoko)
S_2 = AS2D-5M double-pole change-over slide switch, angled wires, PCB mount (Fujisoko)
X_1 = MCE-101 electret micro-phone (Monacor/Monarch)
Re_1 = relay, 5V/380Ω; (Siemens V23100-V4005-A001)

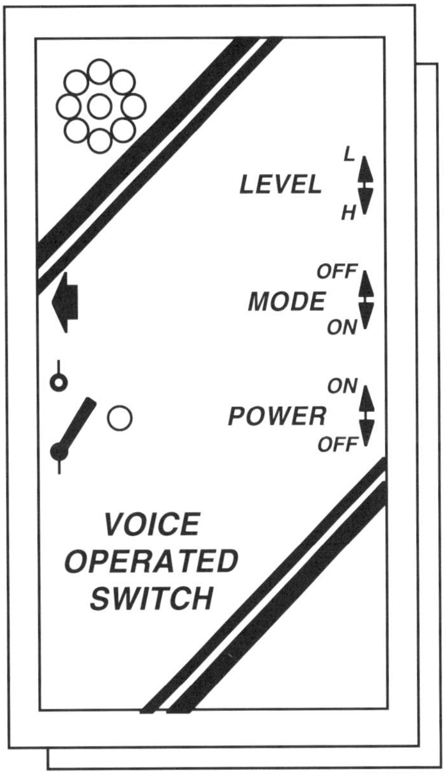

934039-F

1 Plastic case; dim. 61×97×26 mm (Pac-Tec Type HM-kit)
1 9-V battery clip
1 Printed circuit board 934039

Low-noise microphone amplifier

023 With the arrival of matched transistors in a single package it seems that the quest for the zero-noise microphone preamplifier has received a new impetus. The present amplifier is based on PMI's MAT02 transistor pair. The performance of this amplifier is such that even the most demanding audio enthusiast will be impressed by its very low noise contribution. The amplifier is simple to match to a wide range of microphone impedances, and offers a voltage gain of 20 dB or 23.5 dB (amplification of ×10 or ×15 respectively), selectable with a switch. The preamplifier is a two-stage direct coupled type with feedback and an input stage that runs on a very low collector current to minimise the noise contribution. The output impedance is about 70 Ω at a gain of 23.5 dB. The preamplifier is built on a compact printed-circuit board as illustrated. Its low current drain (approx. 2.5 mA) and

9-V battery supply make it ideal for portable use in conjunction with a high-quality dynamic microphone.

Since resistor R_3 is responsible for the input impedance and some other important characteristics of the preamplifier, we will have a look at some measured data with different values of R_3.

With $R_3 = 220$ kΩ, the input impedance is about 30 kΩ. A THD+N (total harmonic distortion plus noise) of 0.045% was measured; the noise level was –65 dB, measured in a bandwidth of 22 kHz, at an output voltage of 15 mV. Source impedance and overall gain were found to have a negligible effect on these figures.

Changing the value of R_3 to 6.8 kΩ resulted in a THD+N figure of 0.042% at an input voltage of 1 mV, and a noise level of –65 dB. At this resistor value the input impedance dropped to about 1 kΩ. At a source impedance of about 600 Ω, the THD is smaller than –95 dB (noise floor). On the down side, however, the lower input impedance results in a reduction in overall gain of about 4 dB. An even lower source impedance of 25 Ω was found to result in a THD of –94 dB.

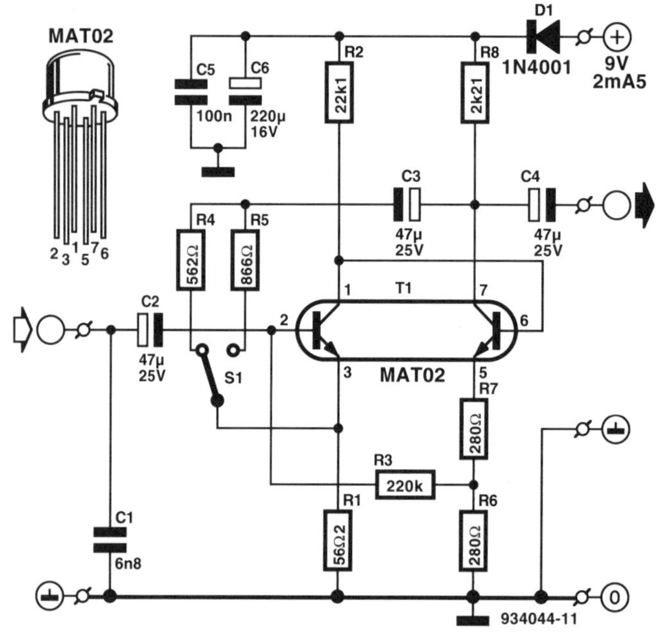

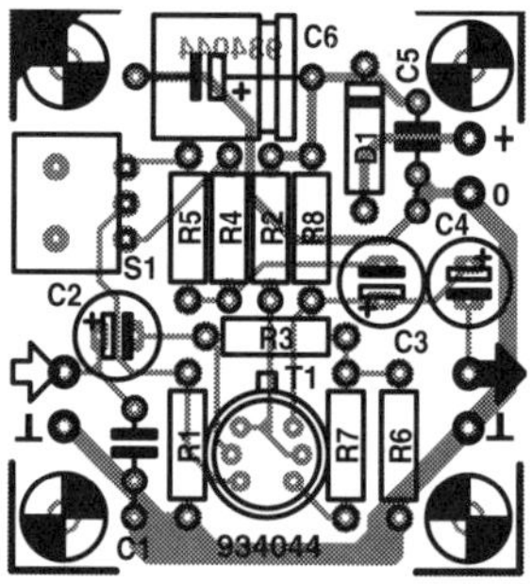

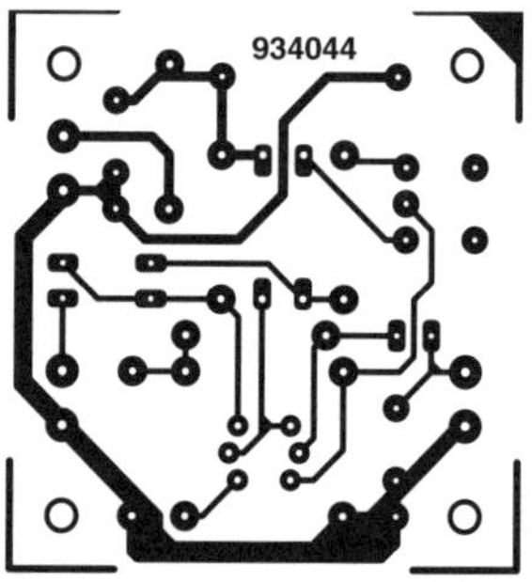

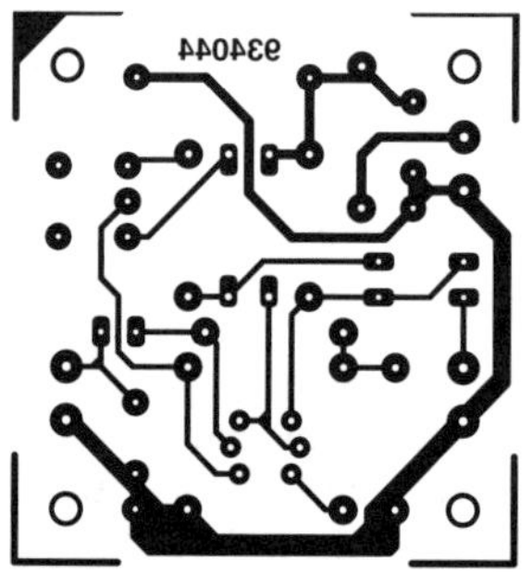

Parts list

Resistors:

(1% resistors: metal film from E96 series)

R_1 = 56.2 Ω, 1%

R_2 = 22.1 kΩ, 1%

R_3 = 220 kΩ

R_4 = 562 Ω, 1%

R_5 = 866 Ω, 1%

R_6, R_7 = 280 Ω, 1%

R_8 = 2.21 kΩ, 1%

Capacitors:
C_1 = 6.8 nF
C_2–C_4 = 47µF, 25V, radial
C_5 = 100 nF
C_6 = 220 µF, 16V

Semiconductors:
D_1 = 1N4001

T_1 = MAT02 (Precision Monolithics Inc.; Analog Devices)

Miscellaneous:
S_1 = PCB mount change-over switch, Fujisoku Type AS1D-5M

4th order single chip filter

024 High-order filters are normally designed by using two or more 2nd-order sections in series. That means that a 4th-order filter needs at least two opamps. The present filter, however, uses only one opamp, which results in smaller distortion, intermodulation, and so on. Also, there is no internal resonance rise, which typifies combinations of 2nd-order sections. Because of this, the peak input signal may be equal to the peak output of the opamp (but consider the common-mode input range).

Drawbacks of the circuit are the rather high ratio C_3/C_4 and the minimum value of the resistors. The resistor values are determined by the load on the output of the opamp (of which the resistors form a part). The maximum load (with large signals) of a TL081 is 2 kΩ. Resistors R_1–R_4 constitute an impedance of 2.5 kΩ, so that the external load must not be smaller than 10 kΩ. If an opamp is used that can handle a load of 600 Ω, it is advisable to give R_1–R_4 a minimum value of 2.5 kΩ. This will lower the noise emanating from the filter, which is generated chiefly by the resistors.

The characteristic of the filter is a 4th-order Bessel polynomial. A Butterworth characteristic is difficult to obtain with this type of filter, since, owing to the unity amplification of the opamp, the ratio C_3:C_4 becomes very high. With component values as shown, the –3 dB point is 1 kHz. Other cut-off frequencies can be obtained by recalculating the component values: these are directly proportional to the frequency. The 4th-order Bessel and Butterworth polynomials and the transfer function of the circuit are given below.

Bessel: $1 + s + {}^{3}/_{7}s^2 + {}^{2}/_{21}s^3 + {}^{1}/_{105}s^4$.

Butterworth: $1 + 2.6131s + 3.4142s^2 + 2.6131s^3 + s^4$.

$$Aj\omega = 1 + j\omega 2C_2C_4 + (j\omega)^2[C_1(C_2+3)C_1(C_4+4)C_2(C_4+3)C_3C_4] + (j\omega)^3[2C_1C_2(C_4+2)C_1C_3(C_4+2)C_2C_3C_4] + (j\omega)^4\, C_1C_2C_3C_4.$$

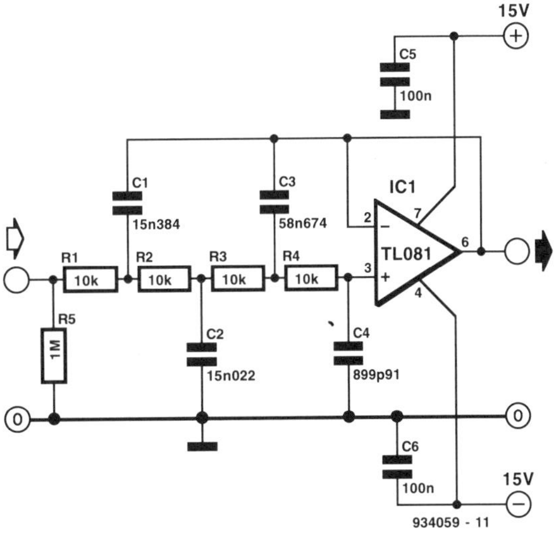

Active bass correction

025 Active loudspeakers offer the only way of obtaining good bass reproduction from inexpensive or small enclosures. The design described does not make use, therefore, of large, heavy enclosures to obtain a good result, but of acoustic feedback. A microphone placed close to the bass drive unit unfailingly registers every movement of the loudspeaker. It is, of course, important that proper attention is paid to the maximum movement of the speaker.

The microphone output is coupled into the negative feedback loop of the output amplifier. In this way, the input signal to the amplifier is compared with the acoustic signal produced by the speaker. In practice, this arrangement appears to work well only with low-frequency signals. Experiments have shown that if the microphone is placed about 10 mm from the cone of the woofer, signals at frequencies of up to 500 Hz are fed back faithfully. To make absolutely certain of correct operation, in the present circuit the upper frequency has been set to 300 Hz; above it, the correcting action gradually ceases. Note, however, that the phase behaviour of the loudspeaker is corrected also for signals above 300 Hz. If the change-over frequency of the cross-over filter of the loudspeaker lies at 300 Hz, it is advisable to make the cut-off frequency of the present circuit, determined by R_6-C_8, lower than 300 Hz.

The gain of IC_2 over the operating range of the circuit is 20 dB, which reduces to 0 dB for frequencies above 300 Hz. This amplifier, which provides the correction up to the cut-off point, also serves as buffer for the microphone signal.

Preset P_1 serves to set the signal level on the basis of the power rating of the output amplifier and the efficiency of the microphone. If this control is set too high,

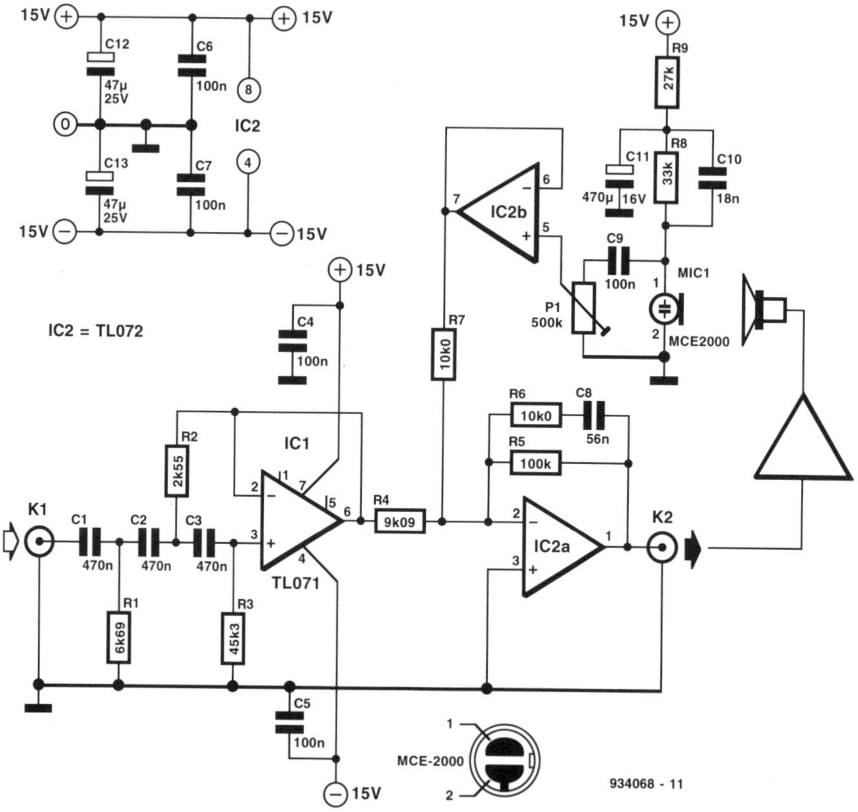

correction is also applied to frequencies above the cut-off point; if it is set too low, little correction will be applied and signals between 20 Hz and 300 Hz will increase along a standard 1st order characteristic.
The choice of microphone is a matter of some experimentation, particularly with high-power amplifiers. That used in the prototype proved to work well with low-power systems with a relatively low efficiency. If another type is used, make sure that the potential across the microphone is about half the supply voltage. This is arranged by R_8 and R_9. Also, make sure that the cut-off point set by P_1-C_9 remains well below 20 Hz (no signal at P_1 results in an increase of the final amplification).
The frequency up to which the microphone signal is compensated is determined by R_8-P_1-C_{10}. This time-constant must be equal to R_6-C_8.
The present circuit can magnify frequencies down to 20 Hz by roughly 20 dB. Since most loudspeakers cannot cope with that frequency, the circuit includes a 3rd-order Butterworth section with a cut-off point of 37 Hz. This frequency may be altered by changing the values of C_1, C_2, and C_3. This filter prevents the loudspeaker being loaded with signals which it cannot reproduce.
The correction circuit is of particular use with active loudspeaker systems. Make sure that the loudspeaker phase is shifted by 180° to prevent positive feedback. This may be done by adding an inverter-buffer in front of K_2.
The circuit draws about ±6 mA, of which only 0.25 mA is drawn by the microphone.

Class A preamplifier

026 Nowadays, the design of preamplifiers is almost invariably based on integrated circuits. However, there are still many constructors who like a discrete design and the one described here is for them.
The design of the amplifier is symmetrical. In the differential input stages dual transistors T_1 and T_2 are used. The differences in current amplification between these make bias correction at the input stage essential and this is provided by T_{12}. The required reference voltage source is provided by D_2. Preset P_2 sets the output voltage to 0 V. The only drawback of the correction stage is that an input coupling capacitor, C_1, becomes essential. However, with the value of this capacitor as shown, the cut-off frequency is only 1.5 Hz, so in practice this is no real problem.
The input stages drive a push-pull amplifier that consists of T_5 and T_6. To increase the stability of this amplifier, its gain is limited by R_{18}-R_{19}.
The output stage is a conventional emitter follower, T_9 and T_{10}. A current source, consisting of T_7 and T_8, ensures constancy of the quiescent current through the emitter follower. For best performance, the plane sides of T_7 and T_9, and of T_8 and T_{10}, should be joined together after some heat conducting paste has been applied to them. A brass clamp to make the mechanical coupling secure is advisable, but not absolutely essential.
The quiescent current through T_9 and T_{10} is set to 20 mA with P_1. The value of the current may be measured on the basis of the voltage drop across R_{22} and R_{23}.
Capacitor C_2 serves to improve the response of the stage and to suppress overshoot. The effective bandwidth is thus reduced to 2.4 MHz, which is more than ample for audio applications.
For optimum performance, all transistors operate in Class A, which results in the relatively large quiescent current of 40 mA.
To ensure correct feedback, the earths of input and output, R_1, R_2, R_{18}, and R_{19}, as well as the negative supply line must be taken to one earthing point only.
Some measuring results (with U_b= ±15 V, 1 V r.m.s. output and a 1 Ω load):
Gain: 16 dB
(input sensitivity 150 mV)
Slew rate 200 V μs^{-1}
Signal-to-noise ratio: 100 dB (unweighted)
Third harmonic distortion <0.00025% (20 Hz–20kHz).

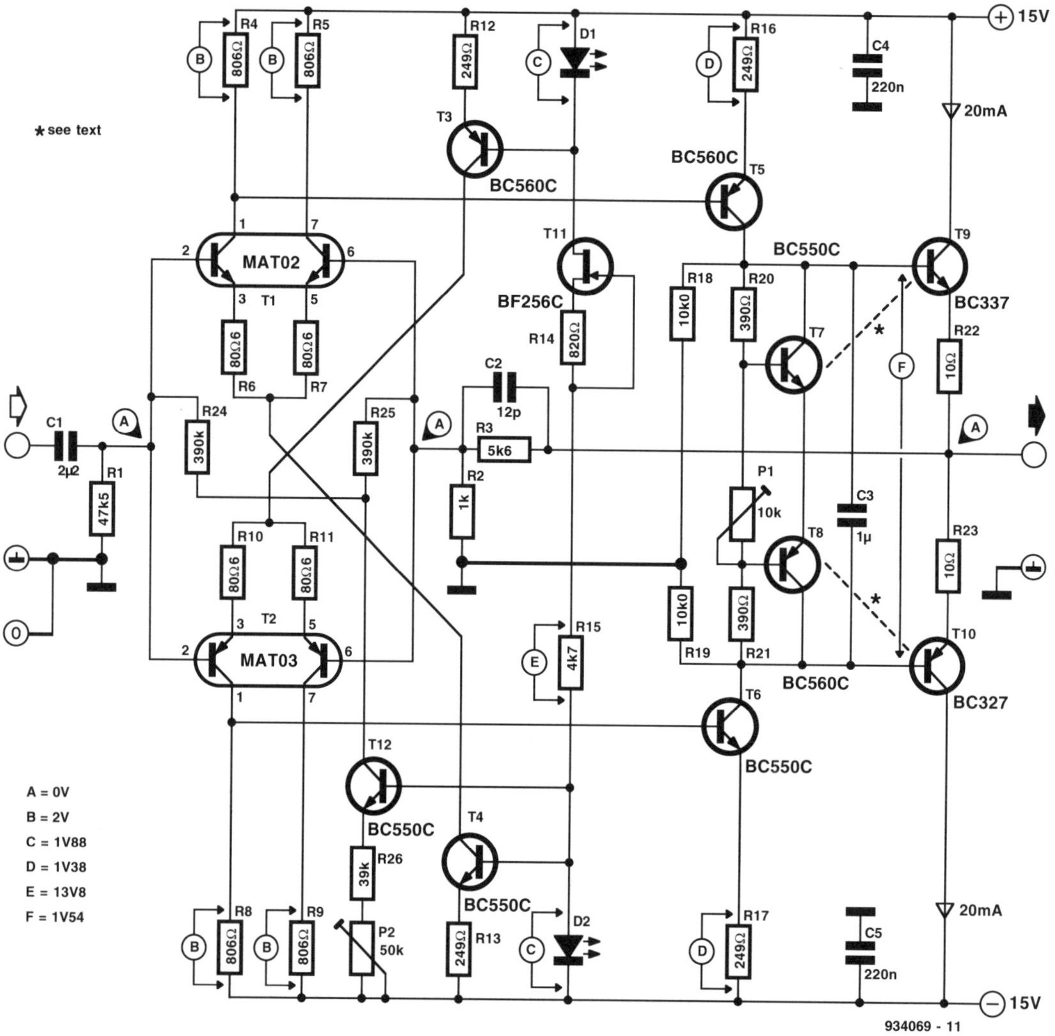

Special band-stop filter

027 Double T filters are used in many circuits. This type of filter can be made into a perfect band-stop filter, at least in theory. In the present circuit (Fig. 1), a double T filter is used in a different manner. It can have the characteristic of the combination of either a band-stop filter and a low-pass filter (switch S_1 in position 1) or a high-pass filter and a band-stop filter (S_1 in position 2). The characteristic curves are given in Fig. 2 and 3 respectively.

The variable K in Fig. 2 and 3 is determined by the setting of P_1. Variable M depends on the setting of P_2, and determines the Q factor.

A drawback of the present design is that the maximum suppression is diminished slightly. At $M = 0.75$, maximum attenuation is (calculated) 50 dB. With components values as in Fig. 1, the frequency at that attenuation is 1 kHz. Other frequencies can be computed easily. The band-stop frequency, f_{bs}, with S_1 in position 1 is

$f_{bs} = 1/2\pi RCK.$ [Hz}

With S_1 in position 2,

$f_{bs} = K/2\pi RC.$ [Hz]

In all cases, $0 \leq K \leq 1$; C is in farad and R is in ohms. Various types of IC may be used for the operational amplifier.
The current drawn by the circuit is around 2 mA.

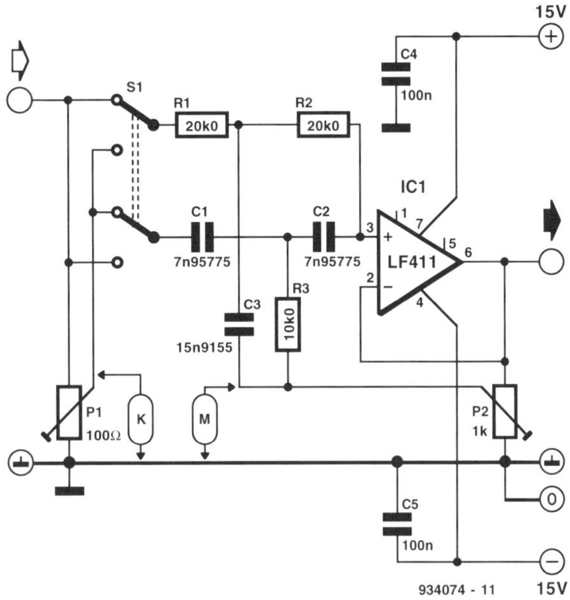

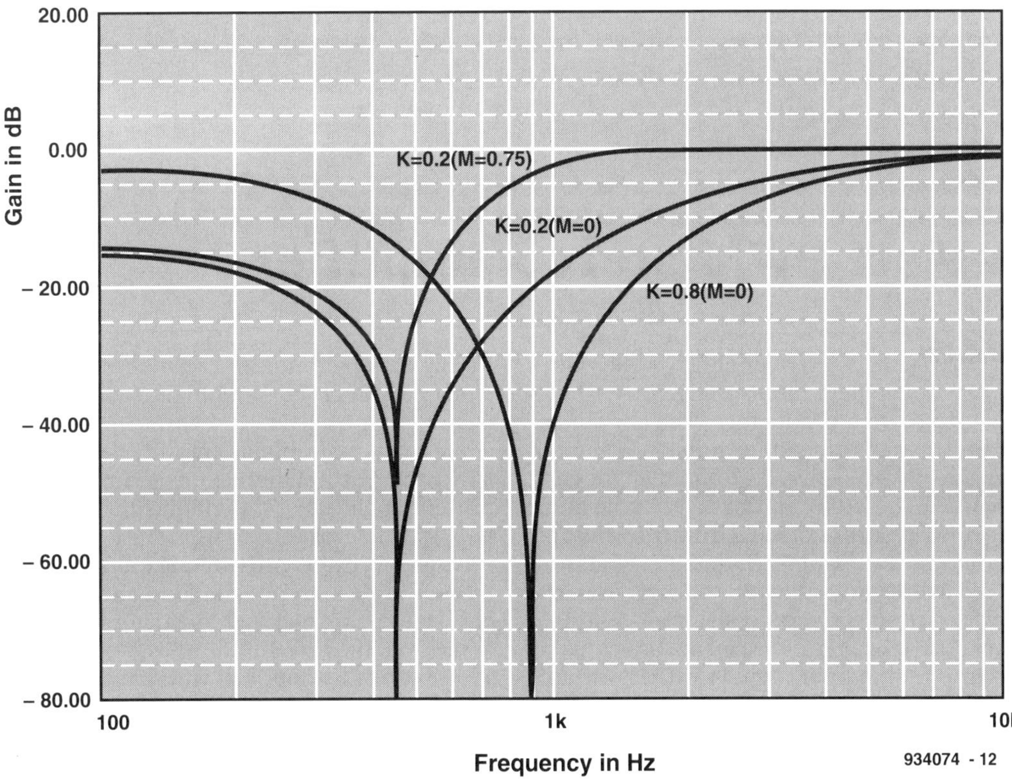

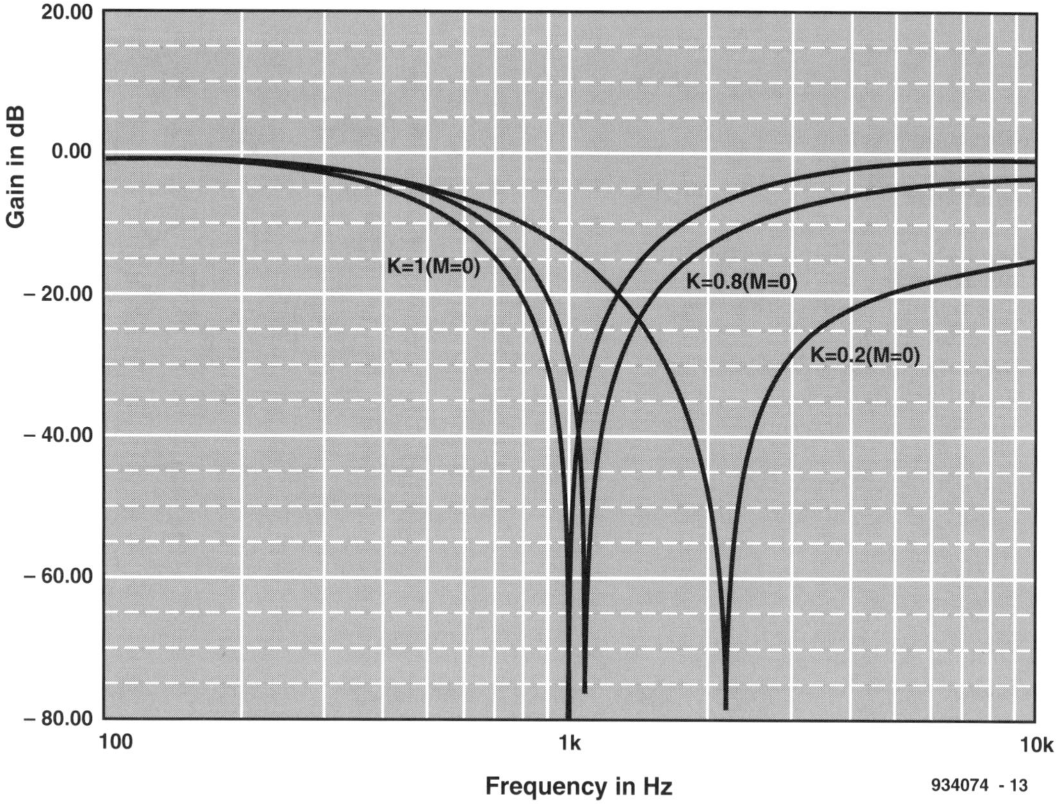

3 Watt PWM stereo amplifier

028 The most remarkable feature of this small stereo amplifier is the use of a stepper motor bridge driver IC as a stereo power output stage. The circuit diagram shows a three-stage pulse-width modulation (PWM) converter/amplifier for each channel. The left (L) and right (R) audio input signals are first converted into triangular waveforms by opamps IC_1 and IC_2, which are wired as integrators, with feedback from the power output stage via resistors R_7 and R_{11}. Next, the triangular waveforms are converted into rectangular signals with a variable pulsewidth by opamps IC_3 and IC_4, which drive the digital inputs of the bridge power amplifier, IC5.

The L6203 from SGS-Thomson is actually a full bridge driver for motor control applications. Its DMOS output transistors have an $R_{DS(on)}$ resistance of only 0.3 Ω, which results in low dissipation and, consequently, high efficiency. Each channel (half bridge) of the device is controlled by a separate logic input, while a common 'enable' (pin 11) is available to switch both channels on and off. In the present application, the channels are enabled by a +5.1 V level derived from the IC's voltage reference output. Although the L6203 contains a bridge circuit, the loudspeakers are connected single-ended for a stereo output arrangement.

The test results obtained with a prototype of the amplifier are not spectacular but none the less worth mentioning.

With a supply voltage of 12 V and an input frequency of 1 kHz, a maximum output power of 2 W was supplied into a load of 4 Ω. The nominal input level required for this output

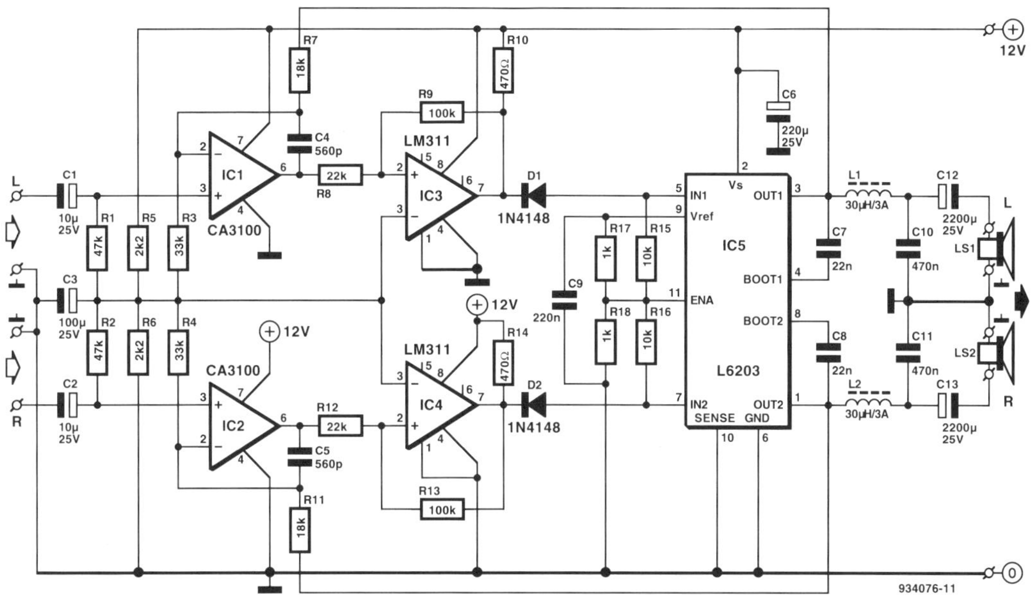

power was 2 V_{rms}. A damping factor of 20 was measured, while the distortion was about 1.5%.

These figures change slightly when the supply voltage is increased to 14.4 V. Maximum output power, for instance, increases to about 2.8 W with a distortion of about 1.5%. The input drive level required for this output power is 2.25 V_{rms}. With full drive, the amplifier achieves an efficiency of about 73%, with PWM clock intermodulation products down to –40 dB.

At lower input signal levels, the distortion of the amplifier remains around 0.3%, and the noise level at about –80 dB.

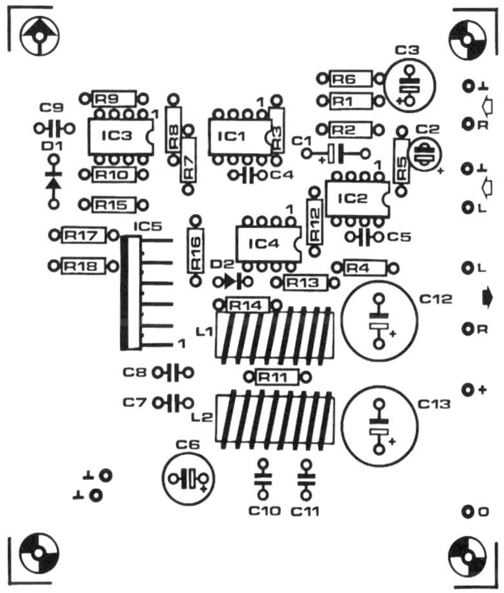

Parts list

Resistors:

R_1, R_2 = 47 kΩ
R_3, R_4 = 33 kΩ
R_5, R_6 = 2.2 kΩ
R_7, R_{11} = 18 kΩ
R_8, R_{12} = 22 kΩ
R_9, R_{13} = 100 kΩ
R_{10}, R_{14} = 470 Ω
R_{15}, R_{16} = 10 kΩ
R_{17}, R_{18} = 1 kΩ

Capacitors:

C_1 = 10 µF, 25 V
C_2 = 10 µF, 25 V radial
C_3 = 100 µF, 25 V radial
C_4, C_5 = 560 pF, polystyrene
C_6 = 220 µF, 25 V radial
C_7, C_8 = 22 nF

C_9 = 220 nF
C_{10}, C_{11} = 470 nF
C_{12}, C_{13} = 2200 μF, 25 V, radial

Inductors:
L_1, L_2 = 30 μH/3 A toroid choke

Semiconductors:
D_1, D_2 = 1N4148

Integrated circuits:
IC_1, IC_2 = CA3100*
IC_3, IC_4 = LM311N
IC_5 = L6203**

* Harris Semiconductor
** SGS Thomson Microelectronics

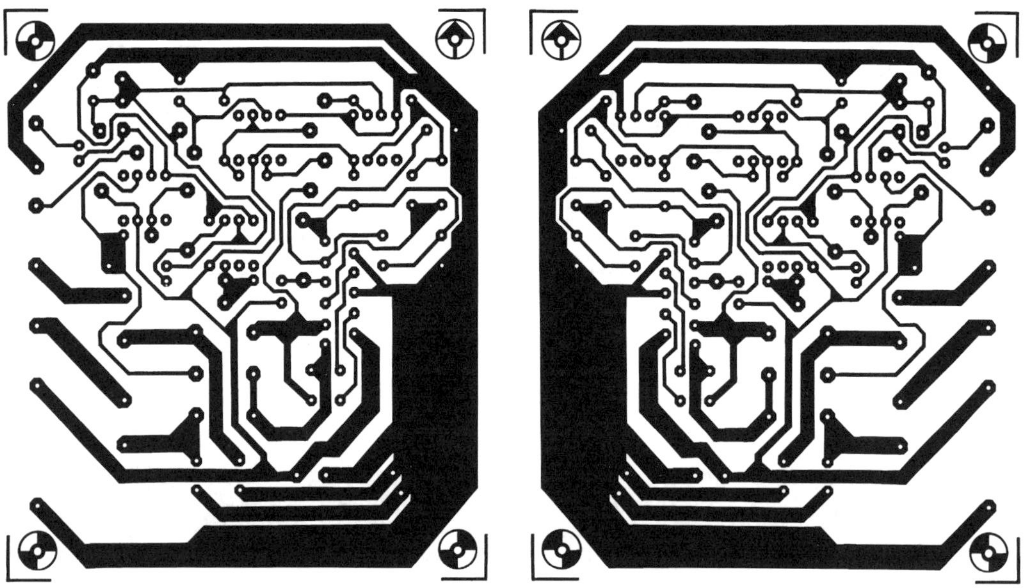

Peak level indicator

029 The indicator shows by means of two LEDs when the signal level in either channel of a stereo audio system exceeds a preset value.

In the diagram, IC_{1a} and IC_{1b} function as comparators. A reference voltage of 0–11 V is applied to their non-inverting inputs by P_1 and P_2 respectively. Resistor R_3 ensures that the reference voltage cannot exceed the common mode range of the opamps.

The signals from the two channels are rectified (half-wave) by D_1 and D_2 and the resulting direct voltage is applied to the inverting inputs of the opamps. Since the input impedance is high, the drop across the diodes is only 200–300 mV. Resistors R_1 and R_2 serve to limit the input current if the drive levels exceed the common mode range.

When the peak value of the input signal rises above the reference voltage less the drop across the diode, the output of the relevant opamp changes state (goes low), whereupon the associated LED lights.

The circuit may be used at frequencies up to 20 kHz (allowing for variations in peak levels of ±0 25 dB).

The indicator draws a current of only 0.25 mA when the LEDs are out and of 24 mA when both LEDs light. This drain may be reduced by the use of high-efficiency LEDs (which draw only 2–3 mA). The value of R_4 and R_5 should then be increased to 3.3 kΩ.

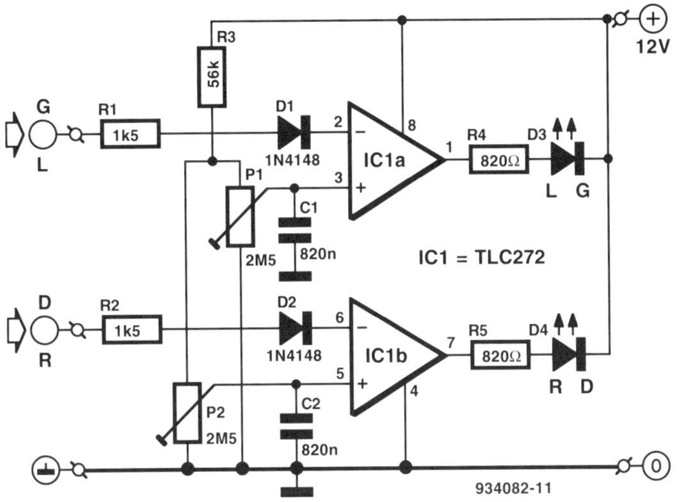

Solid-state voice recorder

030 Integrated circuits ISD1012, ISD1016, and ISD1020 from chip maker ISD enable sounds to be recorded and played back. The final two digits in the type coding show the length of a sound message in seconds. The difference in recording and play back time is caused by the sampling frequency, which is highest (10.6 kHz) for the shortest time.

The circuits have a non-volatile memory that can store the recorded sound for not less than ten years without any supply voltage. The circuits also contain a microphone amplifier and an output amplifier.

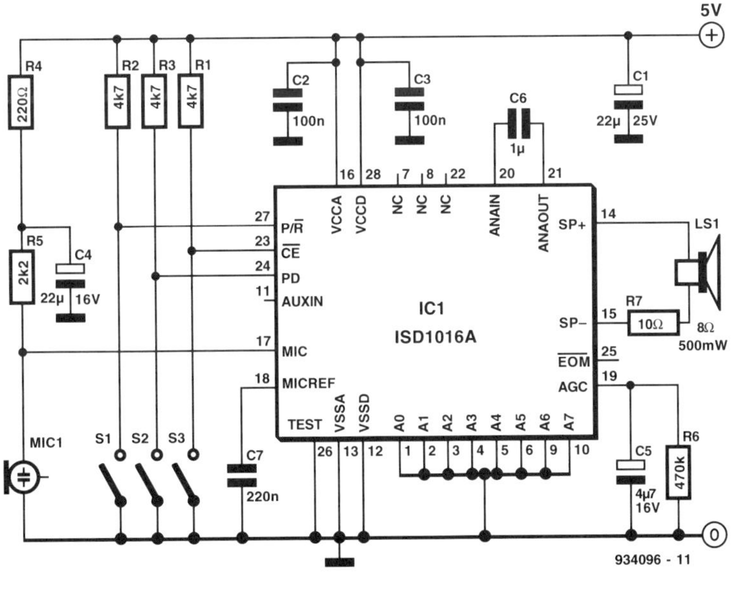

The drawing shows the simplest circuit for using one of these ICs: an electret microphone, a loudspeaker, three operating switches and a power supply.
Switch S_1 is the recording/playback switch: for recording, it must be closed.
Switch S_2, when open, sets the IC to the power-down mode. The larger part of the circuit is then switched off, which reduces the current drain. Switching to power down also serves as a reset when, for instance, during recording an overflow occurs (message too long). The IC indicates this by making the end-of-message (EOM) output low during the recording.
Recording or playback is started by closing S_3. This switch must remain closed during recording and should preferably be a push-button type.
The circuit is kept simple, although it is possible to connect several ICs in cascade if longer messages, or a number of short ones, are to be recorded.
The bandwidth of the circuits is similar to that of telephones: ISD1012 – 4.5 kHz; ISD1016 – 3.4 kHz; ISD1020 – 2.7 kHz. The current drain during playback is about 25 mA.

Copying in MS-DOS with one drive

031 MS-DOS is a program developed for controlling a computer system, which it can do with a variety of procedures. One of these is the making of batch files in which a number of commands can be combined to form a new instruction. Batch file acopy copies files from drive A to drive A. A small alteration makes it possible for the instruction to to be used with other drives as well. The batch file contains several infrequently used procedures and is intended for systems that have only one diskette drive or have two incompatible drives.

The nature of the alteration can be seen in the listing. It has been assumed that the computer has a hard disk or a RAM disk with sufficient free space. This is because the batch file stores all files to be copied temporarily on the hard disk or RAM disk,

The file starts as usual with **echo off**, and then checks whether the temporary directory already exists and whether it contains any files. If so, the user is warned and asked for a decision. This is effected by the first section of the batch file from **if** to the label **:endif**. The **if** command checks whether the directory is empty with **not** and **exist**. If that is the case, the commands up to **:endif** are ignored. If the directory is not empty, its content is shown on the monitor with **dir**, followed by a message directed to the screen by **echo**. The **pause** command gives ample time for a decision to be taken.

After the decision to continue, the content of directory c:\copy.tmp is erased automatically. The question 'are you sure?' is diverted to the zero-device (and is, therefore, not displayed) and replied to automatically by 'y(es)'. The zero device is a fictitious peripheral, which, as far as the computer is concerned, acts in the same way as a monitor or printer. It does nothing else and is eminently suitable for use as a waste basket. The automatic reply 'y(es)' is effected by the **echo** instruction, whose output (y) with the 'pipe' symbol as input is passed on to the **delete** instruction.

The next step is the creation of sub-directory c:\copy.tmp, which, if all is in order, is not present. Should this directory exist after all, it will be empty and be therefore usable.

Next, the diskette with the files to be copied is put into drive A. The files that must be copied are identified in the same way as with the MS-DOS **copy** instruction. No drive letter should be input, because that is already in the file.

```
@echo off

if not exist c:\copy.tmp\*.* goto
endif
dir c:\copy.tmp /w
echo.
echo directory C:\COPY.TMP already
exist
echo press ctrl-C to abort ACOPY
echo press any key to delete
C:\COPY.TMP\*.*
echo and continue ACOPY
pause > nul:
echo y | del c:\copy.tmp > nul:
:endif
ctty nul:

mkdir c:\copy.tmp
ctty con:
echo.
echo Insert SOURCE diskette in drive
a:
pause
copy a:%1 c:\copy.tmp

echo.
echo Insert TARGET diskette in drive
a:
pause
copy c:\copy.tmp\. a:%2 > nul:
echo y | del c:\copy.tmp > nul:
rmdir c:\copy.tmp
```

880191-11

The batch file cannot be used for $3^1/_2$ in **and** $5^1/_4$ in drives: each of these needs its own batch file. The only difference between these two files will be the drive letter in the **copy** instructions.
Once the files to be copied are stored in the temporary directory, the disks can be exchanged after which the copying can be completed.
Finally, the folders in c:\copy.tmp must be erased and the directory removed
The second **copy** instruction contains a full stop where a *.* might be expected. This full stop, in MS-DOS, means 'the present directory' and may be used in many cases instead of *.*. For instance, **del a:.** is the same as **del a:*.***.

Echo instructions followed by a full stop provide an blank line on the screen. This is so even with some other versions of MS-DOS, although the manual does not state this.
The syntax of the ACOPY command is

ACOPY pathname [pathname].

Bear in mind that with this method subdirectories must not be copied together with the files.
A last remark: the **pathname** must be given from the root directory.

2764 EPROM emulator

032 The emulator enables a Type 2764 EPROM in an existing circuit to be replaced by a static RAM. It is a very compact circuit: together with the stand-by power supply, it fits on a 105x40 mm (4 $^1/_8$ x 1 $^9/_{16}$ in) board.
To all intents and purposes, the action of the circuit is indistinguishable from that of an actual 2764 EPROM. The programming voltage may be 12.5 V or 21 V. An additional advantage of the emulator is that programming and erasing are not necessary during a development phase , thus saving much time.
The position of switch S_2 determines whether the circuit is actuated or inactive. When it is closed, the circuit is inactive and the memory cannot be influenced externally: it is then in the stand-by mode. This mode should be selected when it is expected that the circuit will not be used for some time or when it is removed from, or placed into, another circuit.
When S_2 is open, the content of the memory is protected by IC_2, T_4 and T_1. The only way the memory can be erased is by placing it in an EPROM programmer, setting switch S_1 to'erase', and actuating the function 'Blank Check' or 'Read Out' on the programmer. When this routine has been completed, the entire content of the emulator is set to FF-H. Switch S_1 must then be set to position 'program' again, whereupon the emulator can be programmed in the traditional manner.
When programming has been completed, S_2 must be closed and the emulator can then be inserted into its proper place in the circuit. Before that circuit can be operated, S_2 must be opened again.
The battery supply is switched on and off with switch S_3: when it is switched off, the content of the RAM is lost, of course.
As can be seen in the photo, two PCB terminal strips with extra long pins are used to replace the EPROM pins.

Parts list

Resistors:
R_1, R_8 = 22 kΩ
R_2, R_{15} = 100 Ω
R_3, R_4 = 560 kΩ
R_5, R_{11} = 68 kΩ
R_6 = 39 kΩ
R_7 = 3.3 kΩ
R_9 = 330 Ω
R_{10} = 560 Ω

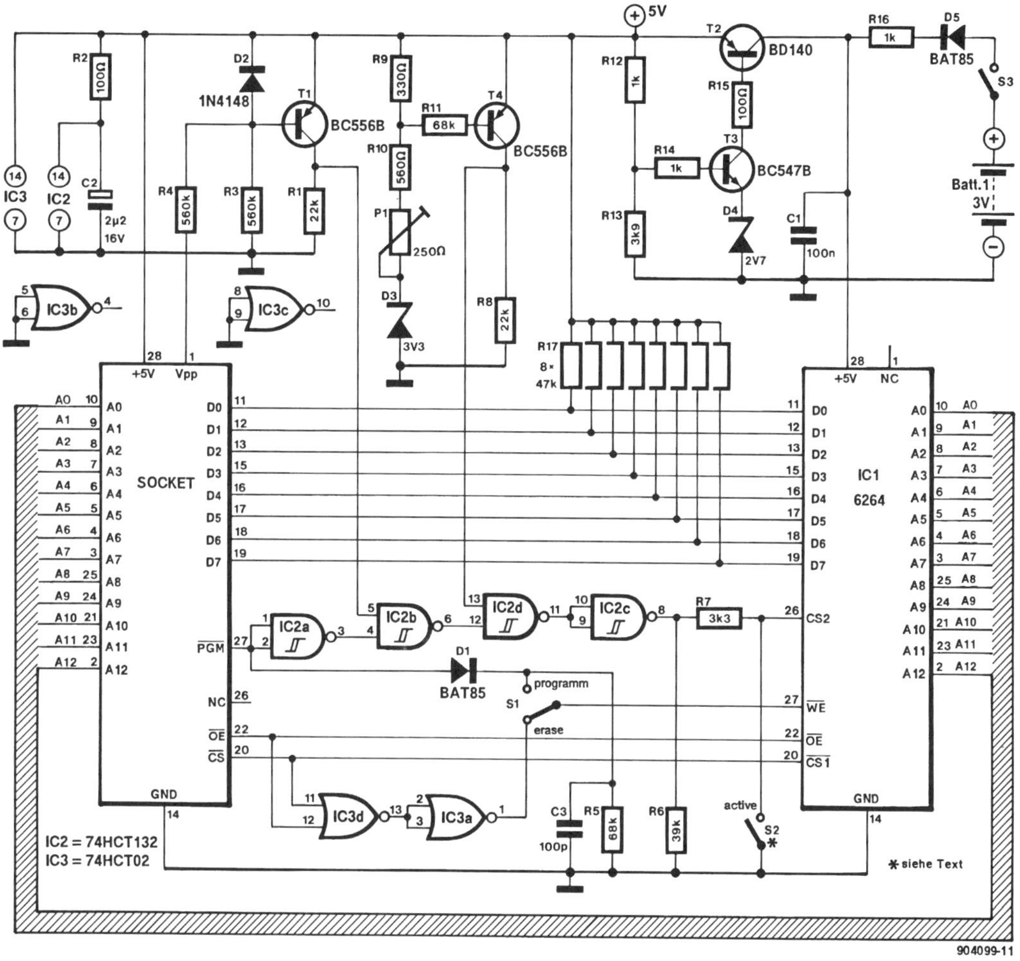

R12, R14, R16 = 1 kΩ
R13 = 3.9 kΩ
R17 = array 8×47 kΩ
P1 = 250 kΩ preset

Capacitors:
C1 = 100 nF
C2 = 2.2 μF, 16 V
C3 = 100 pF

Semiconductors:
D1, D5 = BAT85
D2 = 1N4148
D3 = zener diode 3.3 V, 400 mW
D4 = zener diode 2.7 V, 400 mW
T1, T4 = BC556B
T2 = BD140
T3 = BC547B

Integrated circuits:
IC1 = 6264
IC2 = 74HCT132
IC3 = 74HCT02

Miscellaneous:
S1–S3 = PCB slide switch;change-over contact
Batt1 = 3 V lithium battery

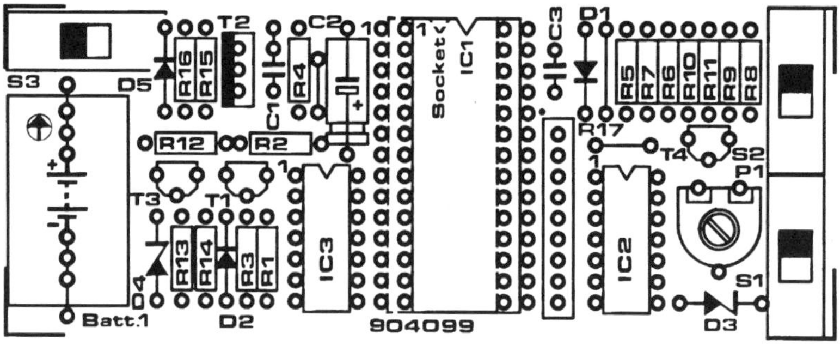

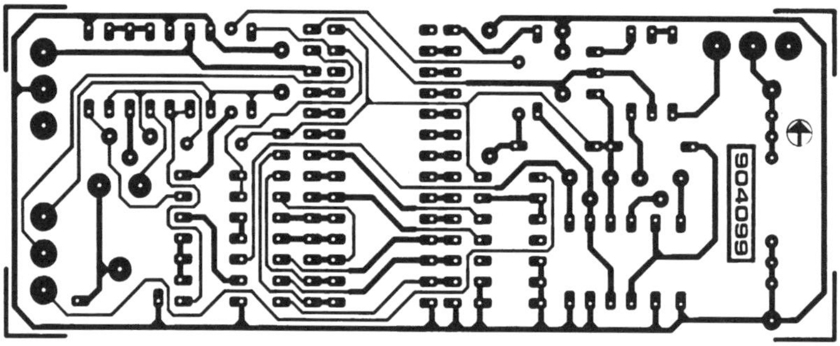

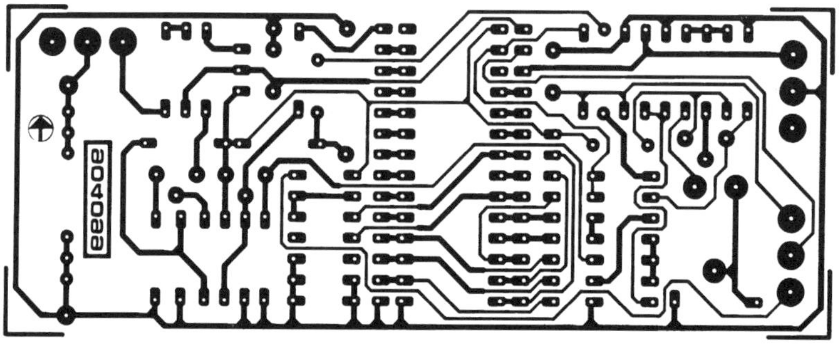

Change-over switch for C64 control port

033 Many people still use the old C64 computer to play games on, but get frequently annoyed by the constant need to change over the joystick connectors. This is because normally only one joystick is available, while some games are controlled via port 1 and others via port 2.

The cause of the annoyance may be removed by the circuit shown here, which uses eight analogue switches packaged in two Type 4066 ICs. Switch S_1 enables pins JOY0–JOY3 to be connected to either port 1 or port 2.

The +5 V supply is derived from the C64.

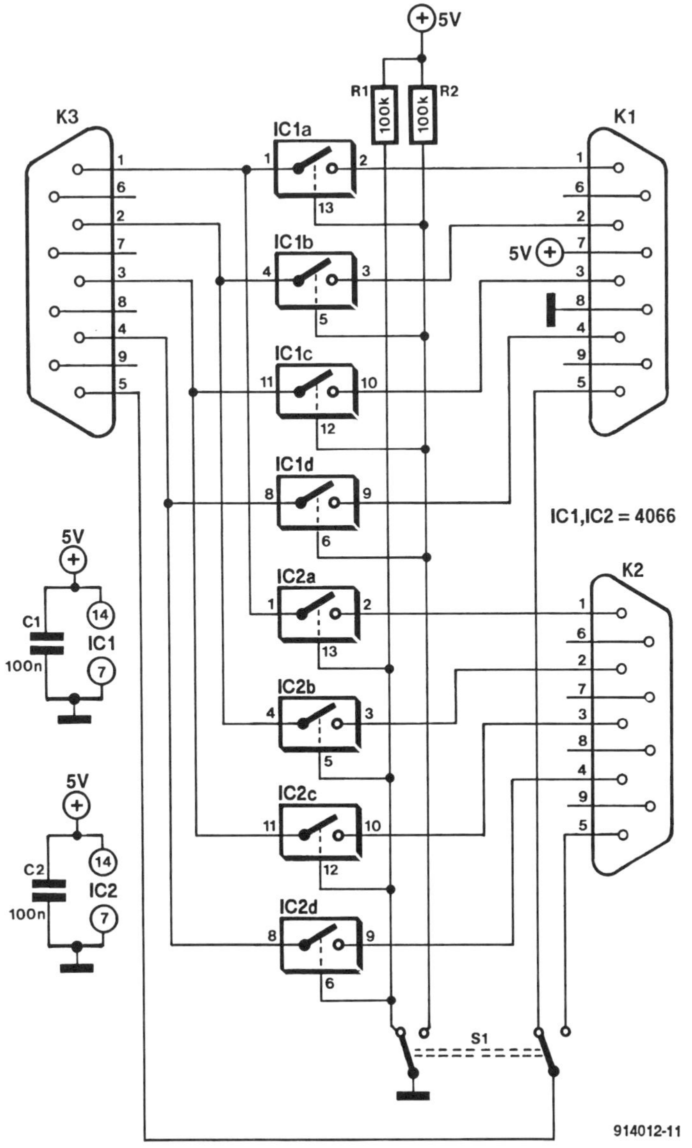

Keyboard change-over switch

034 If you have a non-qwerty keyboard and would like to use this with your computer without having to relearn where the deviating keys are, this simple circuit will help. It receives two keyboards on K_2 and K_3 respectively and connects these via switch S_1 and connector K_1 to the computer. Check the keyboard connections at your computer, because some PC compatibles have a slightly different pin layout. The connection between the circuit and the computer is via a standard 5-way DIN cable; you can, of course, make your own cable as shown at the bottom of

the diagram (K_4 and K_5).
Switch S_1 is a four-pole change-over type, either rotary or toggle. Since the supply line is switched also, the additional keyboard does not increase the load. The additional LEDs that indicate which keyboard is in circuit increase the current drain by about 10 mA.
So much for the hardware; now for the software. Whatever keyboard you connect, the codes it generates do not change. Advising the computer that a different keyboard layout is used is the task of the keyboard driver. In MSDOS versions up to 3.2, this driver is called KEYB??.COM, where in place of the question marks an abbreviation for the relevant country is given. Normally, the correct version of this program is executed in AUTOEXEC.BAT, after the change-over you have to start the program manually. From version 3.3 onwards, there is the file KEYBOARD.SYS (and in some old versions), KEYB??.SYS), and then you have no choice but to restart (alt-ctrl-del) the computer every time the keyboard has been changed with a system disk in drive A on which you have stored the relevant data in CONFIG.SYS and pass these to KEYBOARD.SYS. Alternatively, you can install the correct KEYB??.SYS.

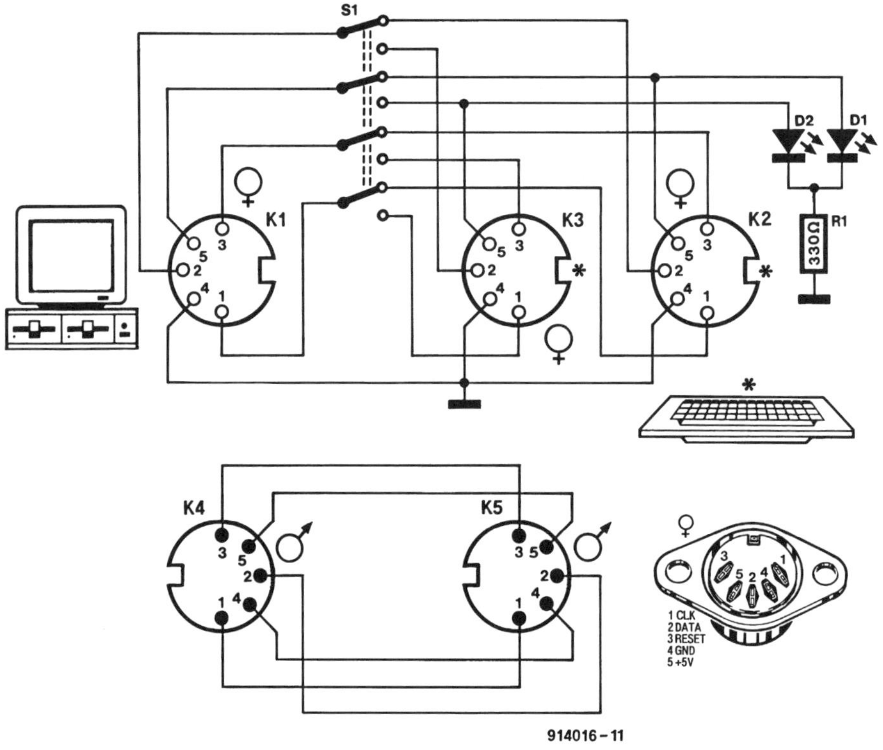

Communication buses

035 There are so many different standards for buses and networks that an overview of the most current types will be found useful by many readers. Note that each bus needs suitable software to transmit data. For instance, the well-known Ethernet network operates with Novell and Lantastic.
Ethernet and Thin-Ethernet buses are intended for use as a LAN (Local Area

Network) between computers and computers or between computers and peripheral equipment like printers and plotters.
The Integrated Service Terminal (IST) bus is used in LANs for offices. It complies with the ISDN norm. It is intended for communication with telephones, view phones, computers, and alarm systems.
The Domestic Digital Bus (D^2B) bus is intended for interconnecting audio and video equipment. It is found on most up-to-date radio tuners and television receivers.
The Controller Area Network (CAN) bus is intended primarily for use in control systems in a noisy environment. It needs only two wires for the distribution of power and information.
The Futurebus is a new standard for parallel processing of data within a computer. Path widths vary from 32 bits to 256 bits. A number of processors can exchange data at very high clock speeds along these paths.
The Inter IC Sound (I^2S) bus is designed for the exchange over short distances only of digital audio (16-bit stereo) between ICs in a digital audio system. The data are transmitted serially.
The Inter IC (I^2C) bus is also designed for communication between ICs. It handles not only data, but also commands. In contrast to the I^2S bus, the I^2C bus is fairly slow and not suitable for the rapid transmission of large quantities of data.

name of network	max. length in metres	kind of information	data format	type of connection
Ethernet	2500	data	serial	1 coax cable
Thin „	925	data	serial	1 coax cable
IST	300	data	serial	2 wires
D^2B	150	control	serial	3 wires
CAN	100	control	serial	2 wires
Future	systemlevel	data	parallel	
I^2S	boards	data	serial	
I^2C	boards	data/ control	serial	2 wires

Microprocessor supply regulator

036 Motorola's TCA5600 is a versatile supply chip for battery-operated microprocessors; its internal circuit diagram is given in Fig. 1.
Diodes D_1 and D_2 protect the circuit against reverse polarity of the supply voltage and voltage peaks respectively.
The 2.5 V reference is powered independently to enable it remaining switched on while the remainder of the circuit is on stand-by.
The d.c.–d.c. converter and presettable voltage regulator, A2, form one entity. The converter arranges for the input voltage of A2 to be of a level that ensures correct and stable operation of the regulator. The level depends on the level of the signals at INH1 and INH2 that control the regulator. Six modes are possible as shown in the table. Potential V_{out2} is used, for instance, as the programming voltage for an (E)EPROM and is set with the aid of resistors R_4 and R_5. Since the converter is on until the level at pin 9 reaches 33 V, it is possible, for example, prior to programming an (E)EPROM, to set the input potential to A2 to the required level in advance of the programming pulse, so that the switching from 5 V to the required level is not subject to any delays. This may also be seen from the timing diagram in Fig. 2.
The 5-V regulator is of conventional design. A differential amplifier, A1, compares the

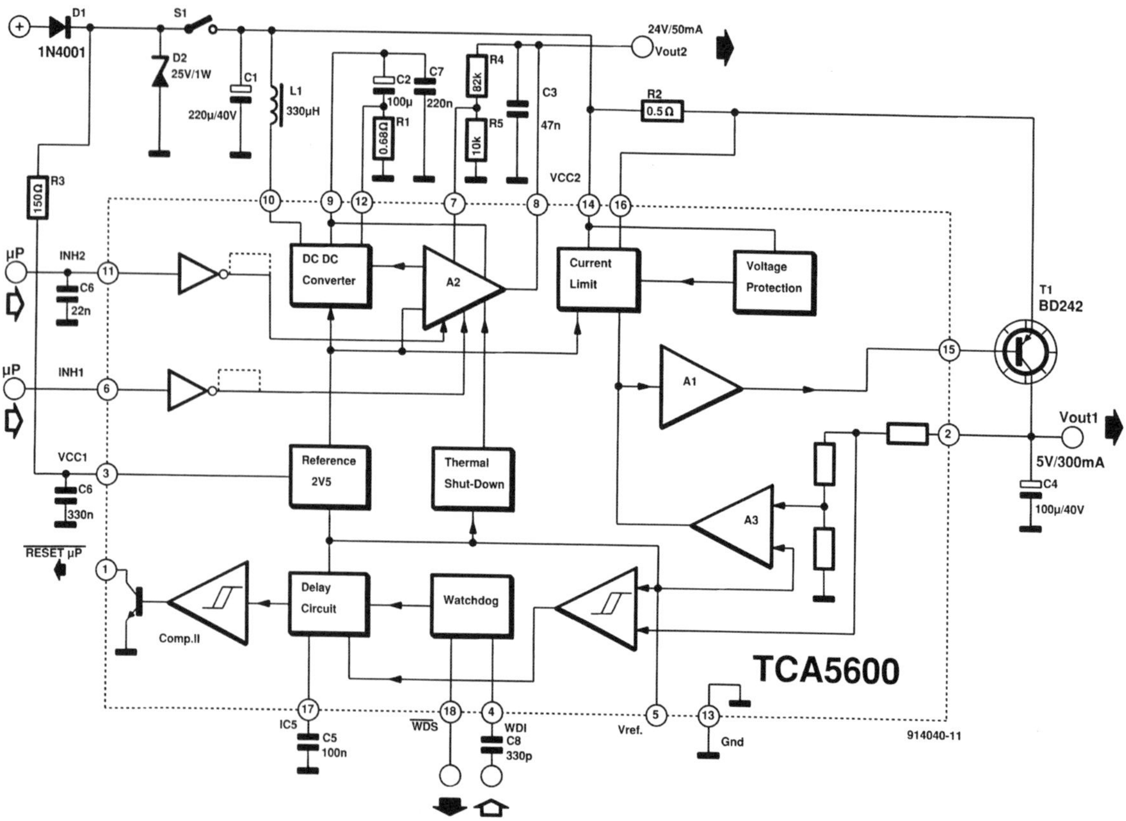

output level via a voltage divider with the reference; the difference drives an external power transistor, T1.

Current limiting is provided by a series resistor in the emitter circuit of the output transistor.

The output of the 5-V regulatr is also applied to a Schmitt trigger that functions as an under-voltage detector. If the output drops below 5 V, the delay circuit is actuated to make the reset output of the IC low. This is intended to prevent spurious operation of the microprocessor at too low a supply voltage. Such operation is also monitored by the watchdog. By programming the processor to emanate regular pulses and provide these to the WDI (watchdog inhibit) pin, the watchdog can monitor whether the program is running normally. As soon as a pulse is delayed, the watchdog actuates the delay circuit, which in turn resets the processor. The watchdog function may be switched on and off via the logic level at pin

TRUTH TABLE

Mode	INH1	INH2	V_{out2}	d.c./d.c.
1	0	0	off	int
2	0	high Z	V_{out2}	on
3	0	1	V_{out2}	int
4	1	0	off	int
5	1	high Z	5.0 V	on
6	1	1	5.0 V	int

int: intermittent operation of the converter means that the converter operates only if $V_{CC2}<V_{out2}+2.5$ V.

on: the converter loads the storage capacitor to its full charge (V_g=33 V), allowing fast response time of the regulator V_{out2} when addressed by the control software.

off: high impedance (internal 10 kΩ resistor to ground).

18 (WDS); if this level is low, the watchdog is enabled.
The delay circuit also provides a power-on reset, so that the processor will always start at the correct point in the program at power-on.

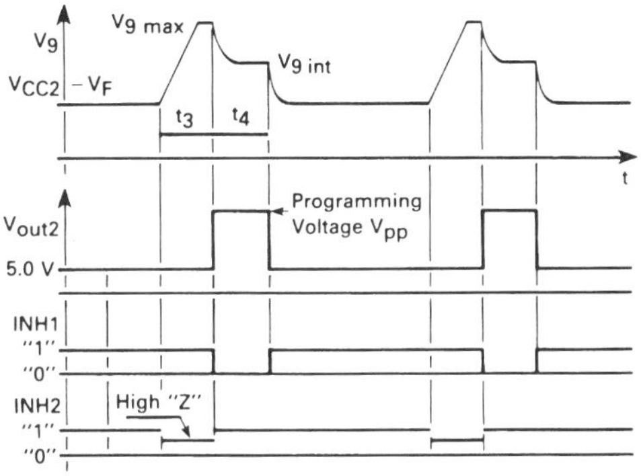

914040-12

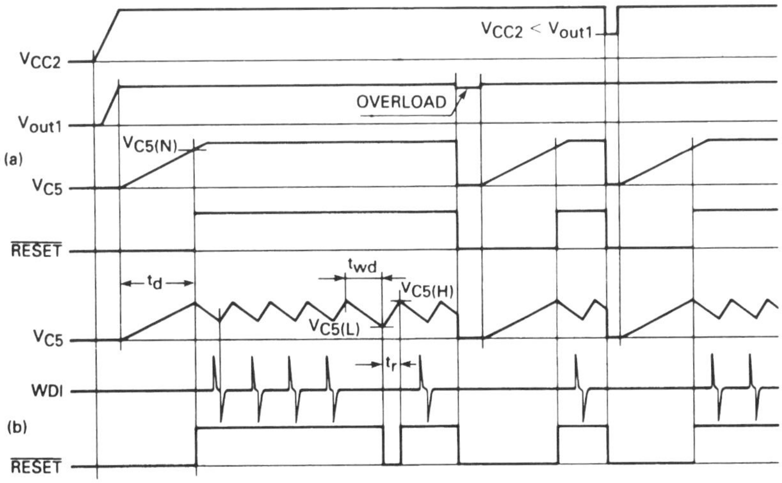

(a) Watchdog inhibited, $\overline{WDS}$ = "1"
(b) Watchdog operational, $\overline{WDS}$ = "0"

914040-13

Angled bus extension for PCs

037 This 8-bit bus extension card for IBM PCs and compatibles allows you to connect and test insertion cards without having to open the computer. The printed-circuit board shown here is angled, and has a 62-way slot connector to accept external boards.
The pins of the bus connector are soldered straight to the copper tracks at the edge of the board.
Since the tracks on the extension card pass through the metal frame at the rear of the PC, it is recommended to insulate them locally with PVC tape. Also, for mechanical stability the extension card must be secured to the frame with the aid of a support bracket.
Finally, take care to fit insertion cards the right way around in the slot connector of the extension card. If necessary, put the PC on a couple of books to create room at the underside.

Parts list

K_1 = 62-way IBM bus slot connector
PCB Ref. 914030

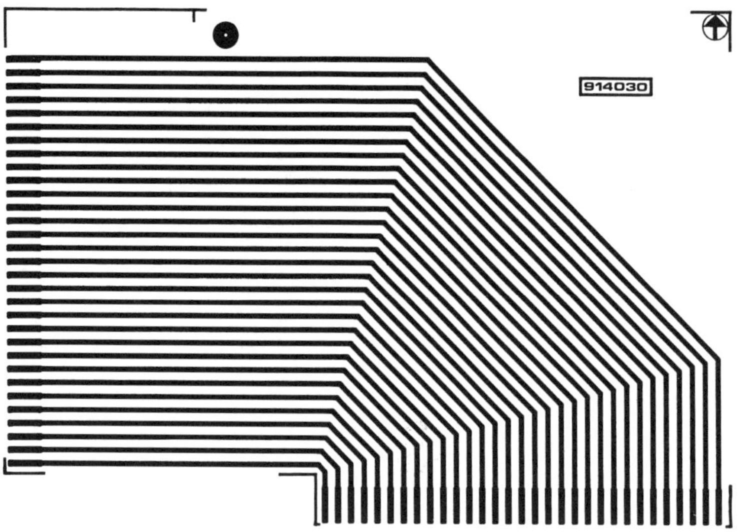

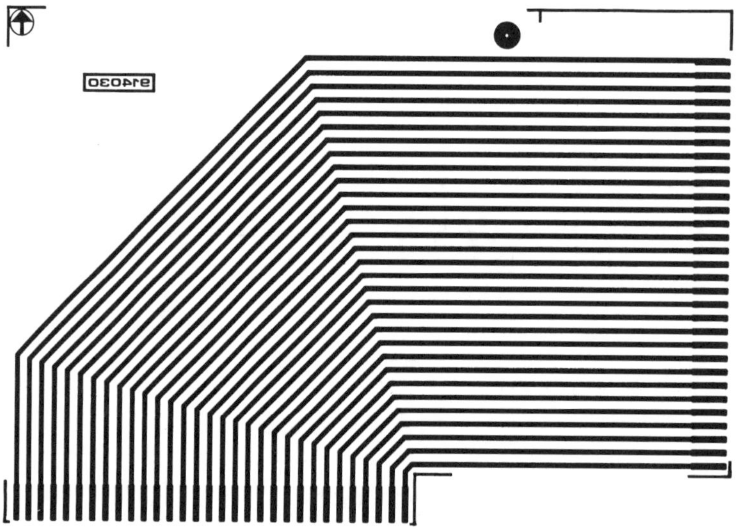

Video enhancement for Acorn Archimedes

038 The Acorn Archimedes, well-known for its speed and good graphics facilities, has a video interface that allows programmers to design a variety of screen modes with the aid of a programmable controller. As so often, this versatility has a draw-back: since the controller uses a fixed clock of 24 MHz, the frame frequency decreases the more pixels are used in the screen image. As a result, high-resolution screen modes have a tendency to cause flicker. Fortunately, this drawback may be eliminated or nearly so by increasing the clock to 36 MHz: this has already been incorporated in the new A540 computers.

All that is necessary to increase the clock is an integrated crystal oscillator and an IC Type 7400. To ensure that the circuit in the diagram is at all times compatible with the existing software, it may be arranged to be switched on (by software) only when the screen mode requires a higher clock.

The design allows the circuit to be fitted simply to the existing connectors in the computer. The TTL oscillator and two small connectors are fitted at one side of the board (not available ready-made), while the other side houses a surface-mount version of the 74HCT00.

The board itself is fitted to the four pins of PL_3 in A300 computers or PL_4 in A400 computers.

The connection with the I/O line that arranges the switching between the two clock frequencies is made with a short length of circuit wire soldered to pin 3 of PL_{10} on the mother board.

The supply for the board is obtained by soldering two short lengths of circuit wire to the supply lines on the back plane (the card with the extension connectors).

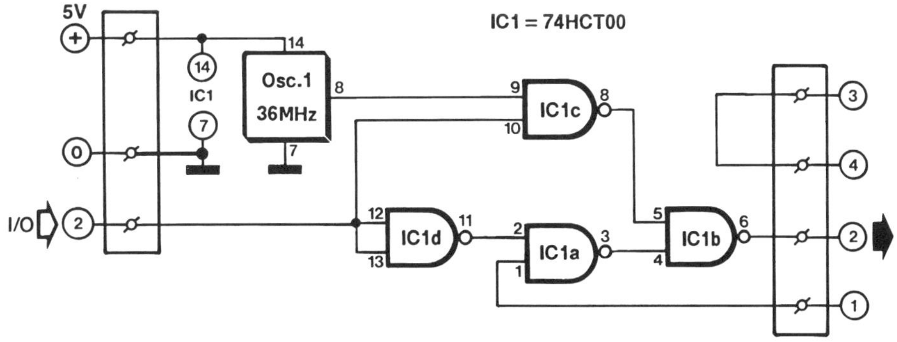

914051-11

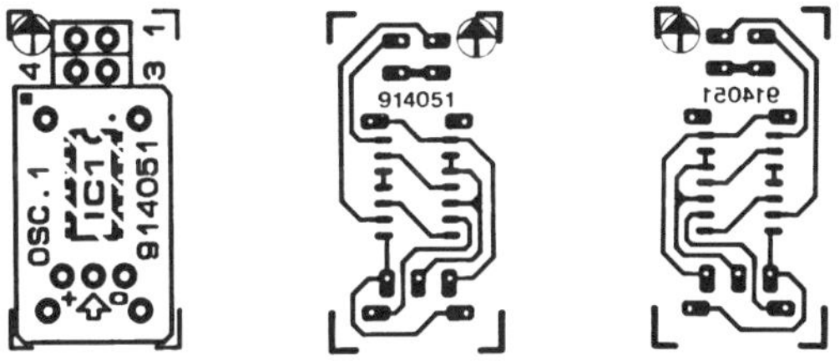

PC interrupt handler

039 The program and the circuit presented here are an introduction into practical interrupt handling in IBM PCs and compatibles, a subject fraught with pitfalls to the average PC user. The program (shown in the listing) is a memory-resident utility that monitors one of the PC's interrupt request lines, IRQ2 through IRQ7. It produces a short beep on the PC's loudspeaker when a request occurs. Interrupt requests may originate from insertion cards, and serve to inform the PC that a particular event has occurred that requires the action of a servicing routine, which interrupts the currently running program. Interrupts may be used to signal the activity of, for example, a telephone ringing circuit, a temperature

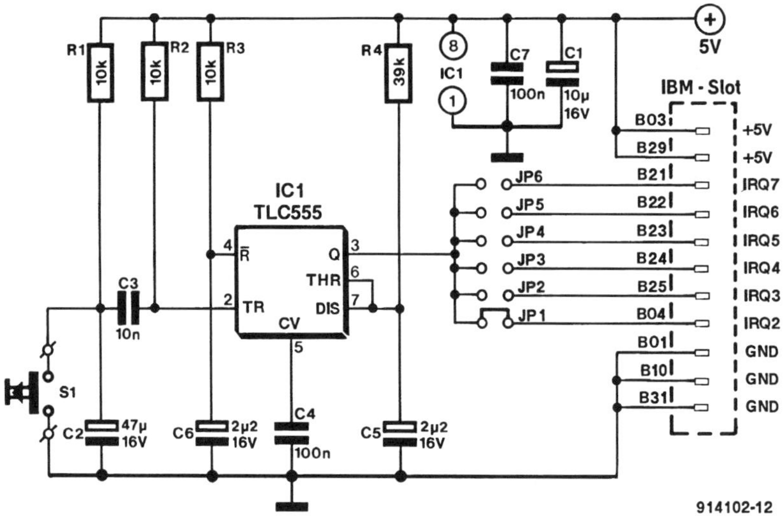

monitor, a voltage level monitor, or a watchdog. To help you understand the use and basic operation of interrupts, a simple circuit is given that generates an interrupt when a push-button is pressed. Taking good care to avoid conflicts with cards already installed in the PC, set the push-button interrupt line with the aid of a jumper, JP_1 through JP_6. In the program, this interrupt line should be defined accordingly by assigning the correct value to the constant 'IRQ' (see the listing).

When the push-button is pressed the TLC555 supplies a 100-ms long interrupt request pulse, which is transferred to the 8259 interrupt controller in the PC via an extension bus slot. Components R_1 and C_2 form a switch debouncing network. The circuit is readily built on the prototyping board for computer extensions described in Ref. 1.

At the end of the listing are two comment lines that serve to prevent the uninstall routine being called. If you intend to make changes to the program, we recommend that you do not attempt to make it resident straight away since the utility can then be removed from memory only by rebooting the computer. Therefore place the KEEP instruction in between braces, and temporarily add REPEAT UNTIL KEYPRESSED and UNINSTALL INTERRUPT-HANDLER. This will save you a lot of time and trouble debugging your own application.

Reference:

1. 'Prototyping board for computer extensions'. *Elektor Electronics* July/August 1988

RS232 with single power supply

040 In most personal computers, the power supply provides +5 V and ±12 V lines. The positive 12 V line is needed for the disk drives, and the ±12 V for the RS232 interface. Over the past few years, ICs have become available (such as the MAX232) that can drive serial channels (which need ±12 V) from a single +5 V line.

In the diagram, a Motorola MC145407 and four electrolytic capacitors provide ±10 V (the supply for RS232 connections may lie between ±5 V and ±15 V). The circuit also provides three input buffers and three output buffers (the MAX232 provides two of each). If more buffers are needed, the IC can supply a 145406 via its Vdd and Vss pins to give a total of six input buffers and six output buffers.

The 10-volt potentials are generated by an integral 20 kHz oscillator and two voltage doublers. When the supply is loaded, these voltages drop a little, but remain well within the RS232 requirement.

The open-circuit current drawn by the IC is only 1.5 mA, but this increases, of course, under load conditions.

It is advisable to keep the construction as compact as possible and to locate the electrolytic capacitors very close to the relevant pins. The 330 nF decoupling capacitor must be soldered with the shortest possible leads between pins 2 and 19. Preferably, do not use an IC socket, because fairly high peak currents flow when the electrolytic capacitors are being charged.

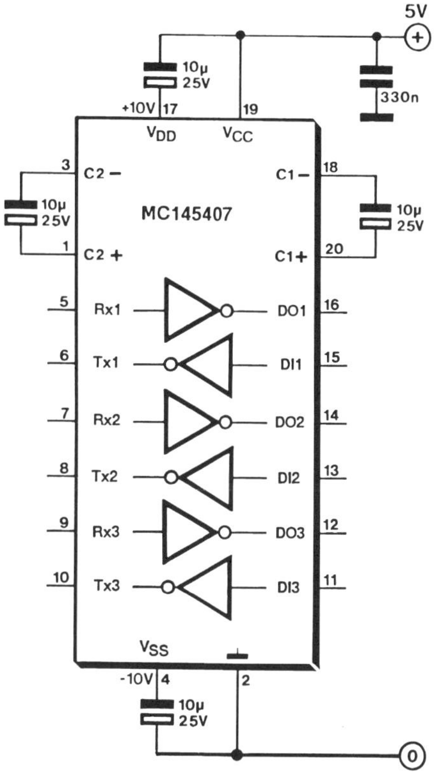

914055-11

Mouse/joystick switch for Amiga

041 Here is an interesting circuit for all Amiga owners who object to having to disconnect the mouse every time a second joystick is required for a video game. The switch is all-electronic, and can be connected permanently to the joystick-1 port on the Amiga. Extremely simple to build from a minimum of components, the circuit detects automatically when the joystick or the mouse is used.

The mouse supplied with the Amiga computer supplies four signals, apart from the two button signals. Signals H and HQ (or V and VQ) indicate the direction, and X0 and X1 the speed (see Fig. 2).

The joystick supplies similar signals, although it must be noted that joystick activity is simpler to detect than mouse activity. The joystick is active whenever one of the four direction signals goes low. By contrast, mouse activity can only be detected by comparing the current state of the H and V lines with the previous state. The mouse is active when these states are different. Hence, a bistable and a clock would be required to implement an activity detector.

Here, a much simpler solution has been found by making use of the propagation delay of the logic functions contained in the GAL. In practice, the actual delay is uncritical, so that even the slowest GALs can be used—a few nanoseconds are sufficient for a reliable mouse activity detector.

The left-hand mouse button is combined with the button on the joystick via diode D_1. Pull-up resistors are fitted at the joystick and mouse inputs to prevent non-defined signals when one of these devices is not connected.

Those of you with access to a GAL programmer can use the source listing in

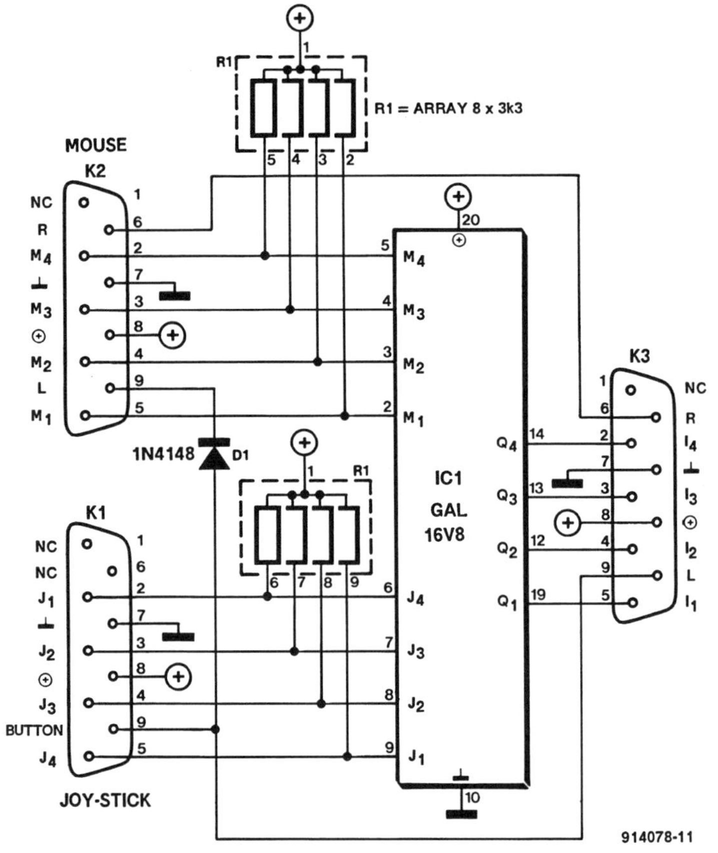


```
*IDENTIFICATION
ELEKTOR;
*TYPE GAL16V8;
*PINS
M1=2, M2=3, M3=4, M4=5, J1=9, J2=8, J3=7,
J4=6, Q2.T=12, Q3.T=13, Q4.T=14
HOUT.T=15, VOUT.T=16, SWM.T=17, DIF.T=18,
Q1.T=19;
*BOOLEAN EQUATIONS
SWM = /DIF & SWM & J1 & J2 & J3 & J4
    + DIF;
Q1 = SWM & M1
    + /SWM & J1;
Q2 = SWM & M2
    + /SWM & J2;
Q3 = SWM & M3
    + /SWM & J3;
Q4 = SWM & M4
    + /SWM & J4;
VOUT = M1;
HOUT = M2;
DIF = /VOUT & M1
    + /M1 & VOUT
    + /HOUT & M2
    + /M2 & HOUT;

*END                                  914078-13
```

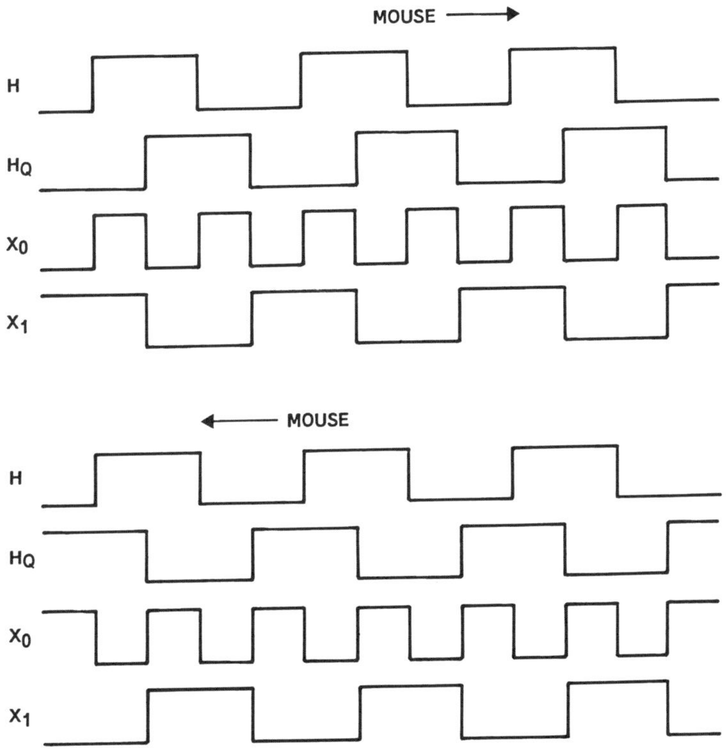

H	HQ	X0
0	0	0
0	1	1
1	0	1
1	1	0

HQ	X1
0	1
1	0

914078-12

Fig. 3 to produce a JEDEC file and program their own device. If you do not have a GAL programmer, you may like to know that IC_1 is available ready-programmed from Elektor Electronics (Publishing).

Parts list

Resistors:
R_1 = array, 8×3.3 kΩ

Semiconductors:
D_1 = 1N4148

Integrated circuits:
IC_1 = GAL 16V8 (ESS6003)

Miscellaneous:
K_1, K_2 = 9-way male sub-D connector, angled, for PCB mounting
K_3 = 9-way, female sub-D connector, angled, for PCB mounting
PCB Ref. 914078

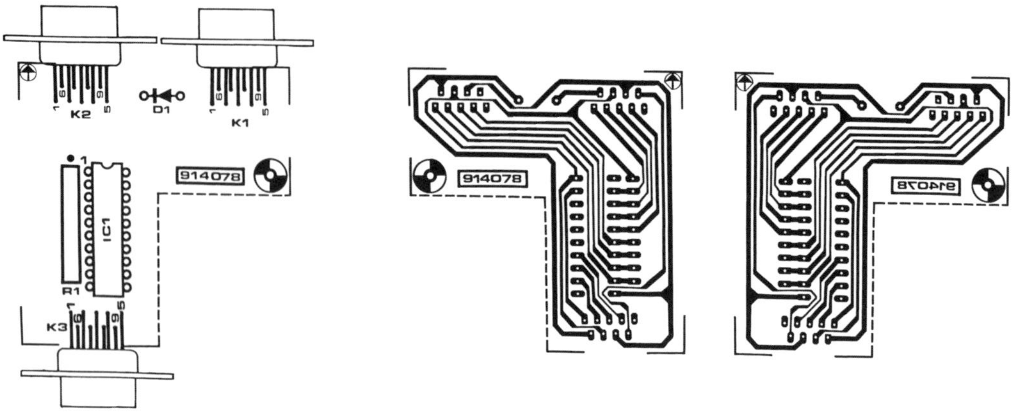

RS232 for Sharp pocket computers

042 Programming pocket computers is normally a tricky job, owing to the small keys and the limited memory.ocket computers from Sharp, which can be programmed in BASIC, have an interface that can be connected to a special cassette interface. The signals at that interface are very similar to those of an RS232 interface, but they are inverted and have different logic levels. It would, of course, be useful if the pocket computer could be linked to a larger computer via this interface, because the writing, changing and storing of the software could then be done much m ore conveniently.

It only requires the small circuit dhown to make that possible. The single 5 V supply voltage is converted into ±10 V in IC_1. With these voltages, the buffers in IC_1 can convert the logic signals of the pocket computer into RS232 levels. Inversion of the levels is effected by four inverters in IC_1. The circuit draws a current of only 30 mA, so that it can be supplied from the larger computer.

The connector to link the interface to the pocket computer presents a small inconvenience, because its pitch (1.27) is rather unusual. Connectors with that pitch are normally too long for this application; the only solution is to cut one to size.

The circuit has been tested with an XON/XOFF protocol, 2400 baud, even parity, 8 data bits and a stop bit. At higher speeds, small problems arose, but that need not always be the case, depending on the software.

The interface of the Sharp pocket computer is set with:

OPEN" COM:2400, E, 8, 1, A, L, &H1A, X, N": CLOSE.

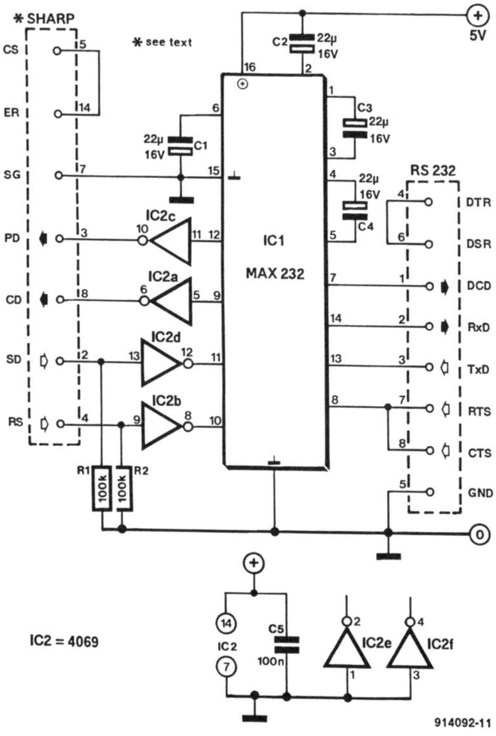

SCSI active termination

043 The SCSI bus (small computer systems interface) is a high-speed bidirectional parallel bus used mainly for connecting 'intelligent' peripheral devices to a CPU. The bus is the *de facto* standard in the Apple computer environment, where it is used to connect devices like CD-ROM readers, scanners, and internal as well as hard disks. The SCSI bus has also found wide acceptance in the IBM PC world, where the main application is with large hard disks. SCSI hard disks feature a 'software' type identifier, and require an appropriate hard disk controller and associated software.An SCSI bus network should have only one bus termination device. These devices are usually supplied with the PC, or the SCSI peripheral, and consist of resistors only, connected between the TERMPWR (+5-V) line, the transmission (signal) lines, and ground (SCSI drivers are open-collector types). The termination resistance 'seen' on every signal line is critical to ensure the highest possible data speed on the network.The circuit shown here is an active termination device, which eliminates the 220 Ω and 330 Ω resistors required for each signal line with a passive termination scheme, reducing significantly the continuous system power drain. When placed in series with 120 Ω resistors, the L200 voltage regulator matches the impedance level of the transmission line, and eliminates reflections. Here, the L200 supplies a fixed voltage of about 2.75 V across R_1.The active termination device should replace the single, passive, termination device already fitted in the SCSI system. Construction is easy on the PCB shown. The voltage regulator, IC_1, may be bolted direct to the PCB without a heat-sink.

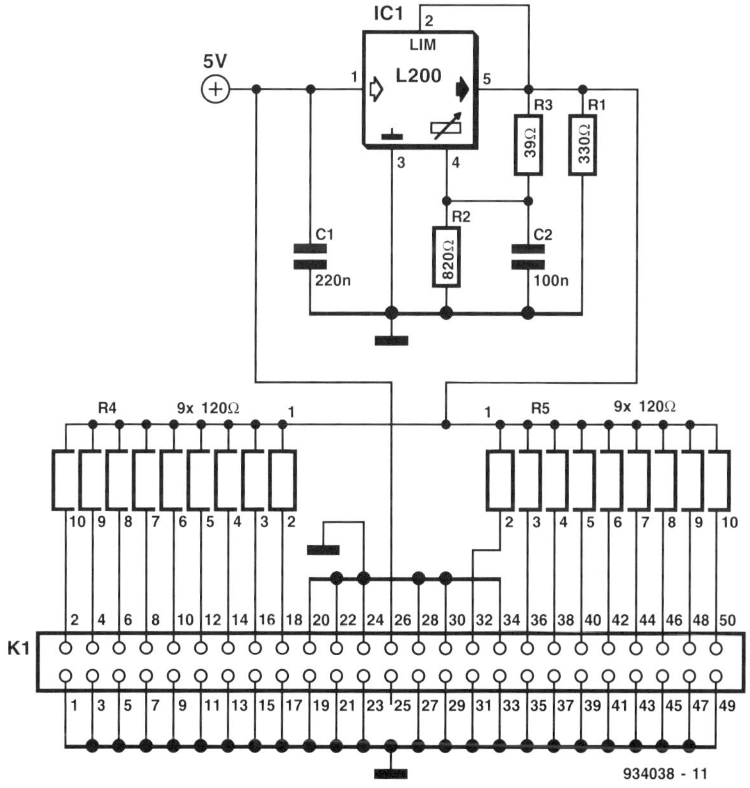

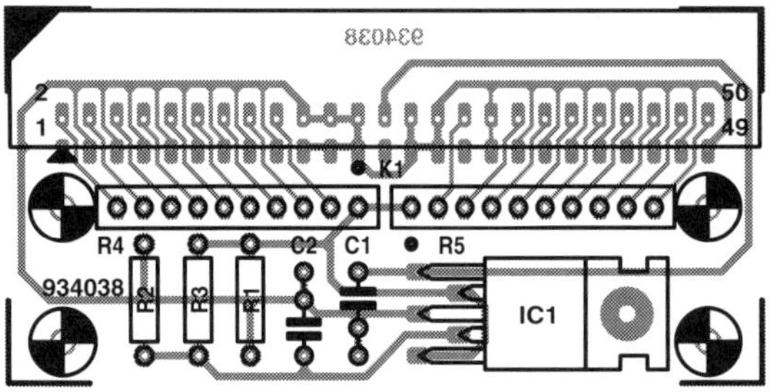

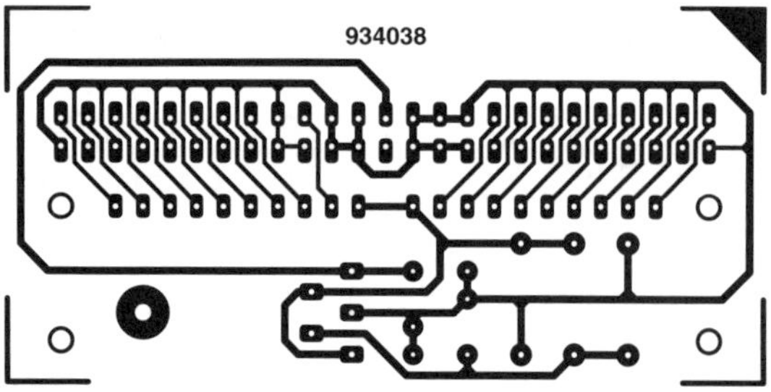

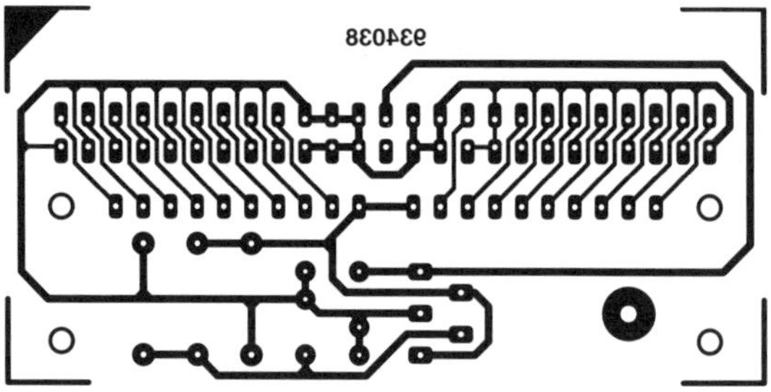

Parts list

Resistors:
R_1 = 330 Ω
R_2 = 820 Ω
R_3 = 39 Ω
R_4, R_5 = 9-way 120 Ω SIL array

Capacitors:
C_1 = 220 nF
C_2 = 100 nF

Integrated circuits:
IC_1 = L200

Miscellaneous:
K_1 = 50-way box header for PCB mounting with angled solder pins
50-way IDC socket

General purpose 64-bit output

044 Since the proposed circuit makes use of the Centronics interface, it is suitable for virtually all types of computer: even older types with a seven-bit Centronics interface, although in their case the possibility of giving an overall reset is not available.

The data from the interface are clocked in IC18 with the strobe signal. The three least significant bits are applied directly to the three address inputs of eight-bit addressable latches IC9–IC16. These latches are addressed by the next three data bits via address decoder IC17. In that way, the six least significant bits form the number of the output bit to be addressed (0–63).

Whether the addressed output is high or low depends on data bit 6. Since drivers IC1–IC8 invert, the logic level at the output is inverted with respect to bit 6. The driver ICs can switch up to 500 mA per output.

All outputs are reset (the open-collector outputs of the drivers become high-impedance) when data bit 7 is high and data bit 6 is low. The state of the three address bits is irrelevant.

When it is required to set one bit and reset the others rapidly (for instance, by a running light), data bits 6 and 7 can both be made high, whereupon the wanted bit is addressed with the other six bits.

Most loads can be driven directly by the driver ICs, as long as the levels at the output do not exceed 50 V and 500 mA. If it is required to switch the mains, the outputs can be expanded with a solid-state relay.

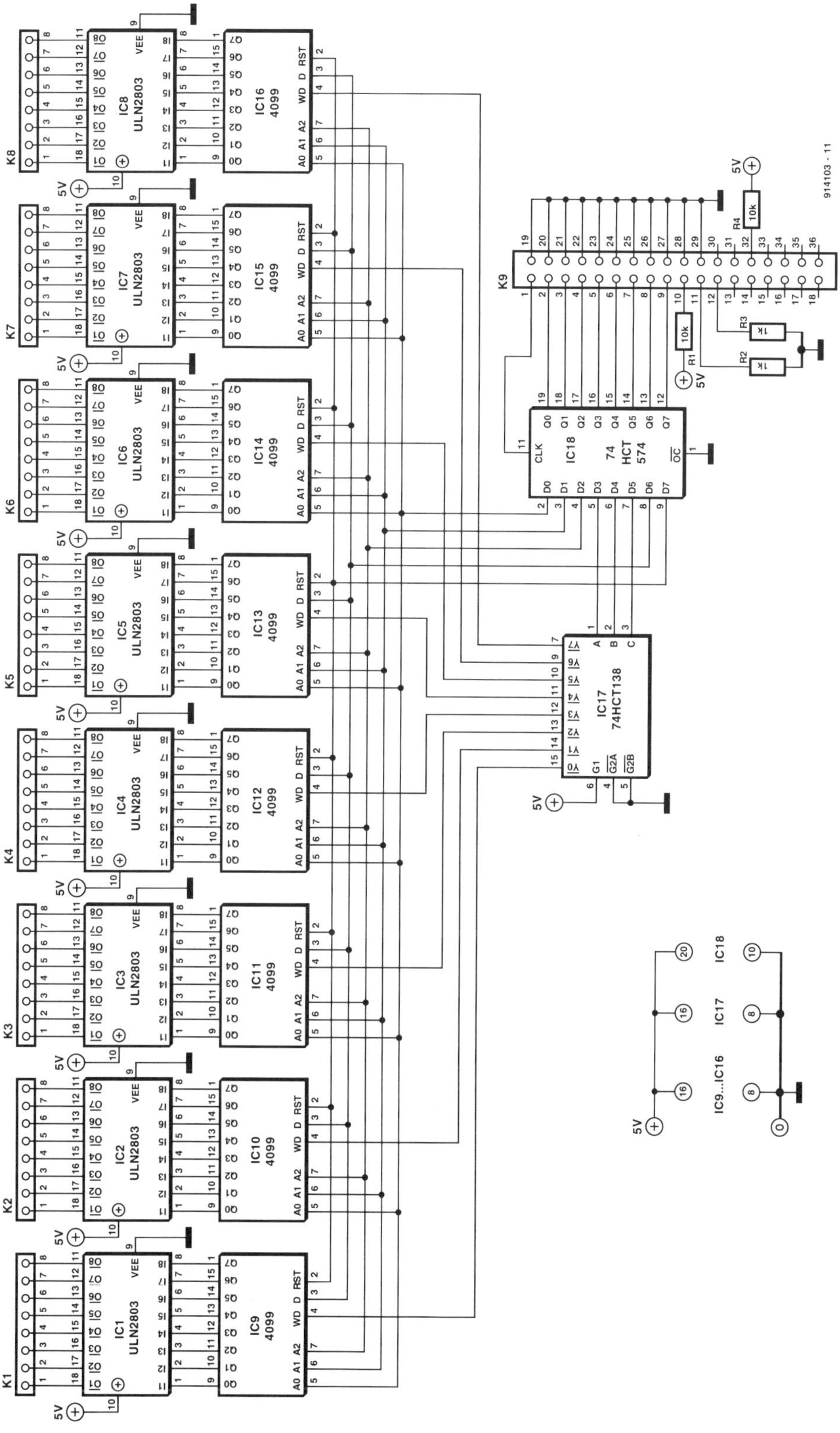
K1
K2
K3
K4
K5
K6
K7
K8
K9
IC1 ULN2803
IC2 ULN2803
IC3 ULN2803
IC4 ULN2803
IC5 ULN2803
IC6 ULN2803
IC7 ULN2803
IC8 ULN2803
IC9 4099
IC10 4099
IC11 4099
IC12 4099
IC13 4099
IC14 4099
IC15 4099
IC16 4099
IC17 74HCT138
IC18 74 HCT 574
R1 10k
R2 1k
R3 1k
R4 10k
5V
IC9...IC16
IC17
IC18
914103 - 11

PC fan control

045 It is an unfortunate and well-known fact that most PCs of the 'IBM and compatible' type make a lot of noise, which is both undesirable (as regards noise in the working environment) and unnecessary (as regards the actual power consumption, which is often quite low). Many fans in PC power supplies are over-rated, noisy, and run at a constant high speed. Fortunately, the authors found that such fans can do their protective job just as well at far lower speeds.

The speed controller presented here consists of (1) a temperature monitor based on a LED as the sensor device, and (2) an idle speed regulator. The combination of the temperature monitor and the idle speed regulator results in a linear relation between temperature and fan speed. In other words: the fan will never run faster than strictly necessary. By virtue of the two separate regulator circuits, there is no interaction between idle speed and temperature regulation, as with many other (less sophisticated) fan controllers. The result is low fan speed (low noise) at low temperatures, as well as good cooling and a safe start-up at all times.

When the temperature inside the PC rises, the voltage drop across the **red** LED, D_2, decreases by approximately –2 mV K^{-1}. This results in a higher output voltage of opamp IC_{1A}. Preset P_1 is used to set the start level, while P_2 determines the slope of the regulator characteristic. The adjustment range of the idle-to-full speed regulation is about 2 °C to 30 ∞C.

The idle speed is set to the desired value by adjusting P_3. Capacitor C_2 ensures a 4-second full-speed start up period, while diode D_3 restarts the fan after a short interruption on the mains.

The outputs of the temperature monitor and the idle speed regulator are 'joined' by two diodes, D_4 and D_5, at the base of T_1. The regulator is stabilized by feeding a

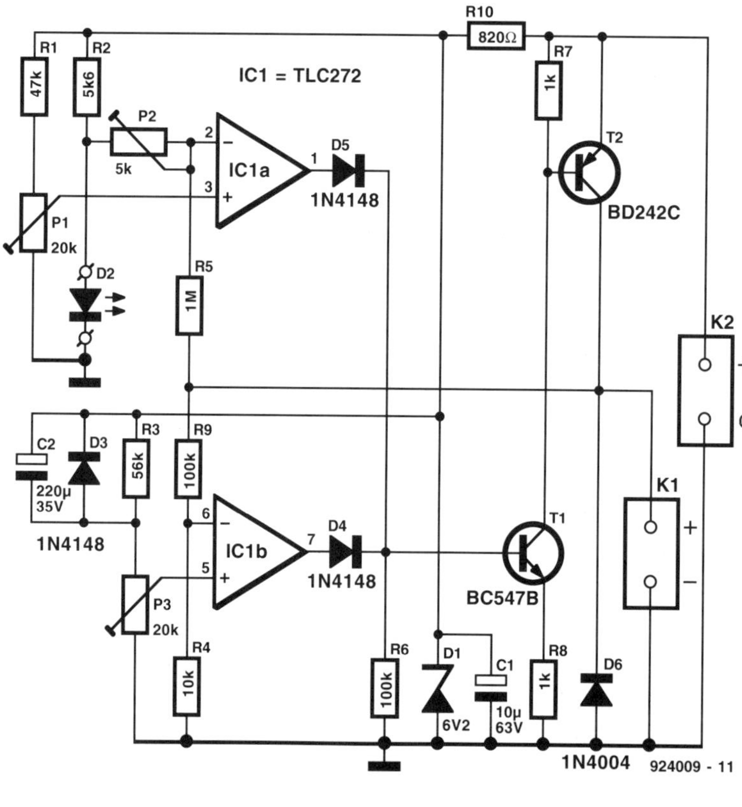

small portion of the fan voltage back to the -inputs of the opamps.

The circuit is built on a small printed circuit board that can be fitted into the PC's power supply. The LED may have to be moved off the board, and connected with wires, to enable it to be fitted in a position where temperature changes are best noticed.

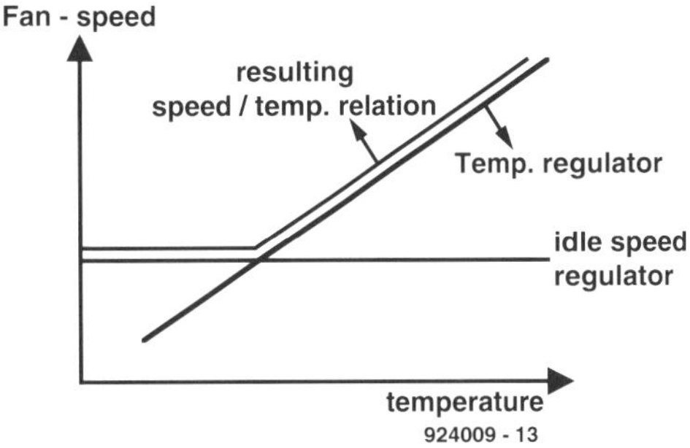

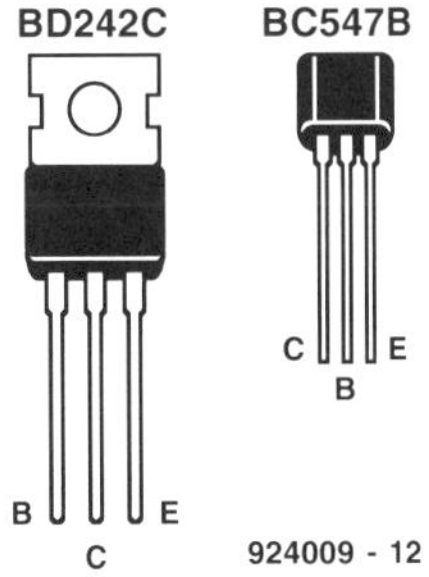

Parts list

Resistors:
R_1 = 47 kΩ
R_2 = 5.6 kΩ
R_3 = 56 kΩ
R_4 = 10 kΩ
R_5 = 1 MΩ
R_6; R_9 = 100 kΩ
R_7; R_8 = 1 kΩ
R_{10} = 820 Ω
P_1; P_3 = 20 kΩ multiturn preset
P_2 = 5 kΩ multiturn preset

Capacitors:
C_1 = 10 μF, 63 V radial
C_2 = 220 μF, 25 V radial

Semiconductors:
D_1 = 6.2 V, 0.4 W zener diode
D_2 = LED, red, 5 mm
D_3; D_4; D_5 = 1N4148
D_6 = 1N4004
T_1 = BC547B
T_2 = BD242C

Integrated circuits:
IC_1 = TLC272

Miscellaneous:
K_1; K_2 = 2-way PCB terminal block, pitch 5mm

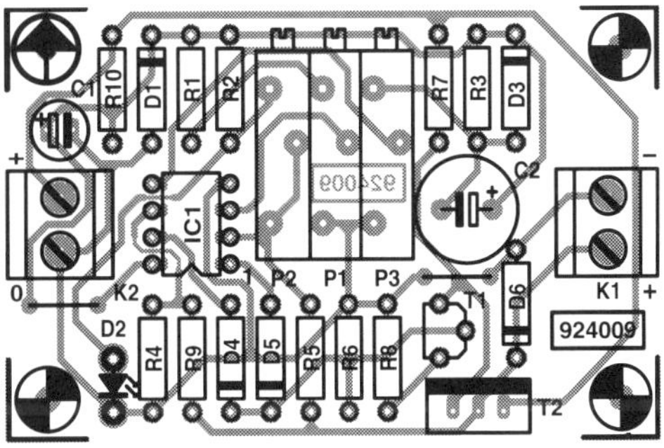

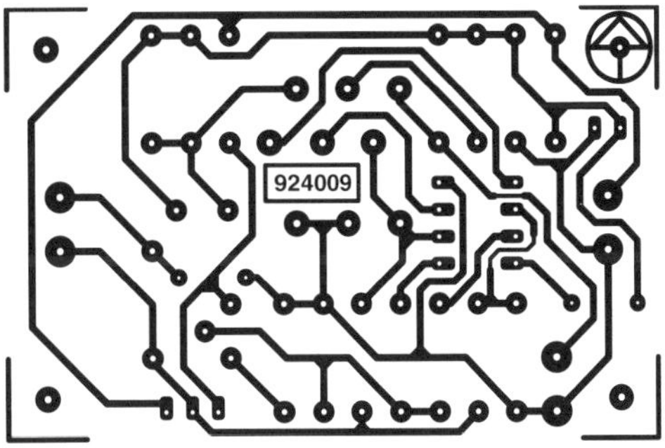

Power-on delay for Atari ST

046 WHEN the Atari ST is provided with an external hard-disk drive, it has to be reset after about 15 seconds, as otherwise the drive is not enabled. The delay circuit presented here obviates that inconvenience by ensuring that the computer is not switched on until a (presettable) period of time has elapsed. The circuit may also be used with the combination of MS-DOS computer and HP DeskJet printer, since the latter can be switched on only after the MS-DOS machine is powered. If that type of printer is used with an Atari ST, the printer must be switched on before the computer is. No doubt, there are other situations where this delay may prove useful.

Operation of the circuit is fairly simple. After the mains has been switched on, capacitor C_3 is charged via R_5 and P_1. When, after a period of time set with P_1, the potential across C_3 has reached 12 V, zener diode D_1 conducts and switches on thyristor Th_1. That device ensures that bridge B_1 provides an a.c. connection between the mains input and the mains output. At the instant the thyristor is switched on, owing to the anode-cathode current, its gate voltage rises slightly. This results in C_3 being charged a little more and thus serves as gate-current buffer to ensure that the thyristor remains on during the mains zero voltage crossings. Therefore, once the thyristor has been switched on, it stays on.

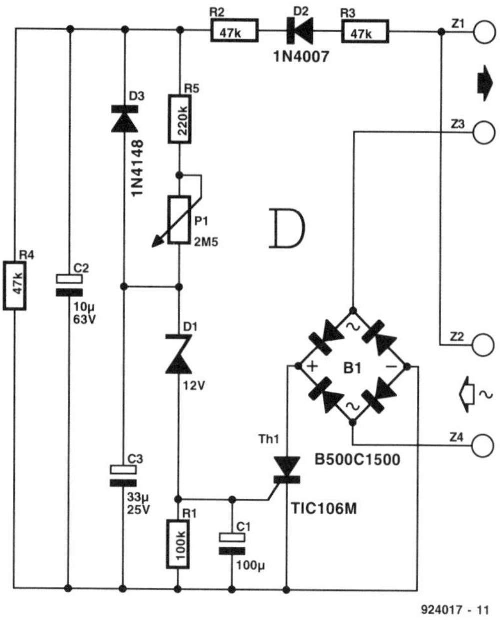

The delay is best built into a small man-made fibre box with integral mains connectors.

The circuit is best built on a small piece of veroboard or other prototyping board, bearing in mind that it will carry the full mains voltage. This means that tracks carrying the mains must be separated by at least 3 mm and preferably 5 mm. This probably entails the removal of tracks between the mains-carrying ones.

Floppy-side chooser

047 This small auxiliary circuit enables either of the two sides of a 3 1/2 in. floppy disk used in an Atari ST drive to be chosen. Some LEDs show which side has been selected, while others indicate whether a read or a write operation is in progress.

In Fig. 1, bistable IC_{1b} serves to choose one of the two sides of the floppy disk. A sensor, formed by a small disk of conducting material, is connected to its clock input via inverter IC_{2f}. The input of the inverter is linked to the +5 V rail via two 10 MΩ resistors.

When the sensor is touched, the input of IC_{2f} is pulled low via the skin resistance. This results in a positive leading transition (edge) to the clock input of IC_{1b}, whereupon the levels at pins 12 and 13 of the bistable change state.

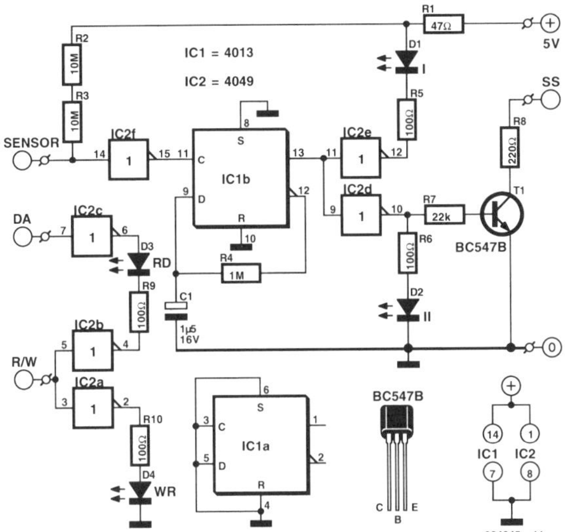

924045 - 11

Network R_4-C_1 suppresses spurious pulses during operation and prevents the bistable continually changing state if the sensor remains touched.

Pin 13 ($\overline{Q}$) of IC_{1b} is linked to inverters IC_{2d} and IC_{2e}, each of which controls an LED: D_1 for side I and D_2 for side II. Since D_1 is connected to the +5 V supply rail and D_2 to ground, only one of the two diodes can light at any one time.

Transistor T_1 serves to alter the level on the Side Select–SS–line of the Atari.

Inverters IC_{2a}–IC_{2c} control D_3 and D_4. The inputs of IC_{2a} and IC_{2b} are linked to the R/W signal of the Atari, while that of IC_{2c} is coupled to the 'disk LED' above the keyboard. This ensures that the 'read' LED lights only when the drive is in motion.

The supply for the auxiliary circuit may be taken from the Atari, since a current of only a few mA is drawn.

Populating the small PCB is simplicity itself. Most of the work will involve fitting the board inside the Atari. The board must be located in the upper part of the keyboard beside the drive LED.

A small front panel as shown in Fig. 2 will be found useful. Note that the LEDs protrude through the small holes in this panel. This means that five holes need to be drilled in the upper panel of the Atari: four for the LEDs and one for the connection to the sensor.

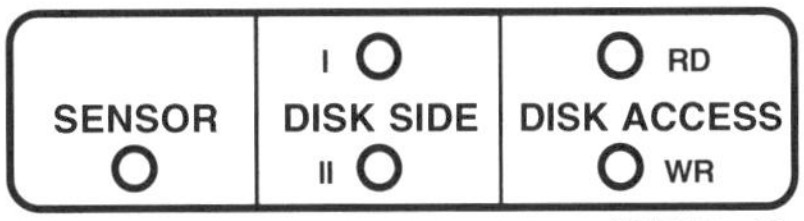

924045 - 12

The following connections must be made:

- earth of the PCB to pin 10 of the 74244 underneath the keyboard;
- 5 V line from pin 20 of the 74244 underneath the keyboard to the present circuit;
- SS line to pin 21 of the YM2149;
- R/W line to pin 25 of the DMA controller adjacent to the YM2149 underneath the drive;
- pin DA to the upper terminal of the resistor in series with the drive LED, just to the left of the 74244;
- sensor to the associated pin on the present PCB.

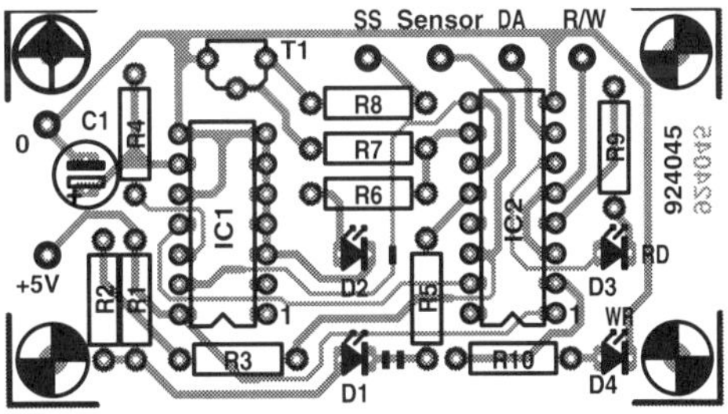

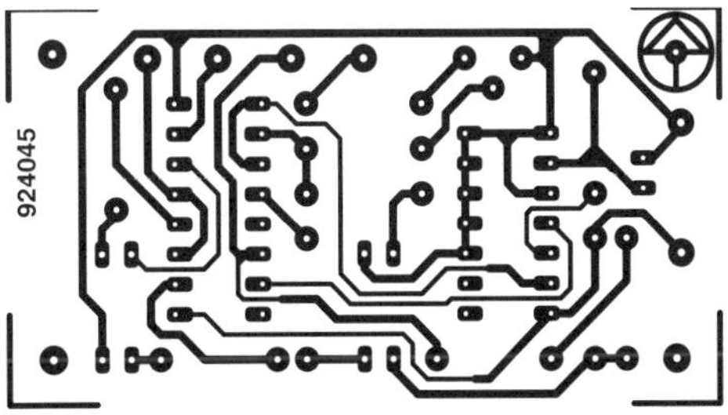

Mini keyboard for Z-80 card

048 Depending on the application of the Multifunction Z80 card (Ref. 1), a basic set of switches to control the program flow may be required. Also, in some cases, the PC/XT keyboard that can be connected to the card may prove a little too bulky to be carried around. The keyboard plus LED indication described here is connected to PIOA on the multi-function Z80 card. The 10 keys are arranged horizontally (see PCB layout) on a board fitted behind the enclosure front panel, and serve to control functions determined by the user software. With the exception of S_1 and S_{10}, each key has an associated LED to indicate the key status.

The number of the key pressed is conveyed to the PIO in the form of an inverted 4-bit binary number supplied by IC_2, a 10-to-4 line priority encoder Type 74HCT147. No key pressed gives output value 15 (binary), while, for instance, key 2 gives 15–2=13.

Bits 6 and 7 are normally logic high, but can be strapped low if desired by fitting jumpers. They may be used as required by the application. A suggestion: the Z80 software, on reading the key number, causes the PIO on the Z80 card to pull line PA4 high, which enables shift register IC_1. Next, the PIO causes one of the LEDs to light by outputting the appropriate number of clock pulses via line PA5. Since the LED activity is controlled by the keyboard decoder software, rather than directly by the keys, the user has a good indication that the selected function has been accepted by the system.

In principle, K_1 on the keyboard unit can be connected to any of the three PIOs on the Z80 card: K_4, K_5 or K_6. The keyboard/LED routines in the BIOS EPROM (ESS6121), however, are based on connection to K_4 (PIOA). The routines 'READEXTRAKEY' and 'LEDOUTPUT' provided in the BIOS, and a demonstration program on the project diskette (ESS1710), make life easy for the Z80 programmer by offering simple ways of reading the keys, and controlling the LEDs, respectively.

The user is, of course, free to determine

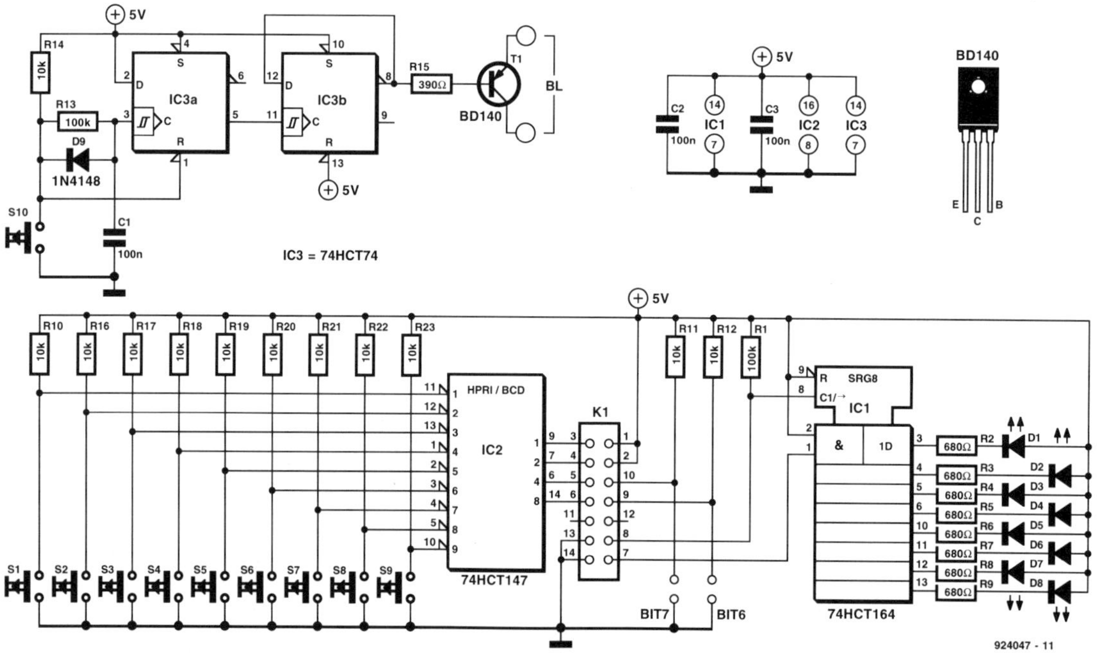

the functions of the keys and LEDs, or, indeed, to omit one or more keys or LEDs not required for his application.

Switch S_{10} has a special function, and is not read by the Z80: it is the LCD back light control switchthat uses two bistables (IC_{3a} and IC_{3b}) to provide a toggle function. Transistor T_1 is effectively connected in series with the back light supply and the back light input of the LCD (more details on this may be found in Ref. 1). The outputs marked 'BL' are connected to the jumper marked 'LCD' on the Z80 card. If the back light is very weak, the BL connections should be interchanged. In this application, the BD140 will happily work in both directions, but its current gain will be much lower when the collector functions as the emitter.

Construction of the keyboard unit is simple. As indicated by the dashed outlines on the component overlay, **the keys and LEDs are fitted at the solder side of the board**.

Finally, the unit is connected to the Z80 card via a length of 14-way flat-cable fitted with IDC sockets. Connector K_1 may be either a 14-way boxheader or a 14-way male IDC connector. The latter may be soldered permanently to the PCB, if the other end of the cable is fitted with an IDC socket. Alternatively, if a boxheader is used in position K_1, the flatcable will have IDC sockets at both ends. Power is supplied via this cable by the Z80 card.

Parts list

Resistors:
R_1;R_2 = 100 kΩ
R_2–R_9 = 680 Ω
R_{10}; R_{11}; R_{12}; R_{14}; R_{16}–R_{23} = 10 kΩ
R_{15} = 390 Ω

Capacitors:
C_1; C_2; C_3 = 100 nF

Semiconductors:
D_1–D_8 = LED, 3 mm, red
D_9 = 1N4148
T_1 = BD140

Integrated circuits:
IC_1 = 74HCT164
IC_2 = 74HCT147
IC_3 = 74HCT74

Miscellaneous:
K_1 = 14-way box header or male IDC connector (see text).
S_1–S_{10} = PCB-mount push-to-make button, Type 3CTL3 (Amroh).
PCB Ref. 924047.

Reference:

1. "Multipurpose Z80 card", *Elektor Electronics* May and June 1992.

80C552 microprocessor system

049 Here is something for microcontroller enthusiasts to grind their teeth on. The 80C552 microcontroller from Philips Compo-nents is an upgraded derivative of Intel's 8032. It costs little more than the 8032, yet offers the following extras: (1) eight analogue inputs; (2) a 10-bit A-D converter; (3) a 'Timer2' with many extra features; (4) a 'Timer3' watchdog function; (5) an on-board I²C interface; (6) 16 I/O lines; (7) two pulsewidth modulation outputs; and (8) a 16-MHz clock.

Here, the 80C552 is used in a single-board microcontroller application, which is intended as an experimental system rather than a replacement for an existing processor. The board is aimed at versatility, and accepts EEPROMs, EPROMs, or RAMs, or a combination of these, as memory devices,

The 80C552 board offers a multitude of I/O connections for your own applications. All I/O lines, except P0 and P2, are accessible via connectors K_1, K_2, K_4 and K_5. The I²C lines, $P_{1.6}$ and $P_{1.7}$, are taken to a

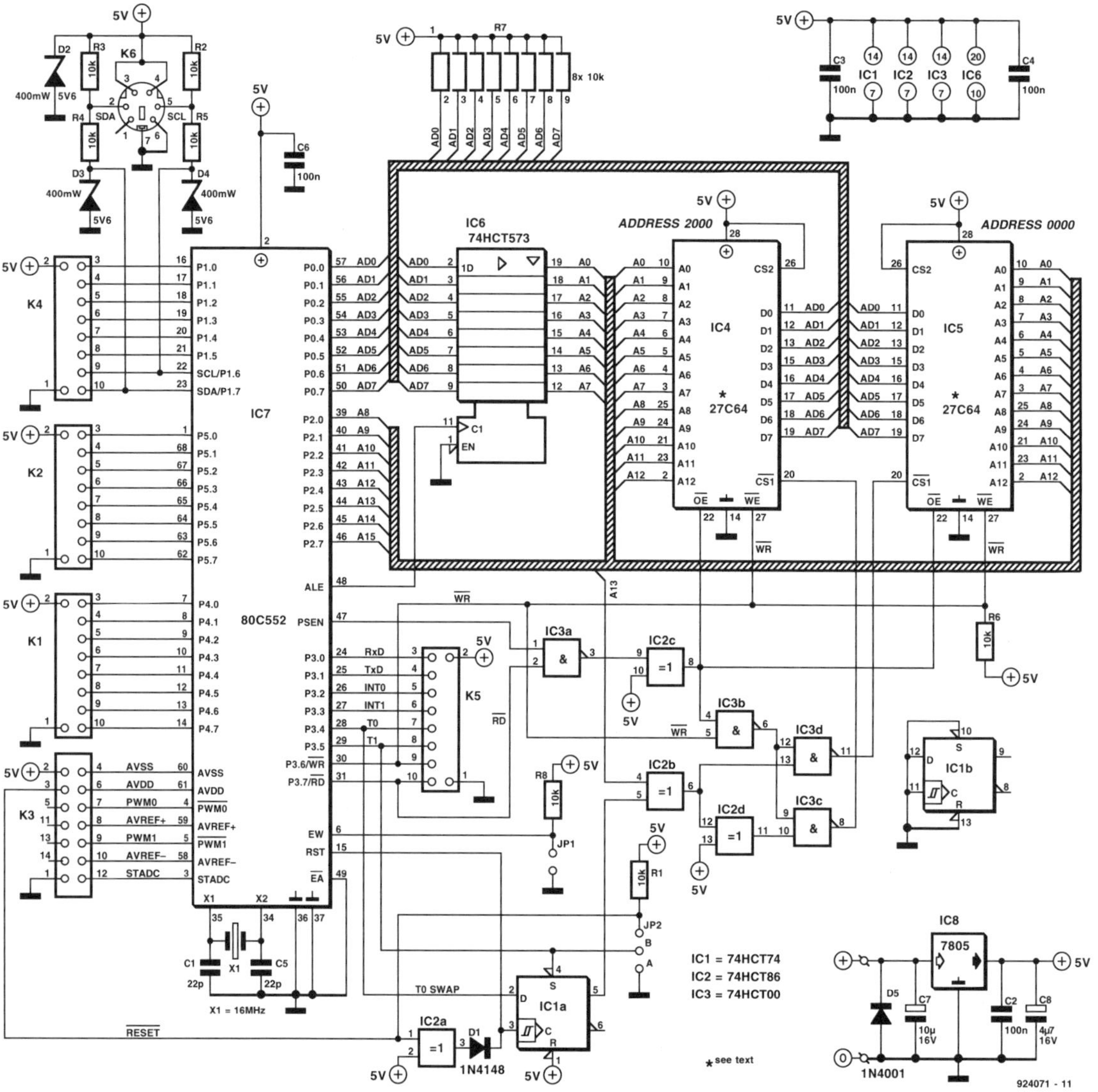

5V
K6
SDA
SCL
R2 10k
R3 10k
R4 10k
R5 10k
D2 400mW 5V6
D3 400mW 5V6
D4 400mW 5V6
K1
K2
K3
K4
K5
C6 100n
IC7
80C552
P1.0
P1.1
P1.2
P1.3
P1.4
P1.5
SCL/P1.6
SDA/P1.7
P5.0
P5.1
P5.2
P5.3
P5.4
P5.5
P5.6
P5.7
P4.0
P4.1
P4.2
P4.3
P4.4
P4.5
P4.6
P4.7
AVSS
AVDD
PWM0
AVREF+
PWM1
AVREF–
STADC
X1
X2
P0.0
P0.1
P0.2
P0.3
P0.4
P0.5
P0.6
P0.7
P2.0
P2.1
P2.2
P2.3
P2.4
P2.5
P2.6
P2.7
ALE
PSEN
P3.0
P3.1
P3.2
P3.3
P3.4
P3.5
P3.6/WR
P3.7/RD
EW
RST
EA
RxD
TxD
INT0
INT1
T0
T1
C1 22p
C5 22p
X1 = 16MHz
R7 8x 10k
AD0
AD1
AD2
AD3
AD4
AD5
AD6
AD7
A8
A9
A10
A11
A12
A13
A14
A15
IC6
74HCT573
1D
C1
EN
ADDRESS 2000
IC4
27C64
ADDRESS 0000
IC5
27C64
CS1
CS2
OE
WE
WR
RD
C3 100n
C4 100n
IC1
IC2
IC3
IC6
IC3a
IC3b
IC3c
IC3d
IC2a
IC2b
IC2c
IC2d
IC1a
IC1b
R6 10k
R8 10k
R1 10k
JP1
JP2
T0 SWAP
RESET
D1
1N4148
IC1 = 74HCT74
IC2 = 74HCT86
IC3 = 74HCT00
* see text
IC8
7805
D5
1N4001
C7 10µ 16V
C2 100n
C8 4µ7 16V
924071 - 11

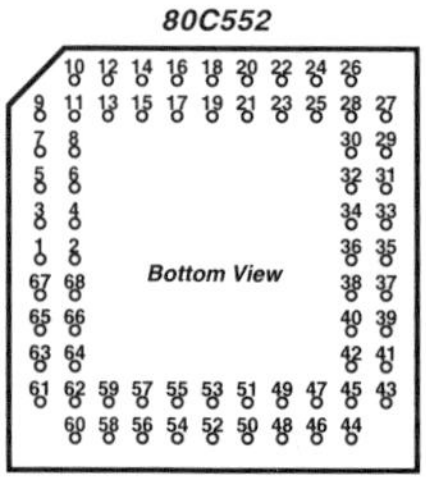

80C552
Bottom View

simple I^2C interface around a 6-way mini DIN connector, K_6. The TxD and RxD lines are available for TTL-level serial communication via two pins on connector K_5, which also carries read, write, interrupt and timer signals.

Those of you familiar with the 8032 processor will find that the address decoding circuit used here is fairly extensive. Also, unusually, the reset input of the CPU is connected to a bistable. As to the address decoding, this must meet a number of special requirements: it must be possible to read instructions (using PSEN\) as well as read data (using RD\), and write data (using WR\). Further, it is possible to swap the address ranges of the IC_4 and IC_5 positions, with the CS1\ (chip enable) inputs of the respective EPROMs (or EEPROMs) used to select and de-select them, so that the de-selected EPROM hardly consumes power. When the processor is switched to 'idle' mode, its current consumption is only one-third of the normal value, while the CS1\ inputs of both EPROMs are automatically taken high, thus reducing power consumption even further.

Bistable IC_{1a} and XOR gate IC_{2b} serve to swap the positions, 0000H-1FFFH and 2000H-3FFFH, of the two EPROMS in the address map. Provided EEPROMs are used, this allows an interesting programming trick: reload the 'upper' EEPROM (i.e., the one with the highest address) with the aid of a program run from the 'lower' EEPROM. *Look, no hands*! No opening of cases, no extracting of EPROMs, and no more time wasted on erasing and reprogramming EPROMs. For instance, the program in the 'lower' EEPROM may read the data supplied by the I^2C or the serial input, organize it, and store it into the 'upper' EEPROM. Next, the program causes the T0 line to be pulled low and also stops triggering the watchdog. After some time, the watchdog will force an internal reset, which also pulls RST (pin 15) high for three clock cycles. This works as a clock for bistable IC_{1a}. The bistable copies the level on T0 to its Q output. This enables IC_{2b} to swap the EEPROM address ranges by inverting address line A13, so that the program just loaded into the 'upper' EEPROM is executed.

Because interrupt vectors are always located from address 0000H, address ranges have to be swapped physically rather than in software. However, this swapping is not allowed while the CPU is about its normal business of fetching and executing codes, because an address might change in the middle of an opcode fetch action. At a clock of 6 MHz, this can be done with impunity, but definitely not at a clock of 12 MHz or 16 MHz as used here, which forces us to use the 'hardware reset' trick.

It should be noted that the analogue port, P_5, may function as an input only (it may also be used to accept digital levels). The two I^2C pins, SCL and SDA ($P_{1.6}$ and $P_{1.7}$), may be used as an input or an output. Contrary to the other I/O pins, they do not have internal pull-up resistors. Further, it is recommended not to use $P_{3.7}$ and $P_{3.6}$, as this will interfere with the normal opcode fetch operations.

If you intend to connect an LCD(liquid crystal display) module to the 80C552 system, it is best to use the LCD in 4-bit mode, because that allows the display to be connected to 7 port lines only.

A direct bus connection is not possible at clock speeds higher than 10 MHz.

Parts list

Resistors:

R_1; R_2; R_3; R_6; R_8 = 10 kΩ
R_4; R_5 = 330 Ω
R_7 = 8-way 10 kΩ SIL

Capacitors:

C_1; C_5 = 22 pF
C_2; C_3; C_4; C_6 = 100 nF
C_7 = 10 µF, 16 V radial
C_8 = 4.7 µF, 16 V radial

Semiconductors:

D_1 = 1N4148
D_2; D_3; D_4 = 5.6 V, 0.4 W zener diode
D_5 = 1N4001

Integrated circuits
IC_1 = 74HCT74
IC_2 = 74HCT86
IC_3 = 74HCT00
IC_4; IC_5 = 27C64 (EPROM) or 28(C)64 (RAM) or 64(C)64 (EEPROM)
IC_6 = 74HCT573
IC_7 = PCB80C552-4WP (16-MHz 68-PLCC)
IC_8 = 7805

Miscellaneous:
K_1; K_2; K_4; K_5 = 10-way box header.
K_3 = 14-way box header.
K_6 = 6-way PCB-mount mini DIN socket.
X_1 = 16MHz quartz crystal.
68-way PLCC socket.
PCB Ref. 924071.

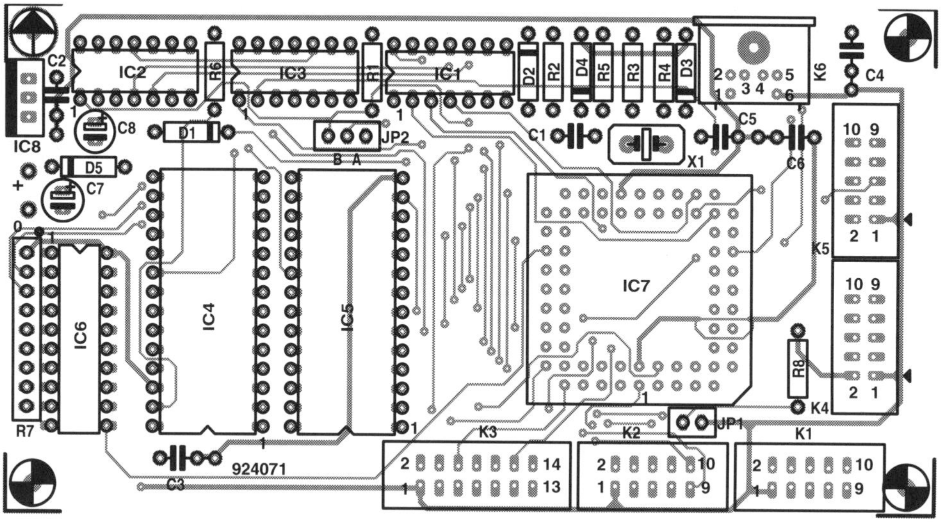

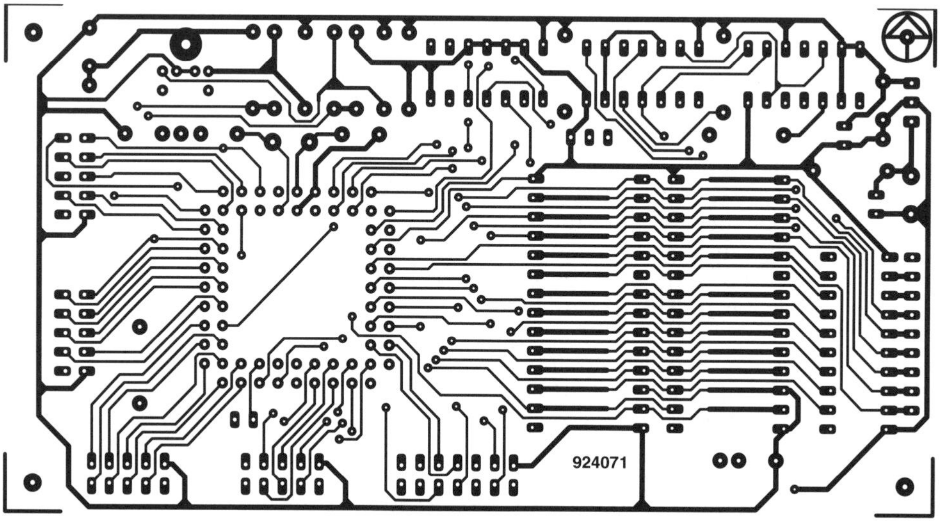

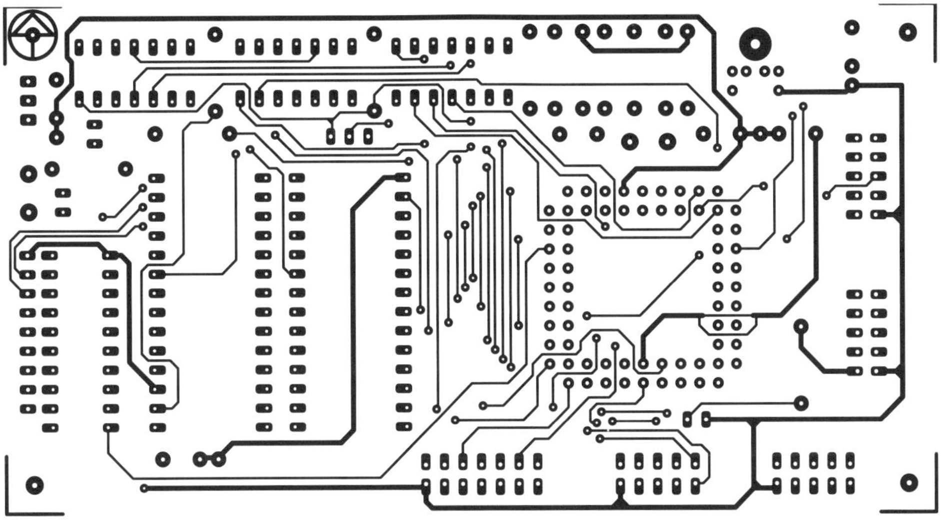

Compact RS232 isolator

050 The MAX252 from Maxim is a special IC for designing an electrically isolated RS232 connection. Such links are particularly important when several pieces of equipment (which may or may not be isolated from the mains) are to be interconnected and they have different earth potentials.

The chip is available in two modes: A (expensive) and B (inexpensive). Note, however, that even inexpensive is relative, because the B type still costs more than £30.

The maximum isolation voltage that the B-type can handle for one minute is 500 V, whereas the A-type can handle up to 1250 V. Moreover, the A-type can have a constant voltage across its terminals of 130 V. The maximum transmission speed of both types is 9600 baud.

The internal transformer of the chip provides the supply voltage for the right-hand side of the diagram. The input and output levels at the left-hand side are TTL compatible. The inputs have a hysteresis of roughly 0.5 V and pull-up current sources of 0.4 μA for the suppression of noise signals. The inputs and outputs at the right-hand side are RS232 compatible. Outputs $T1_{out}$ and $T2_{out}$ deliver a voltage of 7.5 V to a load of 3 kΩ.

With a supply voltage of 5 V, the IC draws a current of 90 mA.

Pin 11 of the MAX252 is an active-low output-enable input and pin 3 an active-high shut-down input. A high level at pin 3 causes the 130 kHz oscillator of the integrated supply to be switched off, $T1_{out}$ and $T2_{out}$ to become high-impedance, the 4 μA pull-up inputs to be deactuated, the power acceptance to be reduced to 50 μW.

RS232 compatible levels at the left-hand side can be provided by the MAX233 as shown in the diagram. This IC also arranges the necessary inversion of the various signals.

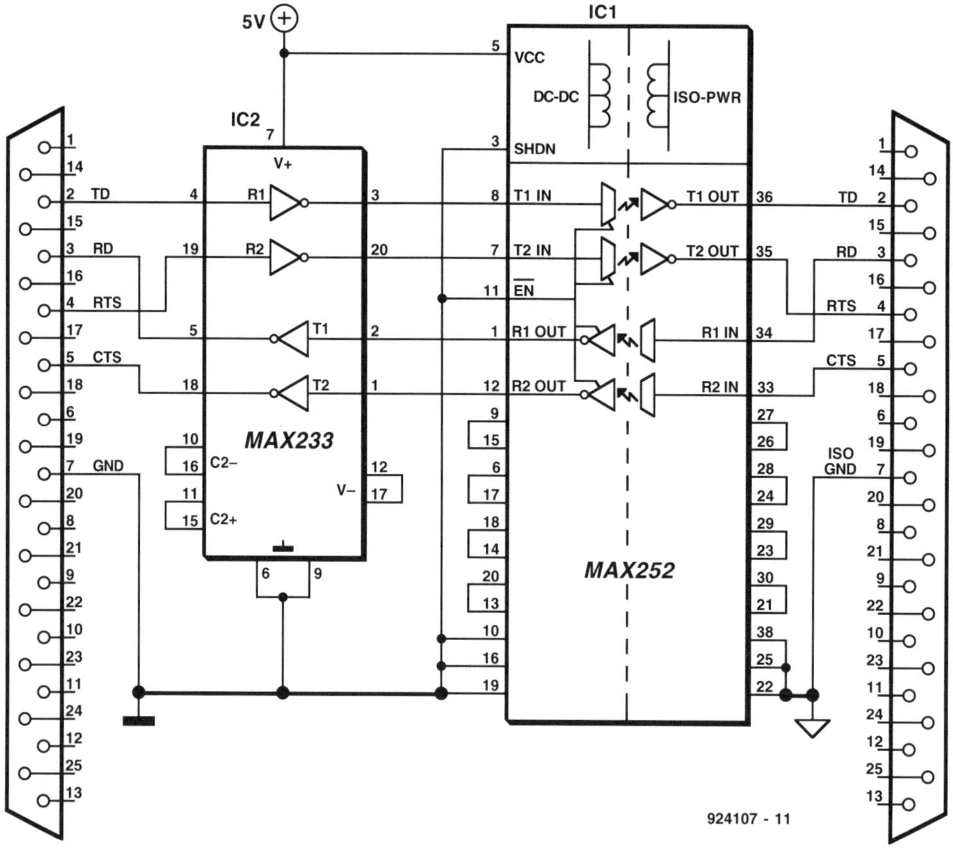

Joystick converter for PCs

051 Although frowned upon by many as addictive and devoid of educational value, computer games can be great fun to play. This little circuit allows you to connect a digital (switch-based) joystick to the analogue inputs on the Game Control Adapter (GCA) which is available on most IBM PC/ATs and compatibles.

A normal joystick interface consists of two inputs, each of which is connected to the two potentiometers in the joystick. Traditionally, the resistance formed by each of these potentiometers is used to charge a capacitor. When the capacitor voltage has reached a certain level, a bistable toggles. This means that the charging time is a function of the potentiometer position. The output state of the four bistables and the four digital inputs is a bit pattern found at address 201H. A software counter then tells the system how long it takes before a bistable bit toggles.

This interface works fine in a closed, unchanging system. Unfortunately, the immense increase in processor speeds that has come about in recent years has caused compatibility problems with regard to the use of the analogue joystick inputs. The reason is simple: the same joystick read routine gives different results on PCs with different clocks speeds. The effect of clock speed differences has been reduced to some extent by making the timing variable with the aid of a switch. In this way, the program still reads the desired value (well, roughly ...).

Most analogue joysticks have a centre position adjustment, which, unfortunately, can be turned by accident in the excitement of a game. Consequently, the program (game) will not respond properly to your signals any more, so that you will be unable to reach the highest score. Fortunately, many programs have a fairly wide margin around the centre position, so that a small misadjustment need not be fatal. This type of software does not follow the above concept of reading the joystick position and turning it into a proportional value. Instead, it just reads left, right, up and down, and no proportional value at all. This circumstance is exploited by the present design.

Arguably, if proportional movement translation is not used, there is no reason why the potentiometers in the joystick could not be dropped altogether and replaced by simple switches, while still using the analogue input on the PC. Fortunately, switch-based joysticks exist – they also go under name 'digital joystick'.

With reference to the circuit diagram, the joystick is in the centre position if it is not operated, and none of the four switches is closed. This means that the PC 'sees' the resistance formed by $R_2//(R_1+D_1)$. The value of R_1 is calculated such that the GCA card 'sees' a joystick resistance of about 50 kΩ, in spite of the diode. If the lever is moved to the right, pin 4 on K_1 is pulled to ground, so that D_1 blocks and the computer 'sees' R_2 (100 kΩ) only. If the lever is moved to the left, pin 3 on K_1 will go low. This causes T_1 to be switched on via R_3, and, consequently, the GCA card to 'see' a joystick resistance of about 100 Ω. In this way, the input signals are converted into resistance values 100 Ω, 50 kΩ or 100 kΩ. The same applies to the up/down section built around T_2.

The 'fire' button is read directly via pin 6 on K_1. If your joystick has a second button, this is read also via pin 5 on K_1. Also note the presence of the +5 V supply voltage on pins 8 and 15 of K_2 (computer's side) – this may be used for inductive joysticks and/or types with an 'auto-fire' button.

Finally, R_1 (R_5) may be replaced by a series combination of a 68-kΩ resistor and a 50-kΩ preset. This allows you to adjust the joystick's centre position, and play 'older' games that lack an up-to-date initialization routine.

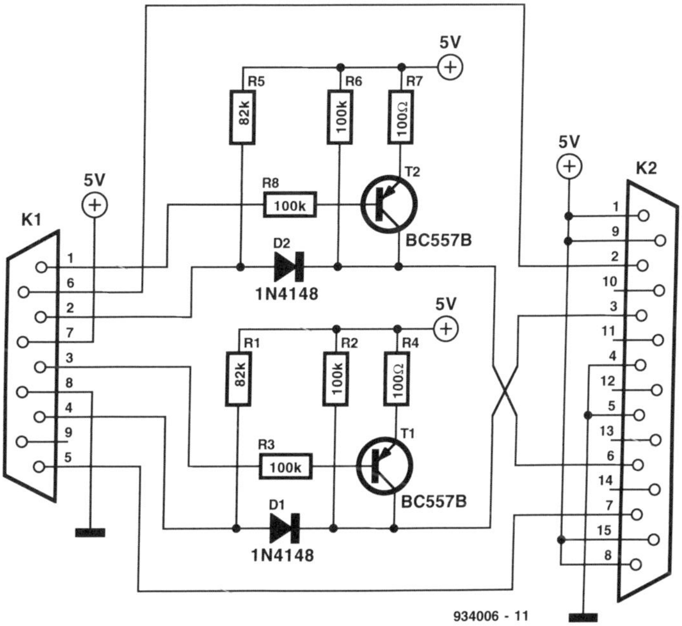

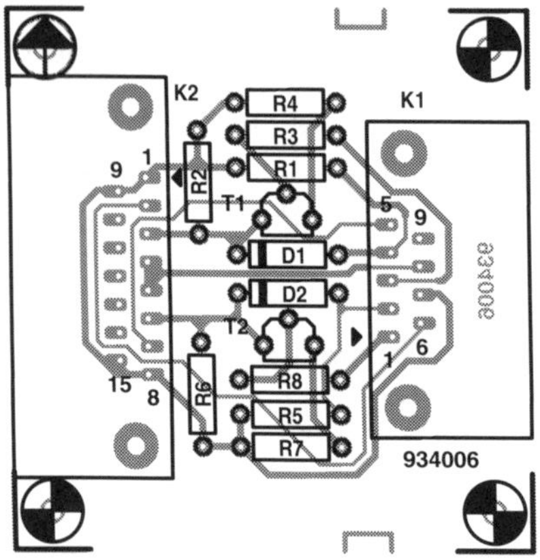

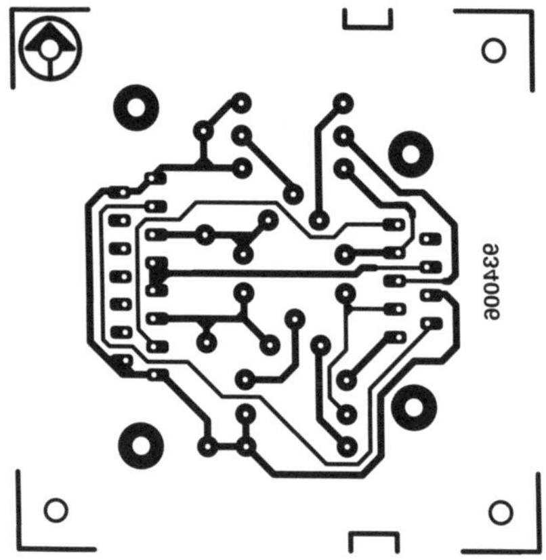

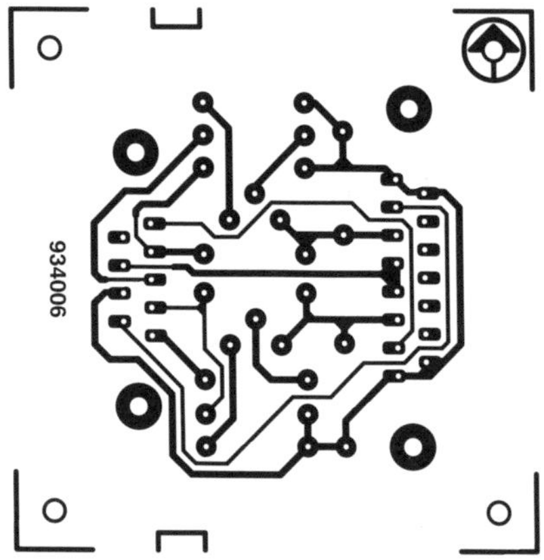

Parts list

Resistors:
R_1, R_5 = 82 kΩ
R_2, R_3, R_6, R_8 = 100 kΩ
R_4, R_7 = 100 Ω

Semiconductors:
D_1, D_2 = 1N41482
T_1, T_2 = BC557B

Miscellaneous:
K_1 = 9-way male sub-D connector, PCB mount, angled pins
K_2 = 15-way male sub-D connector, PCB mount, angled pins

Compact A-D converter

052 Although good and inexpensive integrated analogue-to-digital (A-D) converters are now readily available, it may be instructive to build one from discrete components.

The present convertor is based on a Type TLC274 IC, which contains four comparators. The outputs of these stages also form the output of the converter.

The problem with designing A-D converters is producing a correctly tracking reference voltage. In fact, each bit requires a small digital-to-analogue (D-A) converter to generate the needed reference. This problem is usually solved by assigning the MSB comparator half the supply voltage as reference: this is produced by potential divider R_1-R_2. For each subsequent bit, the preceding bits are then added to the reference voltage, which requires a few resistors. In case of bit B, this means that bits C and D are used; for bit A, bits B, C and D are used. The translating of the level of these bits is effected by IC_2 and R_1–R_{14}.

The regulated 5-V supply voltage forms the basis for the reference voltage. The levels at the outputs of the buffers should ideally be 5 V (logic 1) or 0 V (logic 0). The resistors in the potential divider should be close tolerance types to ensure good linearity. The higher the resolution (the more bits) of the A-D converter, the more accurate the resistors need to be. It should also be borne in mind that the output level of the buffers in IC_2 deviates more and more from the ideal the greater the current that must be provided. In other words, the value of the resistors must be relatively high. The problem is further minimized by connecting two non-inverting gates in parallel to increase fan-out. With values as shown, the voltage drop at the MSB output buffer (that which provides the largest current) was 6 mV in the prototype. Compared with the value of the LSB of 312.5 mV, this is negligible.

The conversion speed depends on the propagation times of the comparators and buffers. Although the TLC274 performs acceptably, for optimum results true comparators should be used.

The input sensitivity of the circuit is set with P_1. Diodes D_5 and D_6 protect the opamp inputs against too high potentials.

The current drawn with all LEDs off is about 7 mA; with all four LEDs on, this rises to about 20 mA.

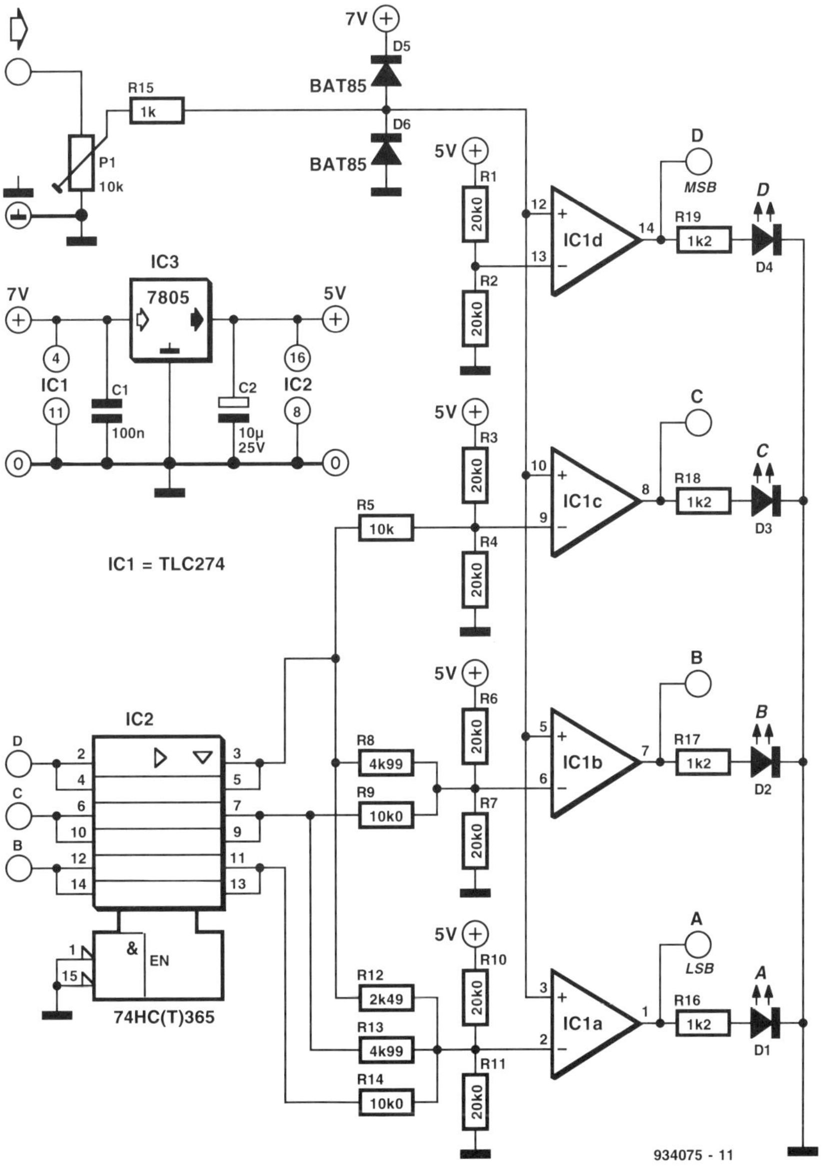

Extended PC drive LED

053 When the disk drive indicator of a computer cannot be read (since the computer is under or beside the table) it may be useful, if difficulties arise, to have an extended indicator alongside the monitor.

The circuit described here is just that. It consists of a p-n-p transistor, an LED, and two resistors, or an n-p-n transistor, an LED, and two resistors, Which is to be used depends on the LED drive of the floppy disk or hard disk.

The LED in the computer must be branched off to a D connector to be fitted at the rear of the computer.

The small circuit is easily constructed on a small piece of prototyping board and built into a small case. The whole is then connected via a cable terminated into a D connector to the computer. The small case can then be located wherever desired.

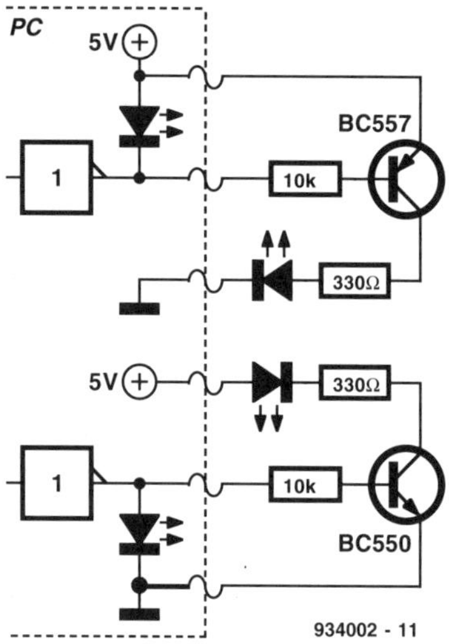

I2C bus fuse

054 All I²C cards published so far in this magazine (see References) are normally powered by the 5 V supply in the PC via the 6-way DIN socket on the I²C PC insertion card. A drawback of this arrangement is that the 5 V supply in the PC is at risk of being short-circuited by a fault on any of the I²C cards connected. Since most PC power supplies are capable of supplying very high currents at the 5-V output (20+ A is typical!), short-circuits can cause a lot of smoke and annoyance.

The circuit shown here protects your costly hardware by inserting a fast 1 A glass fuse in the 5 V supply line between the I²C interface card in the PC and the first I²C card in the chain. Alternatively, for even better security, a fuse card may be inserted into every cable that connects two I²C boards.

Two LEDs are fitted on the fuse card to signal the presence of the 5 V supply voltage ahead of and behind the fuse. The LEDs cause an additional current consumption of about 20 mA.

The printed circuit board shown allows five fuse cards to be built.

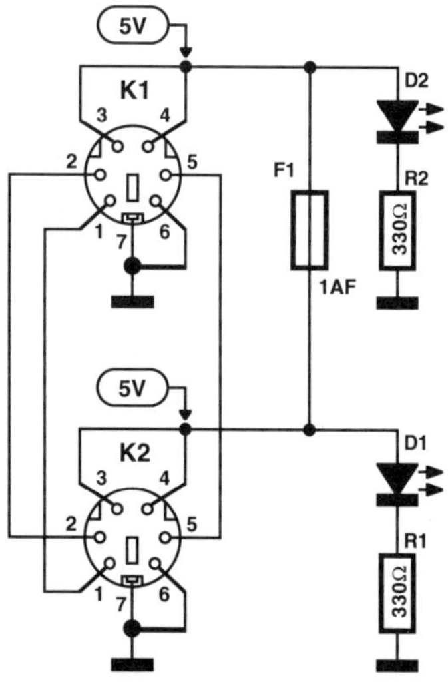

Parts list

Resistors:
R_1, R_2 = 330 Ω

Semiconductors:
D_1, D_2 = LED, 3 mm, red

Miscellaneous:
K_1, K_2 = 6-way mini-DIN socket for PCB mounting
F1 = fuse, 1 A (fast), with holder for PCB mounting
PCB Ref. 934016

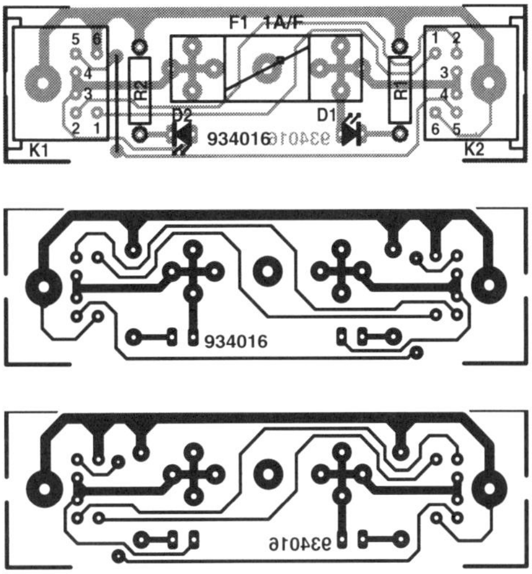

References:

1. I²C interface for PCs. *Elektor Electronics* February 1992.
2. ADC/DAC and I/O for I²C bus. *Elektor Electronics* March 1992
3. I²C LED display. *Elektor Electronics* June 1992.
4. I²C opto/relay card. *Elektor Electronics* February 1993.

Speed indicator for PCs

055 For some reason or other, the principal feature of a PC is reckoned by many to be its CPU clock speed, and that is probably why many PC users have developed a habit of judging a PC by what appears on the 2-digit speed readout. Machines with a display reading below '33' (for 33 MHz), or which lack a display altogether, are not worth talking about. In other words, if your PC is not fast enough, you are 'out' before you know it.

To fool these sufferers of 'PC speed phobia', and to bring you up to par again, we propose a circuit that will feign a much higher CPU speed than actually used in the machine. In this way, you will be able to accredit, say, an 'old' 12-MHz AT with a 'state-of-the-art' CPU turbo speed of 66 MHz, or even 99 MHz! After all, only you will know for sure what's really ticking inside the grey desktop case.

Joking apart, resistors R_1–R_{14} limit the LED segment currents, and their value may be changed as required (within limits, of course) to reduce or increase the readout brightness. The 'turbo' speed indication is switched on via transistor T_2. Applying +5 V to the base of T_2 causes T_1 to be switched off. The reverse situation, T_2 off and T_1 on, applies when the turbo switch is set to the 'normal' position.

Unfortunately there exist differences between PCs as regards the logic levels used to switch between 'normal' and 'turbo' CPU speed. In most cases, turbo speed is selected when the switch contact is a 'make' type that closes to ground. The table shown applies to that configuration. The table is also valid if the 'normal' position of the switch makes contact to +5 V; only R_{17} must then be removed. In case turbo speed is selected when the switch contact closes to +5 V, the table needs to be modified, and R_{17} omitted. These modifications are also required, but R_{17} must be left in place, if 'normal' speed is selected with the switch contact closed to ground.

The modifications to the table are as follows:

(1) in the fourth ('Normal') column, change all entries ×3 into ×2

(2) in the fifth ('Turbo') column, change all 5s into 6s.

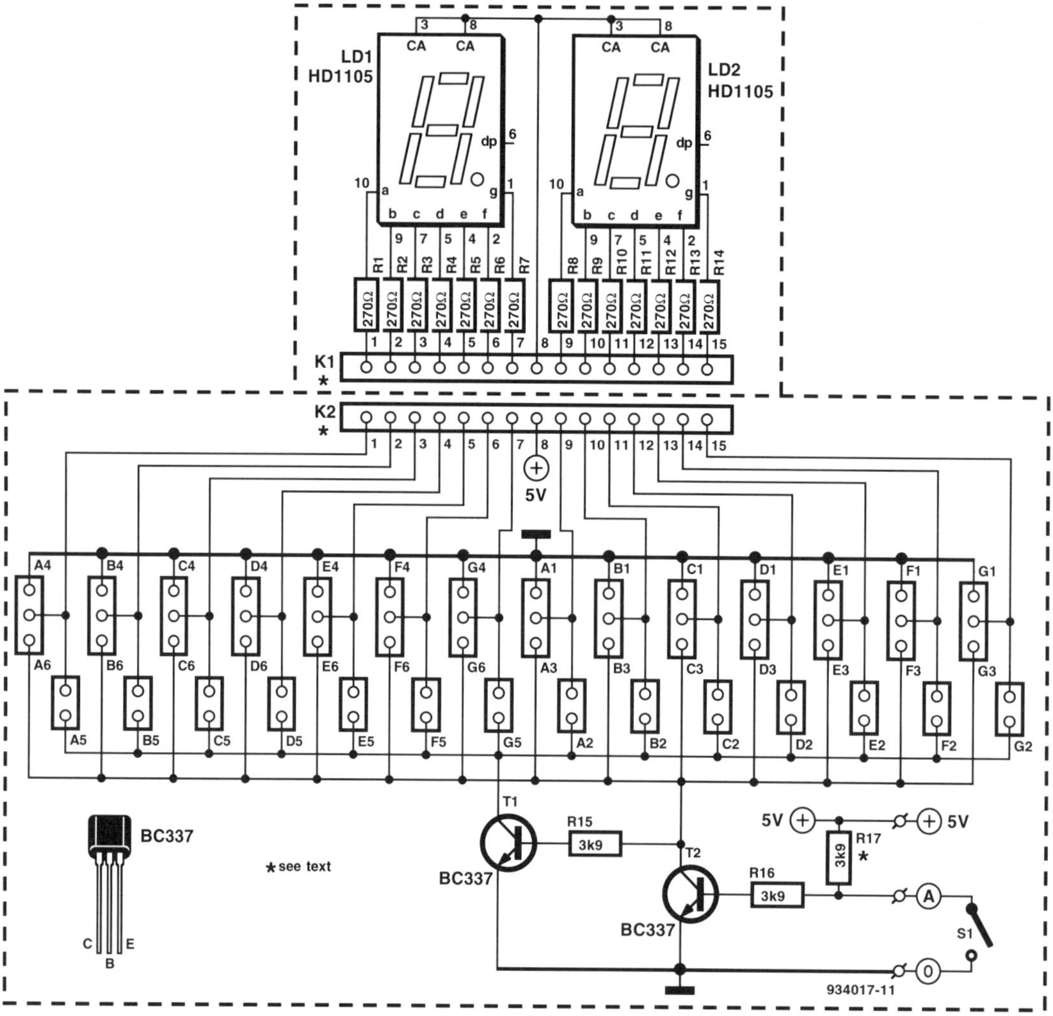

Construction of the readout is simple on the printed circuit board shown (not available ready made). The board is cut into two to separate the jumper section from the display section. The two sections are interconnected by a 15-way SIL socket and a 15-way SIL header. Finally, note the wire link that runs between the speed setting jumpers – if you forget to fit it, the displays will remain dark forever in the absence of their supply voltage.

Normal	Turbo	Always fitted	Normal	Turbo
8	12	A1;B1;D1;E1;G1	C3;F3	B5;C5
8	16	A1;C1;D1;E1;F1;G1	B3	B5;C5
8	25	A1;C1;D1;F1;G1	B3;E3	A5;B5;D5;E5;G5
8	33	A1;B1;C1;D1;G1	E3;F3	A5;B5;C5;D5;G5
8	66	A1;C1;D1;E1;F1;G1	B3	A5;C5;D5;E5;F5;G5
8	99	A1;B1;C1;D1;F1;G1	E3	A5;B5;C5;D5;F5;G5

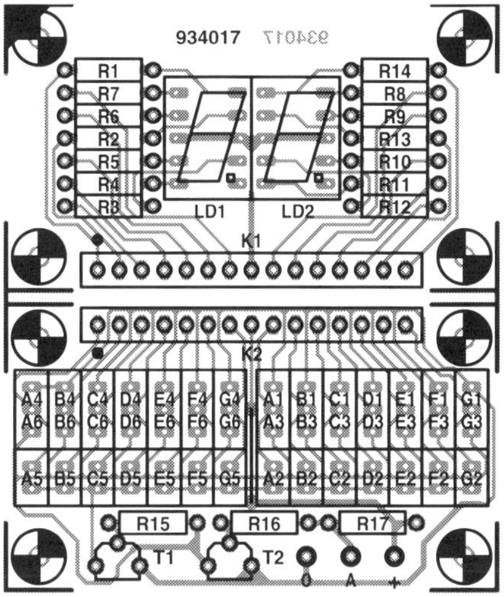

Parts list

Resistors:

14	270Ω	R1-R14
3	3kΩ9	R15;R16;R17

Semiconductors:

2	BC337	T1;T2
2	HD11050	LD1;LD2

Miscellaneous:

14	2-way pin header	
14	3-way pin header	
1	15-way SIL socket	
1	15-way SIL header	
1	Changeover switch (turbo switch in PC)	S1

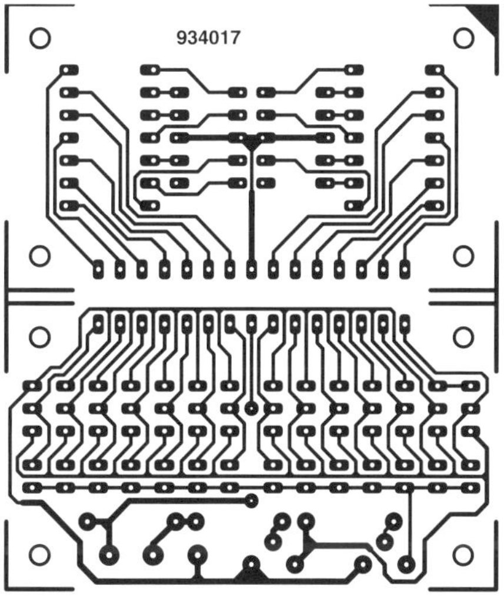

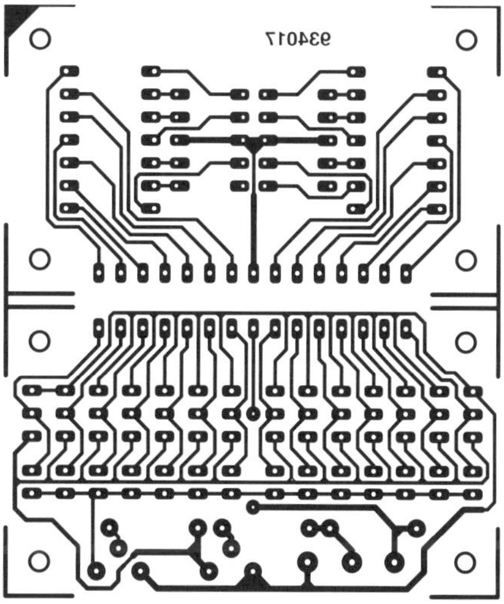

8-channel A-D converter

056 The converter described is controlled by a small I/O unit that has an adequate number of inputs and outputs. It may be built on a small piece of prototyping board and connected to the I/O board via K_1. The 12 V supply and the clock (OSC = 14.318 MHz) which, divided by 16, is used to clock the converter, are derived from the I/O card. The clock may also be drived from a PC, but that on the I/O card is independent of the computer.

How the converter is operated is shown in the BASIC program, except for one action. The EOC output (which goes high at the end of a conversion) needs a minimum of 0 clock pulses and a maximum of eight clock pulses plus 2 μs after the leading edge of the start pulse to go low. This means that with compiled software or machine language routines EOC must be checked in the correct manner (that is looking on line 200 whether EOC is high).

```
10 CLS
20 CTRLWRD = &H99:     'Port A is input, B is output, C is input
30 BASEADDR = &H300:   'base address of 8255
40 PORTA = BASEADDR
50 PORTB = BASEADDR + 1
60 PORTC = BASEADDR + 2
70 CTRLADDR = BASEADDR + 3
80 CHANNEL = 0
90 '
100 OUT CTRLADDR, CTRLWRD:  'initialize 8255
110 '
120 OUT PORTB, CHANNEL:                    'set input channel address
130 OUT PORTB, (CHANNEL  OR &H8):           'B3 HIGH enable address latch
140 OUT PORTB, (CHANNEL AND &HF7):          'B3 LOW
150 '
160 OUT PORTB, (CHANNEL  OR &H10):          'B4 HIGH start conversion
170 OUT PORTB, (CHANNEL AND &HEF):          'B4 LOW
180 '
190 EOC = 0:                               'wait for End Of Conversion
200 WHILE EOC<> &H80 : EOC = INP(PORTC) AND &H80: WEND
210 '
220 OUT PORTB, (CHANNEL  OR &H27):            'B5 HIGH enable ADC outputs
230 LOCATE 1,1:PRINT "Channel ";CHANNEL;" : ";INP (PORTA);"     "
240 OUT PORTB, (CHANNEL AND &HDF):            'B5 LOW disable ADC outputs
250 '
260 PRINT : PRINT "Press N for next channel or S to stop"
270 A$=INKEY$
280 IF (A$="N" OR A$="n") THEN GOTO 310
290 IF (A$="s" OR A$="S") THEN END
300 GOTO 150
310 '
320 CHANNEL = CHANNEL + 1
330 IF CHANNEL = 8 THEN CHANNEL = 0
340 GOTO 110
```

934087 - 12

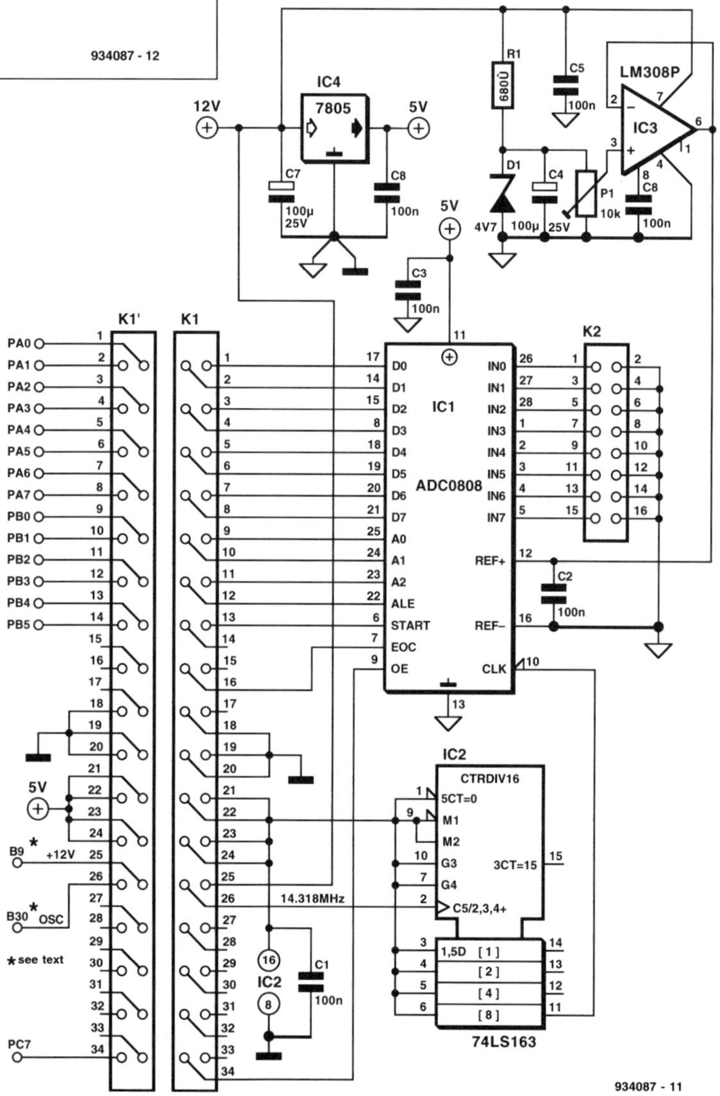

934087 - 11

I²C parallel printer port adaptor

057 The circuit in the diagram permits I²C software programs to communicate via the Centronics output of any IBM compatible PC in conjunction with a buffer IC Type 74LS05. This IC has open-collector outputs to facilitate the parallel switching of I²C applications on the bus. A slight drawback of the present circuit is that everything is done via software, but that is countered by the fact that everything is under full control. Note, however, that some programming languages, such as Visual BASIC, do not allow writing direct to I/O addresses. Most of the 'older' languages do.

The circuit shown is used by Philips in the demonstration packages for I²C applications. Note that the +5 V supply line cannot be taken from the Centronics port, but must be provided by a dedicated power supply. (The 5 V line on the Centronics interface is a supply for the – not present– printer). The circuit draws a current of 20–30 mA.

The base addresses of printer ports LPT_1 and LPT_2 are 378_{HEX} and 278_{HEX} respectively. The table gives an overview of the relations between addresses, bits, pin numbers of the 25-way D-connector, and the signals on the I²C bus. It is best to leave bit 4 of the byte on base address +2 at 0 to prevent interrupts being generated with all the consequences of this.

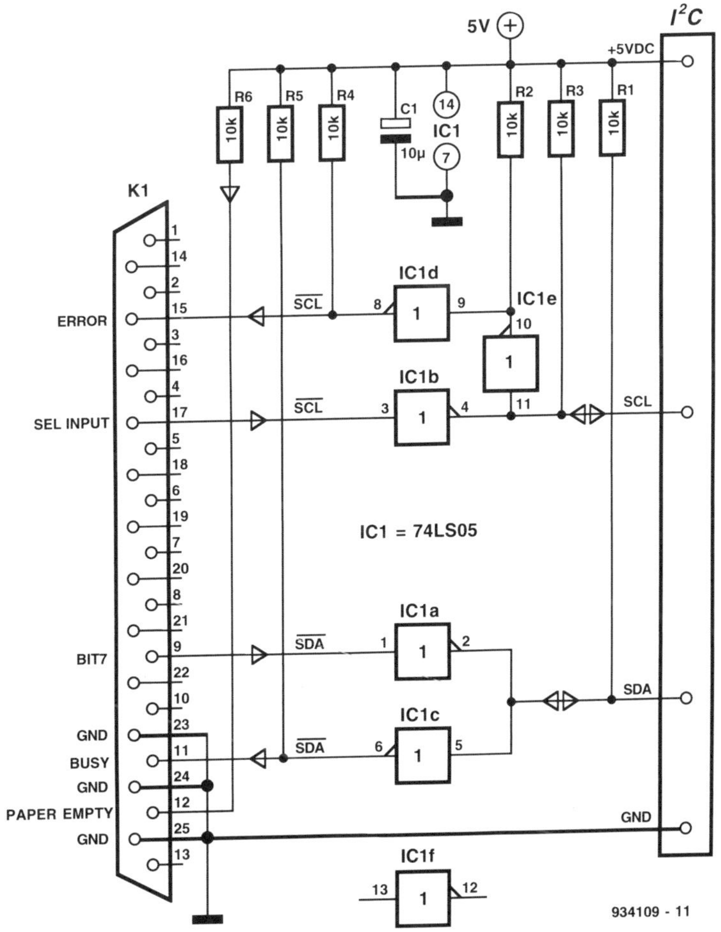

1-Mbit adaptor for EPROM programmer

058 The adaptor allows you to program the 27C1001 EPROM which has a capacity of 1 bit organized as 128 Kbyte ×8. To be able to use the present adapter, your EPROM programmer must be capable of programming 512 Kbit EPROMs such as the 27512 or 27C512 (64 Kbyte ×8). The adaptor programs the 27C1001 in two steps of 64 kbyte each, and is simply plugged into the 28-way (or 40-way) ZIF socket on your EPROM programmer. The 64 Kbyte block selection is effected manually with the aid of a switch.

The circuit of the adaptor has few surprises. Socket 1 connects the adapter to the EPROM programmer, while socket 2 accepts the 27C1001. The «PGM» input of the 27C1001 is actuated (i.e., made logic low) when the programming voltage, V_{pp}, on socket 1 exceeds the zener voltage of D_4, and «CS» on socket 1 is held low by the programmer. The zener diode also enables the programming voltage, 12.5 V, to pull the «OE» terminal of the 27C1001 high, which allows the device to be programmed. The V_{pp} input of the 27C1001 is held either at +5 V (during read operations) or at about +12 V (during programming). To ensure that the 27C1001 is supplied with a sufficiently high operating voltage (nominally 5 V), a Schottky diode type BAT85 is used in position D_2. This diode is marked by a forward drop of 0.2 V only. Switch S_1 takes the highest address input, A16, of the 27C1001 low or high to effect the selection between the lower and higher 64 Kbyte block in the device (A16 = low, and A16 = high, respectively). Diode D_5 lights when the lower block is addressed.

The construction of the adaptor is apparent from the PCB layout. Start by fitting the single wire link on the board. Socket 1 is fitted at the track side of the board, and consists of two 14-pin PCB pin headers. Switch S_1 is a self-locking push-button from ITW with a built-in LED indicator. The PCB accommodates both 28-way and 40-way ZIF sockets.

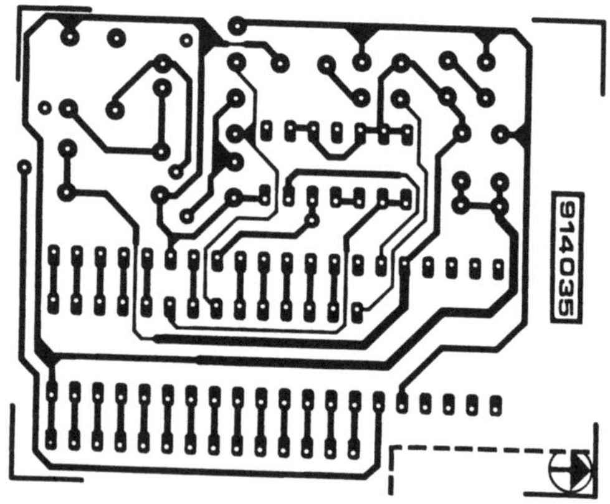

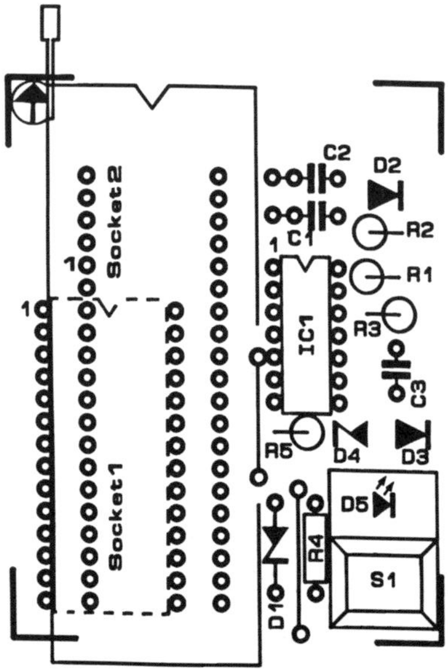

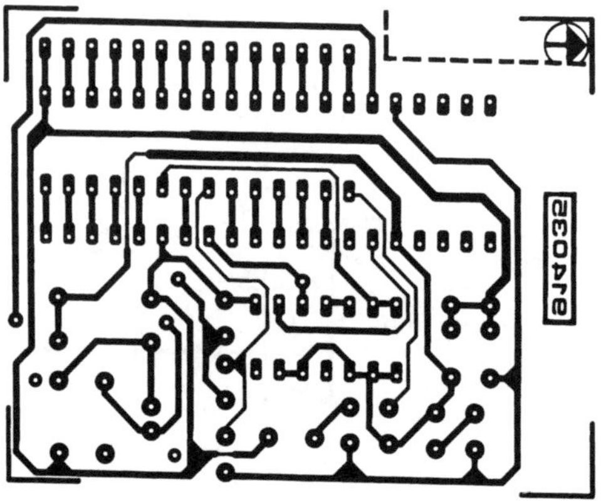

Parts list

Resistors:
R_1 = 10 Ω
R_2 = 10 kΩ
R_3 = 100 Ω
R_4 = 1 kΩ
R_5 = 100 kΩ

Capacitors:
C_1, C_2 = 100 nF
C_3 = 1 μF

Semiconductors:
D_1 = zener, 5.6 V, 400 mW
D_2 = BAT85
D_3 = 1N4148
D_4 = zener, 8.2 V, 400 mW
D_5 = LED, 3 mm

Integrated circuits
IC_1 = 74HCT00

Miscellaneous:
S_1 = self-locking push-button switch with integral LED
2 off 14-way strips of PCB pins (long pins at one side)

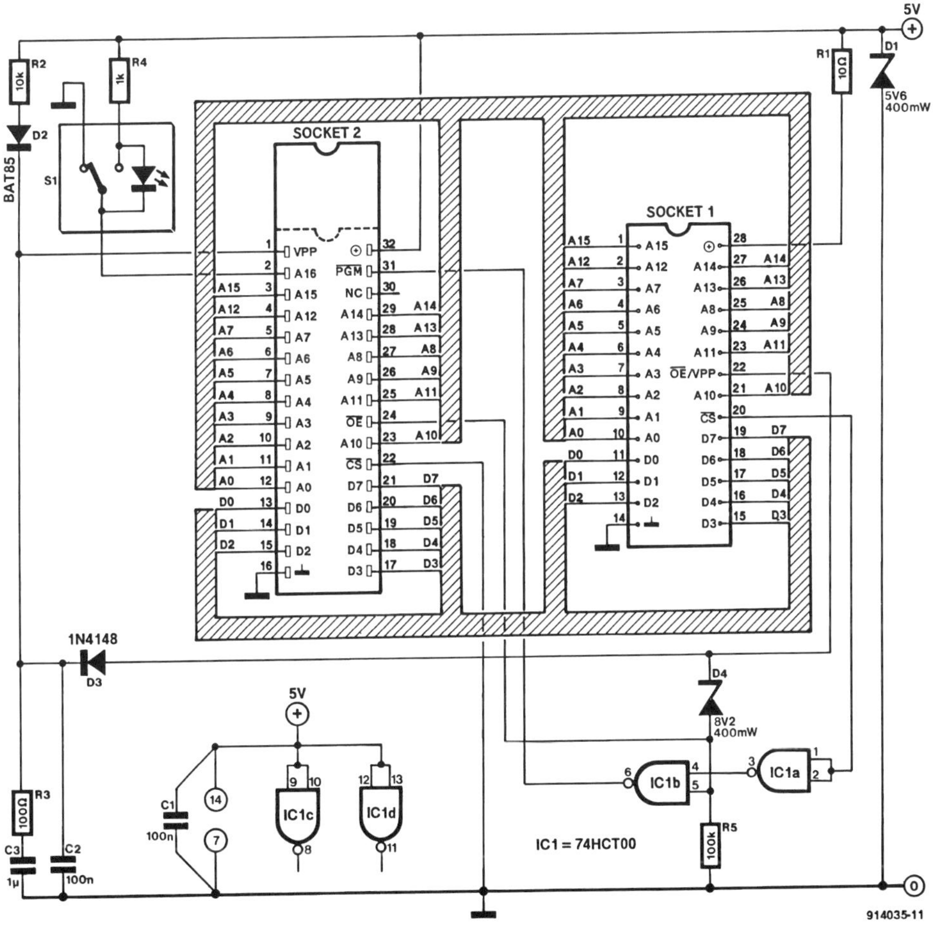

PC cooling fan control

059 Most PCs are provided with a cooling fan to ensure that the internal temperature does not rise unduly. Unfortunately, in many PCs the fan noise soon becomes an irritant. Since for a large part of the time the fan cools the PC more than is necessary, it seems sensible to make the speed of the fan dependent on the ambient temperature. That is the purpose of the circuit shown.

The circuit, designed with discrete component, is intended for the control of 12-V fans that do not draw a current exceeding 200 mA.

To ensure that the fan operates satisfactorily in all circumstances, the supply to it must not drop below its starting voltage. That voltage is equal to the 12-V supply less the 'zener' voltage of T_3-R_6-R_7. With values shown in the diagram, the supply to the fan will be at least 7 V. If the fan does not start at 25 ∞C, replace temperature sensor temporarily by a 1.8 kΩ resistor and lower the value of R_7. If the fan runs too fast, raise the value of R_7.

Transistors T_1 and T_2 compare the fixed potential at junction R_3-R_4 with the temperature-dependent one at junction R_1-R_2. It may be found convenient initially to place 25 kΩ potentiometer meter in the R_2 position, adjust this till the fan runs correctly, measure the resistance and then replace it by a fixed resistor of that value.

Place the temperature sensor in the warm air flow of the fan. When the computer is switched on, the speed of the fan, owing to C_1, will be fairly high, but will soon drop to a minimum. With a thermometer, measure the temperature of the outflowing air close to the sensor. When the temperature has reached a value of about 35 ∞C, the control circuit should come into action, indicated by an increase in the speed of the fan or its supply voltage. If that does not happen, change the value of R_2, or adjust the potentiometer in its place. When the temperature rises, the speed of the fan will increase. The maximum speed will be only slightly lower than that without the control circuit. This is thanks to the fact that T_3 can be driven into hard conduction, so that the drop across it is only some tenths of a volt.

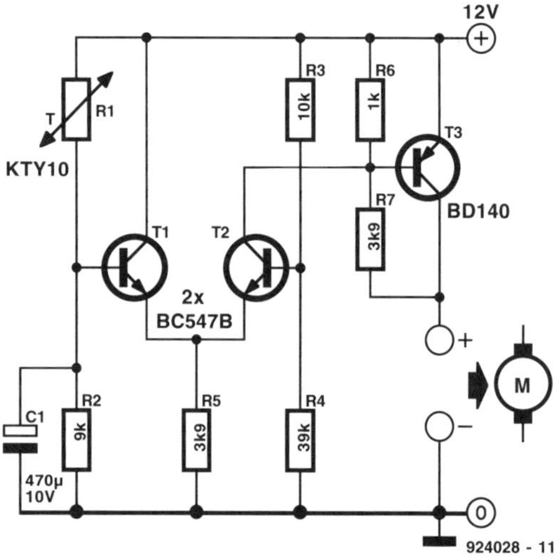

RS232 interface for multi-purpose bus

060 This is an RS232 extension card for the Universal I/O Interface for IBM PCs (Ref. 1). The circuit is based on the COM81C17 UART (universal asynchronous receiver/transmitter) from Standard Microsystem Corporation (SMC). This IC has an on-board programmable baud rate generator, which obtains its clock from a quartz crystal oscillator built around three gates in the 74HCT04 package.

The COM81C17 acts as an 8-bit read/write device on the bus. The RS232 handshaking is limited to the RTS-CTS (request to send – clear to send) pair, which will be adequate for most purposes. The status of these lines is indicated by two LEDs, D_2 and D_3. The INT\ (interrupt request) output of the IC is used as an activity indicator. The INT\ pin goes low when an 'enabled' condition has occurred in the status register, which can be read by the computer system via the bus.

Remarkably, the COM81C17 is housed in a 20-pin DIL enclosure. The fact that it has 20 pins fewer than the well-known AY3-1015D (former 'industry standard') UART is mostly because all status signals normally found on a UART are available as bits in a status register, rather than logic levels on IC pins. These status signals include Tx shift register empty, overrun error, parity error, Rx buffer full, and a host of others. The baud rate selection pins are also omitted and replaced by a 'software' equivalent.

Unfortunately, it is beyond the scope of this short article to cover all programming functions of the COM81C17. For full details, refer to the data sheets supplied by SMC, or to Ref. 3. Some information is, however, given to get you started: an initialization procedure flowchart; mode and status register overviews; the baud rate selection table; and a BASIC program. The latter provides a useful test for the hardware by echoing transmitted characters. This requires K_3 or K_2 pins 3 and 5 (data) and pins 4 and 6 (handshaking) to be interconnected. Two jumpers are provided for this purpose. By the way, K_2 and K_3 are wired such that they can be connected directly to IDC-style D-connectors via lengths of flatcable (a straight pin 1-to-pin 1 connection is used).

A MAX232 (with on-board negative voltage converter) is used in a standard configuration to interface the COM81C17 to the RS232 peripheral connected to K_4 or K_5. As shown in the circuit diagram, these two connectors allow the interface to behave like a DCE (data communication equipment) or a DTE (data terminal equipment). By 'tapping' the connections between the COM81C17 and the MAX232, a TTL-level serial interface is created, which may be used when RS232 levels are not required.

Finally, the addressing of cards connected to the universal bus, and the rules that apply to implementing read/write functions at one address, are discussed in Ref. 2.

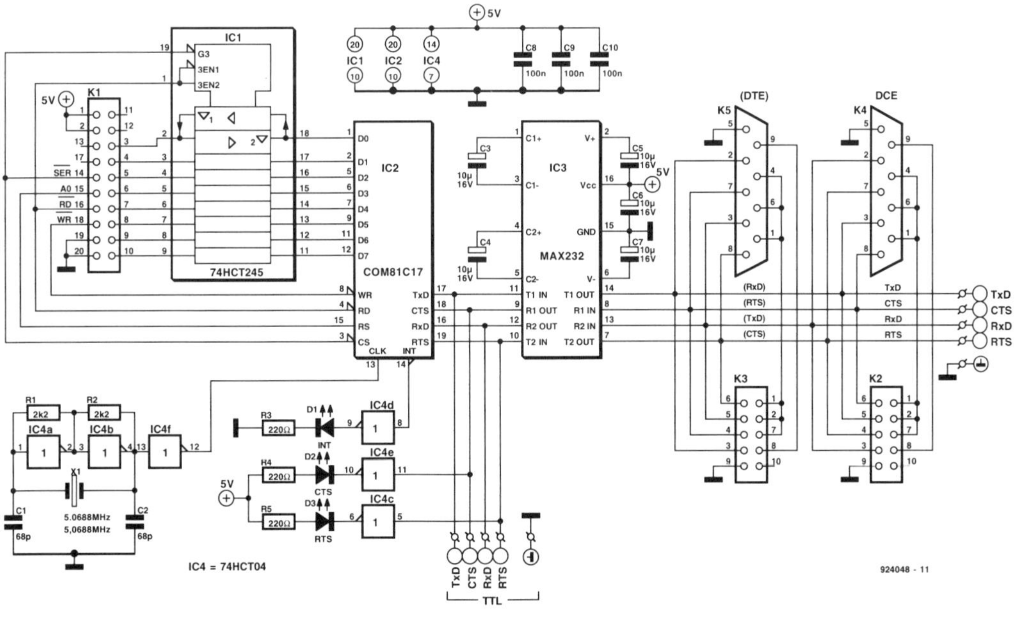

Parts list

Resistors:
R_1; R_2 = 2.2 kΩ
R_3; R_4; R_5 = 220 Ω

Capacitors:
C_1; C_2 = 68 pF
C_3–C_7 = 10 µF, 16 V radial
C_8; C_9; C_{10} = 100 nF
Semiconductors:
D_1; D_2; D_3 = LED, red, 3mm

Integrated circuits
IC_1 = 74HCT245
IC_2 = COM81C17 *
IC_3 = MAX232
IC_4 = 74HCT04

* SMC, head office: 35 Marcus Street, Hauppage, NY 11788, USA. Telephone: (516) 273-3100.
UK distributors: Golden Gate, Maidenhead, Telephone: (0628) 783631, Fax (0628) 71120, or Manhattan Skyline, Maidenhead, Telephone: (0628) 75851.

Miscellaneous:
K_1 = 20-way boxheader.
K_2;K_3 = 10-way boxheader.
K_4;K_5 = 9-way angled PCB mount sub-D connector.
X_1 = quartz crystal 5.0688 MHz.

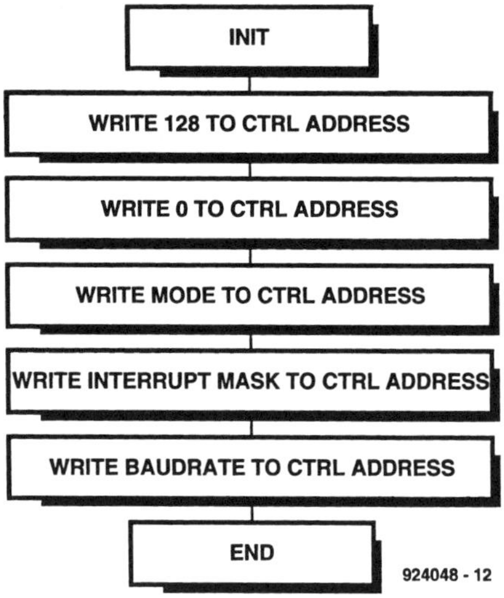

924048 - 12

References:

1. "Universal I/O interface for IBM PCs and compatibles", *Elektor Electronics*, May 1991

2. "Optocard for universal PC I/O interface", *Elektor Electronics*, July 1992.

3. *Databook 4: peripheral chips*, Elektor Electronics (Publishing).

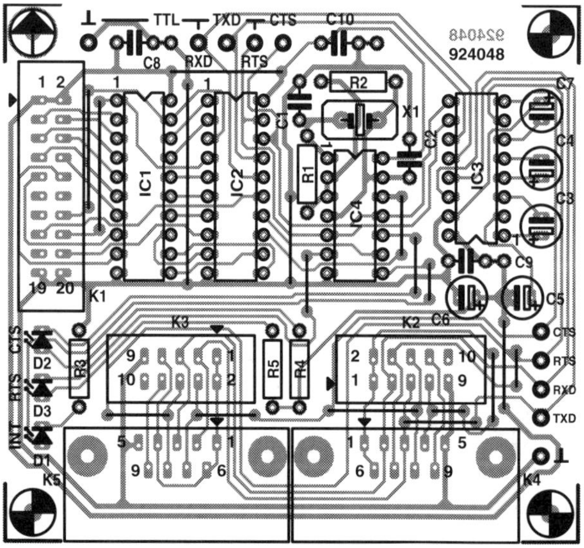

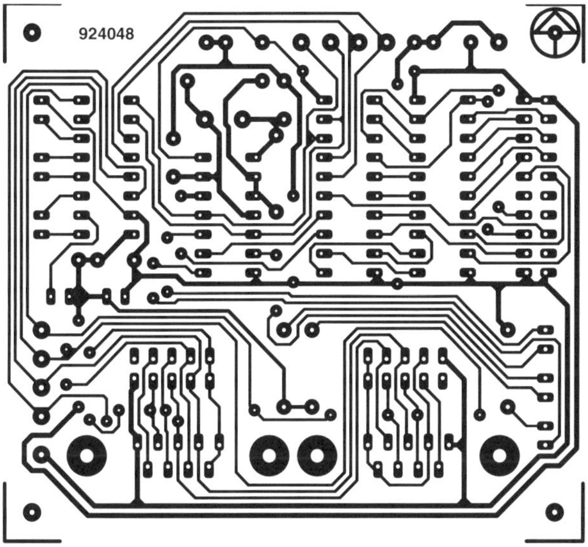


```
10 ADDRESS = &H300
20 TXADDRESS = ADDRESS+0       : REM write transmit buffer
30 RXADDRESS = ADDRESS+0       : REM read receive buffer
40 CTRLADDRESS = ADDRESS + 1   : REM write only
50 STATUSADDRESS = ADDRESS +1  : REM read only
60 CTRLBYTE = 8+16+64
70                               REM reset tpuart
80 CTRLBYTE = CTRLBYTE OR 128
90 OUT (CTRLADDRESS),CTRLBYTE
100 CTRLBYTE = CTRLBYTE AND 127
110 OUT (CTRLADDRESS),CTRLBYTE
120 CTRLBYTE = CTRLBYTE OR 128
130 OUT (CTRLADDRESS),CTRLBYTE
140 CTRLBYTE = CTRLBYTE AND 127
150 OUT (CTRLADDRESS),CTRLBYTE
160                              REM set cts in rts out and internal clock
170 PARITY = 1 : REM 0: no parity
180 ODD = 0    : REM 0: even parity  1: odd parity
190 BITS = 1   : REM # bits per character 0: 7 bits  1: 8 bits
200 STOPBITS = 0 :  REM # of stop bits 0:1 stop bit and 1: 2 stop bits
210 CP1 = 0    : REM cp1 as cts input
220 CP2IO = 0  : REM cp2 as output
230 CP2 = 0    : REM cp2 functions as rts output
240 CLK = 0    : REM internal clock
250 MODE = CP1 * 1 + CP2IO * 2 + CP2 * 4 + CLK * 8
260 MODE = MODE + PARITY * 16 + ODD * 32 + BITS * 64 + STOPBITS * 128
270 OUT (ADDRESS),MODE   : REM define outputs and transmit format
280 OUT (ADDRESS),&HFF   : REM interrupt on all status changes
290 OUT (ADDRESS),5      : REM set baudrate to 2400 Bd
300 OUT (CTRLADDRESS),CTRLBYTE
310 A = INP (STATUSADDRESS)              : REM dummy read
320 CTRLBYTE = CTRLBYTE OR 64            : REM reset error bits
330 OUT (CTRLADDRESS),CTRLBYTE
340 CTRLBYTE = (CTRLBYTE AND (255-8-16)) OR 38 : REM enable communication
350 OUT (CTRLADDRESS),CTRLBYTE
```

```
360 A=INP (STATUSADDRESS) : IF (A AND 128) <>128 THEN 380
370 PRINT A; : A= INP (RXADDRESS) : PRINT A : GOTO 360
380 FOR I=0 TO 255
390 A = INP (STATUSADDRESS) : IF (A AND 64) <>64 THEN 390
400 OUT (TXADDRESS),I
410 A= INP (STATUSADDRESS) : IF (A AND 128) <>128 THEN 410
420 A= INP (RXADDRESS)
430 IF A=I THEN PRINT " ";A;" ";
440 IF A<>I THEN PRINT A;" ";I;" error ";
450 A$=INKEY$ : IF A$ <> "" THEN END
460 NEXT
470 GOTO 380
```

924048 - 16

I²C tip

061 The 8-bit I/O chip Type PCF8574 is very popular: it is used, among others, on the AD-DA I/O module, the opto relay card and the LC display – all published within the past 12 months (see References). Since it is often desirable to connect a number of 8574 modules to the I²C bus, it is possible to give each IC a different address via inputs A0–A2. As therefore only eight addresses can be set, it might be expected that this is also the maximum number of ICs that can be connected to the bus. It is, however, possible to connect 16 PCF8574s, because there are two variants of the IC, each with a different address. The standard PCF8574 has, as is well known, the basic address 40_{HEX}. The variant, the PCF8574A, has the basic address 70_{HEX}. Otherwise the two types are identical and, therefore, completely interchangeable. Thus, eight standard **and** eight variant types of the IC can be connected to one bus.

References:

1. I²C interface for PCs. *Elektor Electronics* February 1992.
2. ADC/DAC and I/O for I²C bus. *Elektor Electronics* March 1992
3. I²C LED display. *Elektor Electronics* June 1992.
4. I²C opto/relay card. *Elektor Electronics* February 1993.

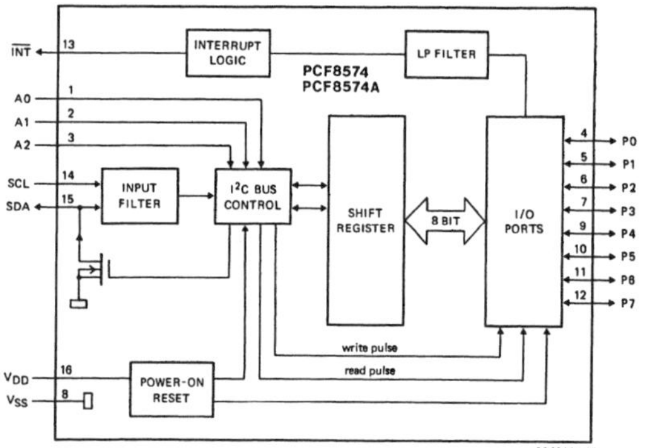

934019-11

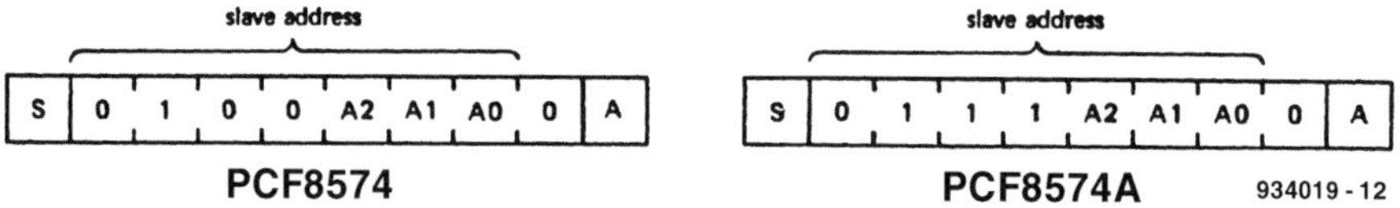

934019-12

Hexadecimal display decoder

062 The byte-to-hex display decoder was designed because nibble-to-hex decodsers are no longer produced commercially. In principle, it is, of course, not difficult to decode a nibble, but from byte to hex is generally more practical and requires fewer ICs. The decoder uses a GAL Type 22V10. This type has an adequate number of inputs and outputs for the present application and may be programmed with the GAL Programmer (incl. the extension) published in Ref. 1. The listing and programming data shown are written in the format that is accepted by the National Semiconductor software and may be converted by a JEDDIC file with the EQN2JED option. Note that although a warning is generated that in the define statements an OR term is used, this is a false alarm, since in the present circuit no fewer than eight OR terms are produced.

The circuit has a drawback in that it requires an external clock. Fortunately, this signal can often be derived from the circuit to which the display is connected. The frequency may be between 100 Hz and 100 kHz.

Apart from clocking the bistables at the outputs of the GAL, the clock also serves the multiplexing of the two displays.

Schottky diodes D_1 and D_2 ensure that T_1 and T_2 switch off fast, so that the transistors are on in turn without any overlap. This prevents 'ghost' segments arising at clock frequencies above 1 kHz.

LT is a lamp-test input with which all segments are controlled. A '1' at input DP0 or DP1 causes the relevant decimal point on the displays to light.

The decoder may serve as an example for the design of a special decoder for another application.

Ref. 1. *Elektor Electronics*, May 1992.

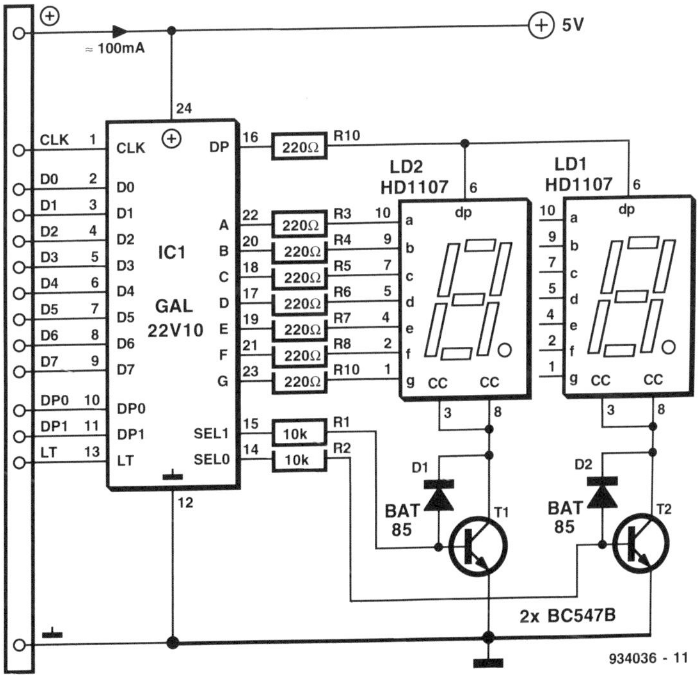

```
Hex Decoder / 7-Segment LED Driver
GAL file for OPAL Junior

CHIP hex_7seg gal22v10

clk d0   d1   d2 d3 d4 d5 d6 d7 dp0 dp1 gnd
lt  sel0 sel1 dp d  c  e  b  f  a   g   vcc

;define product terms for each value and each display

@define zero      "/d7*/d6*/d5*/d4*/lt*sel0+/d3*/d2*/d1*/d0*/lt*sel1"
@define one       "/d7*/d6*/d5* d4*/lt*sel0+/d3*/d2*/d1* d0*/lt*sel1"
@define two       "/d7*/d6* d5*/d4*/lt*sel0+/d3*/d2* d1*/d0*/lt*sel1"
@define three     "/d7*/d6* d5* d4*/lt*sel0+/d3*/d2* d1* d0*/lt*sel1"
@define four      "/d7* d6*/d5*/d4*/lt*sel0+/d3* d2*/d1*/d0*/lt*sel1"
@define five      "/d7* d6*/d5* d4*/lt*sel0+/d3* d2*/d1* d0*/lt*sel1"
@define six       "/d7* d6* d5*/d4*/lt*sel0+/d3* d2* d1*/d0*/lt*sel1"
@define seven     "/d7* d6* d5* d4*/lt*sel0+/d3* d2* d1* d0*/lt*sel1"
@define eight     " d7*/d6*/d5*/d4*/lt*sel0+ d3*/d2*/d1*/d0*/lt*sel1"
@define nine      " d7*/d6*/d5* d4*/lt*sel0+ d3*/d2*/d1* d0*/lt*sel1"
@define ten       " d7*/d6* d5*/d4*/lt*sel0+ d3*/d2* d1*/d0*/lt*sel1"
@define eleven    " d7*/d6* d5* d4*/lt*sel0+ d3*/d2* d1* d0*/lt*sel1"
@define twelve    " d7* d6*/d5*/d4*/lt*sel0+ d3* d2*/d1*/d0*/lt*sel1"
@define thirteen  " d7* d6*/d5* d4*/lt*sel0+ d3* d2*/d1* d0*/lt*sel1"
@define fourteen  " d7* d6* d5*/d4*/lt*sel0+ d3* d2* d1*/d0*/lt*sel1"
@define fifteen   " d7* d6* d5* d4*/lt*sel0+ d3* d2* d1* d0*/lt*sel1"

;define combined productterms to minimize the total number of terms

@define or14or15 " d7* d6* d5*/lt*sel0+ d3* d2* d1*/lt*sel1"
@define or11or15 " d7* d5* d4*/lt*sel0+ d3* d1* d0*/lt*sel1"
@define or12or14 " d7* d6*/d4*/lt*sel0+ d3* d2*/d0*/lt*sel1"
@define or2or3   "/d7*/d6* d5*/lt*sel0+/d3*/d2* d1*/lt*sel1"
@define or4or5   "/d7* d6*/d5*/lt*sel0+/d3* d2*/d1*/lt*sel1"
@define or7or15  " d6* d5* d4*/lt*sel0+ d2* d1* d0*/lt*sel1"
@define or1or9   "/d6*/d5* d4*/lt*sel0+/d2*/d1* d0*/lt*sel1"
```

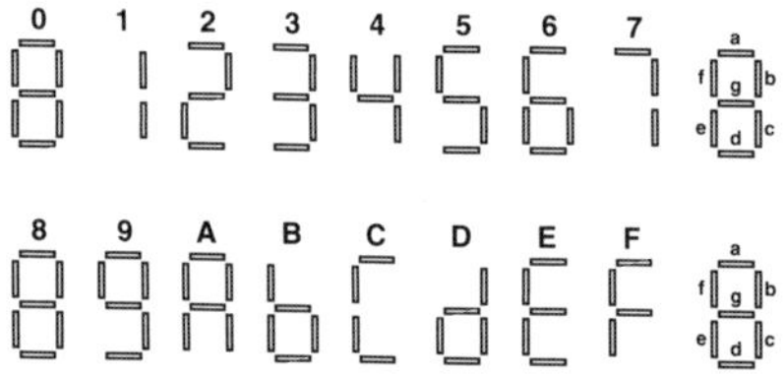


```
EQUATIONS

;To minimize product terms the segment selection is done by using
;the one's that are off, and several options, like 'one+nine', are combined
;into one product term, resulting in one term 'or1or9'.
;All outputs are registered and the multiplex frequency is determined by
;the supllied clock frequency

/a:= one    + four   + eleven   + thirteen
/b:= five   + six    + or11or15 + or12or14
/c:= two    + twelve + or14or15
/d:= four   + one    + ten      + or7or15
/e:= or1or9 + three  + or4or5   + seven
/f:= one    + or2or3 + seven    + thirteen
/g:= zero   + one    + seven    + twelve

sel0 := sel1
sel1 := /sel1

dp:= dp0*sel1 + dp1*sel0
```

934036 - 12

RC-5 infra-red receiver for 80C32 computer

063 'RC-5' is the Philips/Sony standard for infra-red (IR) remote control of audio /video equipment (Ref. 1). The circuit and the listing given enable the 80C32 single-board computer (Ref. 2) to receive and process IR command signals sent by an RC-5 compatible IR remote control unit.

The hardware consists of a Siemens SFH505A IR detector connected directly to the P1.0 port line of the 80C32 microcontroller. The Sharp Type IS1U60 IR detector may be used as an alternative to the SFH505A, with the advantage of higher sensitivity thanks to a passband that is better 'tuned' to RC-5 signals. Note, however, that the SFH505A and the IS1U60 have different pin connections. The IR detector supplies an inverted output signal (high level when no IR signal is detected), which has certain ramifications for the software.

The program (see listing) makes use of the EMON51 monitor program (Ref. 3). Since all timings are based on software, it is essential that the 80C32 computer runs at a clock speed of 12 MHz. It should also be connected to a PC or a terminal, and have the EMON51 system monitor in EPROM. The LOOP routine ensures that the decoding always begins at the start of a new code. COUNT is the count loop proper. If P1.0 goes low during the wait time, R_6 is reset, and the waiting starts again. Next, a number of registers are initialized to prepare for the reception of a new code, and the software waits for the first high-to-low transition. Note, however, that a half bit time of the new word has already elapsed when this H-to-L transition is actually detected. This means that $^1/_4$ bit time of the second start bit has elapsed after waiting $^3/_4$ bit time. Next, the level at P1.0 is monitored twice every bit time — first at $^1/_4$, and again at $^3/_4$ of the bit time. The bi-phase modulation used allows the microprocessor to determine if a '0' or a '1' is meant.

Actually, it is sufficient if a 'check' is carried out only once every bit time. If the synchronization is all right, and the check is made at $^1/_4$ of the bit time, the correct value is established automatically (remember that the input signal is inverted). Checking two times during every bit period, however, enables the software to flag receive errors.

The received bits are shifted into a 16-bit register formed by R4 (low) and R5 (high). The second start bit becomes the MSB. When the received word is complete, the two lowest system address bits are moved from R4 to the LSBs of R5, which leaves the six databits in R4, and the system address bits in the lowest five bits of R5.

Routine CTRL checks if the check bit has changed since the last decoding action. If so, the software decides that a new key has been pressed, or the same key has been pressed again, whereupon the RC-5 system address (that is, the equipment identification number) and the data (that is, the control action required) are sent to the PC. If not, the program jumps back to the start. By omitting the JZ LOOP instruction, the address and data are transmitted on every received code ('auto repeat').

References:

1. Universal RC5 code infrared receiver. *Elektor Electronics* January 1992
2. 8051/80C32 single-board computer. *Elektor Electronics* May 1991.
3. 8051/8032 Assembler course (8 instalments). *Elektor Electronics* February to November 1992.

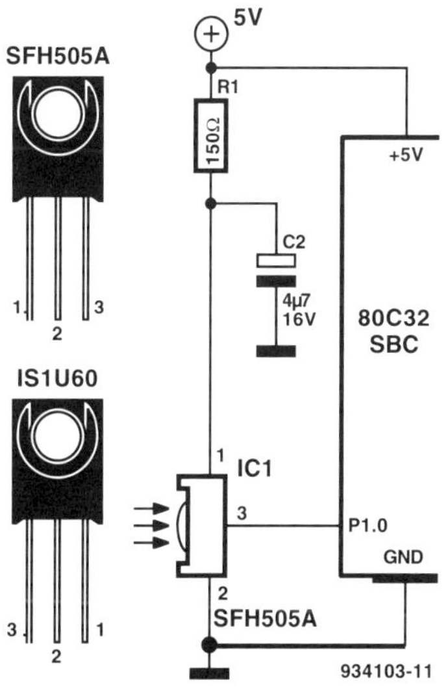

Battery charger

064 The battery charger shown in the diagram may be used to charge a battery or batteries with a total nominal voltage of 12 V (that is, ten NiCd batteries or six 2-V lead-acid batteries. It is small enough to be built into a mains adapter case. Misuse is virtually impossible: batteries connected with incorrect polarity, a short-circuit of the output terminals, or a mains failure have no effect on either the charger or the batteries.

Power is derived from the mains via a transformer with an 18-V secondary. The output of the transformer is rectified by diodes D_1–D_4 and smoothed by C_1, whereupon a direct voltage of 22 V is available across C_1.

Completely discharged batteries are first charged by a current of some 6 mA via R_2-D_5 and R_4-R_6-D_8. Once the battery to be charged has an e.m.f. of 0.3–0.5 V, the base-emitter voltage of T_1 becomes high enough to switch the transistor on.

Charging indicator D_6 then lights and T_2 is also switched on. A charging current of some 60 mA then flows via R_5-R_6. This means that 500 mAh NiCd batteries will be charged in about twelve hours.

If the battery is connected with incorrect polarity or the charging terminals are short-circuited, power transistor T_2 remains off and the charging current cannot become higher than 6–12 mA.

The current drawn by the circuit in full operation is about 80 mA.

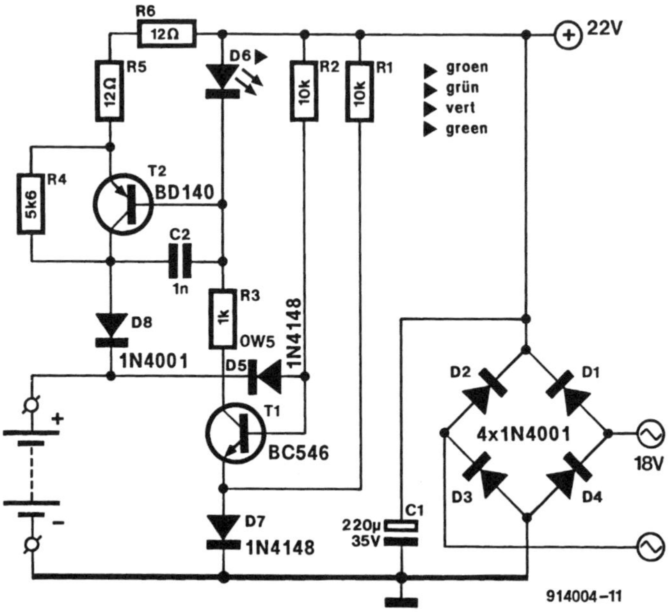

Presettable shunt regulator

065 Depending on their location, linear voltage regulators are arranged traditionally into two sub-groups: series and shunt (or parallel) regulators.
In practical circuits, series regulators, particularly the popular integrated types in the 78xx family, are normally used, since these give good regulation and allow a reasonable output current.
Nevertheless, good shunt regulators are also becoming available, for example, Texas Instruments' Type TL431. The commercial version of this, the TL431C, offers excellent temperature stability and very low dynamic impedance (see table). Although shunt regulators usually function in the same way as a zener diode, the TL431C offers a facility that no zener diode does: the zener voltage may be set anywhere between 2.5 V and 36 V with the aid of two fixed resistors. To function properly, the device needs a cathode current of not less than 1 mA. The voltage across the IC is thenU_{cat} = 2.5(1+R_2/R_3). If the values of R_2 and R_3 are not too high, the current through this reference network is negligible (<4 µA).
A possible application of the device is the compact 5 V power supply shown in the diagram.

SOME TECHNICAL DATA

Cathode voltage, U_{cat}	2.5 V–36 V
Cathode current, I_{cat}	150 mA (1 mA min)
Power dissipation	775 mW (at 25 °C)
Dynamic impedance	0.5 Ω (typical 0.2 Ω)
Temperature coefficient	30 ppm K^{-1}

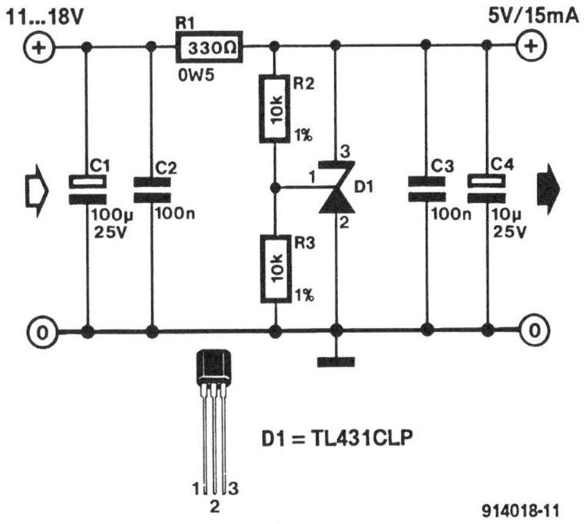

Automatic battery charger I

066 The charger described switches off the charging voltage when the battery reaches its full nominal voltage and switches it on again when the battery voltage drops below a predetermined level.
Part of the battery voltage is taken from across potential divider R_1-R_2-R_3-R_4 and compared with a reference voltage in IC_{2b}. As long as the battery voltage is 0 V, only a small voltage drop is caused across R_5 by the input current of the opamp, so that IC_{2c} toggles at 0 V. The relay therefore remains de-energized. At the same time, the output of IC_{2b} is high, but this has no effect whatsoever owing to AND gate D_4-D_5. When a battery is connected, its small residual voltage ensures that IC_{2c} toggles, diodes D_4 and D_5 are reverse-biased, a reference voltage is applied to the non-inverting input of IC_{2d}, and the relay is energized. The battery is then charged until its voltage has reached the nominal level. Because of potential divider R_1-R_2-R_3-R_4, there is a voltage of more than 3.45 V at the inverting input of IC_{2b}, which causes this opamp to toggle so that its output becomes low (0), the relay is de-energized,

and the charging voltage is removed from the battery.

The (reference) voltage at the output of IC_{1a} is set to 3.45 V. Potential divider D_3-R_6-R_7-P_1 provides a certain hysteresis to comparator IC_{2b}. When the battery voltage drops below the level set with P_1, IC_{2b} toggles again and the charging voltage will be reapplied to the battery.

Calibration is carried out with a voltmeter connected to the output of IC_{2a}, after which P_2 should be adjusted for a reading of 3.45 V. Next, turn P_1 to full resistance. Replace the battery by a regulated, variable power supply and set its output to 6.2–6.4 V (S_1 in position 6 V) or 12.4–12.8 V (S_1 in position 12 V), that is, the voltage at which charging should commence. Adjust P_1 till the relay is energized.

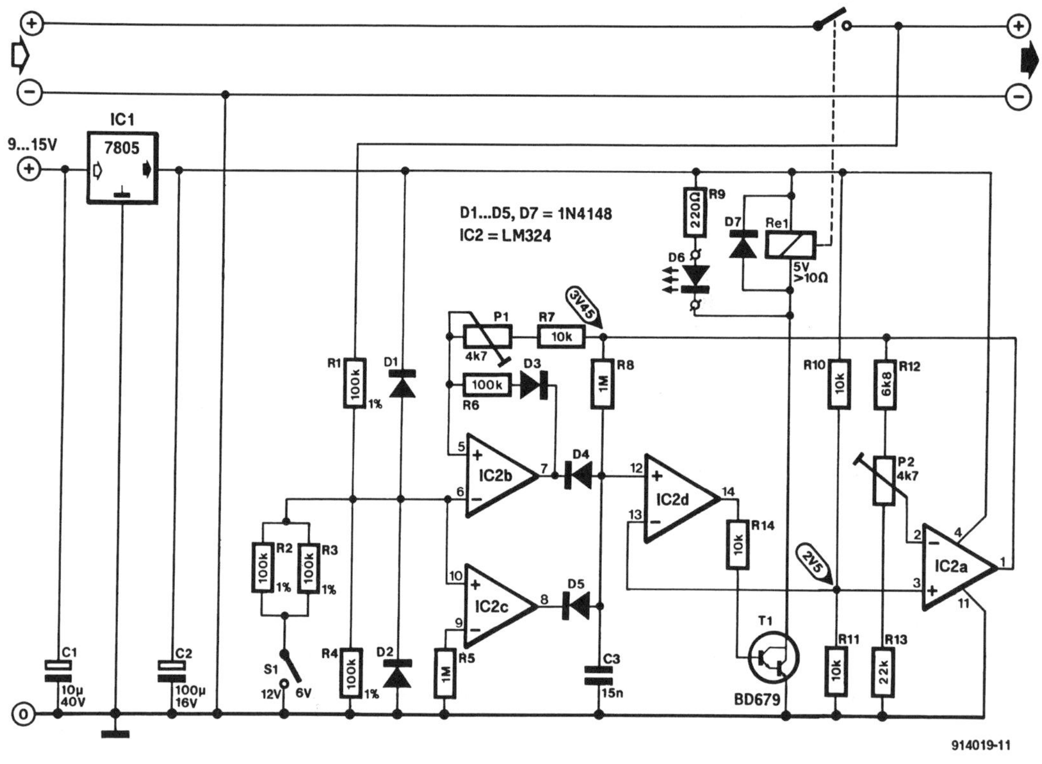

Super voltage regulator

067 A voltage regulator with properties like low voltage drop, 1 A current, protected against reversal and voltage peaks up to 60 V, inexpensive, and simple to design in, is indeed a super device. It concerns here National Semiconductor's LM2941C, which is an integrated device with five pins. Three pins are for the usual connections; the fourth (GND) is required for the low voltage drop; and the fifth provides an additional on-off switching function.

In the diagram, C_1 is required only if the distance to the smoothing capacitor is fairly long. Its value needs to be somewhat larger than is usual with 78xx regulators. This is also the case with C_2. It is advisable to place that capacitor as close to the regulator as possible.

Although it is normal for the quiescent current through low-drop regulators to be rather larger than required by traditional regulators, the LM2941C needs this only at voltage differences of between 0.5 V and 5 V.
The output of the circuit shown is designed to be set between 5 V and 20 V. Since the internal reference voltage is 1.275 V, it should in practice be possible to set it below 5 V. Note, however, that the makers do not guarantee satisfactory operation at such low levels.
The value of resistor R_1 must not be smaller than 1 kΩ. The value of R_2 may be calculated from

$$R_2 = R_1(U_{out}/1.2751-1)\ \Omega$$

where U_{out} is the required output voltage.

Although many three-pin regulators require an electrolytic capacitor at the ADJ output to improve stability, that is not permitted with the LM2941C: in fact, it might lead to oscillations.
A voltage difference of only 0.5 V is sufficient for an output current of 1 A. This difference may be even smaller if the current is smaller.
The input must be actuated by a positive voltage and then requires a current of about 300 µA.
Since the IC toggles with a control voltage as low as 2 V, it may be switched with either CMOS or TTL logic.

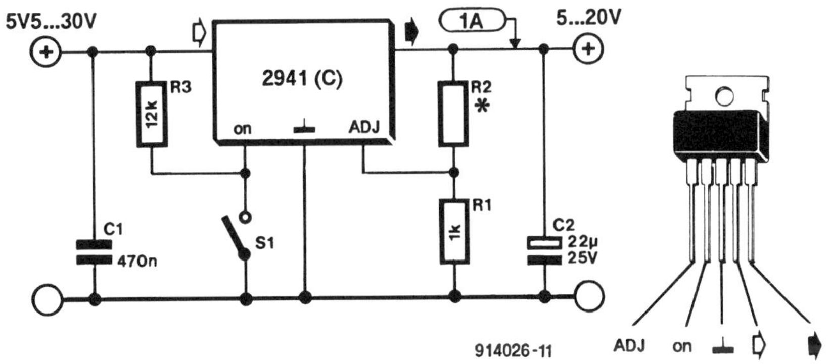

Static d.c.-d.c. converter

068 A Type 555 timer and some passive components can provide a small converter to provide a negative output of 12 V at a few milliamperes.
The 555 is connected as an astable with a rate of 125 kHz. Network C_1-D_2-C_5-D_3 forms a cascade circuit that supplies a negative direct voltage. Since in the design it was required that neither a transformer nor a coil was used, the efficiency of the converter is not high: not more than 16% at an output current of 20 mA. However, in battery-powered equipment requiring a negative supply at only a few mA, that is no hardship. Note, however, that even unloaded, the converter draws a current of about 15 mA.
The output voltage has a ripple of about 0.6 V p-p, which can be suppressed with the aid of a resistor-zener diode network or a low-drop regulator at the output.

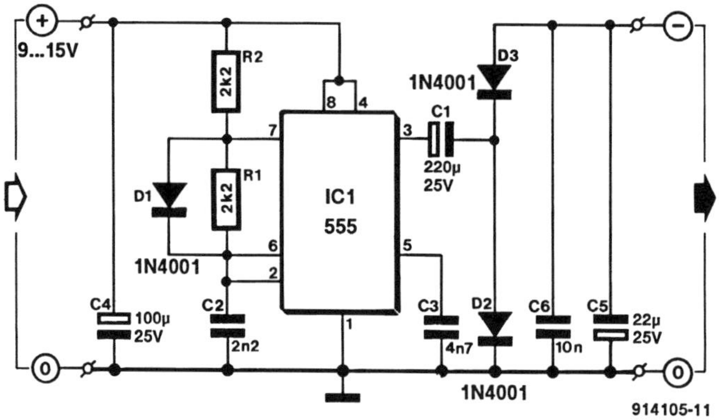

914105-11

Temperature-compensated current source

069 National Semiconductors' LM334Z is a three-pin presettable current source, whose output may be set between 1 µA and 10 mA. It may be also be used 'floating'.

In principle, just one resistor is needed for setting the current. However, the current is then strongly dependent on temperature: about +0.33% °C^{-1}. (This would enable the device to be used as a temperature sensor). Therefore, to obtain a stable current source, an additional resistor and a diode are needed.

For good stability, the diode must be coupled thermally to the IC (the self-heating of the source is then compensated). This is best done by squeezing the IC and the diode, separated by heat paste, into a piece of insulating sleeving asshown in Fig. 2.

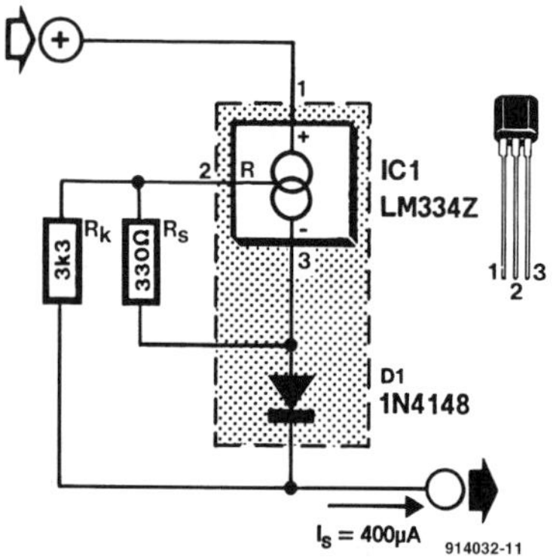

914032-11

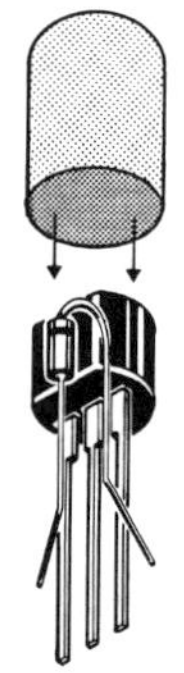
914032-12

Although the current source may be set between 1 µA and 10 mA with the aid of R_s, it is most accurate between 10 µA and 1 mA.

The current provided by the source may calculated from

$$I_s = 2/15R_s$$

Resistor R_k should have a value of $10R_s$.

Set as described and with good thermal coupling between D_1 and IC_1, the prototype showed a temperature drift of not greater than 0.02%°C^{-1} with I_s <1 mA. The largest drift measured, 0.08%°C^{-1}, occurred at I_s = 5 mA. All measurements were carried out with a supply voltage of 9 V.

Relay fuse

070 Simple battery chargers and power supplies are not normally provided with a current limiter. In many cases, however, it would be advantageous if the unit were proof against short-circuits. An electro-mechanical fuse which serves that function and which can be added to the unit is shown in the diagram. There are two variations, one for power supplies (a) and the other for battery chargers (b). The circuit will be described on the basis of (a).

When power is switched on, the relay gets a short energizing pulse of current via C_1. Since the relay contact then changes over, the relay remains energized. When a short-circuit occurs at the output terminals of the power supply, the relay is deenergized and the connection between input and output is broken. The relay is re-energized by a new pulse of current via C_2 after the short-circuit has been removed and S1 is pressed briefly. Capacitor C_2 also prevents an overload if S_1 were pressed while the short-circuit persists. The capacitor is discharged via R_1 when S_1 is opened. Diode D_1 (biased via R_2) shows when the circuit is off.

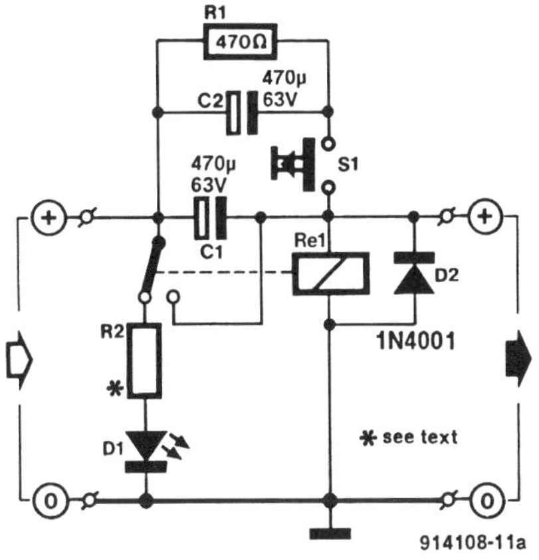

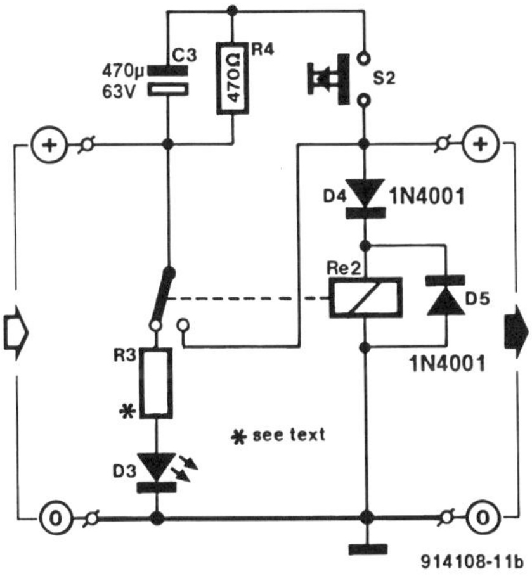

The diagram for battery chargers differs in one respect from that described: on power-on, the relay is not energized via a capacitor, but by the battery on charge via D_4. In case the battery is so flat that it can no longer supply sufficient current to actuate the relay, the relay can be energized, via C_3-R_4, by pressing S_2 briefly.

The value of bias resistors R_2 and R_3 depends on the LED used and the supply voltage. The relay voltage must, of course, also be in accord with the supply voltage.

Single-chip mains supply

071 Harris's Type HV-2405E IC, connected with a few external components as shown in the diagram, enables the direct derivation of a regulated 5–24 V d.c. supply from a 100–260 V a.c. mains supply. The peak output current of the device is 50 mA.

The chip contains a pre-regulator that arranges the charging of C_2, a fairly large capacitance, at the onset of each period of the mains voltage. The charging continues until the potential across the capacitor has reached a level that is roughly the wanted output plus 6 V.

When that state is reached, C_2 provides the voltage necessary for the series regulator also contained in the IC. The output of this regulator can be set between 5 V and 24 V with P_1, and is available at pin 6.

The load current will discharge C_2 to some extent, but the pre-regulator ensures that adequate recharging takes place during every period of the mains supply.

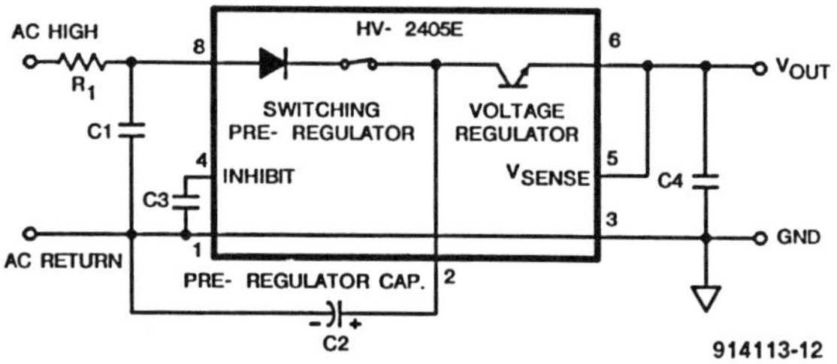

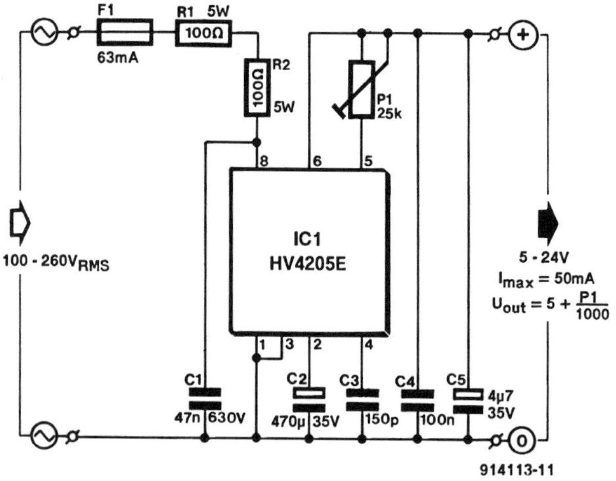

WARNING. Since the circuit and its output is connected electrically to the mains, any apparatus connected to the circuit is, therefore, also at mains potential. Great care should thus be exercised when working on it: touching certain parts is potentially lethal!

Automatic battery charger II

072 Keeping your car battery constantly charged when the car is not in use appreciably increases the life of the battery. Charging is, of course, normally only possible in your garage. The charger described here provides a constant charging current that may, for example, be fed to the battery via the cigarette lighter.

The charger consists of a mains transformer, Tr_1, bridge rectifier B_1 and smoothing capacitor C_1. The charging current through the regulator, IC_1, and the switched series resistors is 107 mA (47 Ω); 230 mA (22 Ω); 500 mA (10 Ω); or 1 A (5 Ω).

Diodes D_1D_4 indicate the position of the switch. Transistor T_1, R_1 and D_5 ensure constant brightness of the diodes.

When the battery is not connected, the relay is not energized and the mains is switched off.

When the battery is connected, C_3 gets charged, T_4 is switched on and the relay is energized. The mains is then switched on and the battery is charged via D_7. The consequent voltage drop across D_7 causes T_3 and T_2 to be switched on, so that the relay remains energized although, since its collector is at +12 V, transistor T_4 is switched off. Resistor R_5 ensures that C_1 is kept charged so that T_4 remains off.

To ensure that the charger works with flat batteries, the relay contact may be bypassed by S_1 which enables the charger to be switched on manually.
Note that during constant charging of lead-acid batteries there is the risk that the water dissolves into hydrogen and oxygen and this will reduce the liquid in the battery. Since sealed batteries can not be topped up, the present charger is not suitable for these types of battery.
Also, do not use a current higher than necessary; in most cases 100 mA is ample. The larger currents are intended for charging large NiCd batteries.

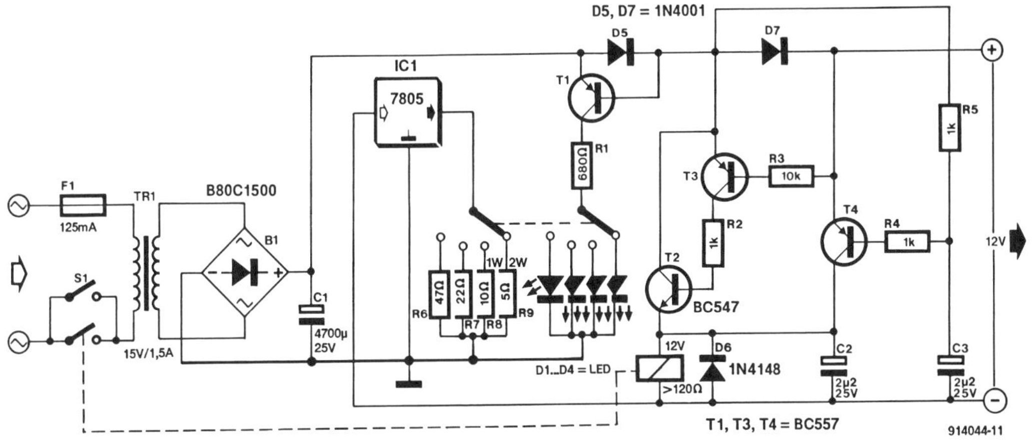

9-volt NiCd battery charger

073 The regulated power supply and the four identical current sources shown in the diagram enable the simultaneous charging of four 9-volt NiCd batteries. The potential at the wiper of P_1 determines to what voltage the batteries will be charged: an unusual, but effective method. The voltage at the wiper is also applied to the non-inverting inputs of four comparators, IC_{2a}-IC_{2d}, via 100 kΩ resistors. When the battery voltage is too low, the relevant comparator toggles, which results in the associated transistor being switched on, whereupon the battery is charged. The rate at which the comparators can toggle is slowed down by a capacitor shunting the opamps (when a battery is being charged, its e.m.f. rises). The charging voltage, resulting from the current through the battery or batteries can rise to the level set with P_1. When that level is reached, the relevant circuit is switched off, and the e.m.f. of the battery drops almost instantly. This might cause the charging current to be switched on again, resulting in a charging voltage rising to the level set with P_1. To prevent this oscillatory action, the capacitor across the opamps enables the battery to stabilize. If, after a short delay, the battery voltage proves to be too low, the current is switched on again. The capacitor then ensures that the current will flow for a while, irrespective of the battery e.m.f. (after all, the battery was found not to be fully charged).
LEDs in the emitter circuits of the transistors give a visual indication of the on and off switching of the charging current. When the battery is nearly flat, the LED will be on continuously; when it is about fully charged, the charging current will be interrupted more and more frequently, so that the LED begins to flicker. The more nearly

the battery is charged, the faster the LED will flicker; when the rate is about 1 Hz, the battery is fully charged.
The circuit requires an alternating voltage of 15–18 V; it draws a current of about 150 mA.

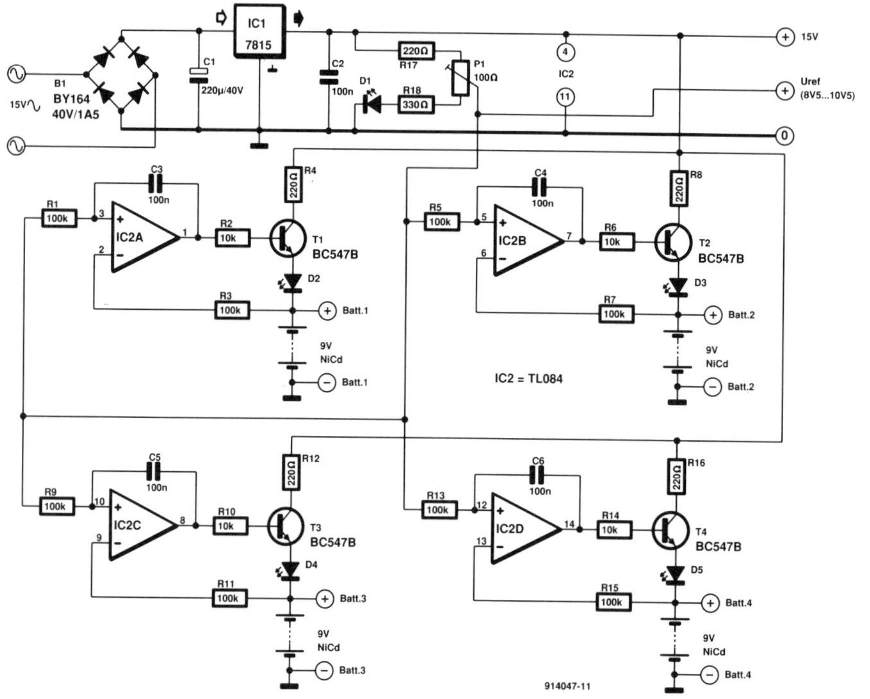

Battery regulator for solar power system

074 Although newer types of solar cell have come to the fore recently, amorphous cells are bound to be with us for some time to come, mainly because of their low cost. Most solar cells based on amorphous technology have a relatively high internal resistance, which results in large differences between the loaded and 'open-circuit' output voltages. Where a rechargeable battery is used as an energy storage device, a voltage regulator circuit arranges charging of the the battery when the cell output voltage is high, and forms a minimum load on the battery when the cell output voltage is low.

For relatively small solar power systems, the parallel (or 'shunt') regulator is a viable alternative. Apart from a single (Schottky) diode, nothing is inserted between the solar cell and the battery. Since the supply voltage is furnished by the solar panel, the regulator works always, even if the battery is fully discharged or not connected. This ensures the best possible protection against overvoltage of all circuits powered by the solar cell (or array of cells).

The heart of the shunt regulator presented here is formed by a Type TL431LP precision voltage regulator from National Semiconduc-tor. When the solar cell output voltage rises above the level set with preset P_1, a current starts to flow through R_3-R_2-D_1. When this current has risen to about 5 mA, transistor T_1 starts to conduct. The transistor used here, a BD642, may be replaced by almost any other, similar, power darlington, for example, the TIP147. The collector of the BD642 is conveniently connected to ground, which means that the device can be bolted direct on to a heat sink.

Since the regulator is capable of shunting quite high currents, it has separate sense inputs, 'A' and 'B', to monitor the battery voltage. The solar cell is connected to terminals '++' and '– –'. Resistor R_4 limits the current through D_3 in the event of the high-power series diode, D_2, breaking down. Not shown in the circuit diagram, but required in the interest of safety, are fuses in series with the battery and the load(s). Also, do not forget to connect a surge arrester in parallel with the solar cell. The Schottky diode used here, an SB530 from Conrad, is capable of passing up to 5 A. For panel output currents up to 3 A, it may be replaced by the more familiar 1N5401 (which, unfortunately, has a slightly higher forward drop voltage). Talking of ratings: the heat sink on to which the power transistor is bolted must have a thermal resistance of 1.5 K W^{-1} or smaller, in which case power dissipation levels up to 40 W can be handled without problems. For very high power applications, connect a number of power darlingtons in parallel, and interconnect their emitters via 0.22-Ω current distribution resistors. Obviously, to cope with the increased power dissipation, the size of the heat sink must be increased accordingly.

The component values shown result in a battery voltage adjustment range of 13.4–17.6 V. This is on the high side for most (lead-acid and gel-type) batteries, which do not fare very well at charging

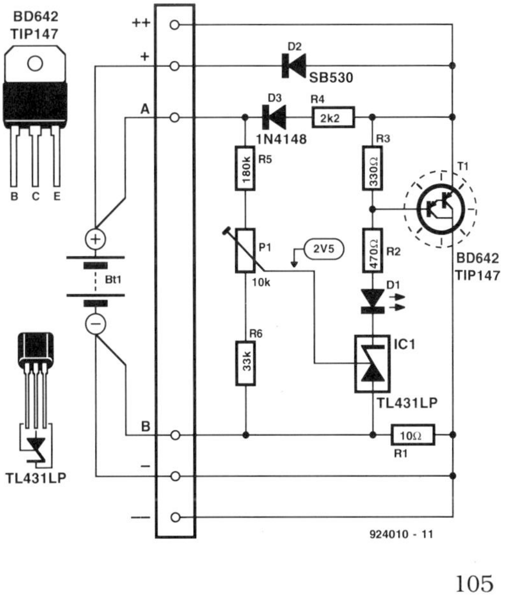

voltages greater than 13.8 V. The circuit can be modified for an output voltage range of about 11–16 V by fitting: R_5=220 kΩ; R_6=47 kΩ; P_1=25 kΩ.

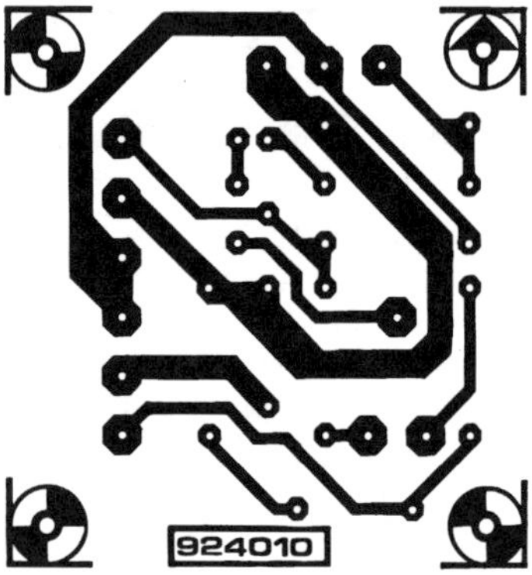

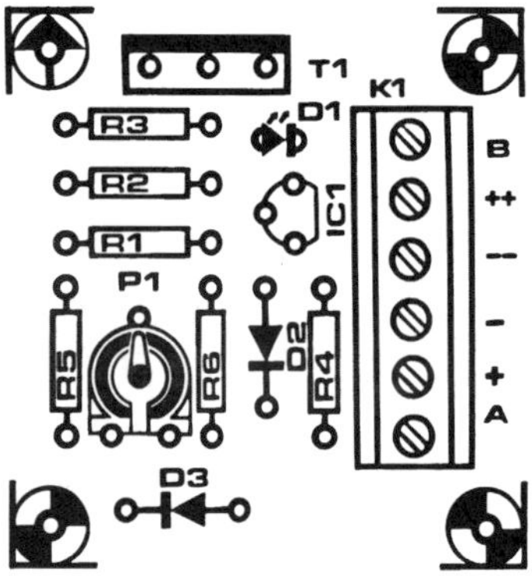

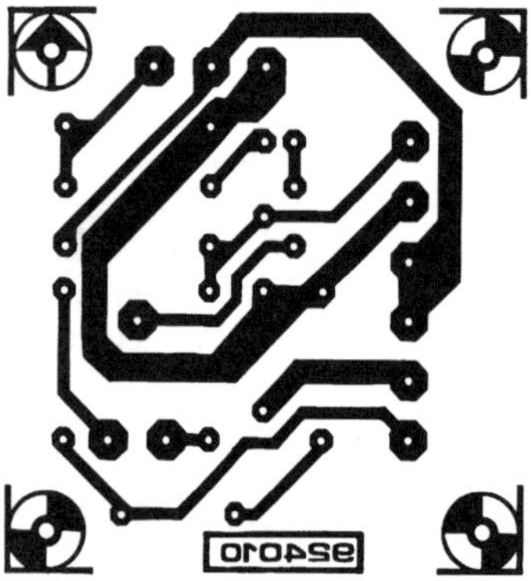

Parts list

Resistors:
R_1 = 10 Ω
R_2 = 470 Ω
R_3 = 330 Ω
R_4 = 2.2 kΩ
R_5 = 180 kΩ
R_6 = 33 kΩ
P_1 = 10 kΩ preset H

Semiconductors:
T_1 = TIP147 or BD642
D_1 = LED
D_2 = SB530
D_3 = 1N4148

Integrated circuits:
IC_1 = TL431LP

Miscellaneous:
K_1 = 6-way PCB terminal block; pitch 5 mm

Low-drop regulator I

075 If a stable 5-V direct voltage is to be derived from a low supply voltage, the 4805 from SGS Thomson is probably the most suitable regulator IC available at present. The popular 7805 does not work too well with supplies lower than about 8 V. The 4805, on the other hand, needs an input voltage that is only about 0.5 V higher than its output voltage. Its data sheets state that its output voltage remains stable so lang as the input voltage does not drop below 5.7 V. That voltage, by the way, is the worst-case voltage when the output current is 400 mA. In practice, therefore, the regulator normally works fine with input voltages as low as 5.4 V.

What to do if the output current is more than 400 mA? Well, in that case, and for output currents up to 1 A, the LM2940T appears to be just about the best available. This IC from National Semiconductor is available in three variations: 5 V, 8 V, and

10 V. The 5-V version, in which we are principally interested here, is type-coded LM2940T-5. These ICs are pin-compatible with the 7805 and 4805, which makes it possible for an existing low-drop supply based on one of these devices to be upgraded fairly simply.

For completeness' sake, the circuit diagram shows a simple 5-V design. The only really important parameter here is the capacitance of decoupling capacitor C_4. According to the relevant data sheets, its value should be not lower than 22 μF to ensure correct stability.

If the input voltage does not rise above about 7.5 V, a heat sink is not necessary. If the regulator is required to operate at a higher output voltage and the full output current, a suitable heat sink is imperative. A heat sink rated at 6.5 °C W^{-1} makes the low-drop circuit suitable for input voltages up to 15 V at full output current or up to 25 V at 500 mA. The worst case input voltage is 5.8 V.

The LM2940T, like the 7805, is short-circuit-proof, but it does draw a rather high quiescent current as shown in the characteristic curves.

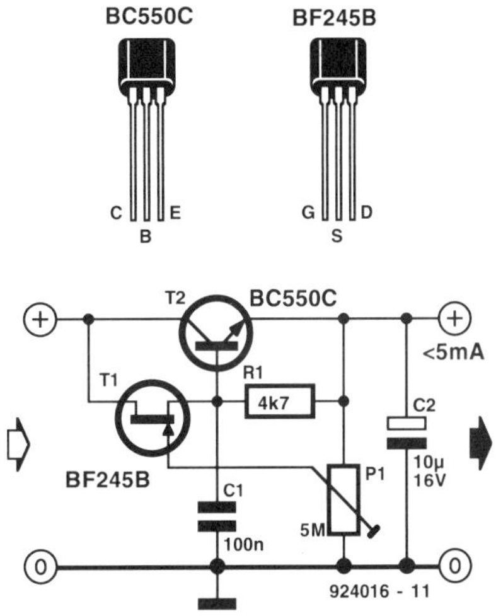

Crowbar protection

076 Quite a few electronic components, particularly active ones, cannot withstand too high voltages. Preventing costly circuits dying a premature death because the supply voltage has risen too high makes overvoltage protector no luxury. Such a protector must, of course, act swiftly, otherwise the deed has been done before it had a chance to act. Therefore, (slow-acting) relays are not suitable for this purpose.

The circuit shown here, a so-called crowbar, contains several fast-acting components. It is intended to be connected between the mains and the appliance to be protected.

The circuit depends, as it were, on brute force: when the supply voltage rises too high, a thyristor short-circuits the output. This means that the too high voltage is immediately removed from the connected appliance and also that fuse F_1 blows. Brute force, indeed!

The voltage at which the crowbar comes into action is set between 5 V and 25 V with P_1. This is done as follows:

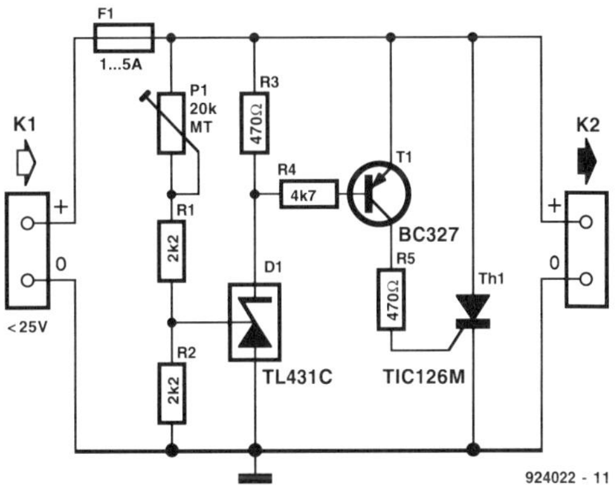

1. Set P_1 to its maximum value.
2. Replace the fuse temporarily by a wire bridge and connect the crowbar to a variable power supply. Set the current limit of that supply to 1 A and the output voltage to the value at which the crowbar is desired to act.
3. Turn P_1 back slowly until the thyristor comes on, that is, when the current limiter of the power supply comes into action.

The crowbar is now set. Replace the wire bridge by the fuse (max. rating 5 A). In quiescent operation, the crowbar circuit draws a current of only about 1 mA.

NiCd temperature monitor

077 THE monitor is particularly intended as an aid for fast charging of NiCd batteries. Most commercially available fast chargers have no temperature sensor, although temperature is an important factor in the fast charging of NiCd batteries.

According to manufacturers' data sheets, temperatures for various states of a NiCd battery should roughly be: (a) 30 °C when half charged; (b) 37 °C when fully charged; and (c) 48 °C when 20% over-charged.

The monitor is connected in series with the battery and the charger. The temperature is assessed by a couple of NTC (negative temperature coefficient) resistors. The output of these devices is compared with reference levels in IC_{1b}–IC_{1d}. The output of comparator IC_{1d} goes low when the temperature—measured by R_{13} and R_{14}—rises above 25 °C, whereupon D_7 lights. The output of IC_{1c} goes low when the temperature reaches 38 °C, whereupon D_8 lights. This signals that the battery is fully

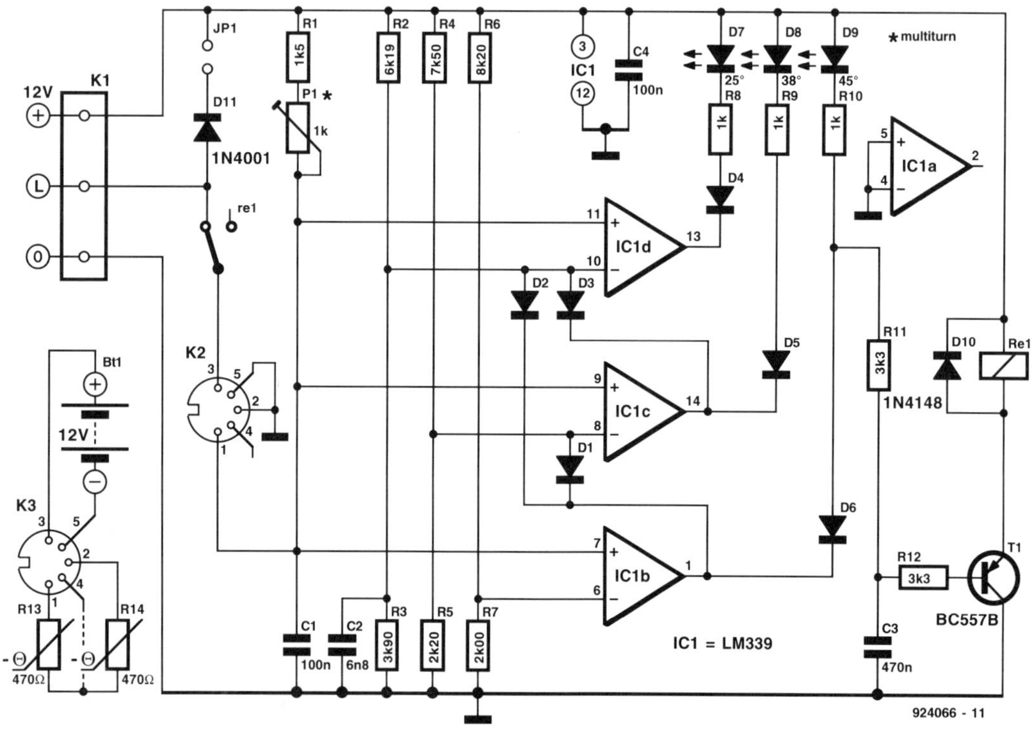

charged. Finally, the output of IC_{1b} goes low when the temperature reaches 45 °C, whereupon D_9 lights. This is an alarm signal calling attention to the battery being over-charged. At the same time, T_1 is switched on, whereupon relay Re_1 is energized and its contact disconnects the charging current. The charging current can be reconnected only after the temperature has dropped to well below 45 °C.

Diodes D_1–D_3 ensure that only one of the three LEDs can light at any one time. The circuit is calibrated with P_1 to make certain that the LEDs light at the correct temperatures.

When the monitor is connected to the charger, it is essential that the charging current, originally flowing via K_1, flows via K_2 and K_3. The batteries must, therefore, be connected to K_2 and K_3. The terminals of the NTC resistors are also coupled to these connectors.

The NTC resistors must be coupled to the batteries in a manner that ensures good contact. It is good practice to fit them permanently to the batteries.

The supply for the monitor may be taken from the charger, either via the 12 V pin or the L pin. In the latter case, wire link JP_1 must be short-circuited. The monitor draws a current of about 15 mA when the relay is not energized.

Variations in the supply voltage do not affect the monitor since the NTC resistors are connected in a bridge arrangement.

As shown, the monitor is intended for use with 12 V batteries. It can be modified for use with 6 V batteries merely by using a 6 V relay instead of a 12 V type.

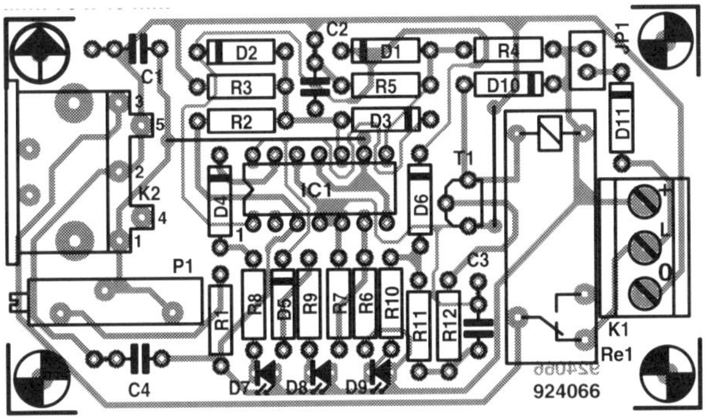

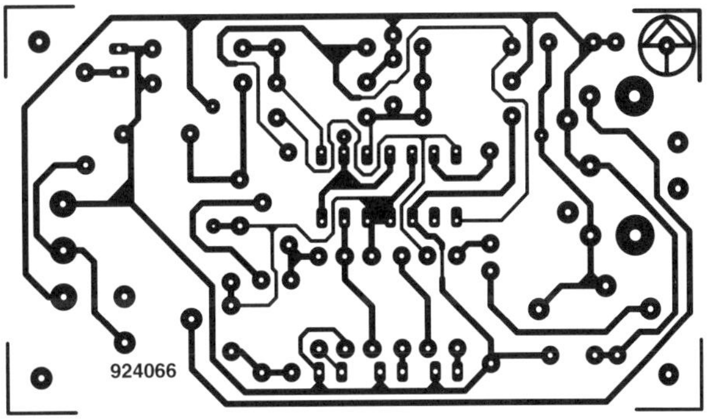

Low-drop regulator II

078 Nowadays the only reasons for not using a voltage regulator in a power supply are: I have not got one; I need an 'odd' voltage; I want to keep the current drain very low.

The regulator shown in the diagram is suitable for currents of 5–10 mA. The two transistors draw only a tiny current. The drop across the regulator depends on the load current and lies between 0.5 V and 1.4 V. The output voltage may be preset between 1.8 V and 8 V.

On power-up, there is no voltage at the source of T_1, so that the FET conducts. Current amplifier T_2 then draws base current and is switched on. This arrangement means that the reference (gate) voltage may be taken from a high-resistance potentiometer. The quiescent current depends on the level of the preset output voltage: at 5 V, it is a mere 1 µA.

When T_2 is switched on, the output voltage will rise to its preset level. The base potential of T_2, and thus the source potential of T_1, remains about 0.6 V higher than the output voltage, so that it rises in step with the output voltage. The gate of T_1 is, however, connected to the wiper of P_1, whose voltage rises more slowly than the output voltage, because the preset is a potential divider. Consequently, the gate of T_1 becomes more and more negative with respect to its source. An equilibrium is soon reached, whereupon the FET reduces the base current of T_2 to a degree that ensures stability of the output voltage.

In normal circumstances, the output voltage vs load current ratio is of the order of 9 mV/mA.

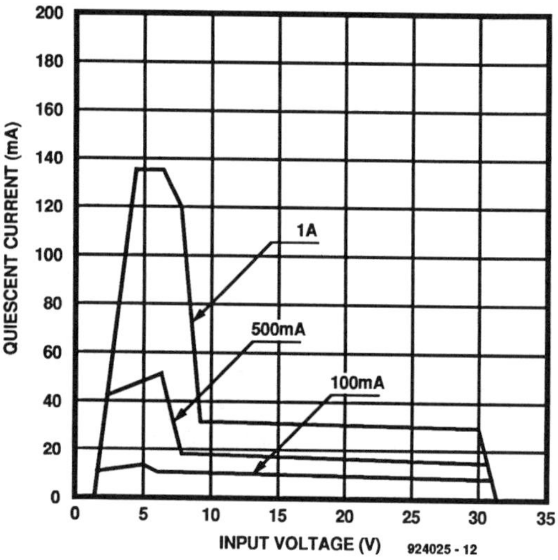

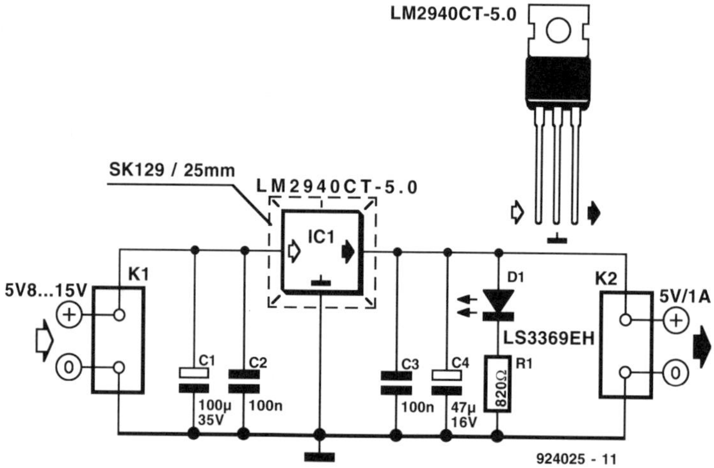

Touch 'on', auto 'off' controller for battery-operated equipment

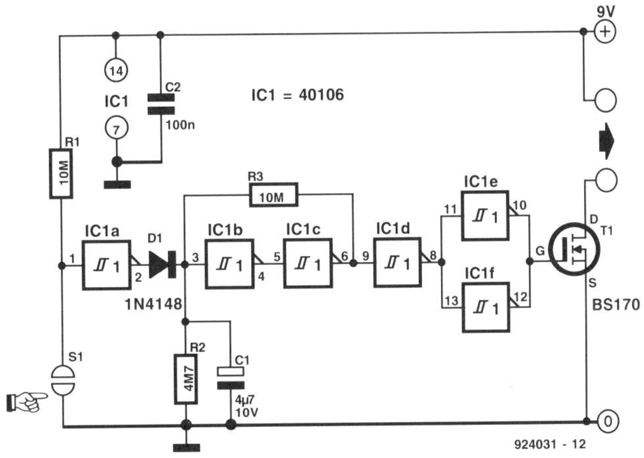

079 This handy circuit is intended for battery-operated appliances. It functions as a touch-operated supply on switch and delayed supply off switch. Figure 1 shows the design for applications that require only a few milliamperes. Figure 2 is identical but for the added FET at the output, which enables up to 300 mA to be switched.

The active electronics is formed by six Schmitt triggers contained in a Type 40106. The touch key consists of two small conducting plates that can be interconnected by our skin resistance. When the key is not touched, R_1 causes a high level at the input of IC_{1a}. That gate is followed by a diode, D_1, which ensures that C_1 can be charged only if the output of IC_{1a} is high. When the key is touched, C_1 is charged rapidly. The capacitor is discharged slowly via R_2: the state of its charge is monitored by N_2-N_3-R_3.

The shunting of gates IC_{1b} and IC_{1c} by resistor R_3 greatly increases the hysteresis at the input of IC_{1b}. That means that the output of IC_{1c} goes high only when C_1 is almost fully charged, and changes state again only when C_1 has been nearly discharged.

Gates IC_{1d-f} serve as output buffers.

The appliance connected to the output terminals will be provided with power as soon as the touch key is operated, because C_1 will then be charged rapidly. The discharge time of this capacitor (and thus the time that the appliance is powered) is fairly long and depends on the value of R_2 and the leakage resistance, and may, therefore, be extended by giving R_2 a higher value. Note that it is not advisable to increase the value of C_1 by much, because that increases the charging time and, worse, simply touching the key may not be sufficient any more. When C_1 has been nearly discharged, the supply to the connected appliance is switched off.

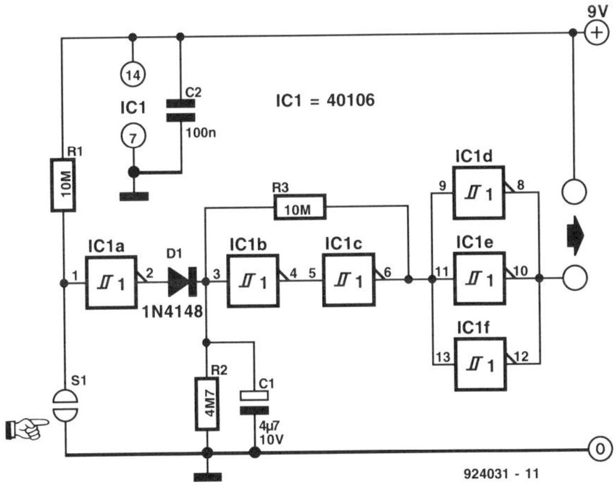

240 V a.c.-to-110 V a.c. converter

080 From time to time one comes across appliances in second-hand goods stores that were designed for operation from a 110 V a.c. (50/60 Hz) supply. If such an appliance is a pure resistive load, such as a radiant fire, a soldering iron, or a melting furnace, the circuit shown may be found useful. Strictly speaking, it is a dimmer set so that the output voltage has an r.m.s. value of 110 V. It is, of course, possible to set it to a different output voltage if so desired.

To obtain an r.m.s. voltage of 110 V across the load, the phase angle at which the triac is switched on must be about 110°. There is no guarantee that this will be met exactly by the present design: owing to tolerances of the various parts, the phase angle may be quite different so that the r.m.s. voltage will be higher or lower than 110 V. It is, therefore, essential, to check the actual voltage across the load. **Bear in mind that the circuit carries mains voltage and is thus potentially lethal**. Checking the phase angle with an oscilloscope cannot be carried out safely without special precautions. The safest and most accurate way of measuring the voltage across the load is with the use of a true-r.m.s. voltmeter (which shows the r.m.s.

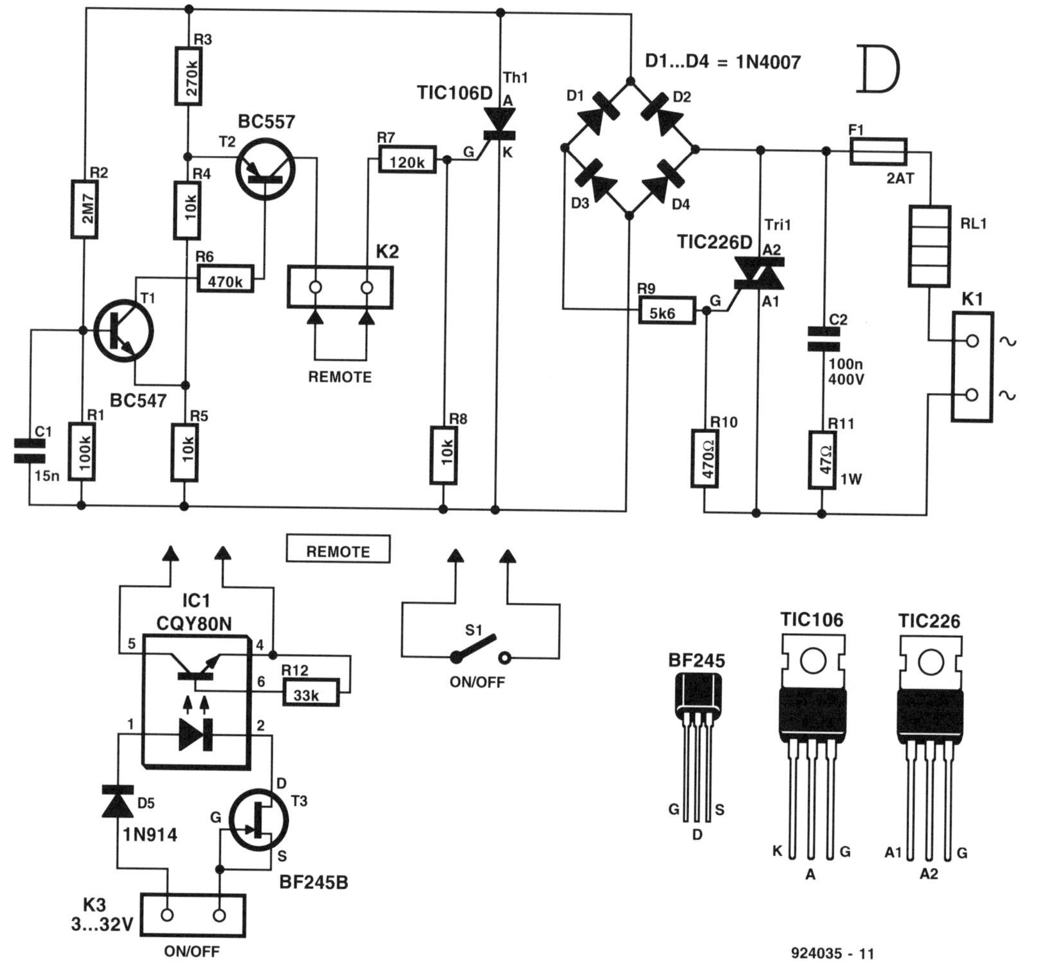

value also of non-sinusoidal voltages). If the voltage across the load is not correct, the value of R_2 must be altered.
If a true-r.m.s. meter is not available, checking may be done in a slightly more primitive way. Use an incandescent 5 W, 240 V bulb as the load and place a thermometer close to it. Switch on the converter, wait till the thermometer gives a stable reading and note that reading (if the thermometer goes off its scale, place it a little further away from the bulb). Do not change the distance between the bulb and the thermometer, and connect a second 5 W, 240 V bulb in series with the first. Once the first bulb has cooled down sufficiently, connect the two in series across the 240 V mains supply (when the bulbs will each drop 110 V). If, after a while, the thermometer has the same reading as before, you may be pretty certain that the converter provides an r.m.s. voltage of 110 V.
The converter may be switched on and off remotely via a direct connection between a suitable 240 V switch to the REMOTE terminals. The wire link should then, of course, be removed. It may also be switched on and off by a voltage of 3–32 V as shown. This optoisolator circuit has the great advantage of isolating the circuit from the mains.
If a triac Type TIC226 is used, the converter can handle currents of up to 2 A. If the triac is mounted on a heat sink, the current may go up to 4 A.

Low-battery indicator

081 The reliability of readings of battery-operated measuring instruments depends, of course, on the state of the battery. Even simple instruments are, therefore, provided with a facility (normally a push button) to check the battery voltage. If, however, the state of the battery must be monitored constantly, that type of check is not very reliable. And then there are instruments that have no battery check facility whatsoever.
The indicator shown in Fig. 1 solves all these problems. With jump links JP_1 and JP_2 in the positions shown, D_2 serves primarily as on/off indicator. This low-current LED lights constantly as long as the battery voltage is sufficiently high and the optional alarm input (a) is not actuated. When the battery voltage drops below a value set by P_1, oscillator IC_{1b} is switched on, whereupon D_2 begins to flash at a frequency of about 0.5 Hz. A second oscillator, IC_{1a}, is actuated when a high voltage level is applied to terminal a. The LED will then flash at a frequency of 10 Hz. This additional indication is often useful to alert the user to an operational error.
When JP_1 and JP_2 are in their other position, there is no on/off indication; the quiescent current then drops from about 4 mA to a few μA. If this condition is selected, the connections of D_2 must be changed over (as shown in dashed lines).
The circuit uses the NAND gates of a Type 74HC132 IC, which is suitable for supply voltages, U_i, of 2–6 V. Since the change-over point of the indicator depends on the supply voltage, connected to U_i, this must be regulated. It is, of course, possible, even preferred, to connect U_i to the regulated supply of the relevant measuring instrument. Here, an independent supply of 5 V is assumed. Since the regulator, IC_2, is a low-drop type, the battery voltage may drop to about 5.5 V before the stability of the output voltage (at U_i) becomes questionable.
The design may be adapted for use with higher supply voltages: if IC_1 is replaced by a Type 4093, the circuit becomes suitable for an input voltage, U_i, of up to 18 V. The series resistor of D_2 must be given a different value, of course: for instance, D_5 should be 3.9 kΩ when U_i = 15 V.
The battery voltage is applied to the input of IC_{1a} via potential divider R_1-P_1. Diode D_1 protects the gate against too high a voltage from a fresh battery.

Although the function of C_6 may not be immediately evident, it is a vital one: the capacitor ensures that the upper switching threshold (about 2.5 V) of IC_{1a} is exceeded briefly on power-on. Oscillator IC_{1b} is then reset. After about one second, the potential at junction R_2-R_3 is a measure of the state of the battery. If that potential lies below about 1.5 V (the lower switching threshold), the oscillator is actuated. If C_6 were not in circuit, the oscillator would not be reset at low battery voltages, resulting in an erroneous low-battery indication.

Preset P_1 (maximum resistance) is adjusted when a variable power supply is connected to U_s+ and set to an output level at which the indicator is to become operational, say, 6 V. Turn P_1 slowly until D_2 begins to flash.

The alarm input, a, may be adapted for use with **negative** battery voltages by the circuit shown in Fig. 2. In fact, the two circuits together monitor a symmetrical battery supply: the positive line via U_s+ and the negative line via the circuit in Fig. 2, connected to terminal a in Fig. 1. When one of the batteries becomes low, D_2 begins to flash: at a frequency of 0.5 Hz in case of the + battery, and at about 10 Hz in case of the – battery. When both batteries become low simultaneously, D_2 flashes with interruptions.

The circuit in Fig. 2 raises the negative battery voltage to above earth level. Note that, because of the loading of its output, R_6 in Fig. 1 must be removed. Connect the ++ line to U_i in Fig. 1, the + line to the negative terminal of the symmetrical battery supply, and 0 to 0 in Fig. 1.

Provided all connections are correct, the total current drain rises by only about 10 μA over that of Fig. 1 (assuming two 9-V batteries in series).

Preset P_1 is adjusted in a manner similar to that described above for P_1 in Fig. 1.

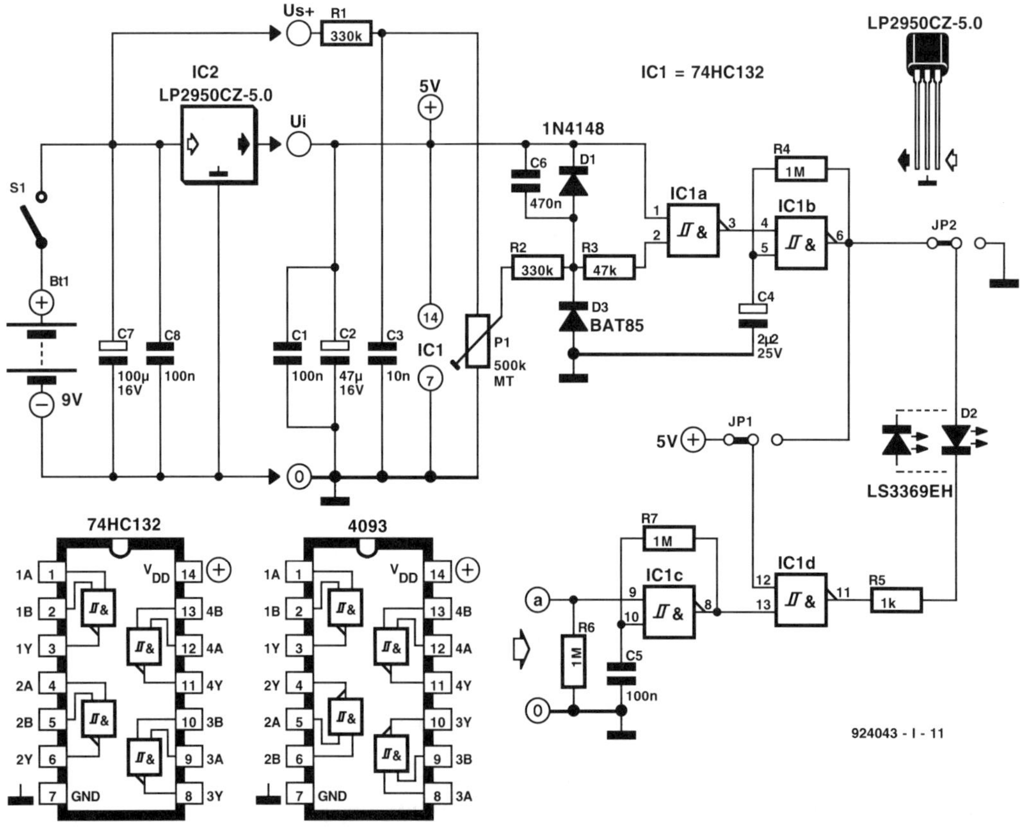

When the upper switching threshold of IC_{1c} is exceeded, the gate oscillates and D_2 begins to flash. The power-on reset (terminal a becomes logic 0) is provided by capacitor C_2. This capacitor is discharged via D_1 and the supply lines.
Diode D_2 protects IC_{1c} against negative input potentials, while capacitor C_1 prevents the circuit being triggered by possible brief noise peaks on the battery voltage.

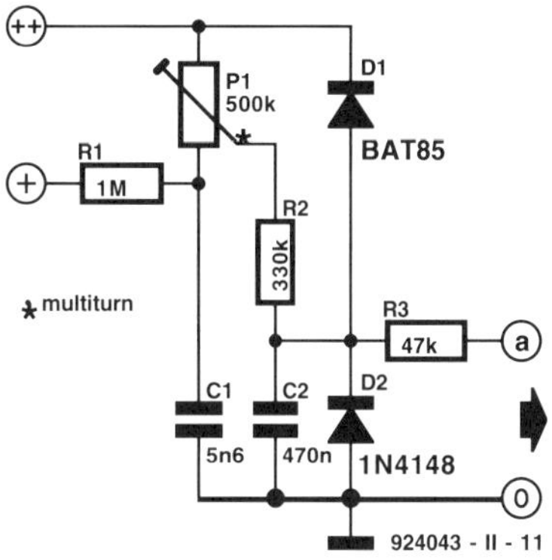

Mains power-on delay

082 This circuit provides a simple means to implement 'soft-start' on heavy mains loads. Basically, this is achieved with the aid of high-power series resistors and a relay. The circuit also enables the load to be switched on with a small (light-duty) switch.
Two relays are used to connect or bypass a series of power resistors inserted between the mains 'live' wire and the load. The 'soft start' is achieved by first connecting a resistor between the mains and the load, and then short-circuiting the resistor again so that the full mains voltage is applied to the load (equipment). The purpose of 'soft' switching is to prevent fuses blowing as a result of very high rush-in currents.
The relay coils are connected in series, and powered by the mains voltage. The reactance of series capacitors C_1 and C_2 keeps the coil current at about 27 mA, while the total direct voltage across the coils is 48 V. When the delay is switched on with S_1, relay Re_2 comes on after Re_1 because its coil is shunted by two large electrolytic capacitors, which take some time to be charged from the current source formed by the mains and the series capacitors. When Re_2 is energized, its contact short-circuits the power resistors.
The simplicity of the circuit gives rise to a minor disadvantage: some time must be allowed between switching off and switching on again. This is necessary to enable C_4 and C_5 to discharge. Depending on the length of the switch-on delay required, the values of C_4 and C_5 may be increased a little. It should be noted, however, that the rush-in current is limited to about 5 A (at a mains voltage of 240 V) by R_3-R_6, and that the resulting dissipation equals about 1200 W in four 5-watt resistors! Therefore, when a longer switch-on delay is wanted, higher-wattage resistors (the PCB can accommodate fairly large types, which may be fitted vertically) must be used.
The current drain of the circuit is practically that of the two relay coils in series, that is, about 27 mA. The current through the neon lamp is negligible.
The specified relays have contacts rated at 16 A. If there is reason to fear that the PCB mounted terminal blocks and their solder connections are not up to the load current, parallel-connected spade terminals ('fast-on' type) should be used instead.
Finally, the values of some components depend on the mains voltage and frequency as follows:

220V, 50 Hz: C_1=330 nF; C_2=220 nF
230 V, 50 Hz: C_1 = 470 nF; C_2 = not fitted
240 V, 50 Hz: C_1 = 470 nF; C_2 = not fitted
(Capacitor working voltage: 630 V d.c. /250 V a.c.)

110 V, 60 Hz: C_1 = 680 nF; C_2 = 470 nF; also R_1 = 120 Ω.
(Capacitor working voltage: 400 Vd.c. /130 V a.c.)

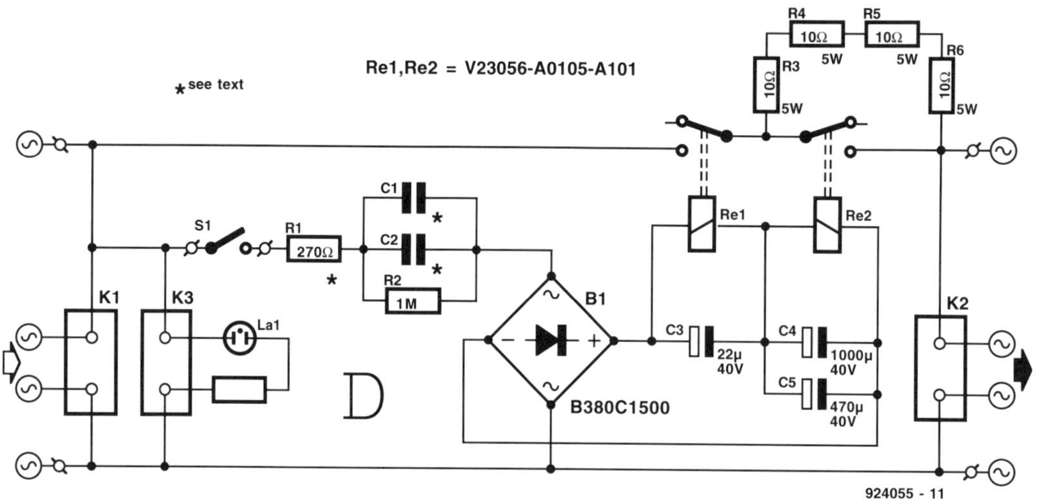

WARNING

This circuit is connected directly to the mains, and carries potentially lethal voltages. Never work on the circuit, or adjust it, while it is connected to the mains. Observe all precautions relevant to working with equipment or components connected to the mains.

Parts list

Resistors:

R_1 = 270 Ω
R_2 = 1 MΩ
R_3-R_6 = 10 Ω, 5W

Capacitors:

C_1 = see text
C_2 = see text
C_3 = 22 µF, 40 V radial
C_4 = 1000 µF, 40 V radial
C_5 = 470 µF, 40 V radial

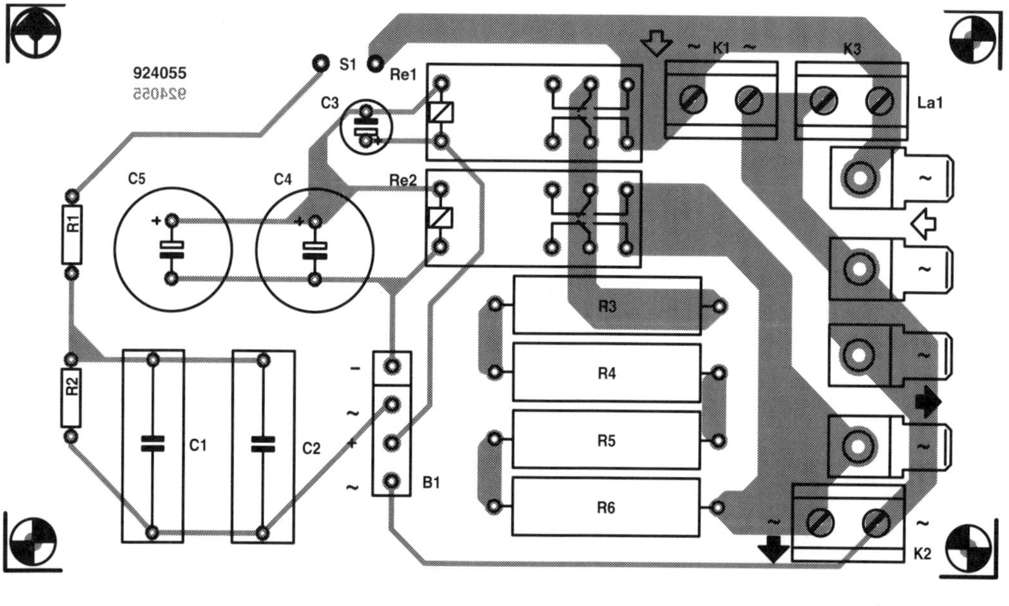

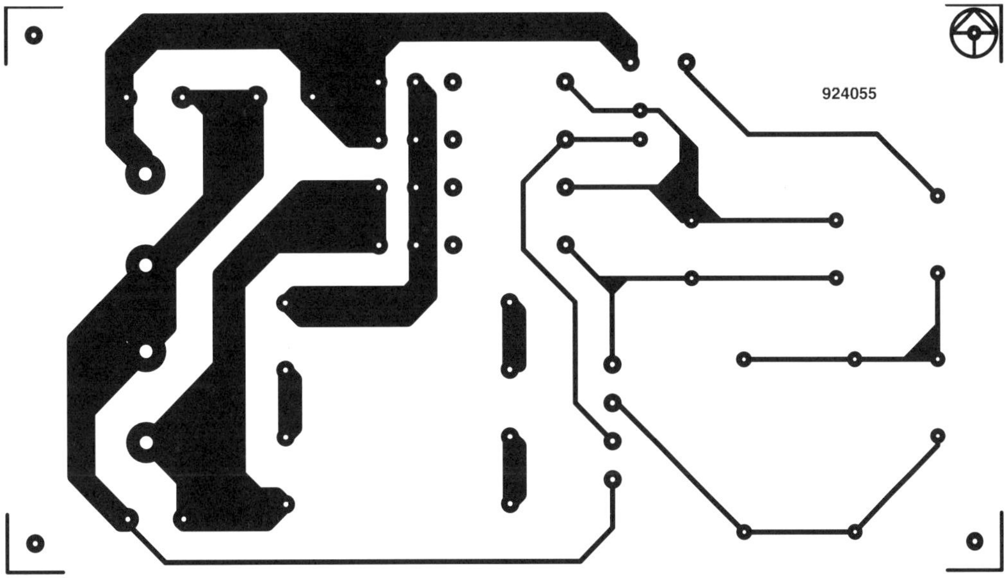

Semiconductors:
B_1 = B380C1500 (380 V piv, 1.5 A)

Miscellaneous:
K_1;K_2;K_3 = 2-way PCB terminal block, pitch 7.5 mm.
S_1 = on/off; mains-rated 1 A.
Re_1;Re_2 = V23056-A0105-A101 (Siemens).
La_1 = chassis-mount neon lamp with integral resistor.
4 spade (fast-on) screw-on type plug for PCB mounting.

Voltage converter I

083 The unit described is capable of converting a 70–260 V r.m.s. alternating voltage into a 180–350 V direct voltage. For that, a full-wave rectifier, contained in an MC34161, is used as a voltage doubler at low input voltages and as a standard rectifier at high input voltages. In this way, an input voltage variation of ×4 is reflected in an output voltage change of not greater than ×2.

The MC34161 has an integral voltage reference source that provides a voltage of 2.54 V at pin 1. The level of a signal applied to pin 2 is compared internally with a potential of 1.27 V.

Voltage divider R_2-R_3 ensures that the internal comparator changes state when the input voltage rises above 135 V: pin 5

Table 1
Input voltage vs output voltage

$U_{a.c.}$	$U_{d.c.}$
70	180
85	218
100	260
125	325
160	212
175	233
200	270
220	291
240	320
260	350

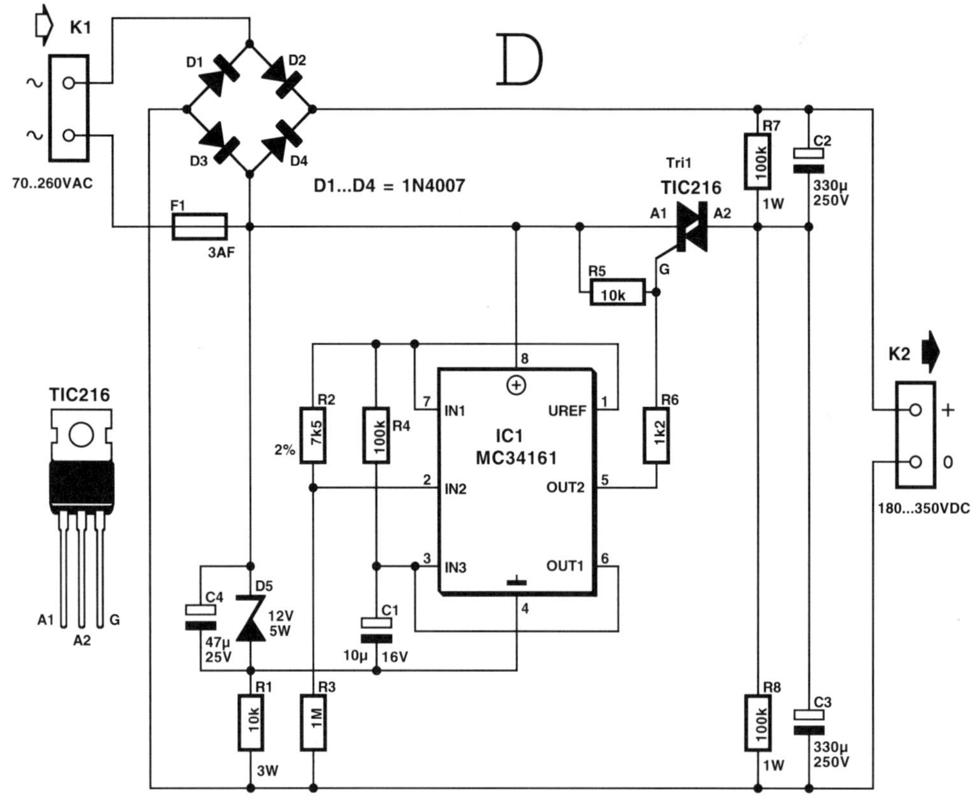

924075 - 11

then becomes logic high. The potential at pin 2 is then lower than 1.27 V. The triac is switched off, and this removes the centre link between the output capacitors, C_2 and C_3, so that voltage doubling cannot take place. Diodes D_1–D_4 then operate as a normal bridge rectifier.

When the input voltage is lower than 135 V, the potential at pin 2 remains higher than 1.27 V: the internal transistor at pin 6 is then off. Because of R_4, pin 3 then carries the reference voltage of 2.54 V. That in turn results in the internal transistor at pin 5 being switched on, so that triac Tri_1 switches from the blocking state to the conducting state. Diodes D_2 and D_3, and capacitors C_2 and C_3, then function as voltage doubler.

Zener diode D_5, in conjunction with R_1 and C_4, ensures that the IC is supplied from a stable 12 V source. The time taken by the circuit to switch from standard rectifier to voltage doubler is determined by R_4-C_1.

The working voltage of capacitors C_2 and C_3 must be ≥250 V.

WARNING: the circuit carries high voltages and must therefore be treated with the greatest care.

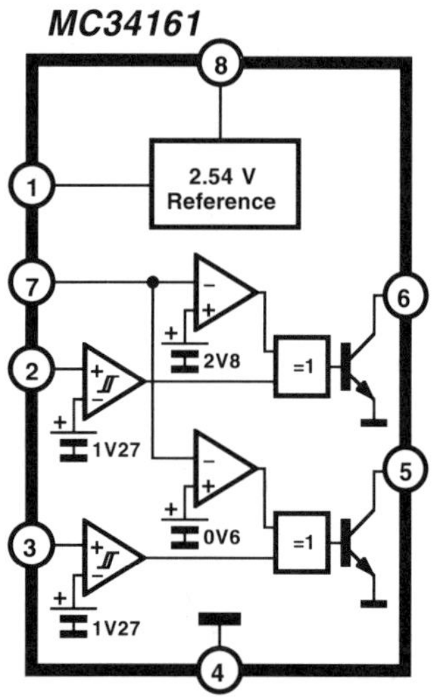

924075 - 12

Voltage converter II

084 Although a 9.5–24 V direct voltage can be brought down to 5 V easily by a standard regulator, the converter described here has the advantage that, since it is a switch-mode type, it hardly dissipates any heat. Its maximum, steady output current at 5 V is 250 mA, although it can cope with peaks of up to 750 mA.

The converter is based on a Motorola Type MC34161 circuit, to which a power stage, T_1 has been added. Inductor L_1, C_5 and D_1 remove any ripple from the output.

The internal comparator at pin 2 of IC_1 is connected to the output of the converter via potential divider R_4-R_5 to monitor the output voltage. The second comparator (pin 3) is used in the oscillator circuit and connected to pin 6 direct and to pin 5 via R_3-C_3.

When the supply is switched on, the output of the converter, and thus the voltage at pin 2, is 0; the oscillator operates normally. Transistor T_1 charges C_5 via L_1. When T_1 is off, L_1 ensures a supply of energy to C_5 via D_1. As soon as the terminal voltage of C_5 has risen to a sufficiently high level, the internal comparator at pin 2 changes state. The oscillator is then switched off via pin 6, so that T_1 is also off. After C_5 has been discharged to an extent that its terminal voltage drops below the preset level, the oscillator is re-enabled, and C_5 is charged again via T_1.

The values of L_1 and C_5 determine the switching frequency; with values as shown, an input voltage of 12 V and a load current of 250 mA, the frequency is 18 kHz. With larger inductances and input voltages, the frequency drops.

It is essential that **all** earth connections are taken to the negative terminal of C_5 as shown in the diagram.

The inductor is a standard triac choke, to which a number of turns have to be added. If the inductance of the choke is L μH and its number of turns is n, the number of turns, n' required for the present inductor is given by $n' = n\sqrt{(470/L)}$.

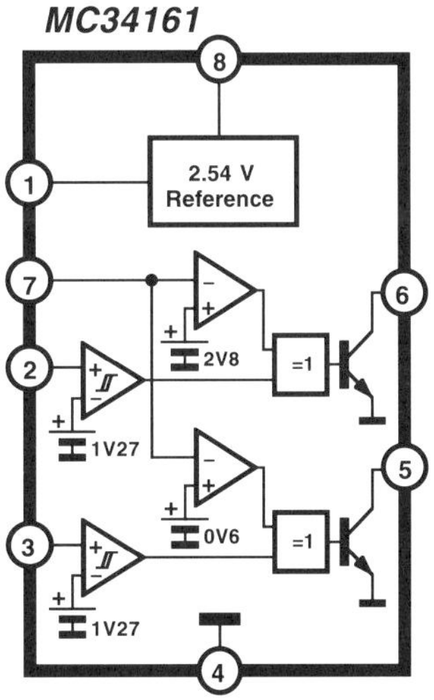

924077 - 12

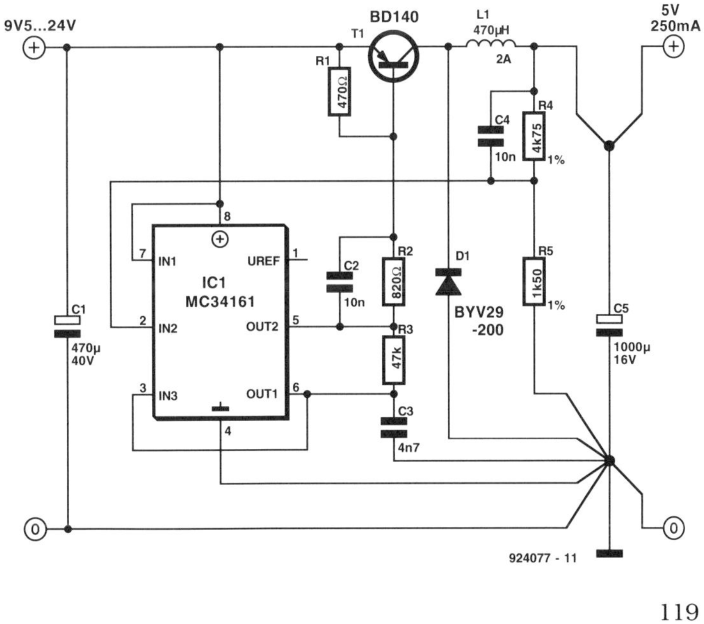

924077 - 11

Voltage inverter

085 A negative supply voltage may be obtained with the use of a special inverter IC, but such ICs are not always readily available. Fortunately, if the current drawn is not excessive, a standard HC-MOS chip may be used. The present circuit is based on a Type 74HC14, which contains six Schmitt triggers, whose combined gates can deliver a fairly high output current. Moreover, such devices can be made to oscillate readily.

Parallel switching is, however, a problem with Schmitt triggers, even if they are contained in the same chip. This problem is particularly acute in the case of slow input signals. In the present circuit this problem is resolved by driving the parallel-switched input via a gate that is not part of the parallel circuits. Since this is also a Schmitt trigger, the output signal has steep edges.

The parallel-switched gates together with IC_{1a} form a rectangular-wave generator, whose output frequency is about 125 kHz. The output signal is converted by a charge pump into a negative supply voltage. The diodes used for that purpose are Schottky types that, owing to their low threshold voltage, do not lower the load voltage by as much as silicon diodes.

Under no-load conditions, the output voltage is about 6 V and the IC draws a quiescent current of around 100 µA. When the load current is about 1 mA, the output voltage drops to 4 V. If this voltage can drop even further (down to half the supply voltage), a load current of up to 10 mA is possible. A higher load current or a smaller drop in the output voltage cannot be obtained by raising the value of the capacitors in the charge pump, since the IC cannot cope with this. Note also that the circuit is not short-circuit proof: the IC will not give up the ghost immediately, but it does not take all that long.

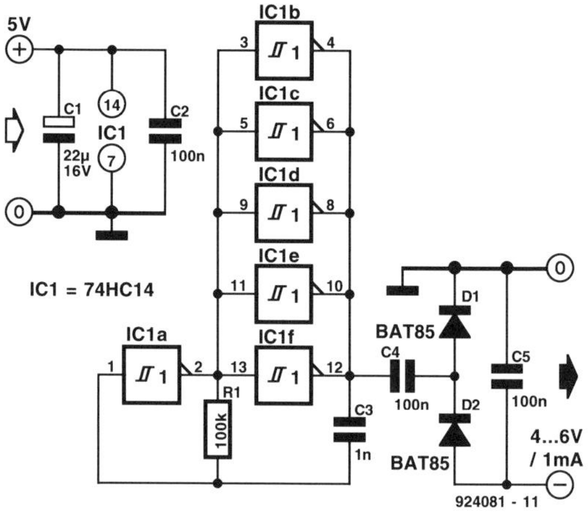

Experimental fast NiCd charger

086 The continuing difficulty in designing fast NiCd chargers is determining when the battery is charged, that is, when to stop charging.

The charger presented here is based on the latest developments as reported by several manufacturers. It is not at all certain whether, and under what conditions, the circuit will always give satisfactory results.

The battery is charged with a current (in mA) that is ten times its nominal capacity (in mAh). That means that, for instance, an HP7 (AA; RG) type battery is charged at 5 A, a current 100 times larger than used in standard charging.

The charging is controlled by a Type 555 timer IC, here connected as an astable. If the IC's output is high, charging takes place. There is, however, a fixed period of time ($=R_6C_3$) during which no charging takes place. As soon as charging stops, C_1 is connected across the battery by electronic switch IC_{3a}. Its terminal voltage is then compared by IC_{1a} with the maximum battery voltage set by P_1. The output of the comparator is integrated by R_3 and C_2 and

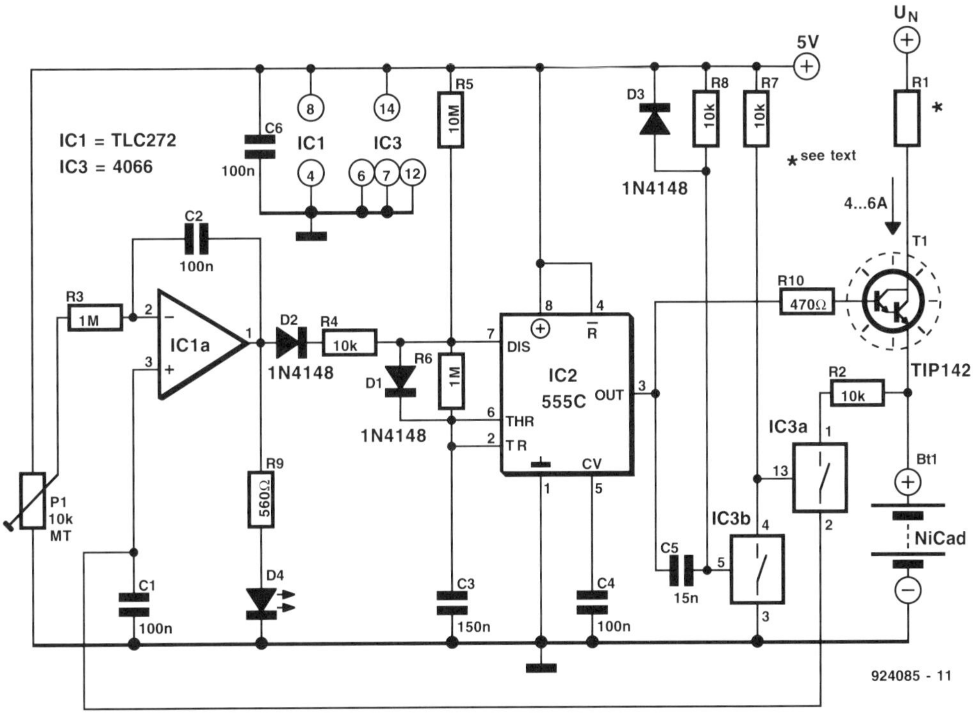

then used to determine the period of the astable. If the maximum battery voltage has not been reached, charging takes place for about 90% of that time. If the maximum battery voltage has been reached, charging takes place for 1% of the time (trickle charging). Do not leave the battery connected to the charger unnecessarily: when the LED lights, the battery is fully charged.

During charging, owing to a variety of resistances, primarily in the supply leads and connections,the battery e.m.f. is an unreliable yardstick for determining the state of charge of the battery. Therefore, the e.m.f. is taken immediately after a burst charge, because then the voltage can be measured exactly. The important question is, of course, to what e.m.f. P_1 should be set; in other words, what is the correct battery e.m.f.? Opinions vary, but with the prototype good results were obtained with a value of 1.42 V at room temperature (21 °C).

The circuit draws a current of only 10–15 mA, which may be obtained with a 7805 regulator.

The proposed charger is intended for charging one 1.5 V NiCd battery in 8–10 minutes. The charging current for a 500 mAh battery is about 5 A, which need not be regulated, since it will be limited by R_1.

The value of R_1 is given by Ohm's law. If, for instance, the charging current is drawn from an 8-V source, and assuming that the drop across the battery and T_1 is 2 V, the voltage across the resistor is 6 V. Its value should thus be 6/5=1.2 Ω. Bear in mind that the power dissipated in it is 6×5=30 W: you will, therefore, have to connect a number of resistors in parallel.

If you want to charge a number of batteries in series, raise the level set by P_1 accordingly (at 1.42 V per battery). Note, however, that you should use only batteries that have already been sorted for equal capacity by the manufacturer. Also, the supply voltage to the charger circuit must always be 2 V higher than the level set with P_1.

Finally, after every five fast charges, give the battery a 'normal' ($^1/_{10}$ mAh over 14 hours) charge.

Automatic NiCd battery charger

087 The charger makes use of the properties of a Type 555 timer. The internal window comparator of the IC is set to 4.7 V by zener diode D_1. If the potential at pin 6 exceeds that level, the output voltage at pin 3 becomes low. If the potential at pin 2 drops below half the reference voltage, that is 4.7/2 = 2.35 V, the output voltage becomes high.

When the voltage of the battery to be charged is low, and the output potential of IC_1 is thus high, the battery will be charged via R_6 and D_2 until the 'fully charged' voltage, set with P_2 is reached. The output of the IC then changes state and charging ceases. However, the IC continues monitoring the battery voltage with the aid of T_1.

When the output of the battery, owing to self-discharge, has dropped below the level set with P_2, the IC causes charging to recommence. Dual LED D_3 lights red when charging is in progress and green during the quiescent periods.

In this way, the battery can be left in the charger for any period you like: it will always be fully charged ready for use when required. It will never be overcharged, nor will it be discharged more than a little.

With values as shown in the diagram, the charger is suitable for 6 V and 9 V batteries. Nine-volt types with 6 or 7 cells are charged at about 20 mA; P_1 should be set to arrange for the charger to switch off after 14 hours. The lower window level is set to about 1 V below that value with P_2.

Six-volt types with 4 or 5 cells are charged at 55 mA. Again, P_1 should be set so that the charger switches off after about 14 hours. The lower window level is set to about 0.8 V below that value.

It should be noted that NiCd battery manufacturers do not recommend 'fully charged' voltage measurements, but rather time or temperature measurements. Nevertheless, the prototype of the present circuit has been used for charging small NiCd batteries to good effect for some time.

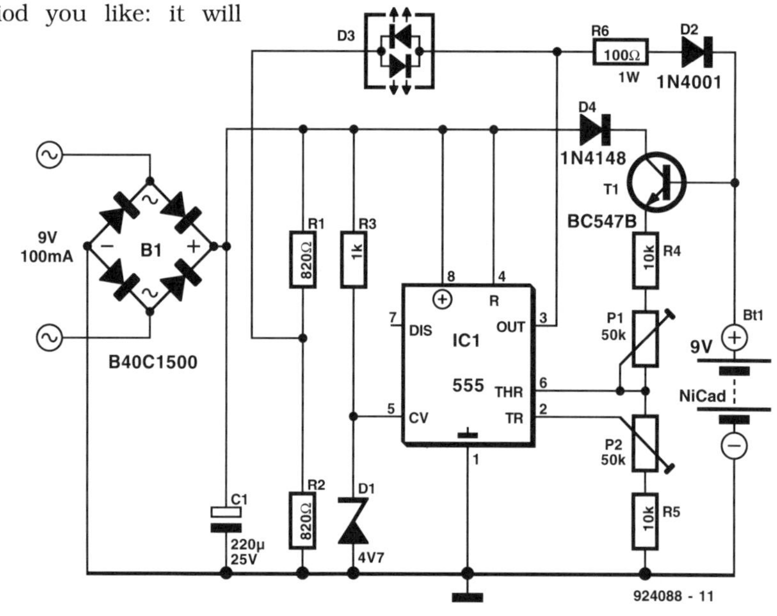

Solar-cell supply

088 The simplest solar-cell supply system consists of three parts: a diode, a solar-cell panel, and a rechargeable battery. The diode prevents the battery being discharged through the solar panel when no or little sunlight falls on to the panel. Although the diode is usually a Schottky type, the forward drop of this type may cause an appreciable loss of energy. The present circuit uses a special low-drop diode.

The opamp draws only a tiny current: on average 20 μA. Although it can operate with a supply voltage of as low as 1.6 V, the gate of T_1 requires at least 3.0 V to be switched on. For that reason, the nominal battery voltage is specified as 4.0 V.

To align the circuit, replace the solar panel by a variable regulated power supply with its current limiting set to a value that is safe for the battery. Set the power supply output to a level that is 0.1 V higher than the current battery voltage. Next, adjust P_1 so that the output of IC_1 just becomes logic high. Finally, check with an ammeter that the battery is not being discharged when the power supply voltage is set to below the current battery voltage.

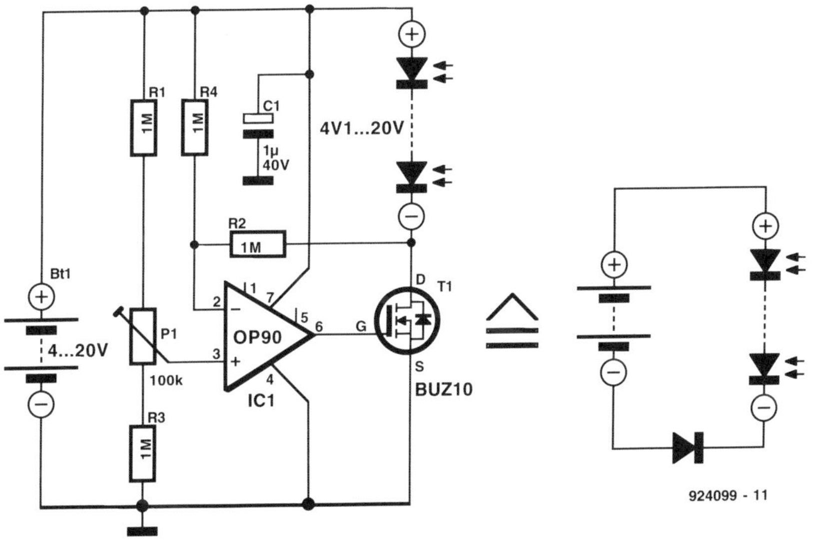

Current limiting for LM317 regulator

089 Although the well-known Type LM317 voltage regulator is already short-circuit-proof, there are cases where limiting the large short-circuit ccurrent may be desirable. As the diagram shows, such a current limiting facility can be provided in a simple manner. Use is made of the fact that the output voltage, U_o, is dependent on the feedback to the control input. As long as the current limiting does not operate, resistors R_2 and R_3, as well as T_1, may be ignored. The output voltage is then:

$$U_o = 1.25(1 + P_1/R_1) + I_{adj}P_1 \text{ (volts).}$$

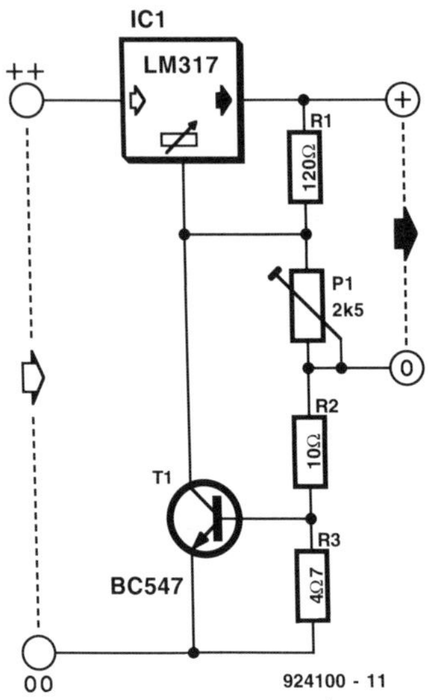

Since the maximum level of I_{adj} is 0.1 mA, P_1 can set U_o to 1.25–27 V.

When the current through the regulator causes a drop of about 600 mV across R_3, T_1 will come on. This will cause a drop in the level at the control input of the regulator, and thus in the output voltage. With the value of R_3 as shown, the current limiting will come into operation at a current of 0.6 / 4.7 = 120 mA.

Improved LM317 regulator

090 Although the properties of the LM317 regulator are excellent, they can be improved by cascading two LM317s. There is then a constant difference between the output voltages of the two regulators and, consequently, a constant voltage across the input and output of IC_2. This arrangements results in an appreciable improvement of the regulating characteristics of IC_2. Moreover, its dissipation is reduced so that the stability of the output voltage with temperature is improved. Other properties, such as the maximum output current of 1.5 A do not change, of course.

The output voltage, U_o, of the circuit depends on the ratio R_3:P_1 as follows:

$$U_o = 1.25(1+R_3/P_1) \quad [V].$$

With values of these components as shown, the output voltage may be varied over the range 1.25–11.5 V. The voltage drop, U_d, across IC_2 depends on the operating point of IC_1 and may be calculated from:

$$U_d = 1.25(1+R_2/R_1) \quad [V].$$

With values of these components as shown, U_d = 3.5 V. It should be noted that this voltage must not drop below 3.0 V. Moreover, the value of R_2 must be about twice that of R_1, and the minimum drop across the entire circuit must not be lower than U_d+3 V.

The circuit is particularly suitable for use as a 5-V power supply. It is, however, important that the direct voltage at the input is not lower than 12 V. This means that the secondary voltage of the mains transformer must be 12 V instead of the usual 9 V.

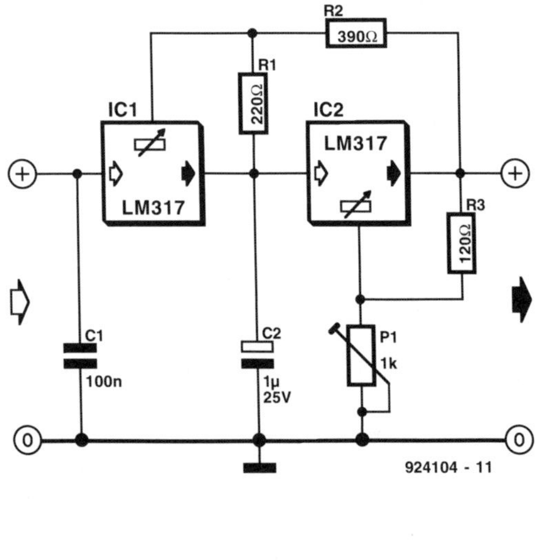

Three-phase simulator

091 Most domestic consumers (in the UK) are provided with a single-phase supply, unless exceptionally heavy loading is foreseen. It may, however, occur that a low-voltage three-phase supply is required for experimental purposes and in such cases, the simulator described can prove useful.

The source signal for the phases, R, S and T is generated with a standard Wien bridge. The sine-wave generator is formed by IC_{1a}. Preset P_1 enables the frequency to be set accurately to 50 Hz; the output level (pin 1) is set with P_2 to 1 V (peak).

Circuit IC_{2a} provides a constant load impedance for IC_{1a}, which is important for the stability of the generated frequency. It

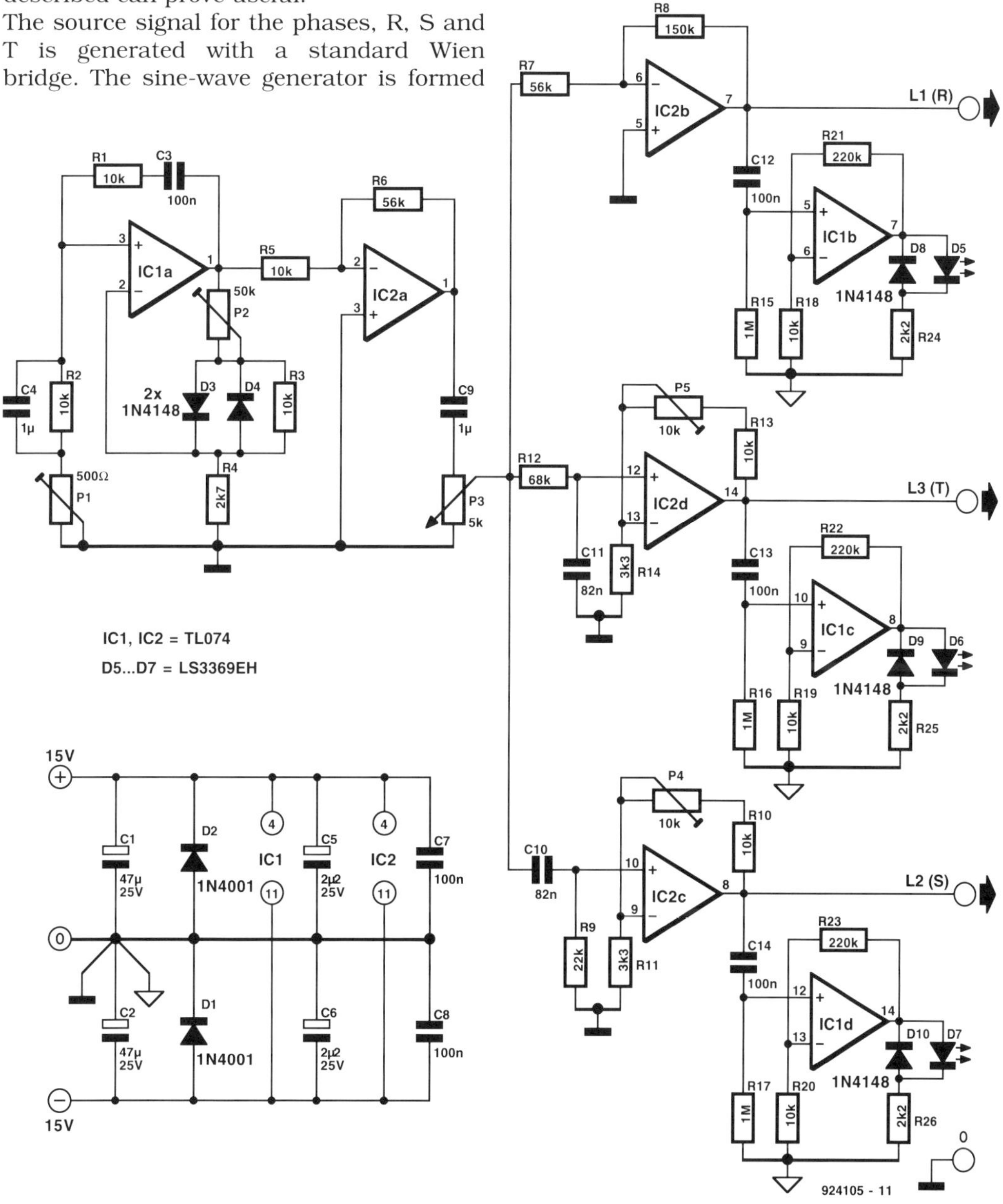

also raises the signal level to 5.6 V (peak). The peak value of the phases is set to 0–12 V with P_3. Series capacitor C_9 prevents the offset voltage of IC_{1a} and IC_{2a} adding direct voltage to the outputs of IC_{2b} and IC_{2d}.

The R phase results from inverting the signal at the wiper of P_3, that is, shifting it by 180°. Owing to low-pass filter R_{12}–C_{11}, the T output lags the signal at the wiper by 60°, while R_9–C_{10} provide a 60° lead at the S output. There is, therefore, a 120° phase difference between any pair of phases.

Presets P_4 and P_5 need to be set only once and that in such a manner that the peak values of the three phases are identical.

Low-current LEDs D_5–D_7 light only if there is an alternating voltage present at the associated output. An accidentally short-circuited phase is, therefore, detected immediately.

The opamps are short-circuit proof and can provide a current of about 10 mA. When a symmetrical ±15 V supply is used, the quiescent current is ±20 mA.

Simple symmetrical power supply

092 Integrated voltage regulators LM317 and LM337 enable a simple, regulated, symmetrical power supply as shown in the diagram to be made. This supply is variable from 0 V to ±15 V, can provide a current of up to 1.5 A, is short-circuit proof, and has thermal protection. Moreover, the regulators have internal dissipation limiting which ensures, irrespective of their cooling, that the dissipation does not exceed 20 W.

To enable the output voltage to be varied from 0 V, auxiliary voltages of ±1.4 V are created with the aid of two zener diodes, D_1 and D_2. Although technically inelegant, this arrangement is the most cost-effective in a simple power supply.

The regulators must be fitted on a heat sink with a thermal resistance of 1.75 K W^{-1}.

The maximum current of 1.5 A is available only if the maximum dissipation of 20 W has not been reached and (with an input voltage of ±18 V as shown), the output voltage is above 5 V. At lower output voltages, the maximum output current drops to 1.1 A (at U_o = 0 V).

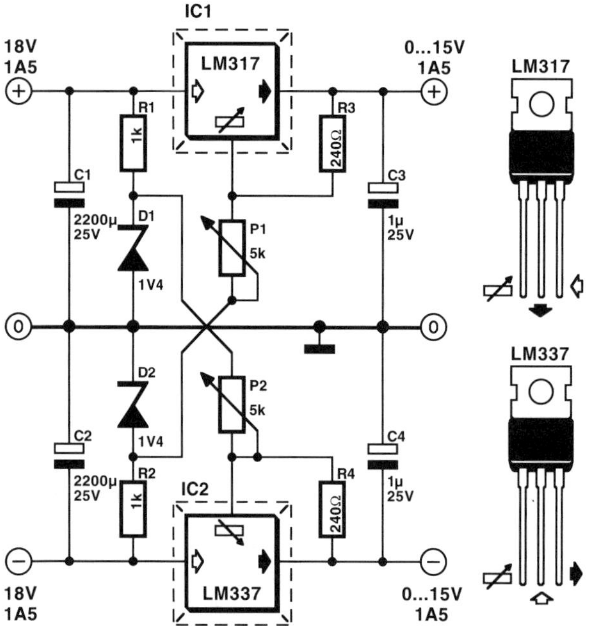

924112 - 11

Switched 5 V supply

093 To derive a 5 V line from, say, a 15–25 V supply, National Semiconductor's LM1575 or LM2575 is ideal. These switched 5-V regulators need only a choke, a freewheeling diode and a couple of electrolytic capacitors. In fact, they go a long way to prove that the industry has overcome most of the problems switched power supplies have suffered from; both the regulators mentioned worked first time without any problems whatsoever.

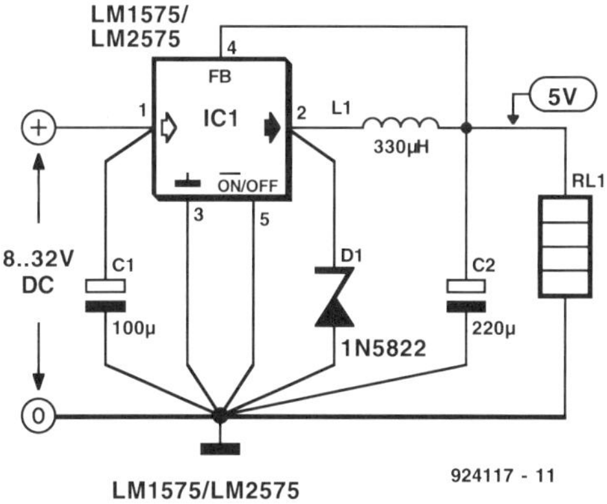

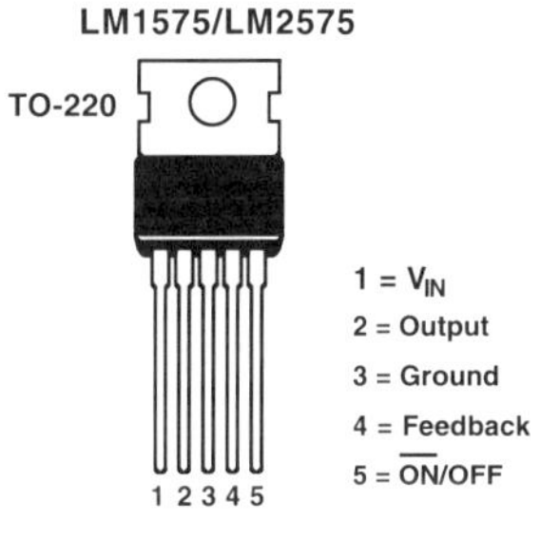

The only thing that is still not as good as in linear power supplies is the ripple: in the the prototypes this amounted to 20 mV_{pp}.

The measured efficiency of the prototypes is 75%, whereas National Semiconductor's give a figure of 82%. This figure does, of course, depend to some extent on the components used.

The diode is a Type 1N5822 or BYW29 from Philips: any fast diode will do, however: do not use a common-or-garden 1N4001 or similar.

Thechoke is a ready-made 330 µH type that must be able to handle currents of at least 1 A.

Build the supply with a central, star-shaped common earthing point as indicated in the diagram. Use good-quality electrolytic capacitors (perhaps two in parllel). The more compact you keep the supply, the better it will work.

The input voltage can lie between 8 V and 32 V. The maximum current the regulator can provide is about 1 A.

Pin 4 is the feedback input.

Pin 5 may be used for switching the supply on/off; if that is wanted, connect this pin to a 2.4–15 V potential instead of to earth.

Simple power supply concept

094 THE best known alternatives to a 'quick and dirty' power supply are the three-pin fixed voltage regulator and the zener-plus-transistor combination. While these basic circuits will suit a good many applications, they do have their limitations, which can be frustrating at times. For example, most types of fixed voltage regulators are limited to an output current of about 1 A only. Where more power is required, a 'current bypass' transistor is often added. However, while this boosts the maximum output current, the regulation of the supply is degraded. Fixed voltage regulators with higher output currents (say, 5 A) are not a good alternative because they are notoriously expensive.

The second alternative, the zener-plus-transistor circuit, has limited use also because of its relatively poor ripple rejection and insufficient stability at output load variations.

The PSU presented here suffers none of the disadvantages mentioned above, and is simple to build as a multi-purpose concept. It is the perfect low-cost supply for a host of applications. At first glance, the circuit looks very much like the familiar zener-transistor combination. However, an essential difference is that feedback is used, which results in a 100-Hz ripple suppression of up to 55 dB—far more than can be achieved with the simple zener-transistor stabilizer.

The voltage reference used here is D_1, a TL431C from Texas Instruments. The internal structure of the TL431C is shown in the diagram. Here, D_1 supplies a base current to T_1 that results in 2.5 V across resistor R_3. This allows you to calculate the supply output voltage, U_o, from

$$U_o = 2.5\,[1+(P_1+R_2)/R_3] \text{ volts.}$$

The specified component values result in an output voltage of 12 V. For other output voltages, simply adapt the output voltage divider, making sure that the current through P_1, R_2 and R_3 is at least 1 mA. This is required to ensure that the current flowing into the reference input of the TL431 is negligible (approx. 2 µA). The power transistor is a darlington with a guaranteed current gain of 1000 or greater at an emitter current of 5 A. This means that only 5 mA of base current is required. Although this is not much, it has to be taken into account when R_1 needs to be given a different value. Also, D_1 requires a minimum cathode-anode current of 0.5 mA, which results in a total, minimum, current of 5.5 mA through R_1. This design infomation, together with the lowest possible input voltage, U_{in} (measured across C_6), and the base-emitter drop of T_1 (approx. 2 V), results in a theoretical value of the current limiting resistor:

$$R_1 = (U_{in} - U_{be} - U_o)/I_{R1}.$$

Because the current gain of the darlington may be up to two or three times the guaranteed value mentioned above, it is often possible to give R_1 a higher value than calculated. Since a higher resistor value results in lower dissipation of R_1 and D_1, some experimenting is certainly worth while.

The PCB designed for the supply accommodates the complete rectifier section, that is, a bridge rectifier, a buffer capacitor and a fuse. The buffer capacitor, C_1, and the on-board heat sink for T1 are large enough for output currents up to 2 A.

As already mentioned, this PSU is a concept. Those of you who do not need the rectifier section may omit it, and connect a d.c. voltage of 16 V to K_1. Note, however, that this requires wire links to be fitted in the positions indicated by dashed lines near the bridge rectifier.

If you require more output current (say, up to 5 A), simply move the power transistor off the board, and fit it on a larger heat sink (see parts list). Also, increase the buffer capacitor to 10000 µF. Since such a capacitor (or array of capacitors) will not fit on the board, connect it as an external part via heavy-duty wires and two spade terminals (marked '+' and '–' on the component overlay). A continuous output current of 5 A also requires the bridge rectifier to be cooled. This is best achieved by leaving it on the PCB, and clamping it on to a side panel of the metal enclosure used to house the supply.

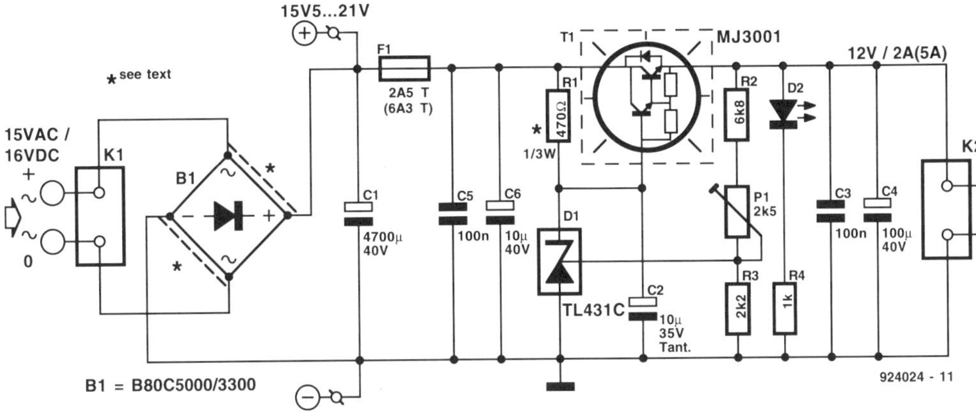

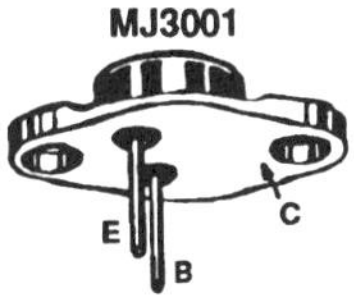

TL431C

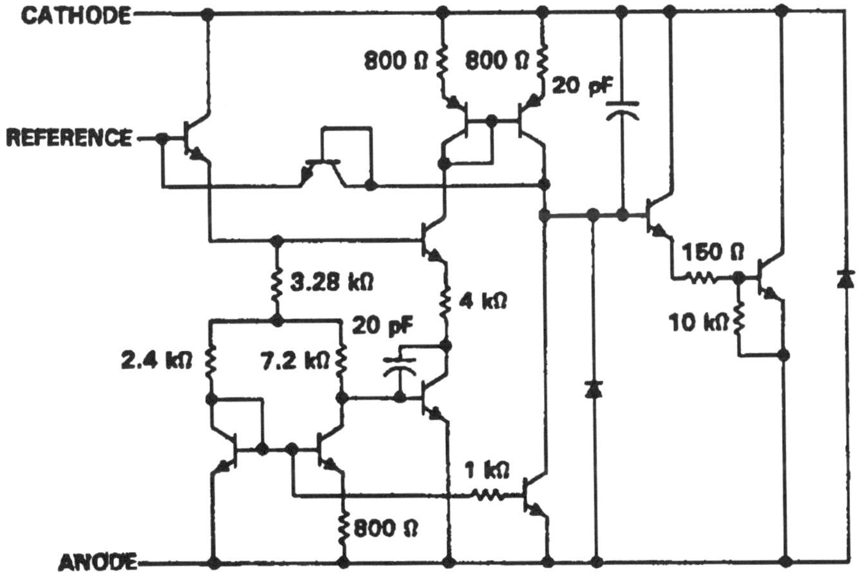

924024 - 12

Parts list

Resistors:
R_1 = 470 Ω, 0.33W (see text)
R_2 = 6.8 kΩ
R_3 = 2.2 kΩ
R_4 = 1 kΩ
P_1 = 2.5 kΩ preset H

Capacitors:
C_1 = 4700 µF, 40V
C_2 = 10 µF, 35 V tantalum
C_3;C_5 = 100 nF
C_4 = 100 µF, 40 V
C_6 = 10 µF, 40 V

Semiconductors:
D_1 = TL431C
D_2 = LED, green, 3 mm
T_1 = MJ3001
B_1 = B80C5000/3300

Miscellaneous:
K_1;K_2 = 2-way PCB terminal block, pitch 5mm.
F_1 = 2.5 A fast fuse (6.3A)* and PCB mount holder.
Heat sink: SK201 (6 K W^{-1}) or SK71/75mm* (1.25 K W^{-1}).
Two 'fast-on' spade terminals for PCB mounting *.
PCB Ref. 924024.

* for 5 A version only.

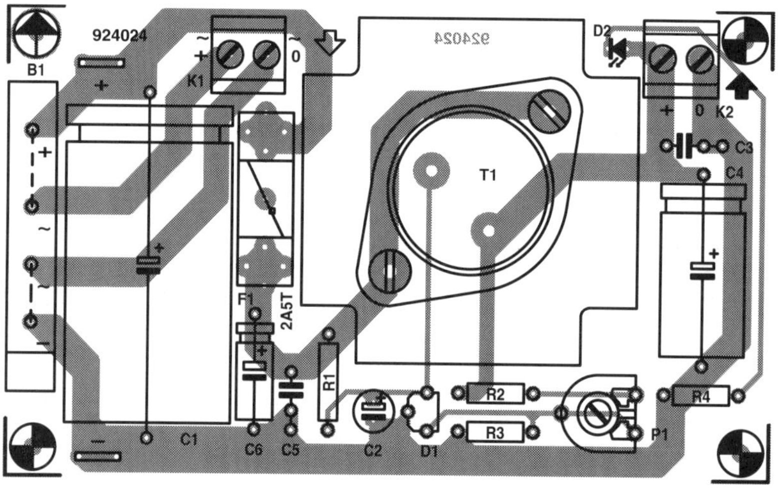

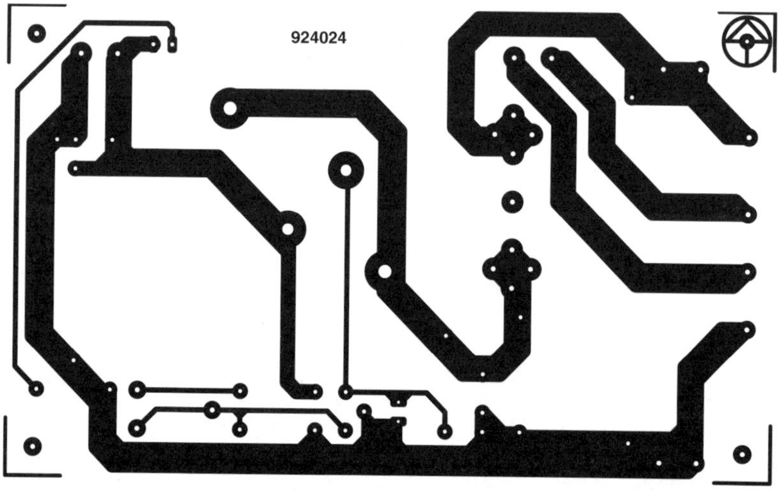

State of battery indicator

095 Many battery-operated appliances have an LED indicator that shows the state of the battery. If the battery is almost flat, the LED flickers. The circuit described enables this facility to be built into apparatus that has no indicator.
The drop across D_1 (U_{LED}) depends on the type of LED and is normally around 2 V. This voltage is applied to the input of a discrete differential amplifier, T_1-T_2. When the base voltage of T_1 is smaller than U_{LED}, it switches on, whereupon C_1 gets charged via R_1. The rising potential on C_1 will cause T3 to switch off, whereupon T_4 begins to conduct. Because of that, D_1 goes off and remains so until C_1 is discharged. Assuming that the base voltage of T_1 is still smaller than U_{LED}, the cycle starts afresh. In this way, the LED continues to flicker until the base voltage of T_1 becomes larger than U_{LED}. Only T_2 and T_3 are then on and T_4 is off, so that the LED lights continuously.
The setting of preset P_1 determines the point at which the LED begins to flicker. The frequency at which it flickers depends on the supply voltage and the value of C_1. This means that the frequency may be changed by altering the value of this capacitor.
The value of R_V determines the current through D_1 and thus the brightness at which it lights. In practice, the LED and R_V are often already present in the appliance.

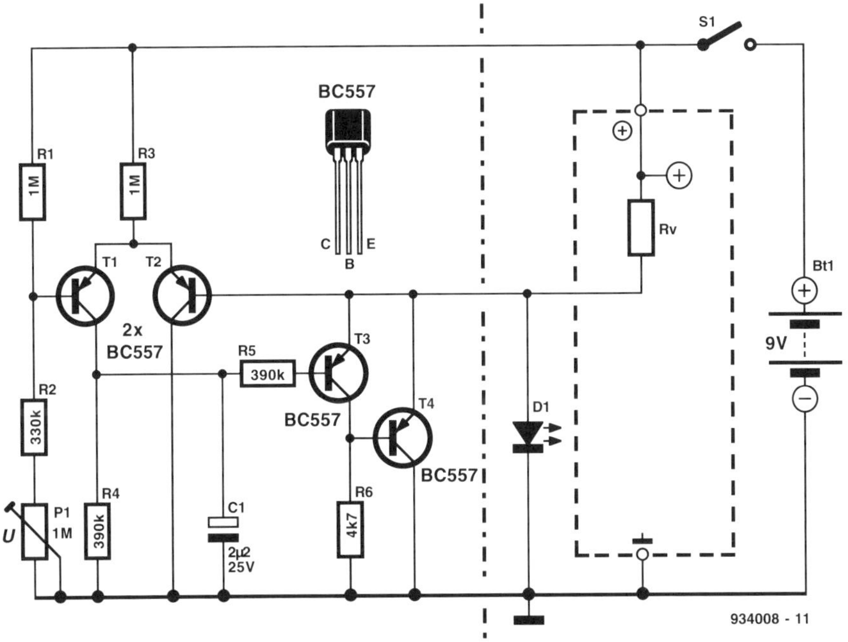

Automatic power off

096 The power off circuit ensures that any load is automatically removed from a battery. It is particularly useful in cars, so that if the driver forgets to switch off the headlights, he does not return to a car with a flat battery a couple of hours later.
The only operating control is a spring-loaded change-over switch with a test position. When the button of the switch is

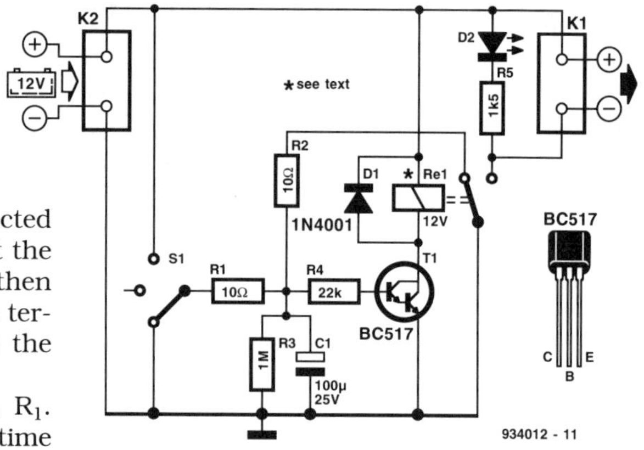

pressed, the +ve supply line is connected to the base of T_1 via R_1 and R_4, so that the transistor switches on. The relay is then energized and its contact links the –ve terminal of the load to ground, so that the load is powered.

At the same time, C_1 is charged via R_1. This capacitor and R_3 determine the time constant of the circuit. When S_1 is set to its centre position, C_1 discharges slowly via R_3. When the base potential of T_1 drops below 1.2 V, the transistor switches off, whereupon the relay contact connects C_1 to ground via R_2. The capacitor is then rapidly discharged completely. The current drawn by the circuit in this situation is nil.

When S_1 is in its lowest position, that is, connected to ground, the discharging process is halted immediately.

Diode D_2 shows the status of the circuit. Resistor R_5 should not have too low a value, otherwise the LED may not show up properly.

The value of C_1 may be increased to 4700 µF: the power-off time is then increased from 10 min to about 30 min. This capacitor should be a type with very low leakage current.

The relay may be 6-V or 9-V type to prevent it clattering when the car engine is started or a large load is switched on. Do not omit protection diode D_1.

Since the circuit has no polarity protection, make certain that correct polarity is observed.

D.C.-d.c. converter

097 The converter raises a direct voltage to almost twice its level and is, therefore, particularly suited to increase the output of solar cells to the level required for charging lead-acid or NiCd batteries. It can deliver a current of up to 3 A. Measurements with a load current of 2 A are given in the table. The open-circuit output voltage is about 1–1.5 V higher.

In the following description it is assumed that the input voltage to the circuit is 12 V and the output voltage 22 V. IC_{1a}, R_2 and C_5 form a rectangular-voltage generator. This signal is also available in inverted form at the output of IC_{1d}. Network R_2C_6 delay the output of IC_{1a}, so that the output of NAND gate IC_{1b} has a duty factor >0.5 (the negative half is shorter than the positive half). The same is true of the output of NAND gate IC_{1c}. (The input signal to this gate is delayed by R_5-C_7).

The output of IC_{1c} is inverted and buffered three times: in IC_{3f}, IC_{3a}, and the four parallelled gates IC_{3b}–IC_{3e}. It is then used to drive power FET T_3.

The output of IC_{1b} drives small-signal transistor T_1. When this transistor is on, junction R_6-R_7 is pulled to 2 V without diode D_1. However, IC_{2a} needs an input signal of 11–22 V, since the supply voltage for this inverter (and, of course, also for inverters IC_{2b}–IC_{2e}), as well as the collector voltage of T_1 is already dervied from the doubled output voltage. The negative supply voltage for this IC is therefore derived from the positive input voltage. Diode D_1 ensures that potential at the input of IC_{2a} does not drop below 10.5 V.

Transistors T_2 and T_3 conduct alternately. When T_2 is on, C_{10} is charged to the level of the input signal via T_3 and D_3. When T_2 is off and T_3 is on, C_9 is charged similarly. Capacitor C_{10} retains its charge, since D_3 prevents its beingdischarged. Since the two capacitors are in series, the output voltage is twice the input potential.

Owing to the multiple inversion of both signals following the delay networks, it is impossible for T_2 and T_3 to be switched on simultaneously.

Capacitor C_1 buffers the input signal, so that its loading is constant in spite of the varying current drawn by the circuit.

It is essential that D_2, D_3, T_2 and T_3 are well cooled. It is best to mount these components on a common heat sink.

The bold lines in the circuit diagram represent heavy-duty wires that should be as short as possible, since they carry a current of 6 A.

A carefully constructed convertor should give an efficiency of 94% (at 22.2 V and 1.8 A).

Input and output voltages with a load current of 2 A are:

U_{in}	U_{out}
10 V	18 V
12 V	22 V
14 V	26.4 V
15 V	28.3 V
16 V	30 V

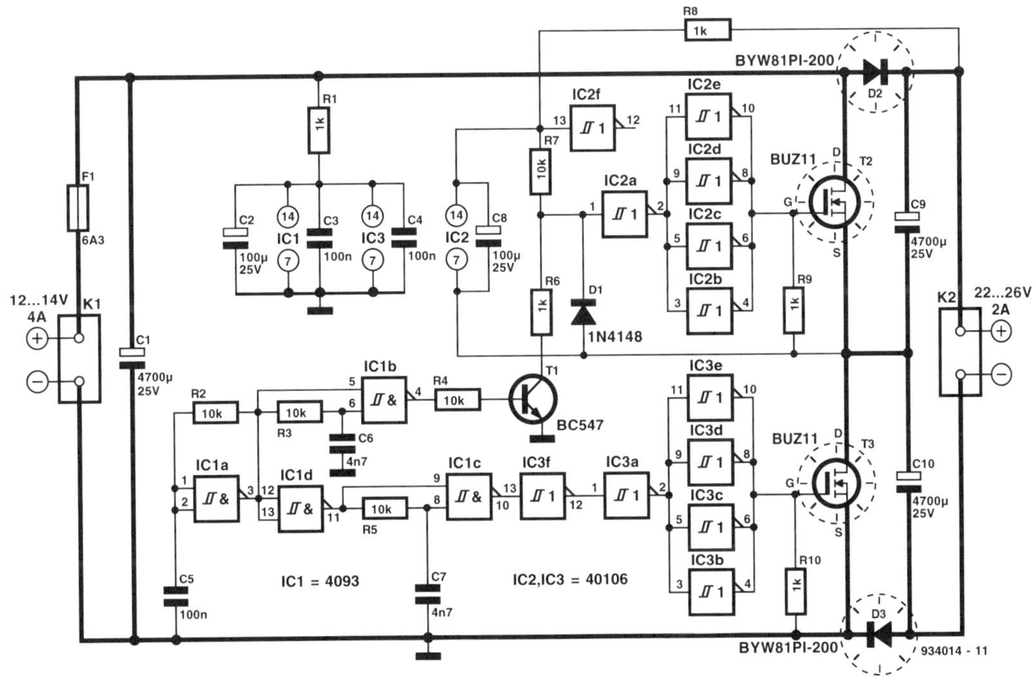

Car battery voltage monitor

098 The monitor uses a two-digit 7-segment LED display to indicate the battery voltage. In addition to this, there are 'low' and 'high' voltage warning indicators. The display range is 9–18 V.The battery voltage is scaled down by a factor 7.83 by resistors R_{20} and R_{17}. This means that the voltage range at the SIG input of the LM3914 (IC_3) is 1.14–2.29 V for battery voltages between 9 V and 18 V. The LM3914 has an internal voltage reference that maintains a voltage difference of 1.25 V between pins 7 and 8. Each of the ten resistors in the ladder network thus drops a voltage of 0.125 V. The voltage difference for each step between 9 V and 18 V is (2.29-1.14)/9, or 0.1278 V, which is close enough to 0.125 V for the resolution required by the present application. Here, the LM3914 is operated in 'dot' mode.The open-collector outputs of the LM3914, L1\-L10\, are connected to 1 kΩ pull-up resistors, nine of which are contained in SIL array R_{19}. The resistors also cause a '1' to appear on the most significant (MS) display for all input voltages greater than 9 V. Only if 9 V is measured will the base of T_1 remain high, so that the MS display is quenched. If the input voltage is higher, the base of T_1 is pulled low, and the B and C segments of LD_1 light, showing a '1'. Output L_2 of the LM3914 is not connected because the priority encoder, IC_1, automatically encodes a '0' if none of its inputs is pulled low. Together with the '1' on the MS display this produces the reading '10'. Inverters IC_{2a}–IC_{2d} ensure the correct input levels for the inputs of the 4511 display driver, IC_1.Inverters IC_{2e} and IC_{2f} function as undervoltage (<8.7 V) and overvoltage (>18.7 V) monitors and indicator drivers. The actual switching thresholds will depend on the IC used, so some experimentation with the value of R_{15} (ovevoltage) and R_{14} (undervoltage) may be called for. The undervoltage LED should light at input voltages between 5 V and about 8.5 V. Lower voltages are not possible because the LED light would not be seen any more. The overvoltage LED should light at an input voltage of about 19 V.The circuit is simple to adjust with an accurate digital voltmeter and an adjustable power

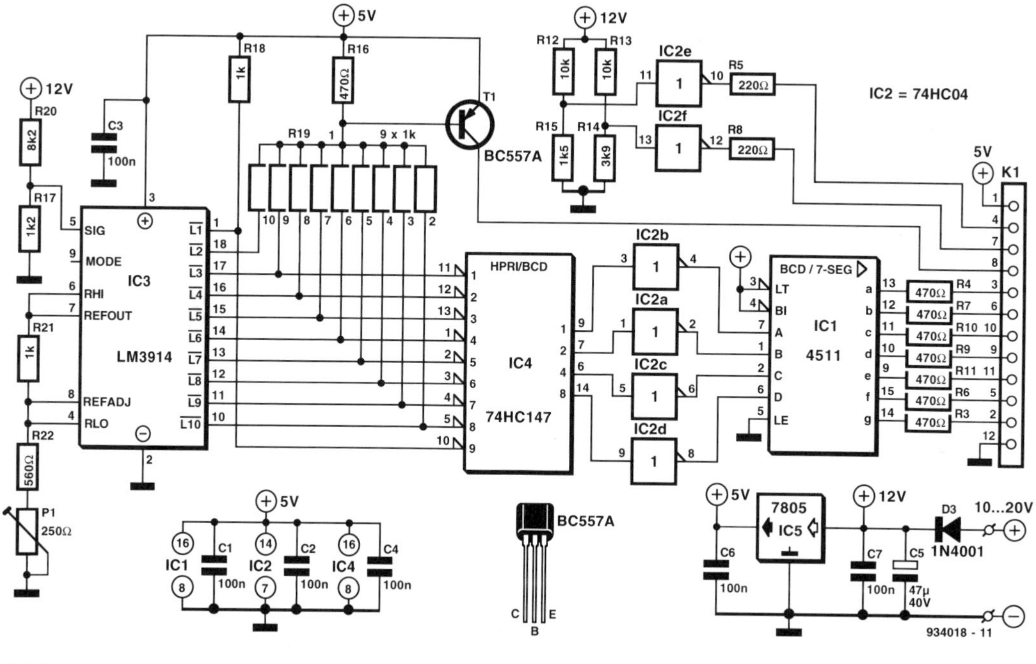

supply. If these instruments are not available, simply set preset P_1 to mid travel, which is a sufficiently accurate setting in most cases. If precision is needed, apply 12.0 V to the input of the circuit. Adjust P_1 until the display reads '12'. This value should remain on the display if the input voltage is varied between 11.5 V and 12.5 V. If not, adjust P_1. Next, reduce the input voltage to 8.0 V. The undervoltage LED should light, and the display read '0'. Slowly increase the voltage to 9 V. At a value of 8.5 V the display should change to '9'. The exact voltage at which this happens is not terribly important. Next, apply 17 V, which should cause the display to indicate '17'. Slowly increase to 18 V, and check that the display changes to '18' between 17.2 V and 17.8 V. If this does not work, R_{21} is out of tolerance, which causes an incorrect step size of the ladder network. The cure is simple: replace R_{21} (and, if necessary, R_{17} and R_{20} also) by a 1% tolerance type.The circuit draws a fairly high current, which causes voltage regulator IC_5 to run hot if it is not fitted on a small heat sink.The printed circuit board consists of two sections (display and voltage monitor proper), which must be separated by cutting before they are populated. The connection between the two units so obtained is formed by a 12-way pin header, K_1, on the main board, and a mating 12-way SIL socket, K_1', on the display board. The display unit is mounted at a suitable position on the car's dashboard, with the displays covered by a bezel. Alternatively, the driver circuit and the readout may be fitted into a small plastic enclosure, for which a suggested front panel layout is shown.

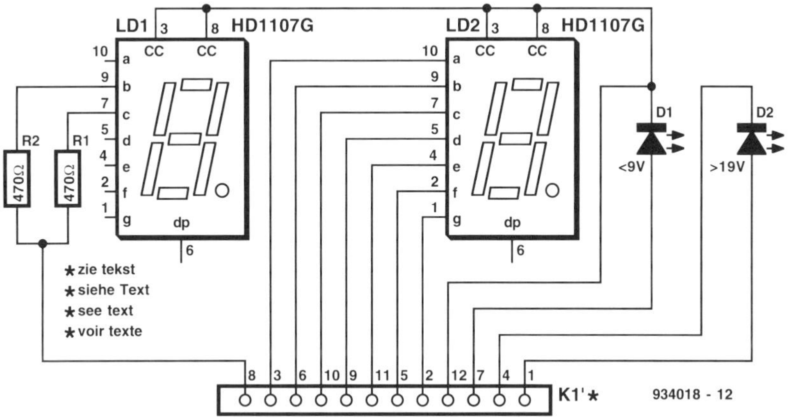

Parts list

Resistors:
R_1–R_4, R_6, R_7, R_9, R_{10}, R_{11}, R_{16} = 470 Ω
R_5, R_8 = 220 Ω
R_{12}, R_{13} = 10 kΩ
R_{14} = 3.9 kΩ
R_{15} = 1.5 kΩ
R_{17} = 1.2 kΩ
R_{18}, R_{21} = 1 kΩ
R_{19} = 9-way 10 kΩ SIL array,
R_{20} = 8.2 kΩ
R_{22} = 560 Ω
R_{23} = 250 Ω preset H

Capacitors:
C_1–C_4, C_6, C_7 = 100 nF
C_5 = 47 µF, 40 V

Semiconductors:
D_1 = LED, yellow, 3 mm
D_2 = LED, red, 3 mm
D_3 = 1N4001
T_1 = BC557A
LD_1, LD_2 = HD1107G (Siemens)

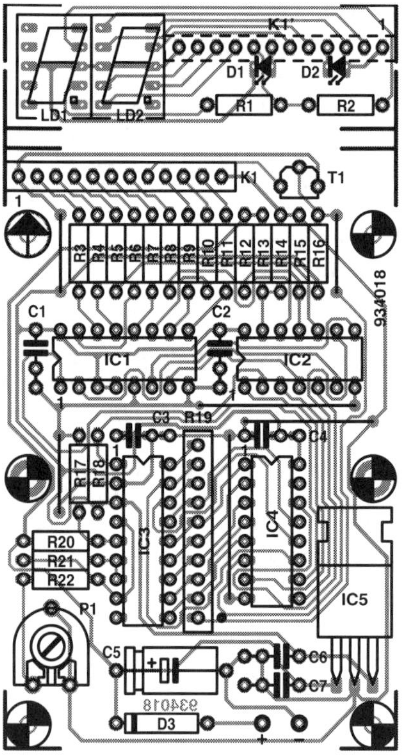

Integrated circuits:
IC_1 = 4511
IC_2 = 74HC04
IC_3 = LM3914
IC_4 = 74HC147
IC_5 = 7805

Miscellaneous:
K_1 =12-way angled SIL header
K_1' = 12-way SIL socket1
Plastic enclosure HM-Kit (Pactec),
61×97×23mm (W×D×H));
order code 6600-902.
Printed circuit board Ref. 934018

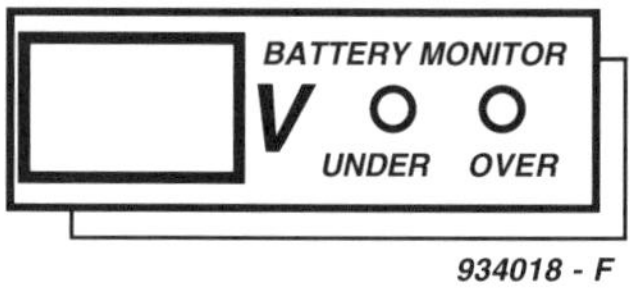

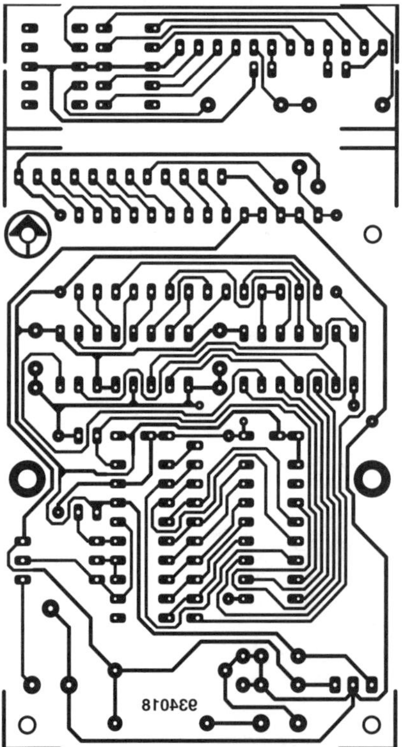

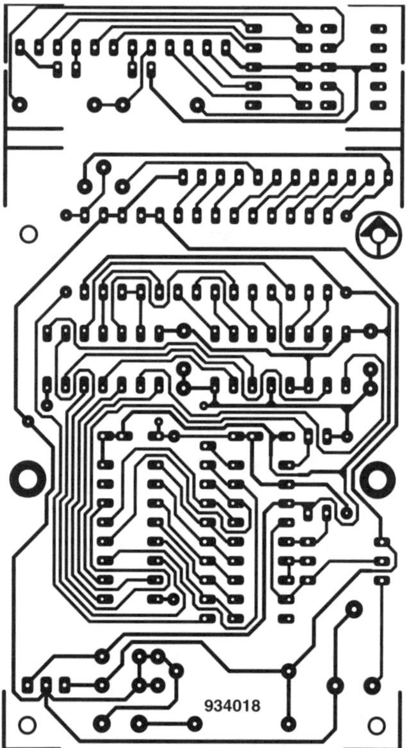

Zener bridge

099 The supply described here uses two zener diodes and two rectifier diodes instead of the usual four rectifier diodes and one zener.

Assuming that the potential at junction D_1-D_3 is positive with respect to that at D_2-D_4 and is greater than the zener voltage (+0.6 V), D_3 will act as a zener and limit the voltage to 10 V. The current flowing from D_1 through C_1 and the load will return to the transformer via D_4 (the voltage across C_1 is then equal to the zener voltage of +0.6 V).

The potential across D_3 (**not** plus 0.6 V) will be applied to C_1, which (in theory) can be charged to 0.6 V below the zener voltage (in practice, without a load, it is slightly higher).

If the voltage at junction D_2-D_4 is positive with respect to that at D_1-D_3, the same process happens, but now D_4 will act as the zener, while D_2 and D_3 are forward biased.

The zener current is determined by R_1. Bear in mind, however, that the diodes act as zeners only half of the time. In other words, the average current through the zener diodes is only half the peak current. The same applies to the power. With the value of R_1 = 47 Ω (as drawn) and an average zener current of 90 mA, the diodes dissipate about 450 mW, but the peak dissipation is nearly 1 W. It is, therefore, not possible to lower the value of R_1.

Since the current through the zeners is not constant (owing to the alternating voltage at the input), the maximum output current of the circuit is rather lower than twice the average zener current. Here, it is around 75 mA. At this current, the output voltage is about 9.4 V; without a load, it is around 10.4 V.

Just as in a traditional circuit, the output voltage of the present circuit rises to the peak value of the input potential (that is, about 18 V) when one of the zener diodes fails.

Also, bear in mind that the circuit limits only the peak value of the output voltage. As soon as the diodes no longer act as zener diodes, the output voltage depends entirely on the charge on C_1. That charge will, of course, drop and so will the output voltage. The present circuit is consequently intended primarily for use with circuits that tolerate ripple or that draw only a small current. With C_1 = 220 μF (as drawn), and the load current 75 mA, the ripple amounts to 7 V. To lower that value appreciably, the value of C_1 must be raised considerably or the load current must be limited to about 1 mA.

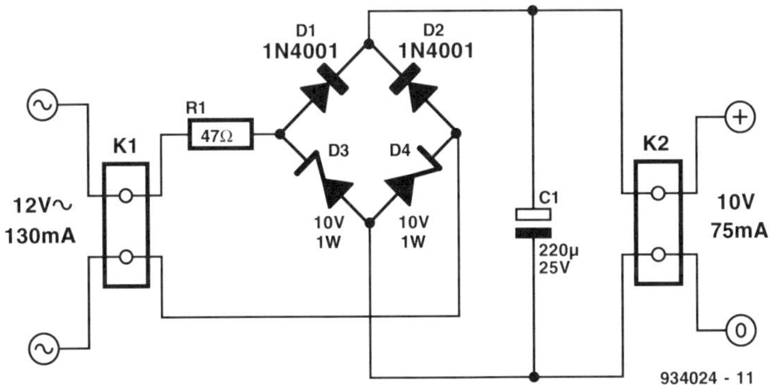

Simple d.c.-to-d.c. converter

100 To keep the converter simple, it contains no preset controls, which means that the level of the output voltage is dependent to some degree on the load. In theory, the output level is twice the level of the input voltage, but, owing to losses in the converter, that level cannot be attained. The main losses occur at the semiconductor junctions of the transistors and the rectifier diodes. Since the drop at these junctions is constant at about 0.6 V, the losses are proportionally larger with an input voltage of 6 V than with one of 18 V.
Oscillator IC_1 generates a signal at a frequency of about 10 kHz. Depending on the output level of the IC, either T_1 or T_2 is switched on. This results in C_2 being charged during one half period; during the other half period, the charge of C_2 is transferred to C_3. This results in an output voltage of twice the input voltage less the losses mentioned.
The circuit contains no critical components; IC_1 may be any version of the 555, bipolar or CMOS, while the transistors may be inexpensive LF types. Although the present circuit uses 1N4004 diodes, Types 1N4001 will do just as well.
Although the switching frequency is of the order of 10 kHz, and the diodes are designed for lower frequencies, no problems were experienced in the prototype, primarily because the voltages and currents are relatively small.
The converter draws a current of 5 mA (555) plus twice the output current.

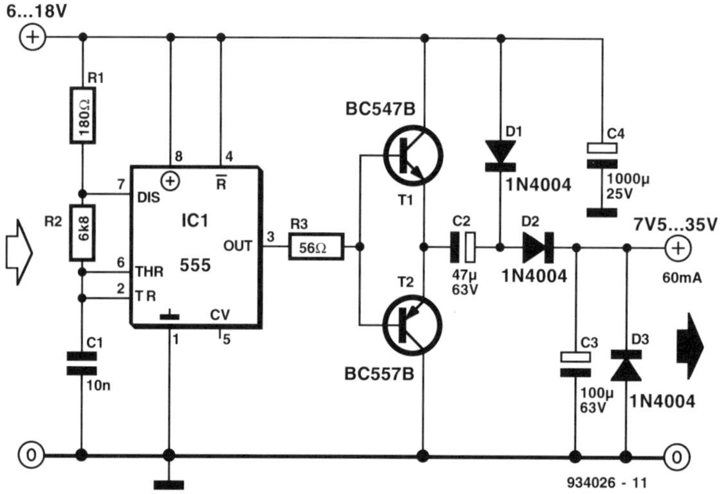

Lead-acid battery conditioner

101 To make sure that a lead-acid battery (whether standard or sealed) is always fully charged, a constant voltage should be applied to it. Chargers that check whether the battery is still charged and only commence charging when it is (nearly) flat, do not guarantee that it is fully charged when it is required in an emergency.
The charger described monitors the battery voltage constantly and tops up the charge if and when required. This has advantages and drawbacks. The battery is always fully charged, but it becomes sluggish, that is, it does not take the charge well. This effect disappears after normal use (a few duty cycles). If the voltage is not too high (≤13.8 V), the battery is charged and has a

long life. It is not advisable to constantly apply a high voltage of, say, 14.4 V across the battery (this is in any case not necessary, because at 13.8 V it remains fully charged). The only reason for applying a high voltage (temporarily) is that this charges a flat battery quicker.

Always use a commercially available charger: this meets safety requirements and is normally relatively cheap. A partly discharged battery may be connected to such a charger without any precautions.

When the battery is fully charged, do not leave it connected to the charger, because that provides quite a high voltage. It is then that the circuit shown here comes in handy. The voltage at which charging should be stopped is preset with P_1. Turn the preset fully anti- clockwise and connect a voltmeter across the battery. The LED will not light. Turn P_1 slowly clockwise till the LED just lights: the battery is then being charged. When the battery voltage has reached the required value, turn P_1 slightly anti-clockwise until the LED goes out. This happens fairly gradually, so it is best to do this in semi-darkness if possible.

The 723 is powered by the battery via D_2 (current about 10 mA). The internal reference voltage is reduced to 2.2 V by R_1 and R_2. This voltage is compared with the divided battery voltage at the wiper of P_1. If it is lower, the output (pin 10) goes high, whereupon D_1 lights and the optoisolator switches on the thyristor. This results in the charger output being applied to the battery: the level of the current is determined by the charger. Since the output of the charger is an unfiltered, rectified alternating current, the thyristor will switch off at every zero crossing, but immediately switch on again if the battery voltage is not high enough.

It is quite feasible to leave the monitor between the charger and the battery. Charging will then take a little longer because there is a loss of about 1 V across the thyristor.

The thyristor must be mounted on a heat sink if the current exceeds 1 A; the maximum current must not exceed 5 A.

The circuit may be built on the PCB shown in Fig. 2. A large part of the board is occupied by the heat sink for the thyristor. Use flat connectors that can be screwed on to the board for the four terminals that carry fairly high currents.

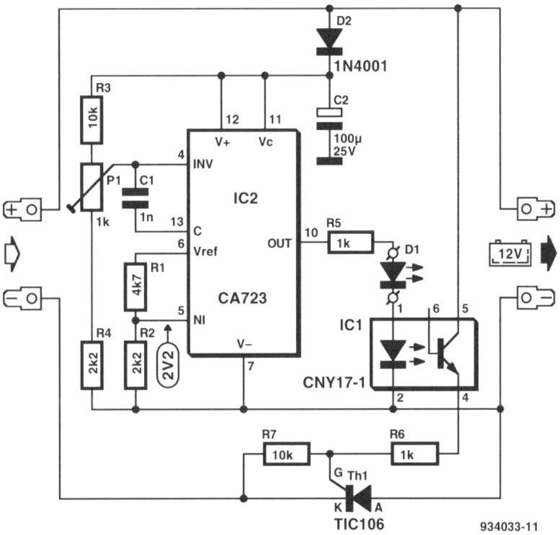

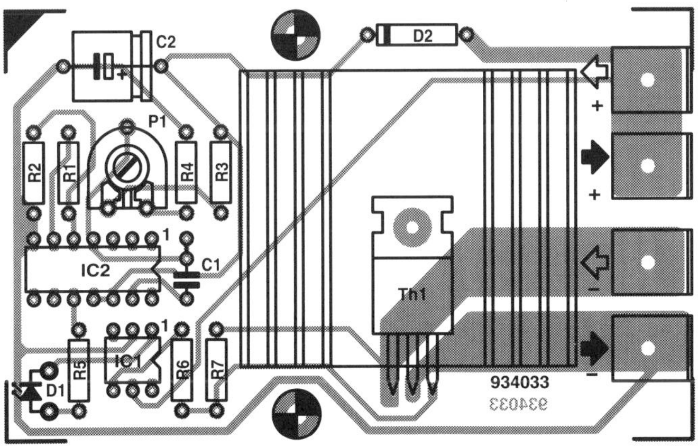

Parts list

Resistors:
R_1 = 4.7 kΩ
R_2, R_4 = 2.2 kΩ
R_3, R_7 = 10 kΩ
R_5, R_6 = 1 kΩ
P_1 = 1 kΩ preset

Capacitors:
C_1 = 1 nF
C_2 = 100 µF, 25 V

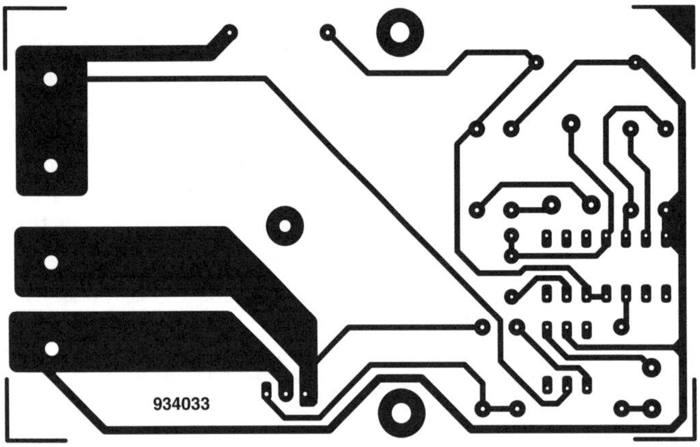

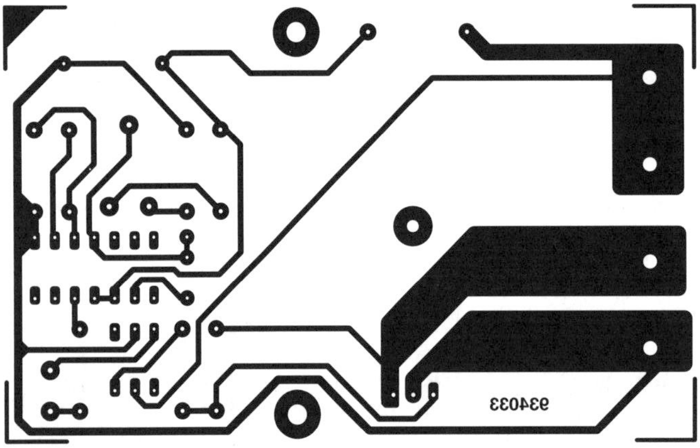

Semiconductors:
D_1 = LED, 3 mm, red
D_2 = 1N4001
Th_1 = TIC106

Integrated circuits:
IC_1 = CNY17-1
IC_2 = CA723

Miscellaneous:
Heat sink for Th_1,e.g., SK59
4 flat connectors with screw fitting
1 case 95×60×24 mm ($3^3/_8$×$2^3/_8$×1 in)
PCB Ref. 934033
Front panel foil Ref. 934033-F

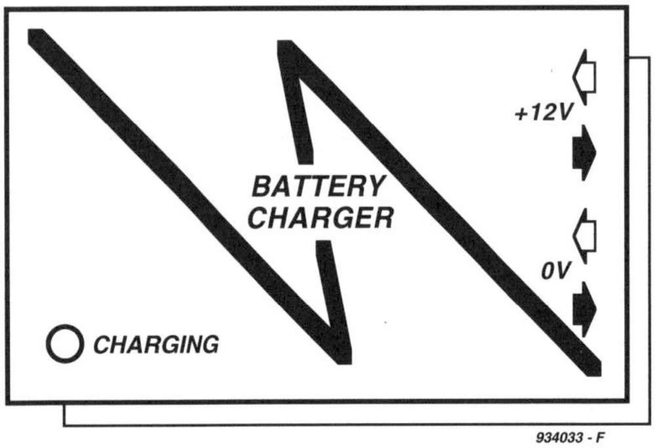

934033 - F

Shunt regulator for solar panel

102 A recurring problem with energy generating systems consisting of solar panels and batteries is that sometimes there is too much sunshine. At those times it is essential to limit the charging current to an already full charged battery or batteries. These will not only use a lot of water (which in itself is no problem), but their life-span will be shortened. Voltage regulators are normally series

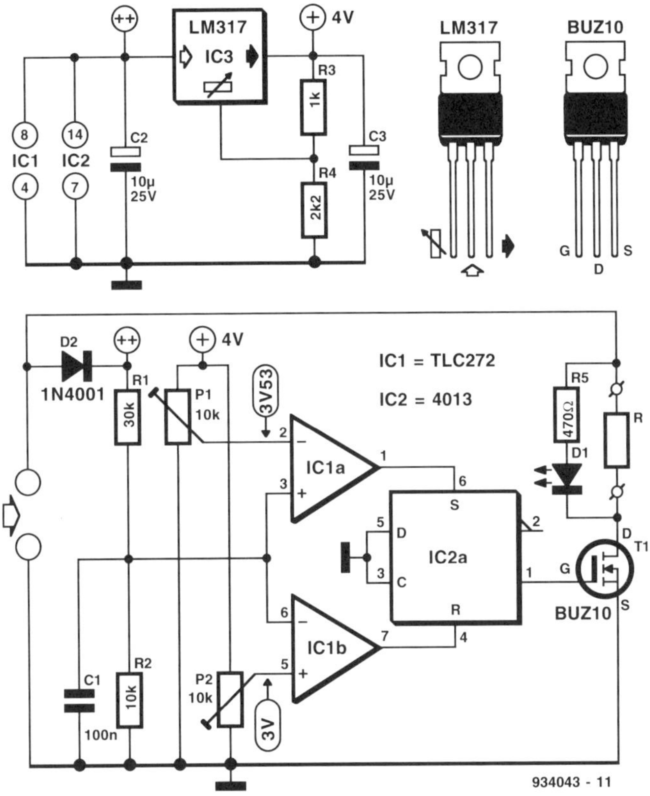

types, that is, the regulator is in series with the load. One of the drawbacks of this is that there is always an additional loss when the battery is being charged. Since the electronic circuits used with solar panel systems already contain a Schottky diode, and solar panels may remain fully loaded even when there is little sun, it is better to use a shunt regulator. These regulators cause losses only when **no** charging takes place. Apart from this, any unwanted energy is then dissipated in a common resistor rather than in power semiconductors on vast heat sinks.

In the diagram, T_1 switches the external load completely on or off. Therefore, the dissipation in this FET is (in theory) nil since the current through, or the voltage across, it is nil. Nevertheless, it is advisable to mount it on a small heat sink (about 5 K W^{-1}).

When T_1 is off, the solar panel feeds all its energy into the battery. When the battery voltage becomes too high, the FET switches on. Provided that *R* is suitably rated, it will dissipate all the energy delivered by the solar panel, whereupon the panel voltage drops and the battery is no longer being charged. The Schottky diode prevents the battery discharging through the shunt resistor. The solar panel can cope well with this load. It is, however, advisable to use a resistor; short-circuiting the panel is not good practice.

The voltages at which the shunt is switched in and out are preset with P_1 and P_2 respectively. Recommended voltages are 14.1 V and 12 V. Since potential divider R_1-R_2 has a ratio of near enough 1:4, these voltages can be set with the aid of a digital voltmeter at the wipers of P_1 and P_2 respectively (3 V at P_1 and 3.53 V at P_2).

The circuit itself draws a current of not more than 4 mA.

The BUZ10 can switch up to 20 A. The value of *R* is calculated with Ohm's law and the data of the solar panel. These data should indicate the output rating of the panel, for instance, 14.4 V at 8 A. In that case, the shunt register should be 1.8 Ω, 115 W. For security's sake, take the value a little lower, say, 1.5 Ω, and double the rating, say, 230 W, otherwise the resistor gets very hot indeed.

Electronic fuse

103 Most modern voltage regulators have an integral current limiter. These are, unfortunately, normally provided with a thermal 'fuse', which in the first instance does not control the current but the dissipation. Since this type of fuse is very slow, in many applications a standard glass fuse is added to protect a number of components against over-current. Also, many small power supplies and mains adaptors have no short-circuit protection. During the design and testing of new equipment, this may have catastrophic consequences.

In all these cases, the fast electronic fuse described here may be found of great value. The fuse is connected between the power supply output and the load. After it has 'blown', and the current has returned to a normal value, it can be reset with S_1.

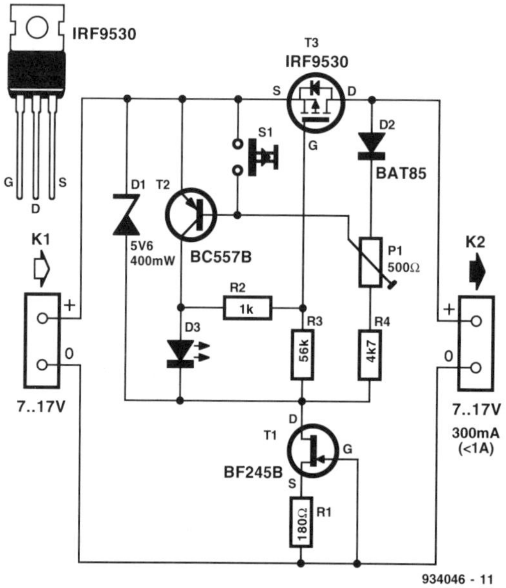

Transistor T_1 and resistor R_1 form a 6 mA current source that, together with zener diode D_1, provides a constant voltage of 5.6 V with respect to the +ve supply line. This gives the gate of T_3 a sufficiently high negative voltage to keep this p-channel MOSFET conducting (and thus providing the link between power supply and load).

When the load current rises to about 1 A, the potential drop across T_3, D_2, and part of P_1 reaches a level at which T_2 switches on, so that the gate-source junction of T_3 is short-circuited. Diode D_3 then lights to indicate that the fuse has 'blown'. In practice, the LED will already light dimly when the critical current is approached. The circuit is reset by pressing S_1.

With values as shown in the diagram, the maximum load current is around 300 mA. The drop across T_3 is only 100 mV. Higher load currents may be accommodated by taking a higher rated p-channel MOSFET for T_3 or connecting two or more FETs Type IRF9530 in parallel. A heat sink is not needed, since the maximum dissipation does not exceed 40 mW.

The LED should be a low-current type, as a current of only 6 mA is available.

The circuit is calibrated by providing a current into K_1 at the level at which the fuse should operate and then adjusting P_1 until the fuse 'blows'.

As already stated, other types of p-channel MOSFETs may be used for T_3. Note, however, that the lower the on-resistance of this transistor, the less precise the value of the load current at which the fuse will act.

Fast high-power zener diode

104 The aim of the design of this 'zener diode' is a fast voltage regulator that can provide high currents. The circuit is based on a Type BUZ10 SIPMOS transistor, which has a high dissipation and can cope with high currents.

From an external power source, whose voltage must be about 2 V higher than the wanted stabilized potential, P_1-C_1-D_1 derive a stable voltage in the range 0–6.8 V. This voltage switches on T_1, which in turn switches on T_2. If U_z drops slightly, the potential at the gate of T_1 also drops, whereupon the T_2 raises its on-resistance, so that the output voltage rises. The control speed is determined by R_4.

The circuit as drawn delivers output voltages in the range 3.9–9.6 V at a supply voltage of 15 V.

In the prototype an open-circuit voltage of 6 V was set (the current drawn by T_2 was 490 mA), which dropped to 5.94 V when a load of 15 Ω (minimum value) was connected across T_2 as shown in dashed lines. The drain current was then 13 mA.

The dynamic behaviour is excellent: with a rectangular input signal at a frequency of 10 kHz, and a load of 15 Ω, the recovery time was smaller than 1 μs.

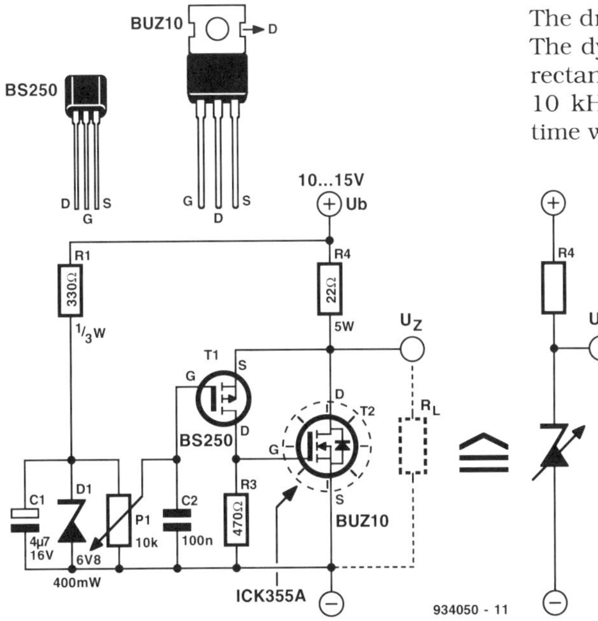

Fast active rectifier

105 The rectifier is intended to improve the bandwidth of most digital multimeters. It is based on a half-wave rectifier in a traditional set-up. Diode D_2 is in series with the output of opamp IC_1. Since feedback resistors R_1 and R_2 follow D_2, the negative halves of the signal at the input of the circuit will appear with the same value at the cathode of D_2. During the positive halves at the input, the output of the opamp is kept at almost 0 V by D_1. Network R_3-C_1 at the output integrates the half-wave rectified voltage, so that the direct voltage across C_1 represents the average value of the negative halves of the input signal. If the original signal was sinusoidal, that direct voltage will be $1/\pi$ times the peak value of the input signal.

The opamp used is a current-feedback

type. Its bandwidth is determined by the values of the resistors in the feedback loop. For a large bandwidth, the value of these resistors should be as low as feasible. To retain an acceptable input impedance (1 kΩ in the prototype), both R_1 and R_2 have a value of 1 kΩ. The bandwidth is then 30 MHz. When the value of both resistors is halved, the bandwidth becomes 60 MHz).

The accuracy of the rectifier is determined not only by the bandwidth, but also by the slew rate. That of the AD844 is not less than 1200 V μs^{-1} (typically 2000 V μs^{-1}). Furthermore, the capacitance and forward bias of D_1 and D_2 play a role. The BAT82 has a capacitance of 1.6 pF,and a forward bias of about 0.5 V at a forward current of4 mA.

In the prototype circuit, the frequencies were measured when the output voltage was dropped by 1% (about 0.1 dB). At 1 V output, the –1% point was at 1.7 MHz; at 100 mV at 400 kHz, and at 10 mV at 45 kHz.

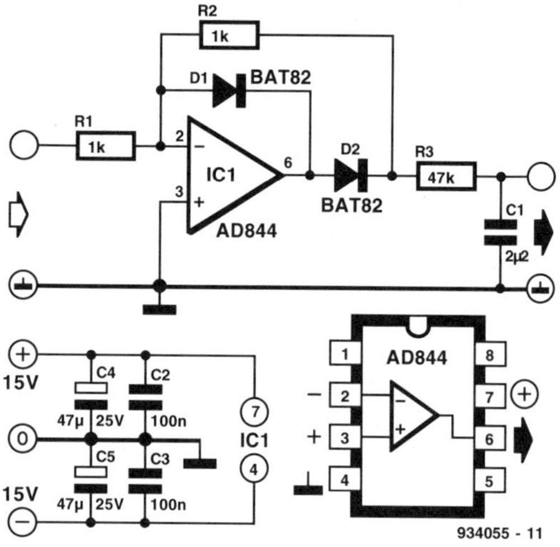

The circuit draws a current of 6.5 mA with a supply voltage of ±15 V. The supply to the AD844 may lie between ±4.5 V and ±18 V.

Button-cell charger

106 In miniature circuits, the trend is away from expensive dry button cells and towards NiCd button types. These batteries are easy to maintain: they may be charged with a constant current for a certain fixed period, normally 14–16 hours.

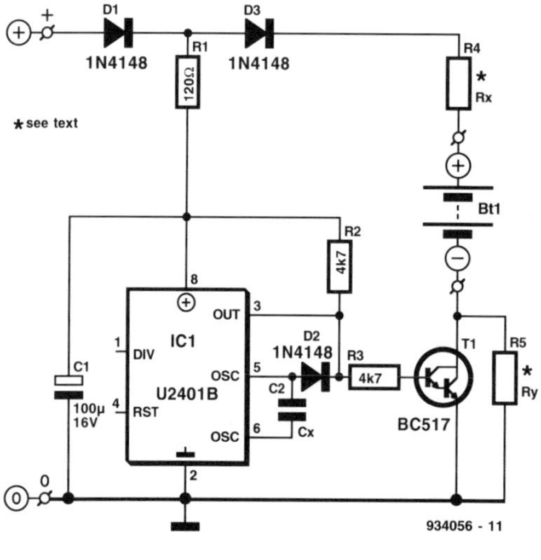

The usual NiCd battery chargers are generally not suitable for charging button cells, as their minimum charging current is too high for these cells. The charger described here operates from a 9-V mains adaptor and can charge from one to five button cells.

D_1 is a protection diode, while R_1 and C_1 decouple the supply line to charging processor IC_1, an economy model of the well-known U2400B. D_3 is a safeguard against polarity reversal of the button cells. R_4 limits the charging current to 5 mA. During trickle charging, a current of 0.5 mA flows through R_4 and R_5. The value of C_2 determines the charging time. After the mains adaptor is plugged in, the cells are charged at full current for a period shown in the table (T_1 is on); after that period, trickle charging takes place (T_1 off). The trickle charging current is always $^1/_{10}$ of the full charging current.

C_2 (nF)	Charging time	C_2 (nF)	Charging time
15	0:58	68	4:27
18	1:10	82	5:22
22	1:26	100	6:33
27	1:46	120	7:51
33	2:09	150	9:49
39	2:33	180	11:47
47	3:04	220	14:25
56	3:40	270	17:41

Capacity (mAh)	Charging Current (mA)	R_4 (kΩ) for number of cells			
		1	2	3	4
8	0.5	15	12	8.2	4.7
15	1.0	6.8	5.6	3.9	2.7
18	1.5	4.7	3.9	2.7	1.8
36	2.0	3.6	2.7	2.2	1.2
75	2.5	2.7	2.2	1.8	1.0
120	5.0	1.5	1.2	0.82	0.47
190	10	0.68	0.56	0.39	0.27
230	15	0.47	0.39	0.27	0.18
310	25	0.36	0.27	0.22	0.12

$R_5 = 0.1R_4$

Battery discharger

107 When batteries are charged rapidly, the manufacturer usually recommends that after every fifth fast charge there is a 'normal' charge, that is, one at 0.1 C(apacity), That can, however, be done only if the battery is discharged. At the same time, the individual cells must not be discharged too much. If, for instance, one cell is flat and the others are still charged or nearly so, the polarity of the flat cell reverses and that must be prevented to save the battery. As a rule of thumb, a battery should be discharged to an average of 1 V per cell. The diagram shows a circuit that is suitable for discharging nickel-cadmium (NiCd) as well as nickel-metal-hydride (NiMH) batteries. It is simple and inexpensive, but the components must be

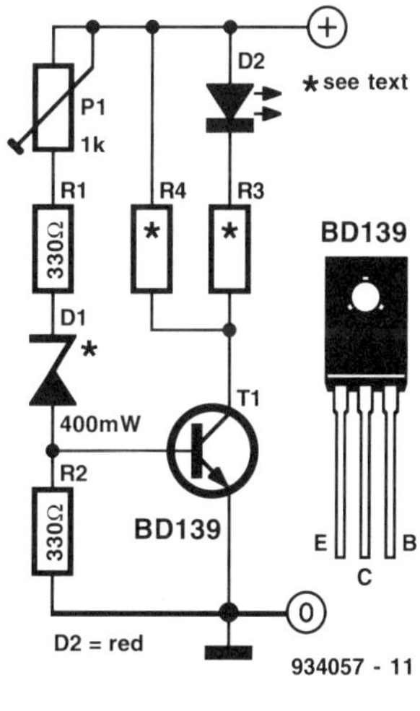

matched to the number of cells the battery contains.
As long as D_2 lights, the battery is discharged via R_4 and T_1. When the battery voltage drops below a value set with P_1, T_1 no longer gets a base current. The LED will then go out and the discharge circuit may be removed. The battery can then be charged.
The circuit may be tested with the aid of a variable power supply with current limiting. Connect the discharger to the power supply: R_4 may be omitted. Set the output of the supply to the wanted level, whereupon T_1 should switch off. Adjust P_1 till D_2 lights dimly. This sets the voltage to which a battery may be discharged,
Since the gain is not very high, there is an area where D_2 diminishes gradually and where, therefore, the discharge current decreases gradually. This does not affect the circuit, however.
If the range is too small, the value of the zener diode should be reduced or increased.
The value of R_4 is calculated for a discharge current of about 0.5 A. This value is not critical. That is, if a resistor of 6.8 Ω proves difficult to obtain, a 4.7 Ω or 10 Ω resistor may be used without detriment. Since the voltage of the cells is monitored, it is not necessary to stabilize the discharge current.

No. of cells	D_1 [V]	R_3 [Ω]	R_4 [Ω]
3	1.2 (2 diodes)	150	6.8 (5 W)
4	2.4 (3 diodes)	270	10 (5 W)
5	3.3	330	12 (5 W)
6	3.9	470	15 (10 W)
7	4.7	560	15 (10 W)
8	5.6	680	18 (10 W)
9	6.8	680	22 (10 W)
10	8.2	820	22 (10 W)

Regulator output delay

108 In this circuit, a Type 317 variable voltage regulator has been provided with a facility whereby, after the input voltage has been switched on, its output voltage rises only slowly to the value set with P_1. For instance, if P_1 was set for an output of 15 V, that value is reached only after 5 seconds. If the output voltage was set at 7.5 V, that value would have been reached after 2.5 second. In other words, the delay time is directly proportional to the set output voltage. The delay may be lengthened by increasing the value of R_3 and C_3. The delay may be disabled by removing the link at JP_1 (which is useful when P_1 is being adjusted).
Diode D_2 protects the base-emitter junction of T_1 against to high a reverse bias. Without the diode, if the output of the regulator were short-circuited, charged capacitor C_3 could cause a U_{BE} that exceeds then permissible maximum 6 V,
Diode D_1 protects the regulator if the input voltage were (accidentally) to drop faster than the output voltage.
The circuit draws a current of only a few mA, although the peak current may amount to 1.5 A. To prevent the temperature protection of the LM317, mounted on a 14 K W^{-1} heat sink, coming into operation, the constant output current should not exceed 500 mA.

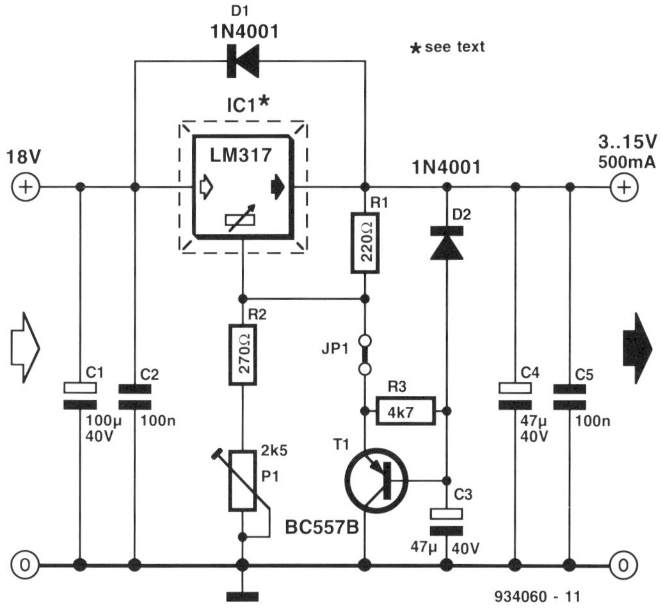

Measuring the capacity of NiCd batteries

109 The capacity tester will give an accurate indication of the capacity of up to six NiCd (nickel-cadmium) batteries connected in series. It is very useful for monitoring the ageing process of NiCd batteries, and for spotting faulty batteries, which are always a source of trouble in battery-powered equipment.

Before the capacity of a NiCd battery can be ascertained, it must be fully charged with the usual charger. Next, the battery is connected to the capacity meter, which draws a constant discharge current. After the battery is connected, and the circuit reset, a counter is started, and stopped again when the tester measures that the 'discharged' level of the battery is reached. The counter is clocked with oscillator pulses, and the number of 'ticks' recorded is an indication of the battery capacity in mAh (milli-ampère-hours). This measurement method works on the assumption that the 'discharged' voltage of the relevant battery is known and predefined as a switching threshold on the tester.

The battery is discharged by a constant current sinking circuit formed by T_1 and IC_1. A MOSFET Type BUZ10 is used as a power regulator because of its very low drain-to-source resistance ($R_{D\text{-}S(on)}$). The discharge current is measured with the aid of R_8, and held at a constant value of 1 A by comparator IC_1. The discharge current is adjusted by preset P_1. Transistor T_2 enables the current sink to be switched on and off under the control of bistable IC_{3a}. The current sink is switched off if IC_{3a} is set.

The discharge cycle is started when the reset input of IC_{3a} is low. The set input is driven by a comparator, IC_2, which compares the battery voltage with the voltage at the pole of rotary switch S_1. This switch selects the number of batteries connected, which can range from one to six. It ensures a supply of 0.5 V nominally per battery connected.

The battery voltage is divided :2 by R_9-R_{10} and then fed to the –input of IC_2. When the voltage per battery drops below

1 V, the voltage at the –input of IC_2 drops below the reference set with S_1. Consequently, the comparator changes state. Its output swings to +5 V, sets the bistable, and switches off the current sink on the battery. Also, oscillator/counter IC_4 is stopped via T_3, so that the count pulses on connector K_1 disappear. Although T_1 is switched off at this point, the batteries are still being discharged at a rate of about 50 mA. This means that they should be disconnected as soon as possible after the test.

The pulses supplied by IC_4 may be fed to any counter capable of accepting a digital input signal with a swing of 5 V. The prototype of the capacity tester was connected to the 4-digit counter module described in Ref. 1. Apart from the supply and clock pulse connections, the reset line is also connected directly. This allows the whole circuit (i.e., capacity tester and readout) to be reset by pressing a single key.

Preset P_2 is adjusted such that a new battery produces a readout that corresponds

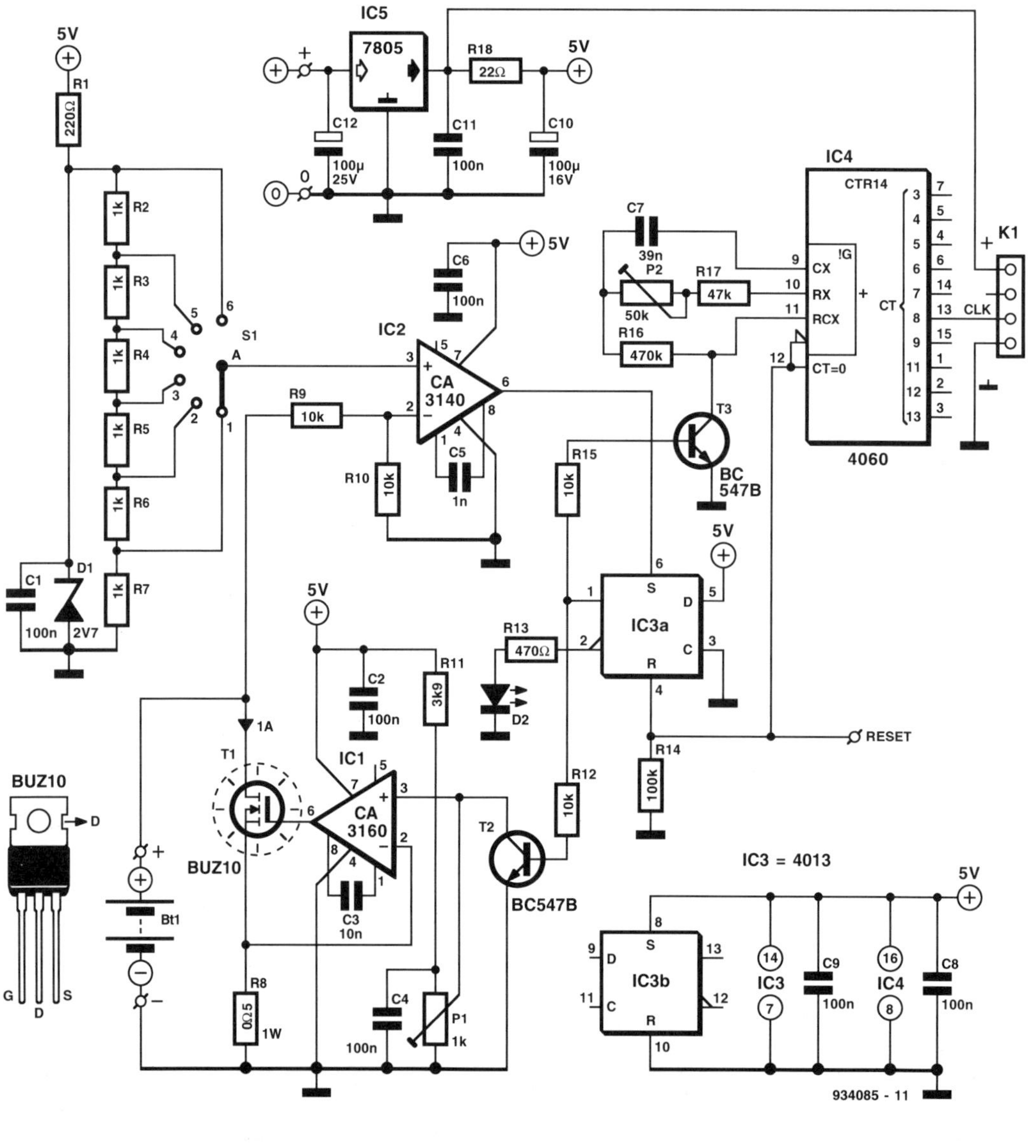

to the nominal capacity of the battery in mAh, e.g., '1200' for a 1.2-Ah battery. Since a battery capacity of 1 Ah equals 1,000 output pulses per hour, the oscillator clock frequency becomes

$$(1{,}000/3{,}600) \times 2^9 = 142.2 \text{ Hz},$$

which can be measured at pin 9 of IC_4. If a frequency meter is not available, P2 may be adjusted until the display counts 100 pulses in 6 minutes.

The printed circuit board allows the tester and the 4-digit readout to be built in one go. The parts list of the readout module is repeated here for convenience. The readout board is fitted on top of the capacity meter board with the aid of four PCB spacers. This allows the pin header on the read-out board to be soldered at the track side, so that it can be inserted into the socket on the capacity meter board. The reset wire must be run separately between the two boards.

Parts list

METER SECTION

Resistors:
R_1 = 220 Ω
R_2–R_7 = 1 kΩ
R_8 = 0.5 Ω, 1 W
R_9, R_{10}, R_{12}, R_{15} = 10 kΩ
R_{11} = 3.9 kΩ
R_{13} = 470 Ω
R_{14} = 100 kΩ
R_{16} = 470 kΩ
R_{17} = 47 kΩ
R_{18} = 22 Ω
P_1 = 1 kΩ, preset (H)
P_2 = 47 kΩ preset (H)

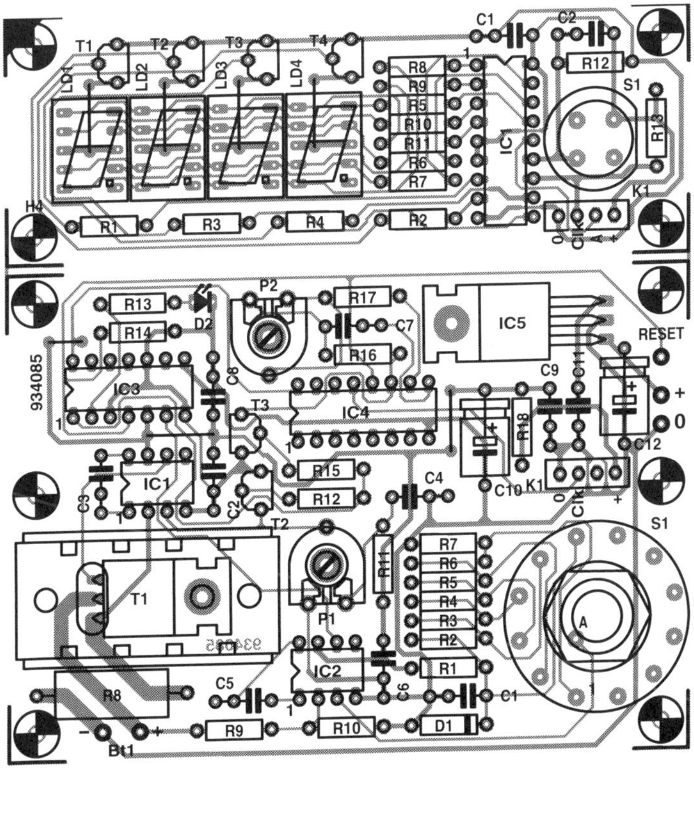

Capacitors:
C_1, C_2, C_4, C_6, C_9, C_{11} = 100 nF
C_3 = 10 nF
C_5 = 1 nF
C_7 = 39 nF
C_{10} = 100 μF, 16 V
C_{12} = 100 μF, 25 V

Semiconductors:
D_1 = zener diode, 2.7 V, 400 mW
D_2 = LED, 3 mm
T_1 = BUZ10
T_2, T_3 = BC547B

Integrated circuits:
IC_1 = CA3160
IC_2 = CA3140
IC_3 = 4013
IC_4 = 4060
IC_5 = 7805

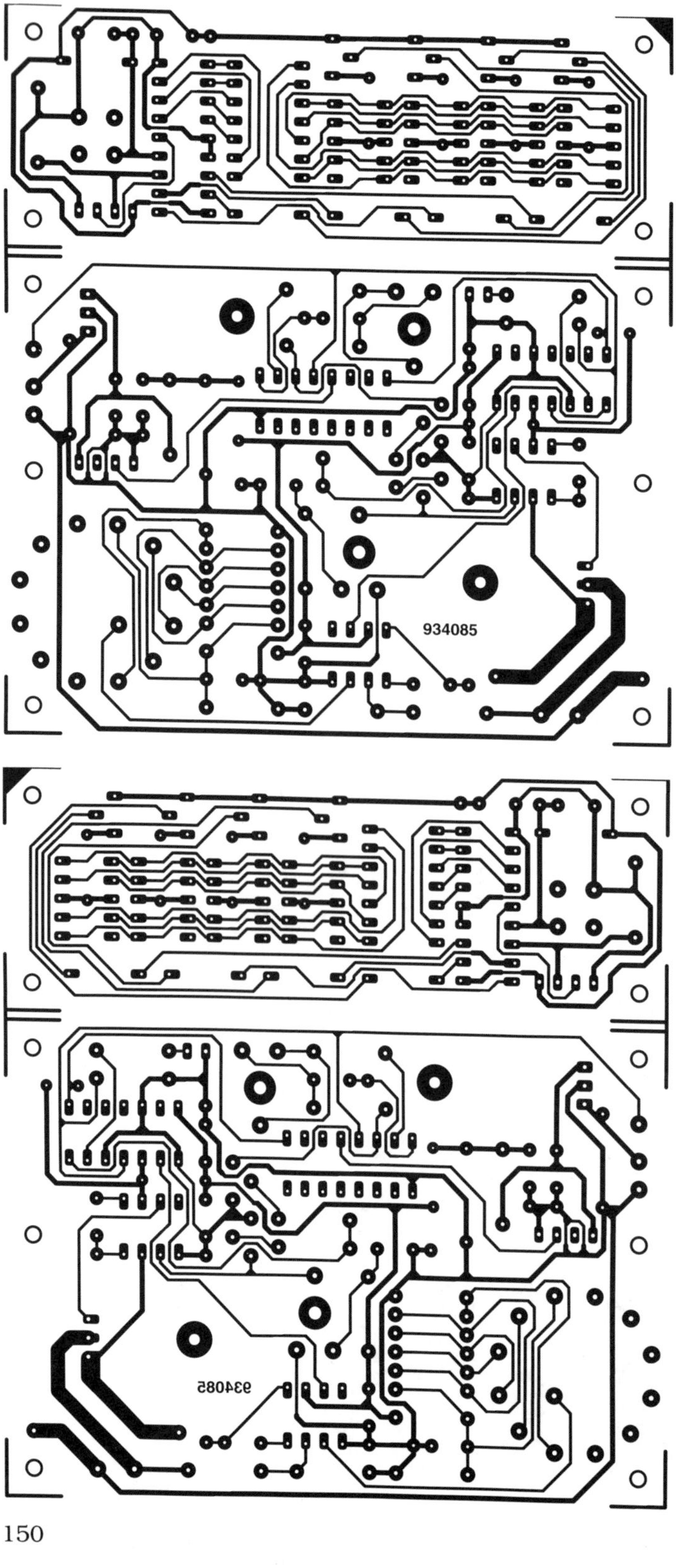

Miscellaneous:
K_1 = 4-way SIL socket
S_1 = single-pole, 12-way rotary switch for PCB mounting
Heat sink for T_1, e.g., Fischer Type ICK35/SA*
PCB REF. 934085)

READ-OUT SECTION

Resistors:
R_1, R_4, R_{12} = 1.5 kΩ
R_5, R_{11} = 56 Ω
R_{13} = 10 kΩ

Capacitors:
C_1, C_2 = 100 nF

Semiconductors:
T_1–T_4 = BC547B
LD_1–LD_4 = HD1107

Integrated circuit:
IC_1 = 74C926

Miscellaneous:
S_1 = push-button switch for PCB mounting
K_1 = 4-way SIL pin header

* Dau (UK) Ltd, 70–75 Barnham Road, Barnham, W. Sussex PO22 0ES. Telephone (0243) 553 031

Reference:
1. 4-digit counter module. *Elektor Electronics*, December 1992.

Battery voltage monitor

110 Philips' IC Type TEA1041T is ideal for monitoring the voltage of 1.8–4.0 V batteries. Internal trigger and timing logic prevent a circuit from reacting to brief breaks in the supply voltage caused by pulses on the load current. One or two LEDs may be used to indicate when the voltage drops below a preset level.

A circuit with two LEDs is shown in Fig. 1. Potential divider R_1-R_2 determines the voltage level below which the LEDs indicate that the battery is charged. The divider should give a voltage of exactly 1.25 V at pin 1. Give R_2 a value between 1 kΩ and 100 kΩ and calculate R_1 from

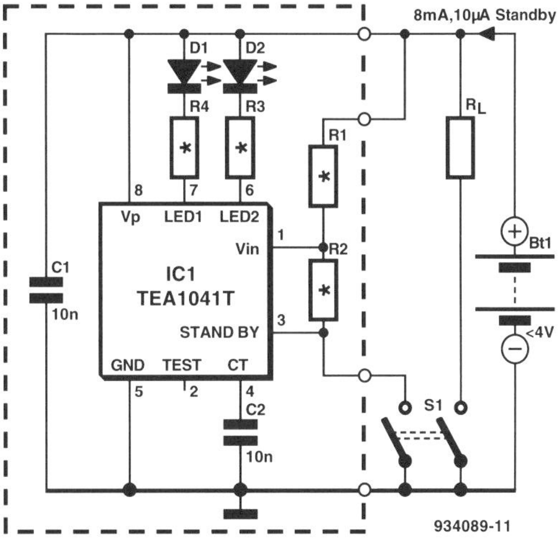

$$R_1 = R_2(U_{th}/1.25-1) \quad [\Omega],$$

where U_{th} is the wanted voltage level. The values of R_3 and R_4 are 100–220 Ω, depending on the battery voltage.

The TEA1041T is enabled when pin 3 is at ground level. If the voltage at pin 1 drops below 1.25 V, a digital counter runs for about 2 s. If the level at pin 1 remains <1.25 V, the IC goes into the alarm state: D_1 lights and continues to do so even when the voltage at pin 1 reaches 1.25 V again. If in this condition S_1 is opened, both LEDs flash for about 4 s. After that, the IC reverts to the standby state, in which it draws a current of about 10 µA, which constitutes only a minute a load on the battery.

Figure 2 shows how the IC can be incorporated in an existing apparatus: D_2 is omitted and pin 3 is permanently at ground level. The monitor is connected across the load as long as S_1 is closed. The IC performs as described earlier, but the LED does not flash after the reset.

If the supply voltage is higher than 4 V, connections must be made as shown by the dashed lines in Fig. 2 (R_1 to R_5 and D_3 and C_3 to earth). The supply voltage is then lowered to 3.3 V by D_3. The potential divider is connected across the full supply voltage.

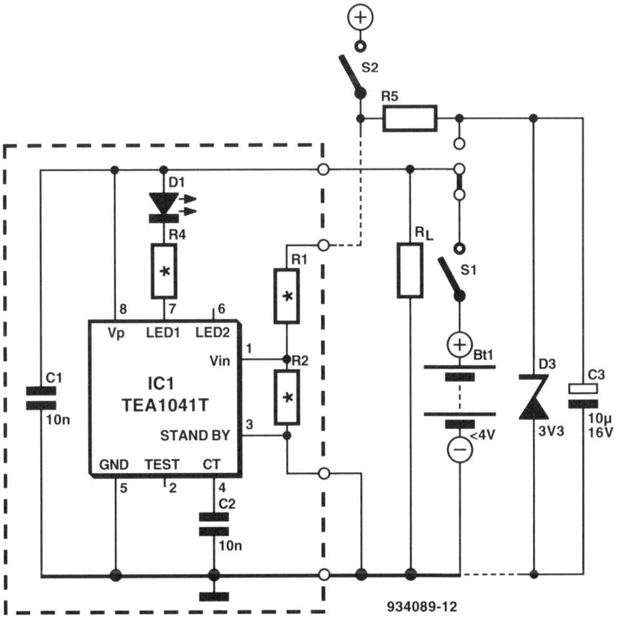

Source: Philips Data Sheet TEA1041T: Battery Low-level Indicator.

Inexpensive voltage doubler

111 Voltage doublers are frequently encountered in electronic circuits. The present one is a variation that is of interest since most digital circuits already have a buffered clock (CLK) available or have a spare Schmitt trigger gate. Since in those cases no new IC is needed, the cost of the doubler is much reduced.
If a buffered CLK signal is available, only four components, C_4, C_5, D_1 and D_2, are required to produce a voltage of 10 V from the 5 V supply. If an oscillator needs to be built from a spare gate, two further components are needed: R_1 and C_3.
The most important parameters of the circuit are given in the table. Note that, owing to tolerances in the clock circuit, these data may be slightly different.

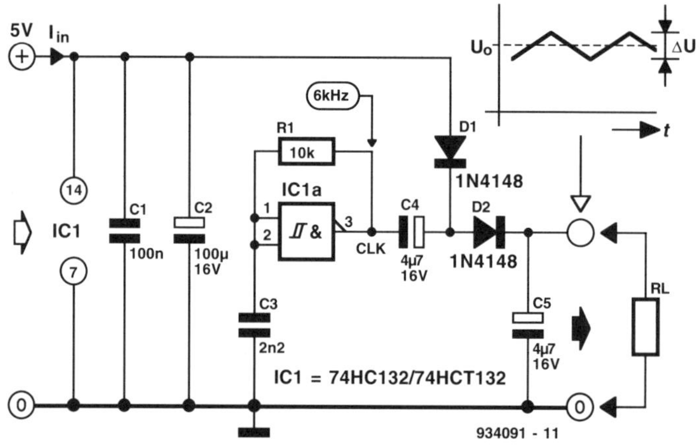

R_L [Ω]	U_o [V]	ΔU [mV_{pp}]	I_{in} [mA]	n [%]
---	9.4	0	2.4	---
2200	8.0	15	20.0	29
780	7.0	40	28.3	45
460	6.0	80	37.6	41

Fast charging of NiCd and NiMH batteries

112 Pin-compatible ICs Type MAX712 and MAX713 from Maxim are intended for fast charging of nickel metal-hydride (NiMH) batteries; the MAX713 may also be used in nickel-cadmium (NiCd) battery chargers (see Table 1). In this table, C is the nominal capacity of the battery in Ah. Note that charging at currents >2 C should be undertaken only if the relevant batteries are suitable (see manufacturer's data sheet),
The ICs fast-charge a NiMH or NiCd battery by forcing a constant current into it (dV/dt). Their output current is always in one of two states: high (fast charge) or low (trickle charge). Once full charge is detect-

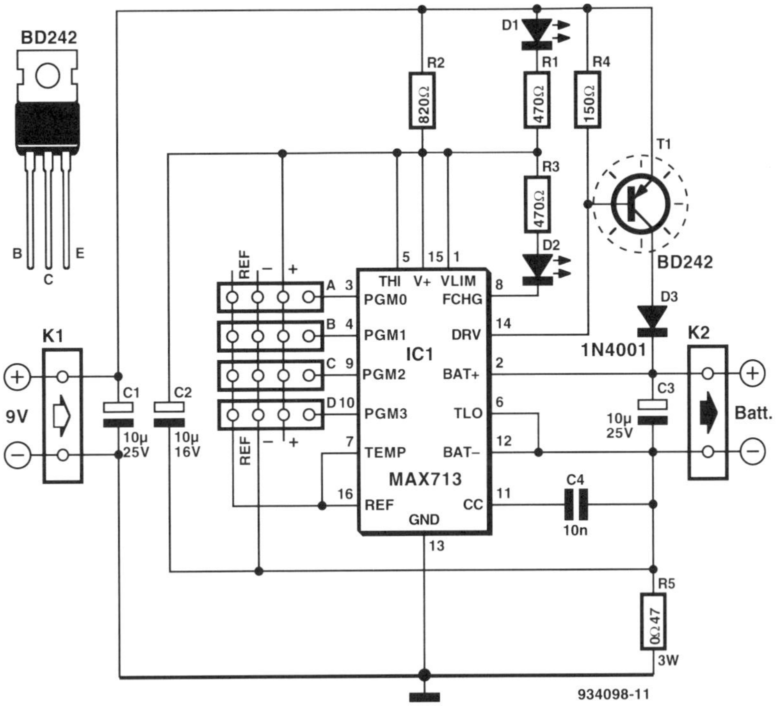

ed, the current is reduced to trickle charge. The ICs monitor three parameters to determine when the battery reaches full charge: voltage slope, battery temperature and charging time.

The charger described here is powered by a mains adaptor with an output voltage of 9 V d.c. and an output current rating of at least 1 A. This allows penlight NiMH batteries to be charged at a rate of 0.5 C, and NiCd batteries at a rate of 1 C. The maximum charging voltage is limited to 1.65 V per battery. Fast charging is discontinued when either the programmed charging time has elapsed, or the battery voltage drops (dV/dt method).

The value of R_5 is calculated from

$$R_5 = 0.25/I_{FAST},$$

where I_{FAST} is the desired charging current. Examples: to charge a 1.2 Ah NiMH battery at a rate of 0.5 C (two hours), a charge current of 1.2×0.5 = 0.6 A is required. This means that R_5 becomes 0.25/0.6 ≈ 0.42 Ω. Roughly the same resistance is needed when a 500 mAh NiCd battery is to be charged at a rate of 1 C (one hour). In practice, the 0.47 Ω resistor shown in position R_5 will work fine for both applications, since the charging current need not be exact, and, in any case, the ICs monitor the individual cells.

The ICs have a built-in timer that can be set to a wide range of charging times. The theoretical charging rate of 1 C is approximated by the '66 minutes' setting. Similarly, 0.5 C corresponds to 132 minutes. When the programmed fast charging period has elapsed, D_2 goes out, and the ICs automatically switch to trickle charging. The trickle charging current depends on the programmed charging time (see Table 2). As a rule of thumb, batteries shoud be given a 14-hour trickle charge once every five 'fast' charging cycles.

Construction of the charger is straightforward on the printed circuit board (Fig. 2). The number of cells and the charging time are set with the aid of wire jumpers, whose position may be taken from Tables 2 and 3. The power transistor, T_1, is mounted on a heat-sink. Power resistor R_5 is mounted a few millimetres above the board to assist in its cooling.

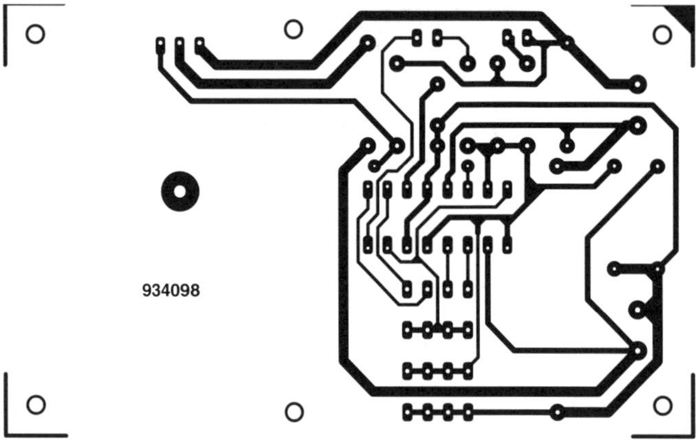

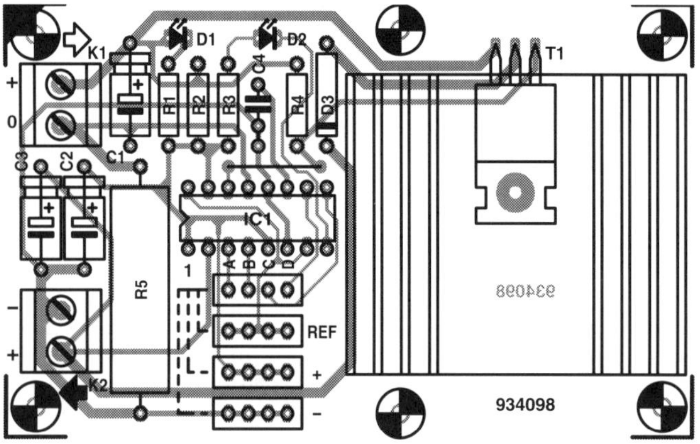

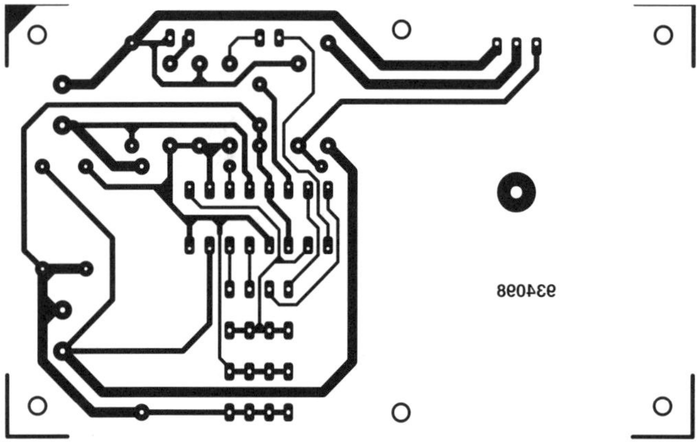

Parts list

Resistors:
R_1, R_3 = 470 Ω
R_2 = 820 Ω
R_4 = 150 Ω
R_5 =0.47Ω, 3 W

Capacitors:
C_1, C_3 = 10 μF, 25 V
C_2 = 10μF, 16V
C_4 = 10 nF

Semiconductors:
D_1 = LED, red
D_2 = LED, green
D_3 = 1N4001
T_1 = BD242C

Integrated circuits:
IC_1 = MAX713 or MAX712*

Miscellaneous:
K_1, K_2 = 2-way PCB terminal block,
pitch 5 mm
Enclosure, e.g., Pactec Model HM,
Type 6600-902**
Heat-sink Type SK59 (Fischer)***
PCB REF. 934098

* Maxim Integrated Products (UK) Ltd., 20A Horseshoe Park, Pangbourne, Reading RG8 7JW. Telephone (0734) 845255, fax (0734) 843863

** OK Industries UK Ltd., Chickenhall Lane, Unit 1, Deacon Trading Estate, Eastleigh, Hants S05 5RR. Telephone: (0703) 619841, fax: (0703) 643279.

* Dau (UK) Ltd, 70–75 Barnham Road, Barnham, West Sussex PO22 0ES, Telephone (0243) 553 031.

Symmetrical power supply

113 A symmetrical power supply may be designed from two standard operational amplifiers, if only a small current (of the order of a few mA) is required.

In the diagram, the upper half provides the positive voltage. A 3.3 V zener diode, connected to the non-inverting (+) input of opamp IC_1, serves as reference. To ensure correct starting of the circuit, the diode is initially powered via R_1 and then, when the output voltage is sufficiently high, via D_1. Part of the output voltage is fed back to the inverting (–) input of IC_1 viaP_1. The lower the feedback voltage, the higher the output. The supply voltage for the TL071 may be 36 V maximum, so that the output voltage can go up to about 30 V. It is, however, safer to make the supply to the opamp rather lower and accept a slightly lower output voltage.

The lower part of the diagram is a mirror of the upper part: it provides the negative output. Opamp IC_2 needs an additional (negative) supply voltage.

The output voltages may be made more stable by connecting a 10 µF electrolytic capacitor across each of the outputs.

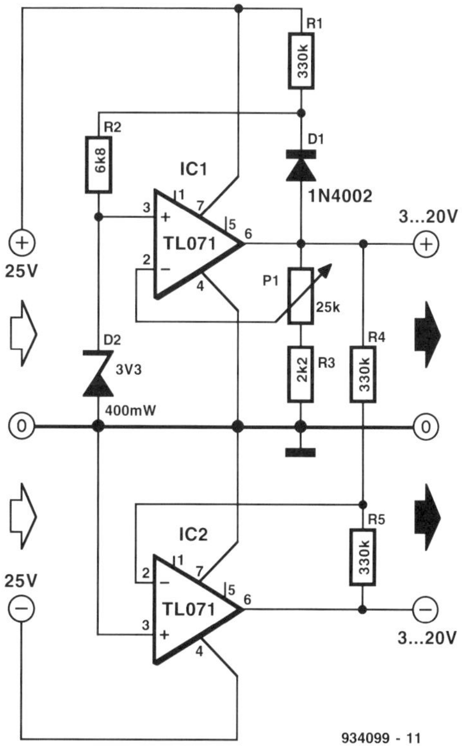

Regulator short-circuit indicator

114 Modern integrated voltage regulators are protected against short circuits, but do not give an indication when a short circuit occurs. In the case of regulators with fixed output (78xx) a short-circuit indicator is easily arranged by connecting an LED and series resistor across the regulator output. The LED will light only during normal operation, that is, input available and output not short-circuited.

If this sort of indicator is used with regulators with a variable output, such as the LM317, the brightness of the LED will vary with the set output. To make the brightness constant, the current through the LED must be made independent of the output voltage. The LED current must then not be limited by a series resistor but by a current source. Such a source can be made with only one extra component: T_1 in the circuit diagram. The design makes use of the available reference voltage across R_1. This voltage is generated by the LM317 and serves in the first instance, in conjunction with R_1, R_2 and P_1, to regulate the output voltage.

In the present circuit, the reference voltage is used to keep the drop across R_3 constant:

$$U_{R3} = 1.25 - U_{eb} = 1.25 - 0.65 = 0.6 \text{ V}.$$

The current through D_3 is then 0.6/180 = 3.3 mA, which is more than ample for the low-current LED used here.

On the prototype, the LED current remains constant with output voltages from 3 V to 25 V. In other words, over that range of voltages, the LED lights with constant brightness.
The base current of T_1 is only 15 µA, so that the operation of the regulator is not affected by the branching off of the reference voltage.
In the present circuit, T_1 dissipates only 100 mW when the collector-emitter voltage is 25 V. Since this transistor can dissipate up to 500 mW (with heat sink), overheating is highly unlikely. Also, its collector-emitter voltage of 25 V is well below the maximum permissible 45 V. These limits should be borne in mind, however, if the power supply provides higher output voltages.
Capacitor C_3 increases the ripple suppression from 65 dB to 80 dB. Diode D_2 protects IC_1 and T_1 against too high a discharge current from C_3 and too high a base-emitter reverse bias (max. 5 V) when the regulator is short-circuited.
Diode D_1 protects the regulator against discharge currents from C_4 and any electrolytic capacitors in the circuit being supplied). Without this diode, these capacitors would be discharged through the regulator should the input of this IC be short-circuited or be connected to a lower voltage.
The quiescent current of the indicator circuit is about 10 mA, while the peak current may rise to 1.5 A. At maximum input voltage (35 V) and minimum output voltage (3 V), the circuit can provide a constant current of about 150 mA.

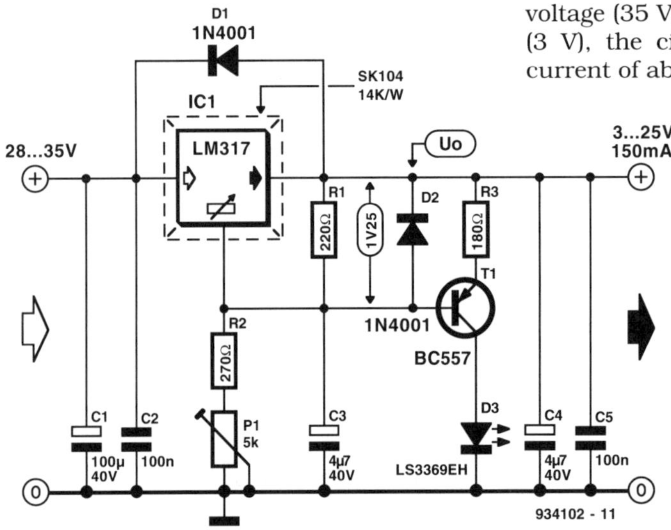

Solar panel switch

115 When a battery is charged by a solar panel during the day, it may become (partly) discharged via that panel after sunset. This may, of course, be prevented by a diode, but that has the drawback of a forward voltage drop of 0.7 V or, in the case of a Schottky diode, 0.4 V. In many applications, this is not acceptable.
The present circuit replaces the diode and connects the solar panel to the battery via a relay contact. It is powered by the solar panel. When the supply voltage is too low, the relay is not energized so that the battery is not connected to the solar panel.
When the panel output is sufficient to energize the relay and light-dependent resistor (LDR) R_2 receives enough light to cause T_1 to conduct, the relay is energized and the battery will be charged.
Owing to its inherent hysteresis, the relay will remain energized even if the potential across the panel drops to a certain degree. A connected and charged battery cannot

actuate the relay when the light begins to fail, because R_2will then switch off T_1. At what brightness this occurs is preset with P_1.
Since the current drain is determined primarily by the relay, it is important that this is a miniature type with a high coil resistance, which can nevertheless switch up to 10 A (e.g., Siemens Type V23037-A0002-A101*).

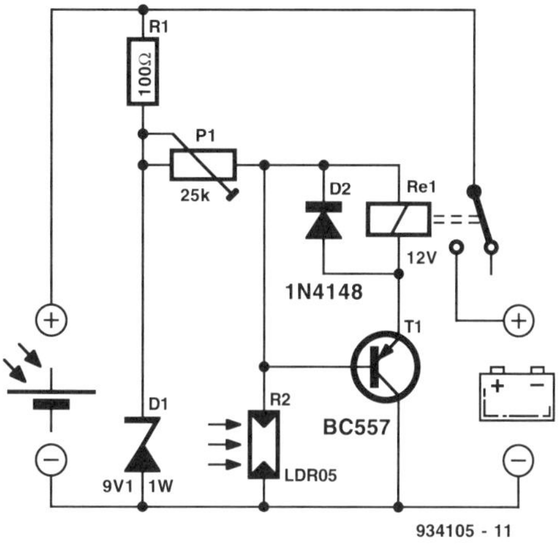

*Available from ElectroValue, 3 Central Trading Estate, Staines TW18 4UX. Telephone (0784) 442 253.

Battery charging regulator

116 Philips' IC Type TEA1100 uses the delta-peak principle to charge nickel-cadmium (NiCd) and nickel-metal-hydride (NiMH) batteries fast and effectively. The task of the delta-peak battery charging regulator is to ascertain and evaluate the **change** in the battery voltage: if the voltage at pin 7 drops 1% or more below the average maximum, charging is discontinued. The voltage at pin 7, according to the manufacturers' data sheet, must be 0.385–3.85 V. Potential divider R_4-R_5 ensures that the battery e.m.f. corresponds with this range. The values of these resistors are given by:

$$R_4/(R_4+R_5)\cdot n\cdot 1.8 < 3.85;$$

$$R_4/(R_4+R_5) < 2.14n;$$

$$R_4/(R_4+R_5)\cdot n\cdot 1.1 > 0.385;$$

$$R_4/(R_4+R_5) > 0.35n.$$

The charging current, I_c, is given by:

$$I_c = 1.25R_{sense}/R_{shunt}\cdot R_{ref},$$

where R_{sense} is R_{10}; R_{shunt} is R_8 in parallel with R_9; and R_{ref} is R_2.
Timeout, TO, is a safety facility of the IC. If, for whatever reason, no maximum can be detected, the IC stops the charging after the timeout. This time constant is given by

$$TO = 2^{26}\cdot 0.93\cdot R_{ref}\cdot C_{osc},$$

where C_{osc} = C_3. Note that both timeout **and** charging current are influenced by R_2. When the charging current and timeout are being determined, the value of R_2 must therefore be fixed and that of C_3 must be variable.
When the battery is fully charged, the TEA1100 switches to the 'maintain charge' mode. In this mode, during $^1/_{10}$ of the normal charging time $^1/_2$ of the usual charging current flows (in other words, the battery is trickle-charged at $^1/_{20}$ of the normal charging current). This trickle-charge current may be altered on the basis of the following formula:

$$I_{trickle} = 1.25R_{sense}/10R_{shunt}R_N.$$

The current is determined from the drop across the shunt resistor and switched by the transistors via pin 2 of IC_1.
Stability of the circuit is ensured by capacitor C_2.
When no battery is connected to K_1 or when a battery is being trickle-charged, D_1 flashes; when the voltage at pin 7 drops, the LED lights continuously.

The supply voltage at pin 12 must be 5.65–11.5 V. The quiescent current (when the outputs are all off) is some 4 mA.
This IC enables many other applications: a number of these are given in Philips' Data Sheet TEA1100 & TEA1100T.

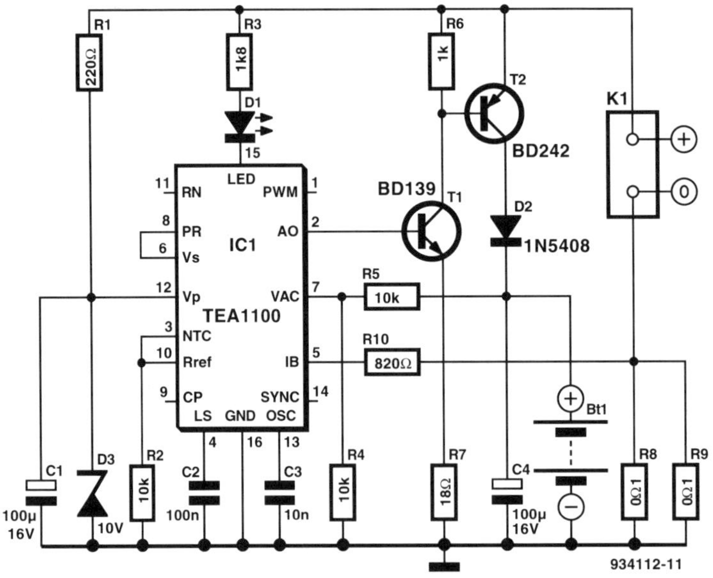

NiMH battery charger

117 Nickel-metal-hydride (NiMH) batteries are now widely available. A typical one is the 120AAH, which has a capacity of 1.2 Ah in size AA (MN1500; LR6). This type does not lose more than 45% of its capacity through self-discharge over a period of 25 days at a temperature of 20 °C. Recommended fast charging instructions are charging with a current of 0.3C for not more than 2.5 hours, or until the cell voltage has risen to 1.49 V, or until the cell temperature rises above 40 °C. The cell is then charged up to 75% of its capacity, after which it should be charged with a current of 0.1C, which it can stand for long periods. The circuit shown enables these instructions to be adhered to.
Output CT13 of IC_2 goes high after $2^1/_2$ hours, whereupon charging stops. At the same time, the output of IC_{1a} goes high when the battery voltage rises above 1.49 V per cell. This output then toggles, which also stops the charging.
The number of cells contained in the battery to be charged is set with the value of R_{14}: for 1 cell, the resistor is omitted; its value for 2 cells is 100 kΩ; for 3 cells, 2×100 kΩ in parallel (= 50 kΩ); for 3 cells, 3×100 kΩ in parallel; and so on.
The third way of ending the charging is by means of the temperature measured with IC_5. When the temperature rises above 40 °C, IC_{1b} terminates the charging.
The three charging outputs are formed into OR gates by D_2–D_4. Since these are low-current LEDs, it is immediately visible on what basis charging has been discontinued.
The charger is switched from 0.3 C to 0.1 C (that is, off); the control input of current source IC_4 is connected to earth via

thyristor Th_1. The output potential of IC_4 then can not rise above 1.2 V. Since this is lower than the battery voltage, D_5 will be reverse-biased. Further charging of the battery is then via R_{15}. The value of this series resistor (in Ω) is obtained by dividing the difference (in V) between the supply voltage and the battery voltage by 0.12. The quotient should be rounded off to the nearest standard value (the exact value is not terribly important).

The level of the supply voltage depends on the number of cells contained in the battery to be charged. Its minimum value is 4 V plus the number of cells times 1.5 V.

It is necessary to mount IC_4 on a small (10 K W^{-1}) heat sink.

Two keys must be pressed to switch the charger on: S_1 resets timer IC_1 and S_2 switches off thyristor Th_1. These switches could be combined, but double-pole key switches are not generally available.

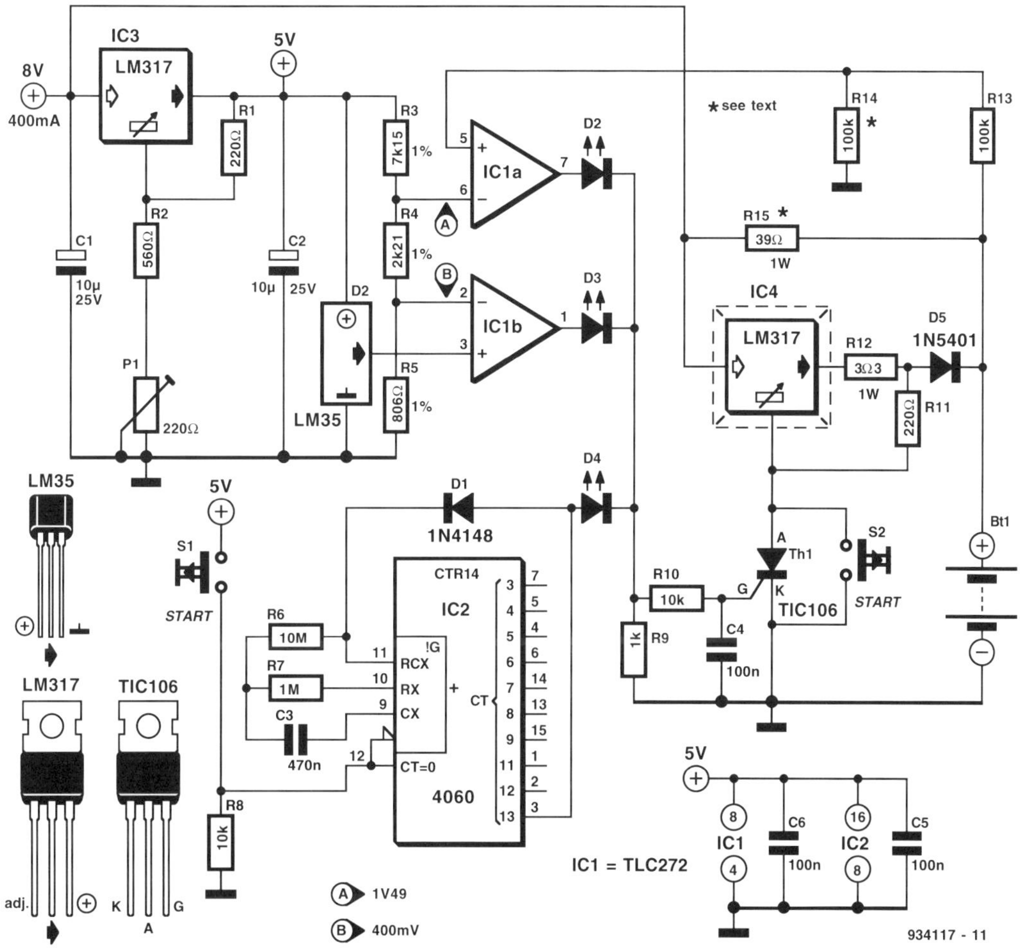

UHF remote control transmitter

118 This low-power UHF FM transmitter is intended for remote control applications such as garage doors and wire-less alarm systems. It is a single transistor design that operates at a frequency reserved for low-power wireless signalling. The operating frequency is determined by a resonator, Fl_1. The following types may be used:

UK: Type R2528 for 418 MHz;
USA: Type R1530 for 315 MHz;
Holland, Belgium, Germany: Type R2554 for 433.92 MHz;
France: Type R2523 for 224.5 MHz.

The transmitter is frequency-modulated by an audio (or digital) signal applied to the junction of a varactors D_1, D_2 via R_3. The varactors effectively change the shunt capacitance of the resonator as a function of the modulation signal, which results in FM. The design may be changed to produce AM (amplitude-modulation) by omitting components D_1, D_2, R_2, R_3 and R_6, and interconnecting points 'A' and 'B'.

Transistor T_1 is fitted at the solder side of the printed circuit board, as indicated by the dashed outline on the component overlay. Also on this side of the PCB is a short wire link (use silver-plated wire) that connects strip line inductor L_1 to the positive supply track that runs in parallel with it. The position of the wire link depends on the transmit frequency. The lower the frequency, the more inductance is required, that is, the further the link has to be 'moved' towards the edge of the PCB. Some experimenting may be necessary to find the best position. Start by setting the trimmer, C_3, about mid-way, and fit the wire link about 'half way' on the strip line. Monitor the received signal, and adjust C_3 until a maximum is found. If you can not find a maximum, try moving the wire link either towards the transistor (smaller inductance), or towards the PCB edge (greater inductance). The best position is that at which C_3 'peaks' when set about mid-way.

Since we are dealing with a UHF circuit, it goes without saying that all component terminals must be kept as short as possible. The transmitter must be housed in a plastic enclosure to enable it to radiate.

Note that radio regulations apply to the possession and use of this transmitter. In the United Kingdom, a licence must be obtained from the Department of Trade and Industry.

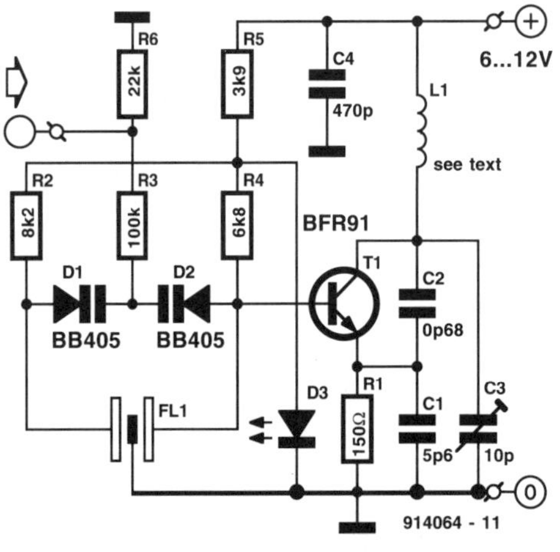

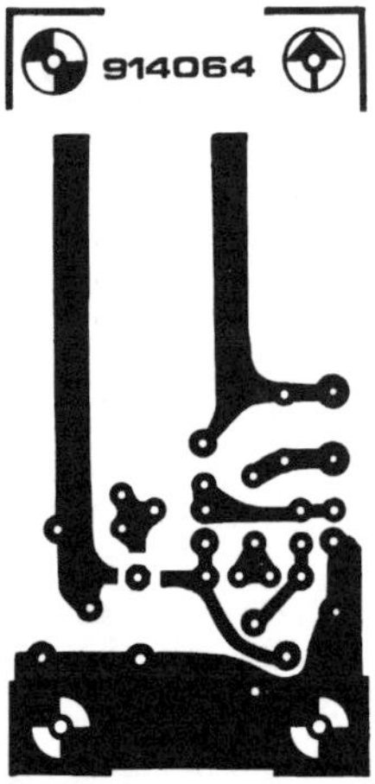

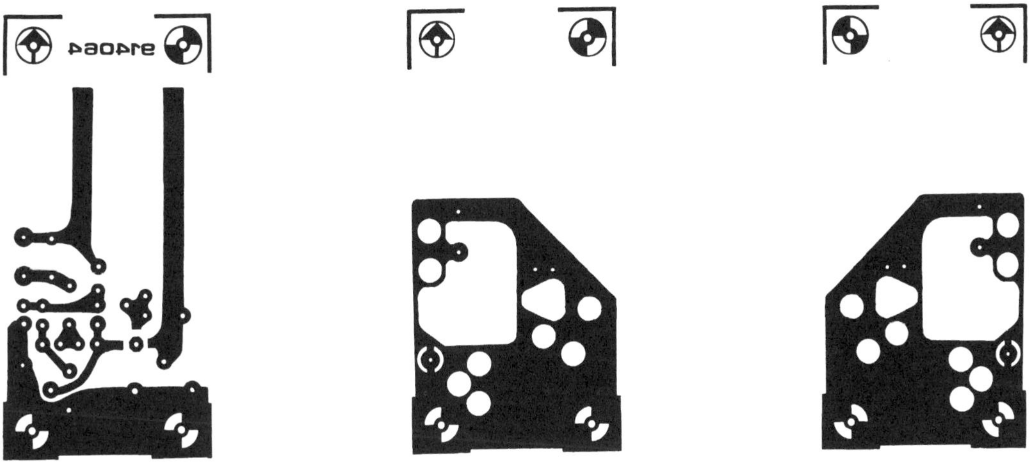

Parts list

Resistors:
R_1 = 150 Ω
R_2 = 8.2 kΩ
R_3 = 100 kΩ
R_4 = 6.8 kΩ
R_5 = 3.9 kΩ
R_6 = 22 kΩ

Capacitors:
C_1 = 5.6 pF
C_2 = 0.68 pF
C_3 = 10 pF
C_4 = 470 pF

Semiconductors:
T_1 = BFR91
D_1, D_2 = BB405
D_3 = LED, red, 5 mm

Miscellaneous:
Fl_1 = ceramic resonator (see text)
Silver-plated wire, 0.8 mm dia.

UHF remote control receiver

119 The receiver is intended primarily for use with the remote control UHF transmitter described in the preceding article. It is a super-regenerative type with an active RF amplifier, T_1. The antenna signal is applied to the input inductor via a BNC socket, K_1. The input circuit is tuned by trimmer C_4.
The amplified RF signal is applied to the input of the super-regenerative stage based on transistor T_2. Although the oscillator is, strictly speaking, not tuned, it will lock on to the amplified RF signal applied via coupling capacitor C_7. The low-frequency modulation component is extracted from the oscillator signal with the aid of low-pass filter, R_6-C_{11}-R_7-C_{12}-R_8-C_{13}. The signal level at the demodulator output is 50 to 800 mV_{pp}, so that further amplification is required before the signal can be applied to a digital input.
The inductors in the RF amplifier input and output are made from 1 mm dia. silver-plated wire. The length of the pieces of wire is indicated by the component overlay. The wires run at a height of about 3 mm above the board surface. Note that the stator terminal of C_4 is bent upwards and soldered direct to the input inductor. The same goes for junction C_6-C_7, which is soldered 'in the air', directly on to the hot end of the inductor wire.

Inductor L_1 consists of 12 turns of 0.6-mm dia. enamelled copper wire. Its internal diameter is 3 mm.
Each of chokes L_2 and L_3 consists of 4 turns of 0.2-mm dia. enamelled copper wire through a 3 mm long ferrite bead.
Capacitor C_8 is a surface-mount technology (SMT) type which is fitted at the solder side of the board, as are the BFG65 and the BF980. The type indication printed on the transistors is legible from the component side of the board.
As indicated by the dashed lines on the component overlay, the super-regenerative section of the circuit must be screened from the rest. To do this, it is best to solder a 20 mm high box, made from tin plate, directly to the PCB copper surface.

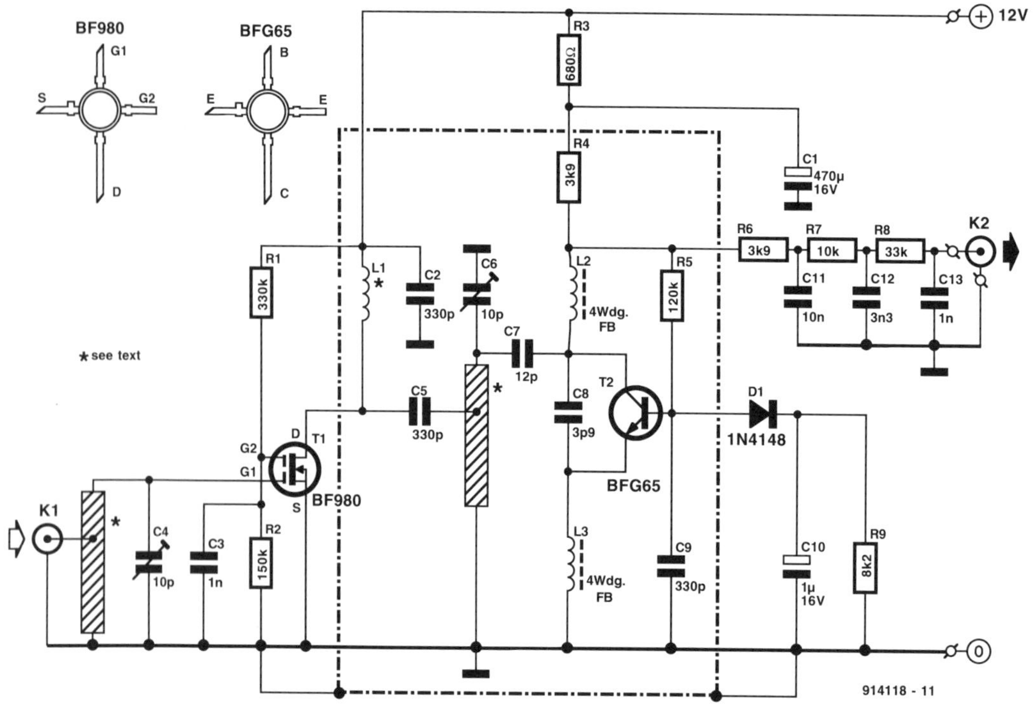

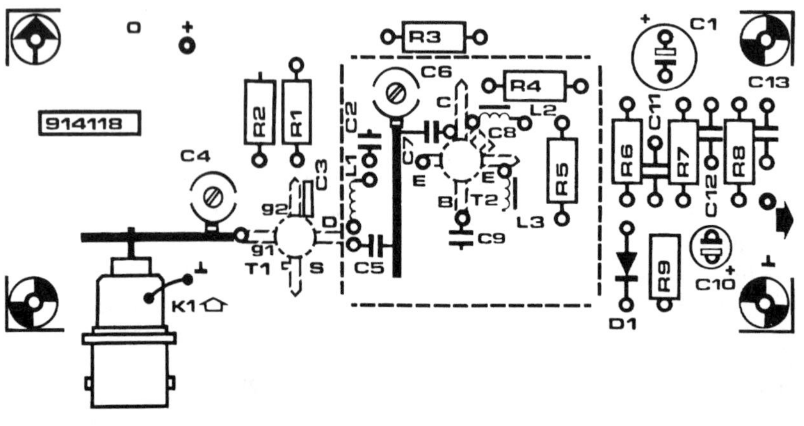

Resistors:
R_1 = 330 kΩ
R_2 = 150 kΩ
R_3 = 680 Ω
R_4, R_6 = 3.9 kΩ
R_5 = 120 kΩ
R_7 = 10 kΩ
R_8 = 33 kΩ
R_9 = 8.2 kΩ

Capacitors:
C_1 = 470 µF, 16 V, radial
C_2, C_5, C_9 = 330 pF, ceramic
C_3 = 1 nF, coffin-style leadless ceramic
C_4, C_6 = 10 pF trimmer
C_7 = 12 pF, ceramic
C_8 = 3.9 pF (surface mount)
C_{10} = 1 µF, 16 V, radial
C_{11} = 10 nF
C_{12} = 3.3 nF
C_{13} = 1 nF

Semiconductors:
T_1 = BF980 or BF966
T_2 = BFG65
D_1 = 1N4148

Inductors:
Construction details are given in the text
2 off 3-mm ferrite bead
0.6 mm dia. enamelled copper wire
0.2 mm dia. enamelled copper wire
1 mm dia. silver-plated copper wire

Miscellaneous:
K_1, K_2 = BNC socket

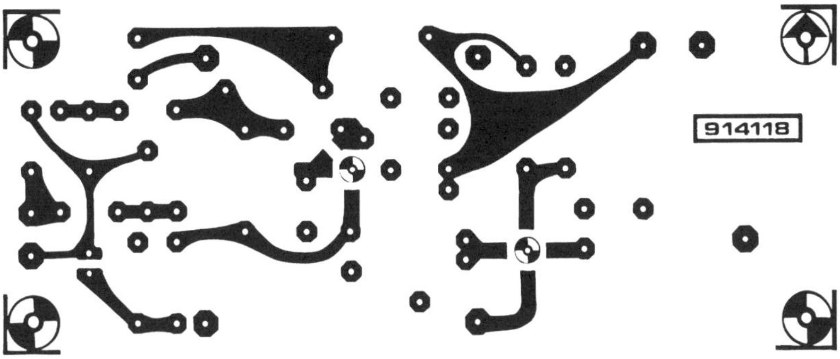

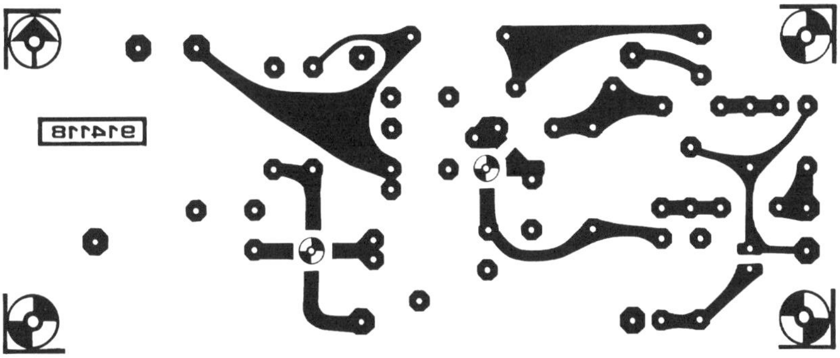

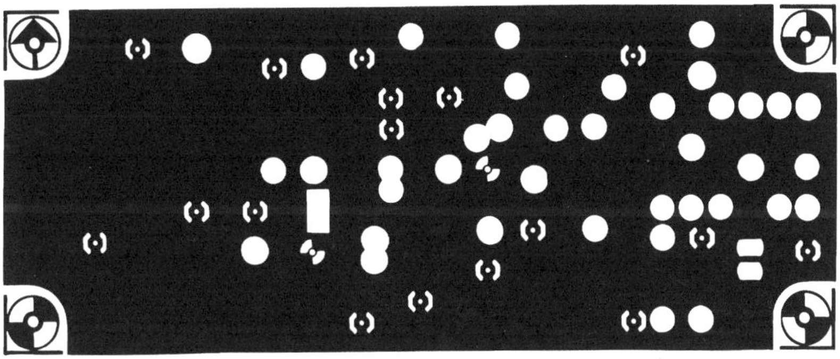

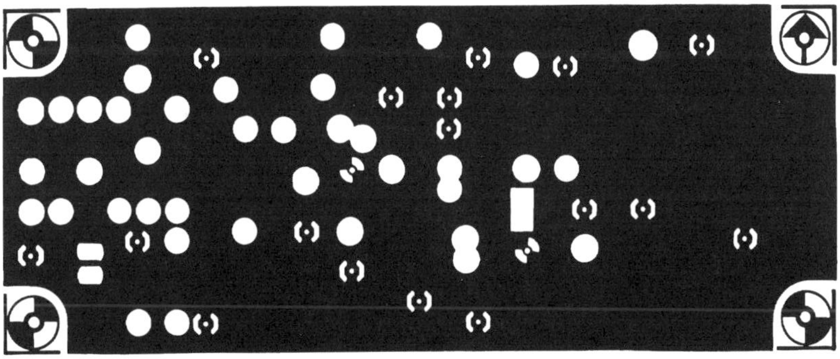

Synchronization separator

120 This circuit forms the missing link between various video sources and, say, a multisync monitor. Based on discrete parts only, it extracts the composite synchronization (i.e., a mix of the horizonal and the vertical component) and the vertical synchronization from a composite video signal with an amplitude of about 1 V_{pp}. The output sync signals are available in true as well as inverted form to suit your monitor.

The positive composite video signal is filtered by R_1-C_2 and clamped by a Schottky diode, D_1, to extract the sync components. The CSYNC signal is fed to XOR gate N2 which functions as an inverter when S_1 is closed.

The CSYNC signal is also fed to a two-stage *L-C* filter which suppresses the line sync component and leaves the raster sync, VSYNC, at the inputs of opamp IC1b. Like CSYNC, VSYNC is available in true or inverted form.

The circuit draws about 200 μA at a supply voltage of 5 V. The output signals are TTL compatible.

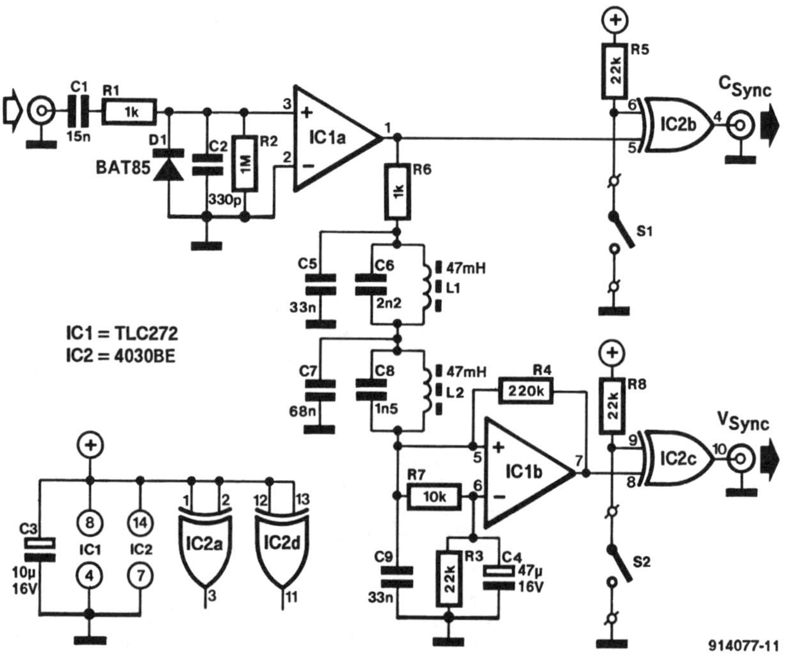

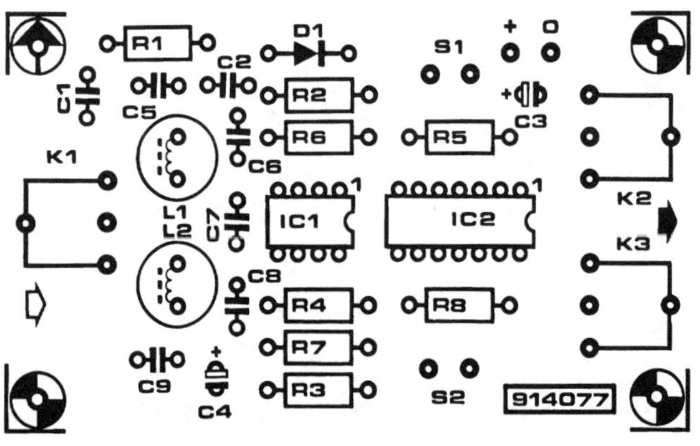

Parts list

Resistors:

R_1, R_6 = 1 kΩ

R_2 = 1 MΩ

R_3, R_5, R_8 = 22 kΩ

R_4 = 220 kΩ

R_7 = 10 kΩ

Capacitors:

C_1 = 15 nF

C_2 = 330 pF

C_3 = 10 μF, 16 V, radial

C_4 = 47 μF, 16 V, radial

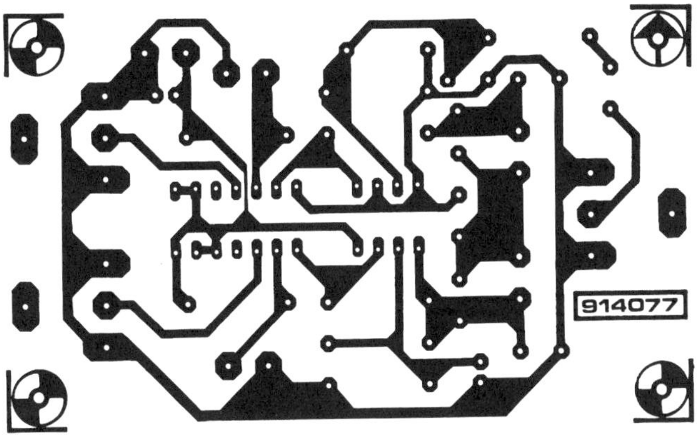

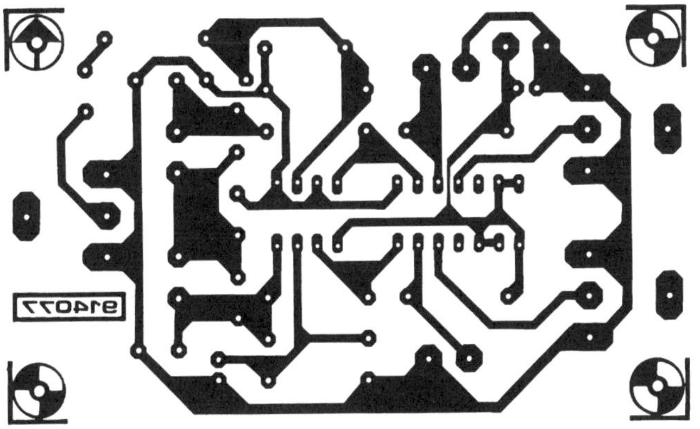

C_5 = 33 nF
C_6 = 2.2 nF
C_7 = 68 nF
C_8 = 1.5 nF
C_9 = 33 nF

Semiconductors:
D_1 = BAT85

Integrated circuits:
IC_1 = TLC272
IC_2 = 4030BE

Miscellaneous:
L_1, L_2 = 47 mH choke, radial
K_1, K_2, K_3 = phono socket for PCB mounting
S_1, S_2 = miniature on-off switch
PCB Ref. 914077

Low power NBFM transmitter

121 The transmitter is a crystal-controlled, battery-powered, one-chip narrowband frequency modulation (NBFM) model for operation in the 27-MHz band, primarily as a wireless microphone*.

The circuit is an application of Motorola's MC2833 single-chip VHF narrowband FM transmitter IC, designed to work in the 27-MHz band. The transmitter output power is about 10 mW (+10 dBm), which, owing to the expected low efficiency of the antenna used, will result in a typical effective radiated power (ERP) of less than 1 mW. The range of the transmitter, therefore, is limited to 10–20 metres (33–66 ft).

The active circuitry contained in the MC2833 includes a microphone amplifier, a voltage controlled oscillator, and two auxiliary transistors which are used for frequency multiplication or RF amplification, depending on the desired output frequency.

Preset P_1 is used to adjust the microphone gain, and preset P_2 to adjust the deviation. Remember, the transmitter will produce NBFM only, with a maximum deviation of 5 kHz. This means that a narrow-band receiver (such as a typical 27 MHz CB unit) is required for sufficient audio output.

The quartz crystal, X_1, resonates in fundamental mode (here, 9 MHz), calibrated for parallel resonance with a 32 pF load. The final output frequency is generated by frequency multiplication (here, ×3) within the MC2833.

Construction of the transmitter follows

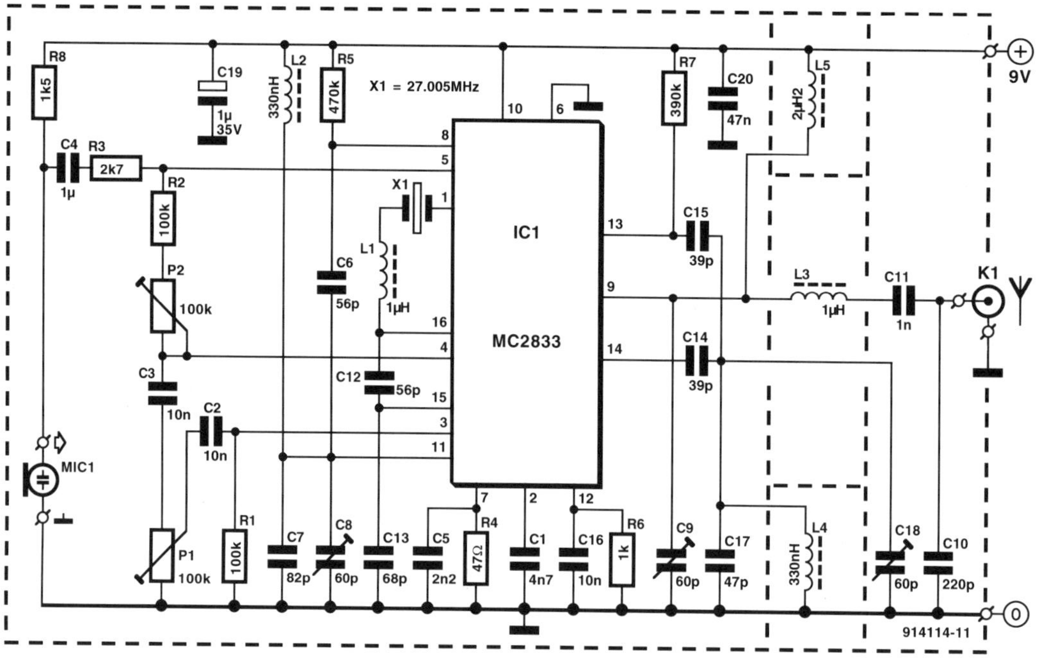

the rules of RF design: keep all component leads as short as possible, fit the screening as indicated on the printed circuit board, and do not use an IC socket for IC_1.

The transmitter is fairly simple to adjust: simply peak the three trimmers, C_8, C_9 and C_{18}, for maximum output power delivered to a 50 Ω dummy load. Alternatively, connect the antenna and an oscilloscope to the transmitter output, and adjust the trimmers for maximum RF voltage. Next, listen to the transmitted signal on a 27 MHz receiver, and adjust the two presets until the best possible modulation is achieved. Do not set the microphone gain too high, since this will easily cause clipping.

If the transmitter is used as a wireless microphone, the antenna will typically be a piece of flexible wire with a length of about 1 m. The transmitter draws a current of about 7 mA, so it is good practice to keep an eye on the state of the battery.

* This transmitter is not licensable as a wireless microphone in the UK.

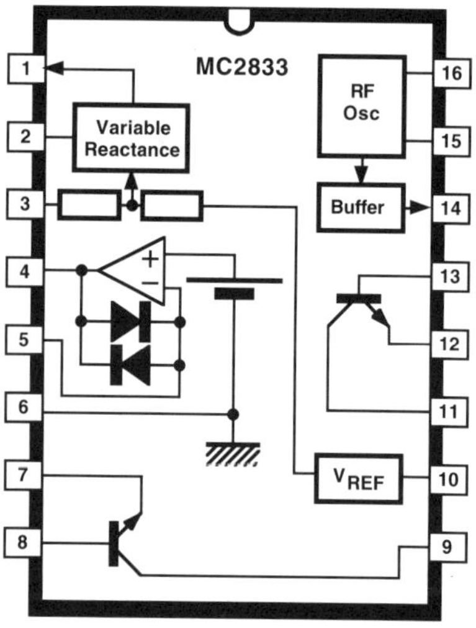

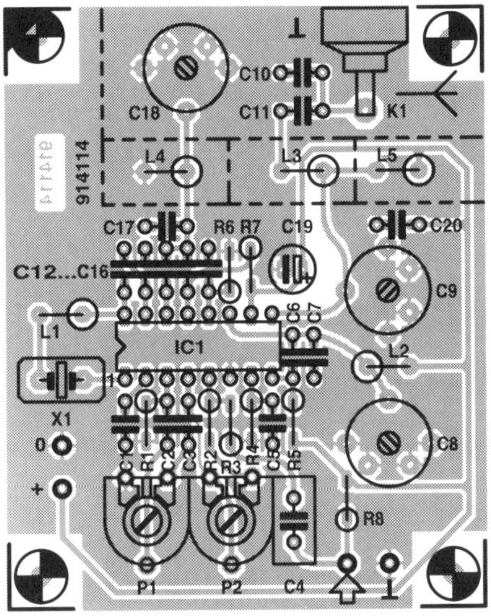

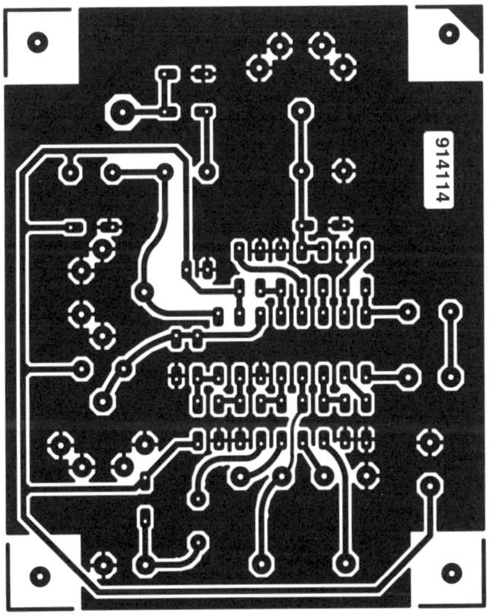

Parts list

Resistors:
R_1, R_2 = 100 kΩ
R_3 = 2.7 kΩ
R_4 = 47 Ω
R_5 = 470 kΩ
R_6 = 1 kΩ
R_7 = 390 kΩ
R_8 = 1.5 kΩ
P_1, P_2 = 100 kΩ preset, H

Capacitors:
All fixed capacitors are ceramic, unless otherwise indicated
C_1 = 4.7 nF
C_2, C_3, C_{16} = 10 nF
C_4 = 1 μF MKT
C_5 = 2.2 nF
C_6, C_{12} = 56 pF
C_7 = 82 pF
C_8, C_9, C_{18} = 60 pF foil trimmer
C_{10} = 220 pF
C_{11} = 1 nF
C_{14}, C_{15} = 39 pF
C_{17} = 47 pF
C_{19} = 1 μF, tantalum
C_{20} = 47 nF

Inductors:
L_1, L_2 = 1 μH
L_3, L_4 = 330 nH
L_5 = 2.2 μH

Integrated circuits:
IC_1 = MC2833P (Motorola)

Miscellaneous:
K_1 = BNC socket
X_1 = crystal, 27.005 MHz
MIC_1 = Electret microphone

Field-strength meter

122 The unit discussed here indicates, by means of a chain of LEDs, in logarithmic ratios the strength of an electric field surrounding the unit. The input signal across the conducting discs is applied to the non-inverting input of IC_1. The amplification of IC_1, $A = R_4/P_1$. In the prototype, P_1 was set at about 210 kΩ to give an amplification of about ×50.

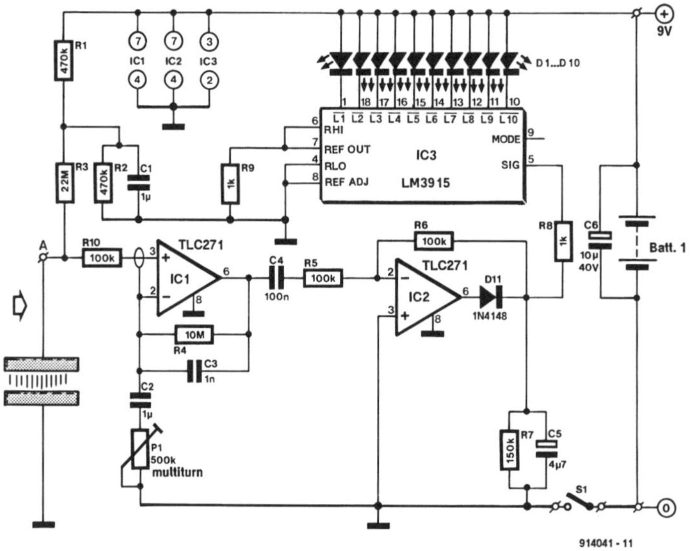

Opamp IC_2 functions as a rectifier: during the negative halves of the input signal, its output goes high and D_1 conducts. During the positive halves, the signal is applied direct to the output via R_5 and R_6. Diode D_1 is then reverse biased and IC_2 does not function.

The level of the output voltage across C_5 is monitored by diode chain D_1–D_{10} via IC_3. The capacitor is discharged slowly via R_7 when the signal ceases.

The conducting discs are formed by two round, double-sided PCBs as shown in Fig. 2. The components are mounted at the track side of one of the boards. Carefully bend the terminals of the components about 2 mm from their ends to make better contact and strengthen the solder joints. Note that C_6, S_1 and the LEDs must be fitted lying down, all the others upright.

The two boards are fixed together with non-metallic screws and 20 mm spacers.

The indicator is calibrated by turning P_1 until the LED associated with a field strength of 10 V m^{-1} just lights.

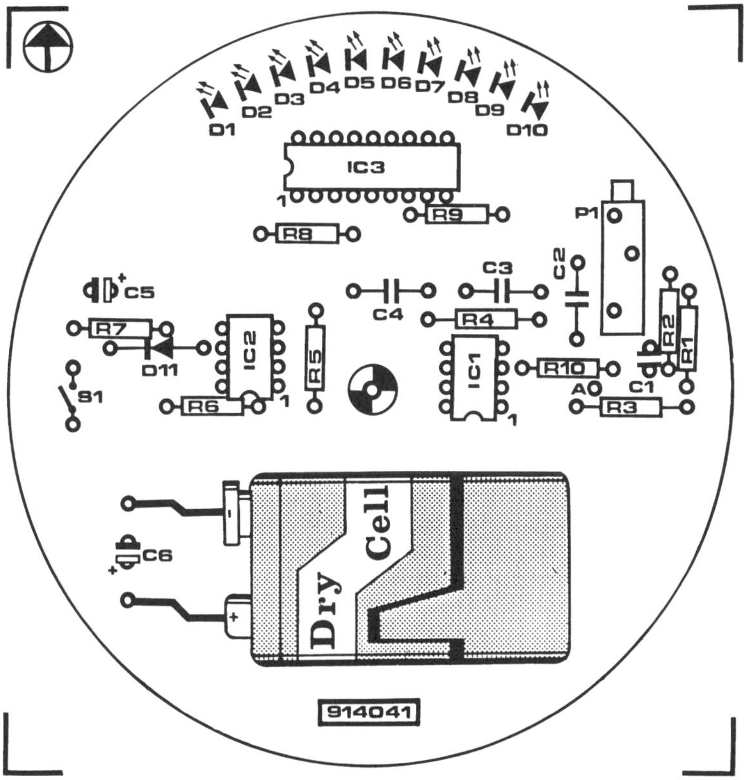

Parts list

Resistors:
R_1, R_2 = 470 kΩ
R_3 = 22 MΩ
R_4 = 10 MΩ
R_5, R_6, R_{10} = 100 kΩ
R_7 = 150 kΩ
R_8, R_9 = 1 kΩ
P_1 = 470 kΩ (or 500 kΩ) multi-turn preset

Capacitors:
C_1, C_2 = 1 μF
C_3 = 1 nF ceramic
C_4 = 100 nF
C_5 = 4.7 μF, 25 V, radial
C_6 = 10 μF, 40 V, radial

Semiconductors:
D_1 = 1N4148
D_2–D4 = LED, 3 mm, green
D_5 = LED, 3 mm, yellow
D_6–D10 = LED, 3 mm, red

Integrated circuits
IC_1, IC_2 = TLC271
IC_3 = LM3915

Miscellaneous:
S_1 = SPST switch
9 V battery

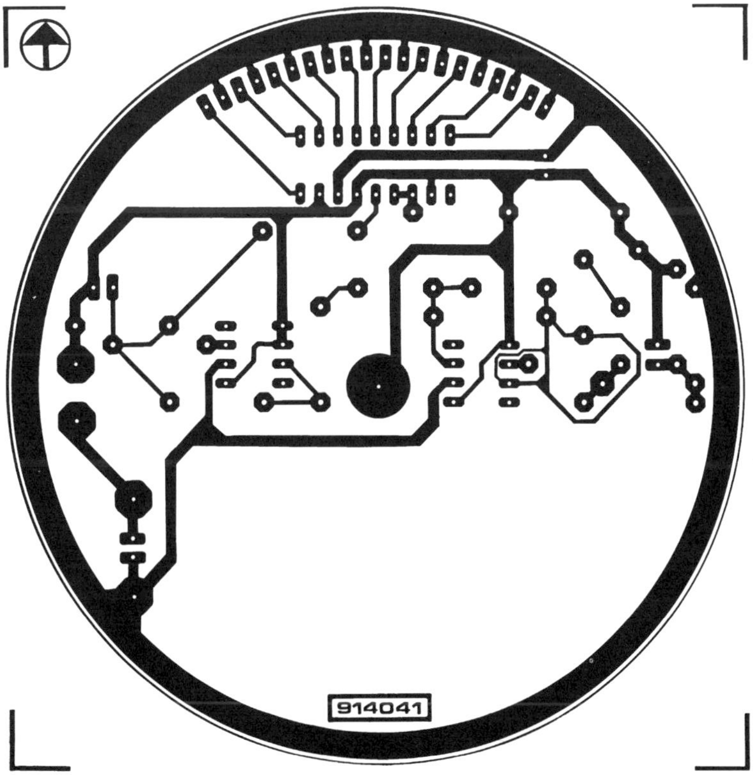

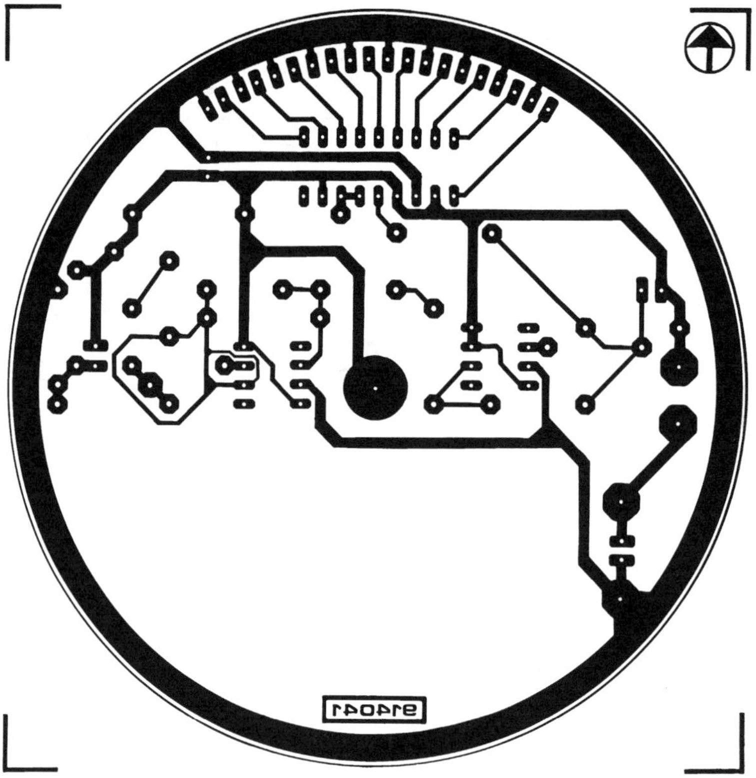

S-meter for short-wave receivers

123 Some radio amateurs are very keen on accurate RST reports, others (mostly the VHF/UHF fraternity) never look at the S-meter on the receiver, and are satisfied as long as they can hear the other station. This circuit is for the first group.

Traditionally, one S-point corresponds to a 6-dB increase in signal strength, while 'S9' is defined as 50 µV into 50 Ω. Unfortunately, very few receivers these days have a calibrated S-meter, hence the confusion among radio amateurs about the interpretation of the signal strength reports they exchange and write on QSL cards.

The logarithmic-to-linear converter contained in the NE604 from Valvo (Philips Components) is used here to build an accurate S- (signal strength) meter for short-wave receivers. The amplifier in the NE604 is tuned to the intermediate frequency (IF) of the receiver with the aid of L1 and C_2. Here, the circuit is dimensioned for an IF of 455 kHz, which is applied to input capacitor C_1.

The output of the field strength detector in the NE604 supplies a current of 0 to 50 µA at pin 5. This current is converted into a voltage of 0–5 V by a 100 kΩ resistance, R_2+R_3. Note that two E96-series 1% resistors are used here, plus a diode, D_1, instead of a single 100 kΩ resistor. This is done to compensate temperature effects which would cause degrading of the linear output voltage. If the specified E96 resistors can not be obtained, R_2 may be replaced by a parallel combination of two 120 kΩ 1% resistors, and R_3 by a parallel combination of a 39 kΩ 1% resistor, and a 1 kΩ 1% resistor. It should be noted that the usable range of the log-to-lin converter in the NE604 is roughly from 5 μA to about 40 μA of output current, corresponding about 70 dB, or 0.5 V to 4 V at pin 6 of IC_2. The lower level is caused by background noise of the IF amplifier in the NE604, and the upper level by limiting and saturation effects. Fortunately, the effective range of the converter is large enough for the present applications, bearing in mind that S-meter readings lower than S_3 are rare and of little meaning in the short-wave bands.

Components R_4, C_9 and C_{10} suppress ripple and noise. Opamp IC_2 is set to provide unity gain, i.e., its output voltage is 0 to 5 V. The moving-coil meter is connected between two presets. P_1 is adjusted until the meter reaches full deflection at a voltage of 4.5 V measured at pin 6.

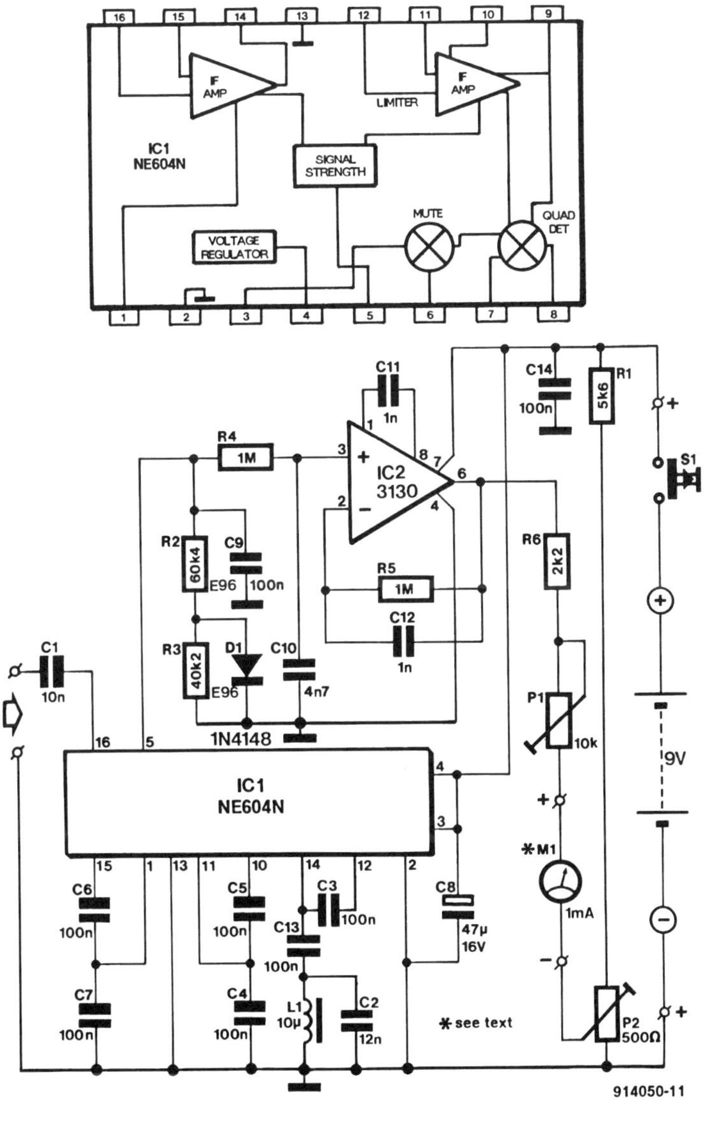

Parts list

Resistors:
R_1 = 5.6 kΩ
R_2 = 60.4 kΩ (E96 - see text)
R_3 = 40.2 kΩ (E96, see text)
R_4, R_5 = 1 MΩ
R_6 = 2.2 kΩ
P_1 = 10 kΩ preset H
P_2 = 500 Ω preset H

Capacitors:
C_1 = 10 nF
C_2 = 12 nF
C_3, C_4–C_7, C_9, C_{13}, C_{14} = 100 nF
C_8 = 47 μF, 16 V, radial
C_{10} = 4.7 nF
C_{11}, C12 = 1 nF

Semiconductors:
D_1 = 1N4148

Integrated circuits:
IC_1 = NE604A
IC_2 = CA3130E

Miscellaneous:
S_1 = on-off push-button
L_1 = 10 μH choke, axial
M_1 = MC meter, 1 mA

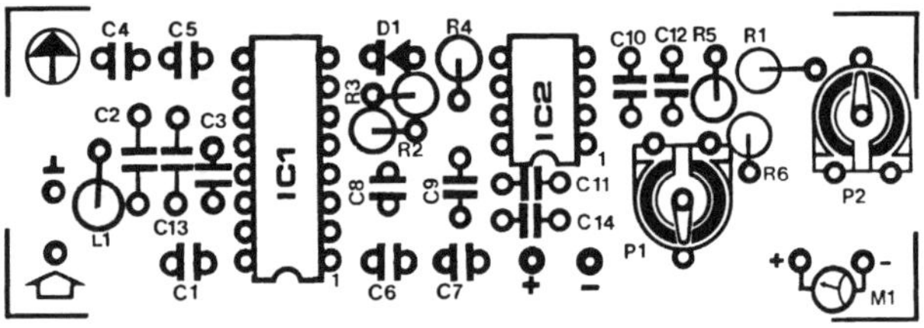

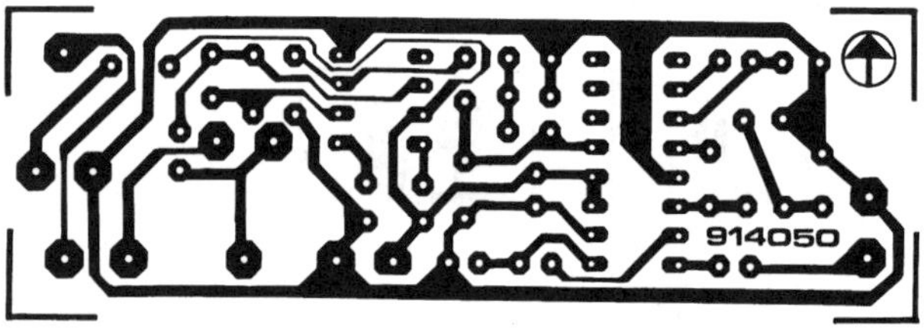

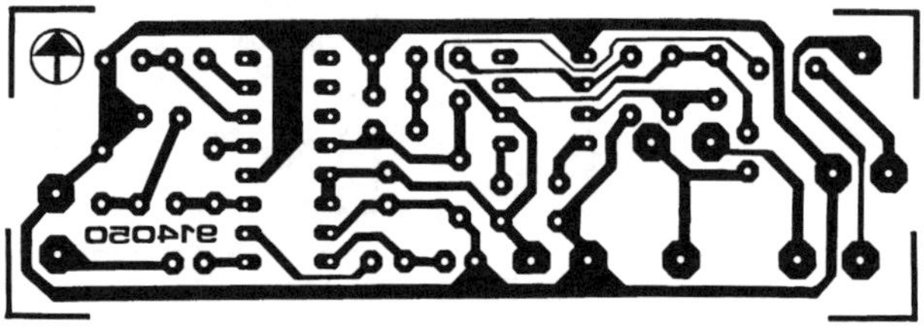

CB-to-SW down converter

124 This converter enables long-distance (DX) reception of AM or SSB stations in the 27-MHz citizens' band (CB) on a short-wave or medium-wave radio (note, though, that AM or SSB modulation in the 27-MHz CB band is no longer allowed in a number of countries).

The converter consists of a prestage, T1, and a mixer/oscillator, IC1. The antenna signal is coupled inductively to the gate of dual-gate MOSFET T1 via tuned circuit L1-C1, which acts as a 27 MHz input filter. The operating point of the MOSFET is determined by resistors R1-R2 connected to the gate-2 terminal. The amplified signal is fed to the mixer amplifier via a coupling capacitor and a second tuned circuit, L3-C2. The oscillator on board the S042P IC (manufacturer: Siemens) works with a quartz crystal, X1. The quartz frequency is selected such that the difference frequency produced by the mixer (also on board the S042P) falls within the tuning range of the radio connected to the output of the converter. For instance, if an inexpensive 26.800 MHz crystal is used, the frequency of the received 27-MHz station, f_{in}, is

$$f_{in} = 26{,}800 \pm f_{dial}$$

where f_{dial} is the frequency read on the radio's tuning scale (in this case, the received stations will appear in the medium-wave band). Other crystal frequencies may be used, e.g., 10 MHz, to move the CB band into the SW range (17 MHz).Construction of the converter is fairly easy on the printed circuit board shown here. Parts shown with a dashed outline are fitted at the solder side of the board. The converter is shielded all around to prevent spurious radiation. The antenna and the IF output are best connected via coax sockets (SO239 or BNC). The two inductors in the converter are simply adjusted for best reception.

The converter is powered either by a regulated 9-V adaptor, or by the radio it is connected to, if this is capable of furnishing 9 V at a few tens of mA.

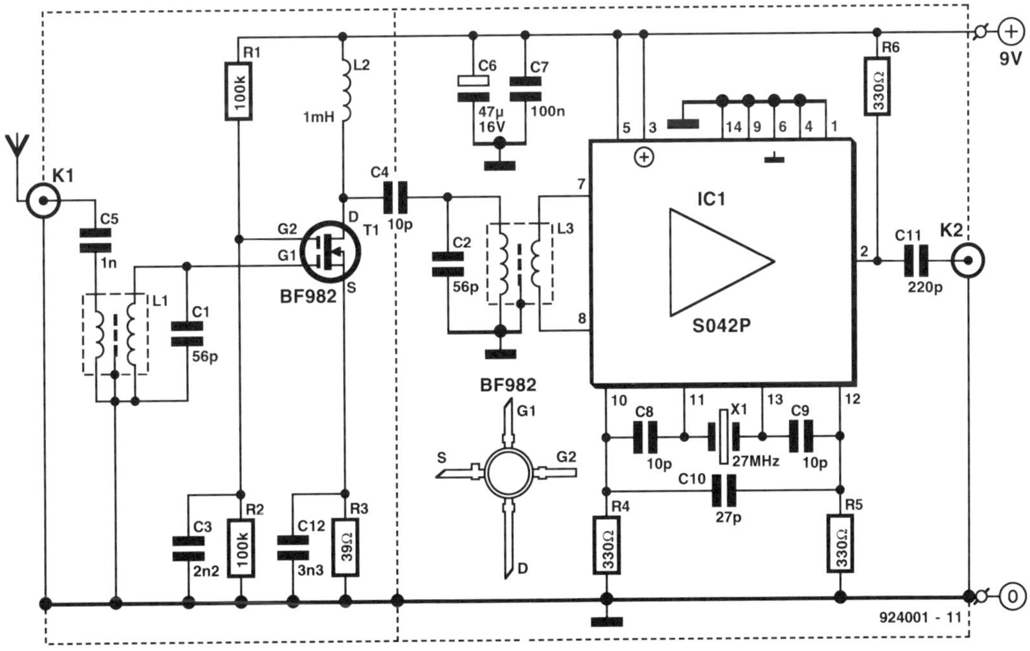

Parts list

Resistors:
R_1;R_2 = 100 kΩ
R_3 = 39 Ω
R_4;R_5;R_6 = 330 Ω

Capacitors:
C_1;C_2 = 56 pF ceramic
C_3 = 2.2 nF ceramic
C_4 = 10 pF ceramic
C_5 = 1 nF ceramic
C_6 = 47 µF 16V radial
C_7 = 100 nF ceramic
C_8;C_9 = 10 pF ceramic
C_{10} = 27 pF ceramic
C_{11} = 220 pF ceramic
C_{12} = 3.3 nF ceramic

Semiconductors:
T_1 = BF982

Integrated circuits:
IC_1 = S042P

Inductors:
L_1;L_3 =
113CN2K50989189ADZ
(Toko)
L_2 = 1mH choke

Miscellaneous:
K_1;K_2 = BNC socket
X_1 = see text

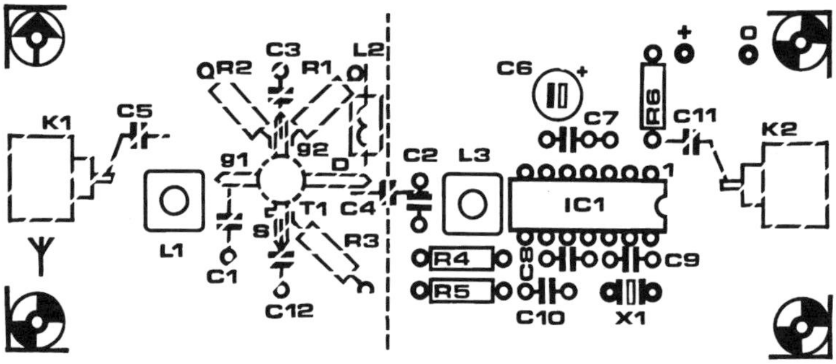

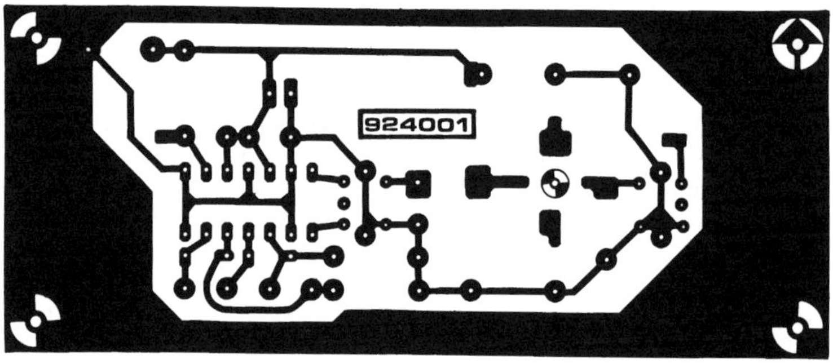

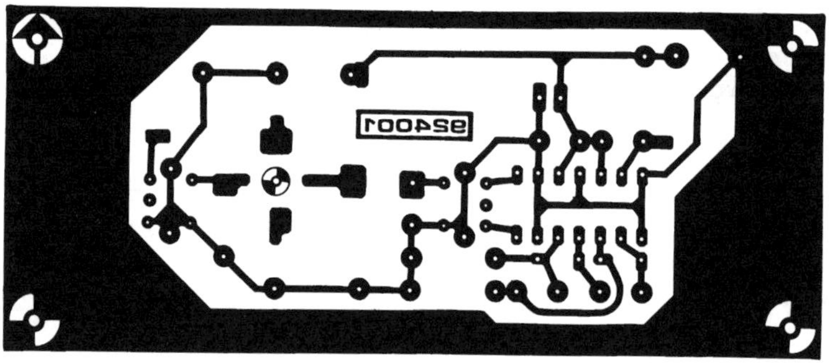

CGA-to-SCART adaptor

125 The adaptor has the great advantages of not needing a separate power supply and requiring so few external components that these can easily be accommodated in the SCART connector.

The signals on the R, G, and B pins of the CGA card are converted from TTL level to SCART level with the aid of transistors R_4, R_5 and R_6 and the input impedance (75 Ω) of the SCART inputs. The output impedance of the so created voltage dividers is not exactly 75 Ω as specified for SCART video inputs, but in practice this does not appear to make much difference.

Resistors R_1–R_3 between pin 6 of the CGA card and the R, G, B pins of the SCART connector ensure that the brightness is brought down to 50% when pin 6 is active.

A composite sync(hronization) signal and a

blanking signal for the TV are derived from the horizontal and vertical sync pulses of the CGA card with the aid of R_7–R_{11} and T_1.
The 5 V supply for T_1 is obtained from the CGA card via pin 7. Since that pin is not normally used, it must be linked to the 5 V line by a short length of wire.

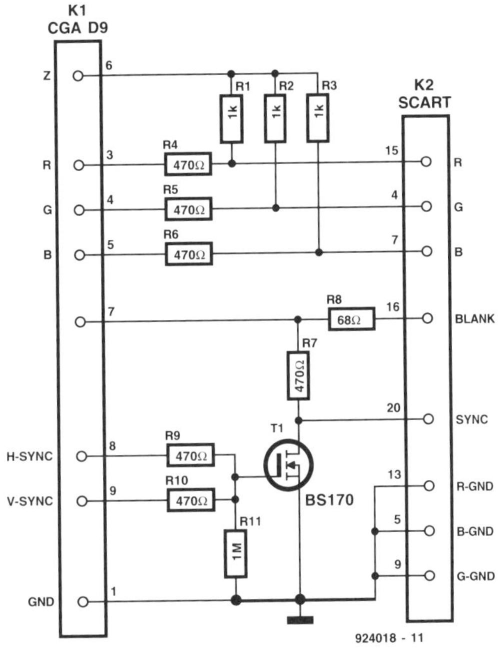

Miniature crystal oscillator

126 Nowadays, with the advent of SMT (surface mount technology), it is possible to construct tiny circuits. In the case of crystal oscillators, however, this may not always be feasible because of the size of the crystal. Fortunately, Statek, a specialist crystal manufacturer, produces crystals measuring only 8×4×1 mm for SMT. Together with a single inverter and four passive components, such a crystal makes it possible to make a truly miniature, stable oscillator as shown.
The circuit works very well up to frequencies of 16 MHz if an HC-IC is used. With an HCT-IC, the maximum frequency was found to be about 8 MHz.
Finally, the circuit works perfectly all right with standard components as well.

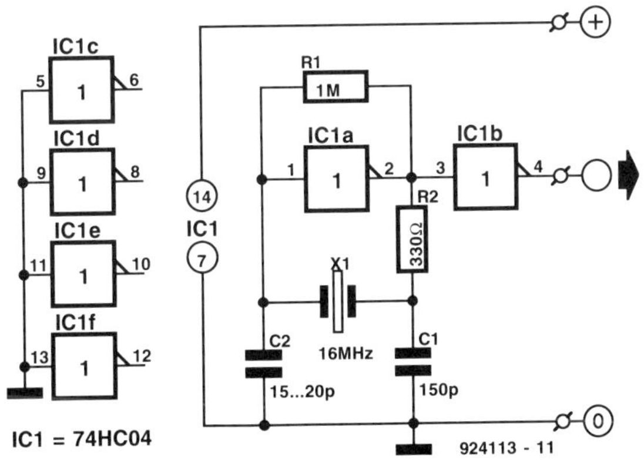

Kalundborg 10 Mhz frequency reference

127 Ever since the long-wave broadcast station Kalundborg changed its frequency from 245 kHz to 243 kHz to comply with the CCIR's recommendations for a 9 kHz raster in the LW and MW bands, it has become possible to use the carrier in the locked frequency standard described in Ref. 1. Kalundborg is a 300 kW long-wave transmitter in Denmark, with a range of about 500 km (300 miles). This article is, therefore, of particular interest to our Scandinavian readers.

Briefly, what is proposed here and in the next item ('preamplifier for Kalundborg frequency reference') is to change the divider in the frequency reference such that a carrier input frequency of 243 kHz can be used instead of 77.5 kHz (the transmit frequency of DCF77 in Germany), for which the circuit was originally designed. The changes are outlined in the simplified block diagram, Fig. 1, which is best compared with Fig. 1 in Ref. 1. In practice, the original circuit of the reference is changed to the extent that a new circuit diagram is required—see Fig. 3. The new circuit is much simpler than the original, mainly because some circuit sections could be omitted, including tthe VLF preamplifier (T_1-T_4 in the original design), the 10 MHz 'locked only' output (IC_7 and T_{12} in the original design) and the 'error' detector (N_2, N_3, N_4 and the beeper in the original design).

In the 'Kalundborg' circuit shown here, the 10-MHz signal supplied by X_1 and T_3 is multiplied by 3 by parallel tuned circuit L_4-C_{39}. The 30-MHz signal is subsequently divided by 100 (IC_5) and again by 100 (IC_6) to obtain the 3 kHz reference for multiplier IC_3. The 240 kHz signal used to heterodyne with the 243 kHz carrier is obtained by dividing the 30 MHz signal by 25 (IC_5) and then by 5 (74LS90).

The antenna, formed by an inductor wound on a ferrite rod and resonated by capacitors C_1, C_2 and C_3, is connected direct to the balanced inputs of the S042P mixer (IC_1). The error signal at the output of the multiplier (IC_3) is filtered and converted into a tuning voltage, which is applied to a dual varicap, D_5. The varicap is capable of detuning (to a small extent) the 10 MHz quartz oscillator, and so closes the phase-locked loop (PLL). Provided Kalundborg is received with adequate strength (rotate the ferrite rod), the LED at the output of IC_8 lights, and a 'rock-steady' 10 MHz reference signal is available at the output of N_2.

Figure 2 shows a detailed diagram of the ×3 multiplier. Inductor L_4 consist of 4.5 turns of 0.5 mm dia. enamelled copper wire on a former with a ferrite core. The former has an outside diameter of 6 mm,

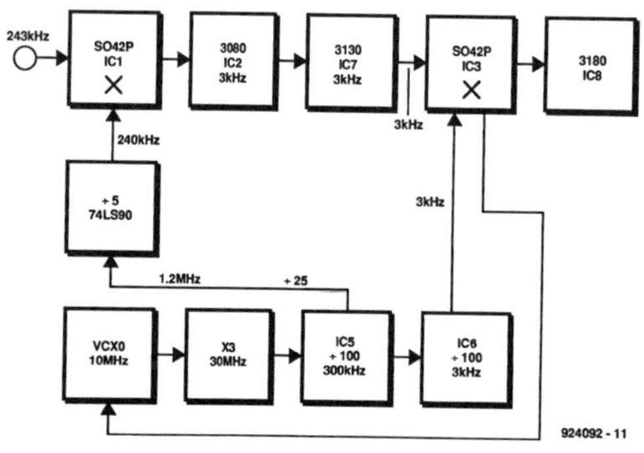

Figure 1

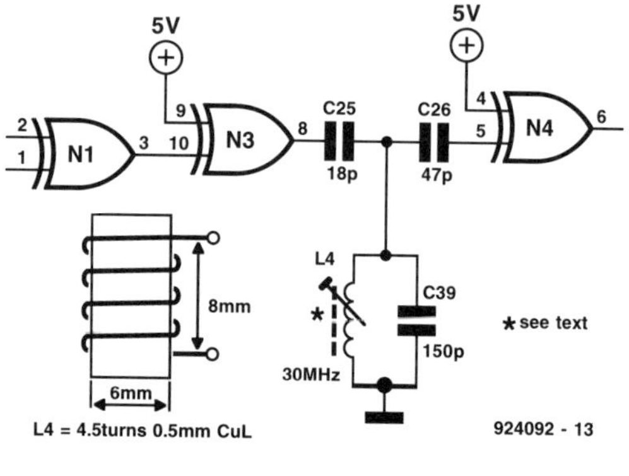

Figure 2

and L_4 is drawn out to a length of about 8 mm. The operation of the multiplier is easily checked with the aid of an oscilloscope and/or a frequency meter, the signal levels being quite high. The core in L_4 is adjusted for the highest 30 MHz level at pin 5 of N_4.

Reference:

1. "DCF77 receiver and locked frequency standard", *Elektor Electronics* January 1988.

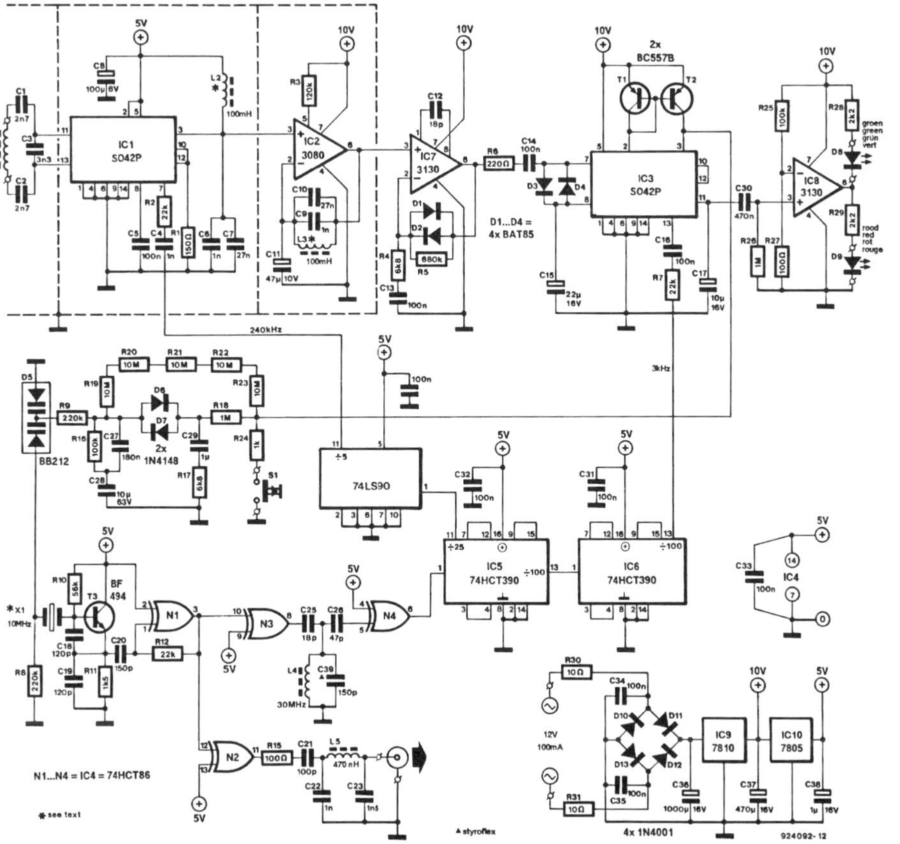

Figure 3

Infra-red headphone transmitter

128 The proposed transmitter provides an optical (infra-red) that is wireless, connection to a headphone. The receiver is described in the next article.

Three infra-red (IR) LEDs are provided with a quiescent current by T_1. The level of that current is set with P_1. When an audio signal is applied to the gate of T_1, the current through the LEDs is modulated. Consequently, the light emitted by the diodes is also (amplitude) modulated.

To prevent overdriving of the gate causing too high a current through the LEDs, a current limiter, consisting of T_2 and R_3, holds the current below 100 mA.

The maximum dissipation of a BS170 is 830 mW at an ambient temperature of 25 ∞C, while the maximum drain current is 500 mA. Therefore, even when the FET is overdriven, those limits are not exceeded.

The optimum quiescent current through the LEDs must be determined in conjunction with the receiver (which must be adjusted for minimum distortion).

The prototype transmitter drew a current of about 60 mA at a supply voltage of 9 V. It is, therefore, advisable to use a mains adaptor, because that current is just a little too high for a PP3 battery. Keep the earths of the mains adaptor and the audio signal separated as shown in the diagram to prevent feedback of the LED current to the input.

The gate-source voltage of a BS170 may be up to 15 V. If you use a signal source that delivers a higher level, it is advisable to incorporate a simple protection circuit (for instance, a 10-V zener diode parallel, and a resistor in series, with the input).

The optical connection is fairly directional, but this can be improved by placing the LEDs at varying angles. Also, the distance of operation can be extended appreciably by fitting reflectors behind the LEDs.

The optimum input level for an operating distance of some metres (4–8 ft) is 100–200 mV.

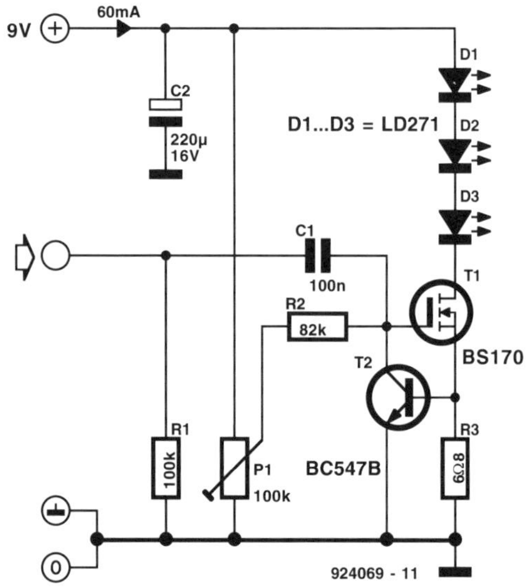

Infra-red headphone receiver

129 This receiver is meant to complement the transmitter described in the previous article. Its design is based on just one FET. This has the advantage that construction is simplicity itself, and the disadvantage that for a sufficiently low output im-pedance the value of drain resistor R_3 has to be fairly low. That results with correct operation of T_1 to a fairly large (certainly for a battery) current. The value of R_3 was chosen, as a compromise, at 560 Ω, which makes driving 600 Ω headphones possible. The load seen by T_1 is then 300 Ω.

Both usable types of receive diode shown in the diagram have a daylight filter and are tuned to the wavelength of the LEDs in the transmitter (950 nm at 25 ∞C). At a couple of metres (5–8 ft) distance, a (no-load) output voltage of 200–300 mV is obtained, which is quite sufficient for most headphones. The circuit then draws a current of 9–10 mA.

The setting of P_1 is fairly critical, but its control range may be reduced by adding a small resistor at either side of the preset. The preset should be adjusted for minimum distortion. This is best done by applying a 1 kHz audio signal at a level of 150 mV to the transmitter and adjusting both circuits for minimum (audible) distortion. This should be done without electric light, because the transmitter does not modulate the audio signal on to a carrier, so that light bulbs, and particularly neon tubes (which emit an appreciable amount of IR light at 950 nm, modulated with 100 Hz) can cause quite a hum. Even normal ambient light causes a deterioration of the signal-to-noise ratio. However, with a little ambient light and a distance between transmitter and receiver of 3–4 m (10–13 ft), the distortion was 1–2%, which is not bad for such a sparse design.

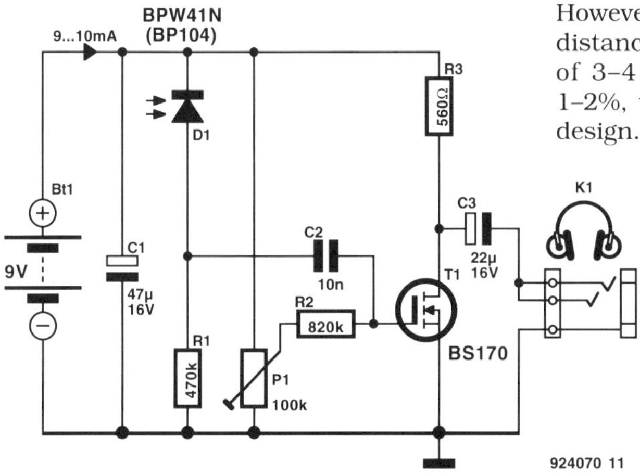

PLL synthesizer for TV receivers

130 Modern TV tuners normally contain means to scale down the VCO (voltage-controlled oscillator) signal. The one used in the proposed synthesizer is a Philips Type UV816/6456, whose scale factor can be set to 64 or 256. It operates over the low VHF band, the high VHF and hyperbands, and the UHF band.

The proposed circuit is based on a Siemens Type SDA3002 frequency synthesizer IC that may be controlled by a computer. It works exclusively with a :64 scaler. This is ensured by leaving pin 15 of the tuner open.

Network R_2-C_2-C_1 forms the filter for the phase-locked loop—PLL. The level of the charging current for the filter is determined by the value of R_1 and bit 14—see Table 1. In the prototype, both low and high levels of current gave stable PLL operation, but the loop reacted faster with a high current.

The SDA3002 has a unique oscillator for providing a reference frequency. The frequency of crystal X_1 is scaled down internally by 4096, so that the reference frequency for the PLL is 976.5625 Hz.

The PLL is set by entering a data word into the SDA3002, for which a clock, CPL, an enable signal, PLE, and a data signal, IFO, are needed. At each trailing transition (edge) of the clock, a bit is shifted into the IC, provided PLE is high—see Fig. 2. Only when PLE has become low will the PLL accept the entered data into a latch.

A total of 18 bits must be shifted into the chip. Fourteen of these contain the frequency setting of the PLL. The 15th bit determines the charging current. When this bit is high, the current is $10I_r$, where I_r is the reference current; when it is low, the

charging current is I_r. The 16th bit controls the NORM output. The final two bits determine the state of the band selector outputs: pins 4–7, see Table 1.
The first 14 bits are computed fairly easily once the desired channel frequency, f_c, the intermediate frequency, *i.f.* (=38.9 MHz), the scale factor, z_a (=64) of the tuner, and the reference frequency, f_r, of the PLL are known. The overall scale factor, z, is then

$$z=(f_c+i.f.)/z_a f_r.$$

The result should be rounded to the nearest whole number, and then split into two parts (IC_1 contains a dual-mode prescaler) to obtain the bits needed by IC_1. The nine most significant bits—MSBs—are calculated by dividing *z* by 32 and ignoring the digits following the decimal point. The remainder of z:32 forms the five least significant bits—LSBs.
As an example, assume that channel 29 is wanted; the carrier frequency is then 535.250 MHz. The scale factor is

z=(535.25+38.9)/64¥976.5625=

=9186.

Dividing that number by 32 gives 287, remainder 2. In binary form, that is (MSB)100011111 00010 (LSB). However, the data must be entered in inverted form, that is, a logic 1 corresponds to low and a logic 0 to high. Taking the LSBs first, the following levels must appear at the data input

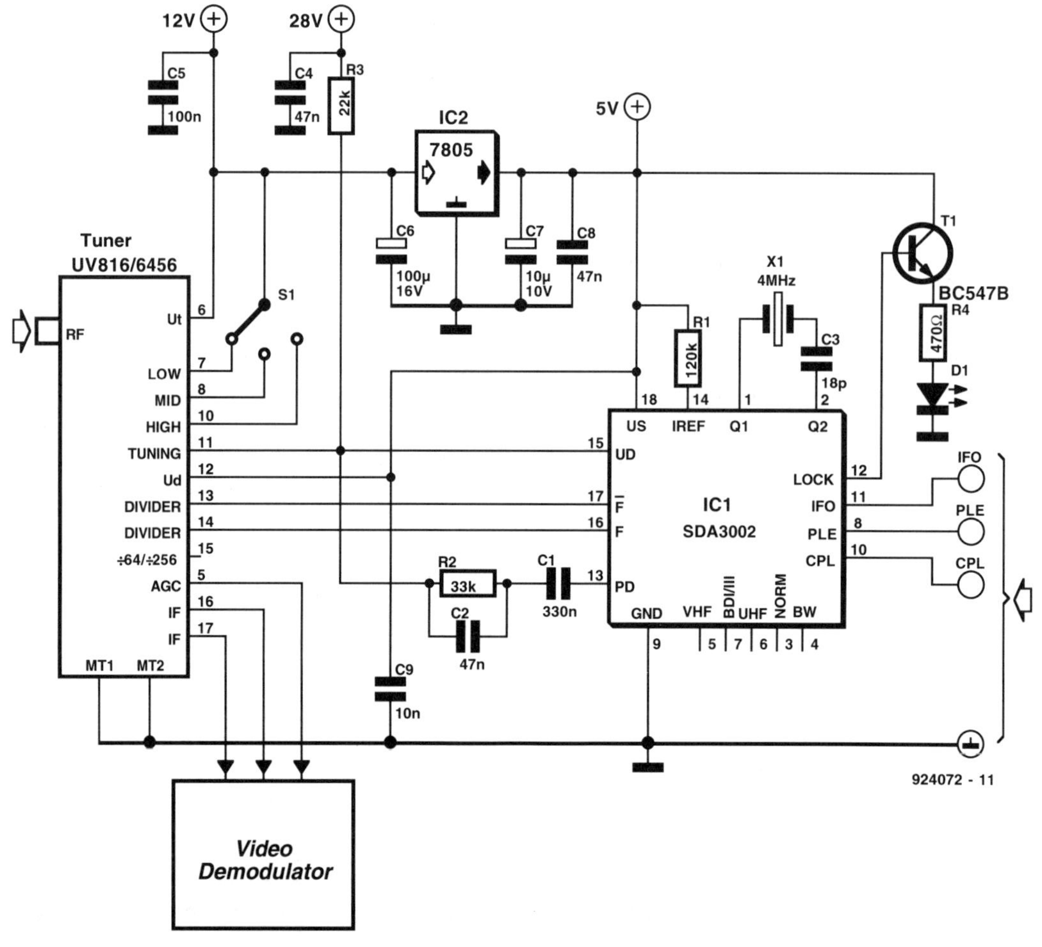

HLHHHLLLLLHHHL.

Diode D_1, which is controlled by the LOCK output via T_1, then lights to indicate that the PLL is locked.

The maximum tuning voltage is determined by the supply voltage, which is connected to the loop filter via R_3. The UV816 needs a tuning voltage of 0.7–28 V. The SDA3002 can handle a maximum voltage of 33 V at its UD output.

Two further supplies are needed: 12 V for the tuner and 5 V for IC_1 and the prescaler in the tuner. Since the combined current drain of IC_1 and D_1 (lighted) is only 29 mA, the 5 V rail is easily obtained from the 12 V supply via 5 V regulator. That regulator can at the same time provide the current—about 25 mA— for the prescaler in the tuner. Additionally, the 12 V supply must provide the current for the tuner, which in case of the UV816 is about 85 mA. A total current of some 140 mA is, therefore, required.

IFO bit 2^{14}	pump current			
L	Ir			
H	$10I_r$			
IFO bit 2^{15}	NORM output			
L	L			
H	H			
IFO bit 2^{16} 2^{17}	band selection output			
	band I/III	VHF	UHF	BW
L L	H	H	L	H
L H	H	L	H	H
H L	L	L	H	H
H H	L	L	H	L

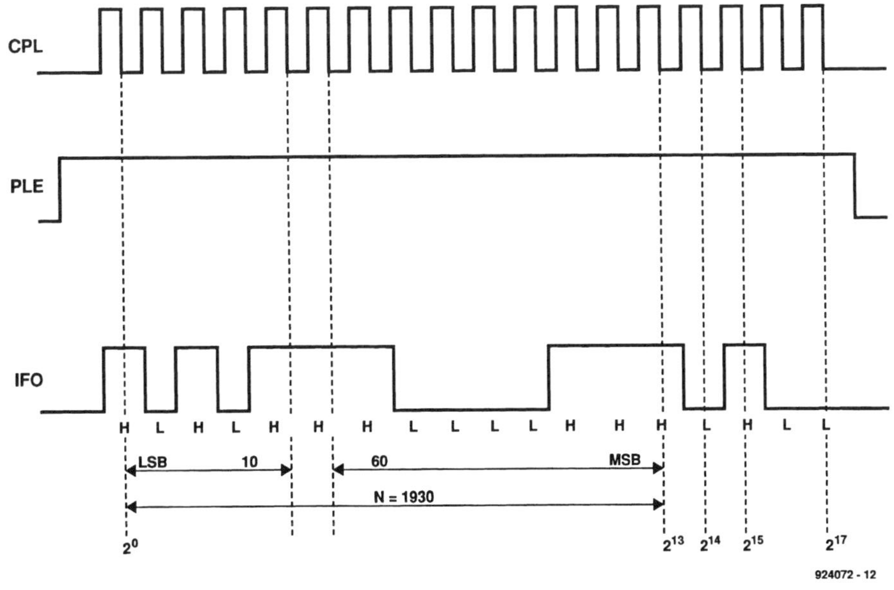

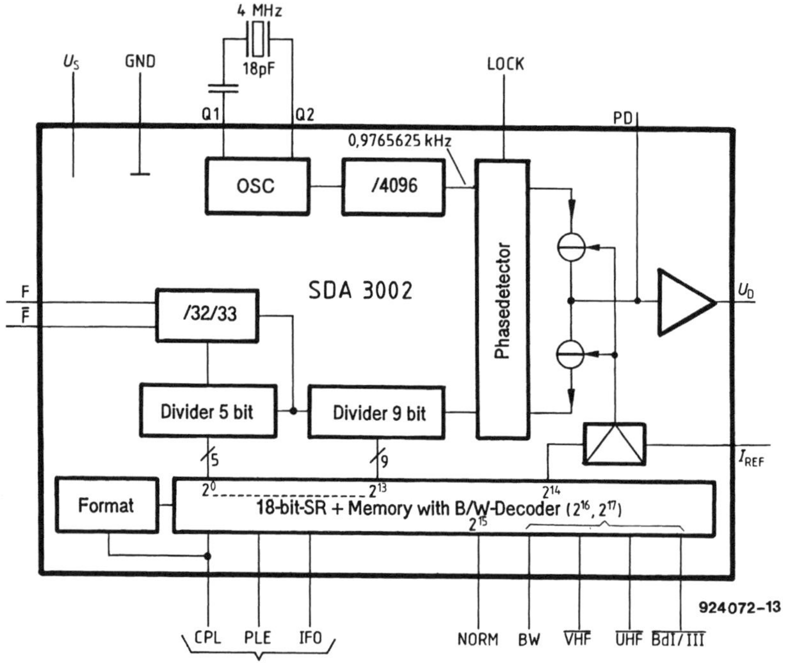

Video demodulator

131 The demodulator is based on a Type TDA8341 chip, which is the successor of the well-known Types TDA2541 and TDA3541. Apart from a demodulator, the chip contains an AGC (automatic gain control) section for tuners whose AGC voltage is directly proportional to the gain, and an AFC (automatic frequency control) facility. In the present design, the AFC is used merely as a tuning indicator.

The control range of the AGC is 67 dB. Because of internal supply regulation, the input sensitivity of the IC is 40 µV.

The IC provides a video signal at a level of 2.7 V at pin 12, from where it is passed to a Toko low-pass filter, Fl_1, which removes any residual carrier frequencies from the video signal. The filter has an attenuation of 6 dB over its pass-band. To obtain a standard video signal of 1 V_{pp} into 75 Ω, the filter may be followed by a video amplifier/buffer, for instance, a Type NE592. Note that the output impedance of the filter is 1 kΩ.

The AGC operating point is set with P_1. The AGC control voltage is taken from pin 4 via R_2. The level of the current through the AGC output is limited to about 10 mA. Network R_7-C_{11}-C_{12} forms an AGC detector, which also provides pulses for a sample & hold circuit. That circuit ensures that no video information is present in the AFC output.

The reservoir capacitor for the S&H circuit is C_6. If pin 6 is connected directly to earth, the AFC is disabled and the potential at pin 5 is then roughly half the supply voltage.

The AFC requires a synchronous demodulator with its own tuned circuit: L_2-C_8.

Because of (parasitic) capacitive coupling (C_9-C_{10}) with the tuned circuit for the reference amplifier, L_1-C_7, the skirt of the AFC characteristic becomes steeper. Since the AFC voltage here provides a tuning indication, a steep characteristic is not really desirable, so that C_9 and C_{10} must be kept as small as possible. The tuning

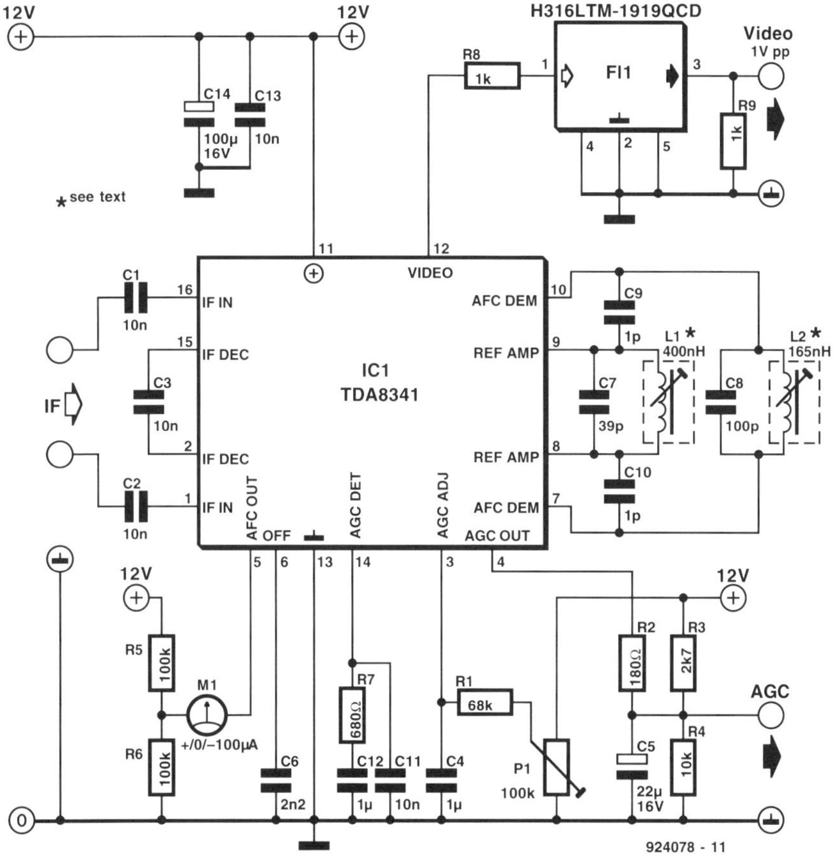

924078 - 11

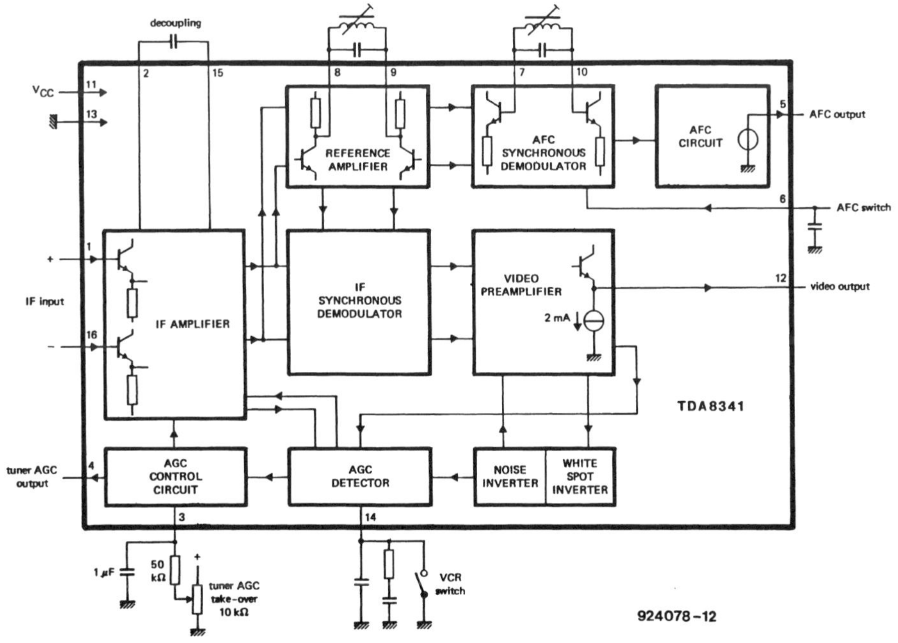

924078-12

indicator is formed by centre-zero meter M_1. Tuning the reference circuit (to remove any residue of the carrier) is tricky and can really be done properly only if an r.f. analyser or modulated r.f. generator and frequency meter are available. If crystal-controlled PLL (phase-locked loop) tuning is used, however, the adjustment can be carried out without this test equipment.

Once the channel tuning and the adjustment of L_1-C_7 are correct, the AFC circuit, L_2-C_8, may be tuned for centre-zero reading of the meter (half supply voltage at the AFC output). Resistors R_5 and R_6 hold the other terminal of M_1 at half the supply voltage and at the same time limit the current through the meter to about 100 µA.

Because of the circuit layout, there may be cross-talk between the video output and the i.f. input. This may, however, be cured by connecting a 6.8 µH choke in series with pin 12 (as close to the IC as feasible).

The demodulator draws a current of about 45 mA.

The input signal may be taken from a SAW (surface acoustic wave) filter, which is readily available nowadays. Most tuners – although this must be checked – are capable of driving a variety of SAW filters.

Video enhancer

132 The enhancer amplifies the high frequencies of a video signal, resulting in a sharper picture. It may be inserted between, say, the video recorder output and the SCART input of a television receiver.

The simple design is based on only three transistors. The first, T_1, is a buffer. Resistor R_1 ensures that the input impedance is of the order of 75 Ω. The signal is then applied to amplifier T_2, whose gain is determined by the setting of P_2.

The frequency characteristic of the signal at the base of T_2 is shaped by P_1, R_6, and C_6, and is, therefore, to a certain extent under the control of the user (by P_1).

Buffer T_3 provides sufficient current for correctly driving most 75 Ω loads.

Preset P_2 must be set to give an output voltage of 1 V_{pp} (terminated output; for an open-circuit output, the level should be 2 V_{pp}).

The enhancer draws a current of about 50 mA. Note that the 12 V supply should be regulated.

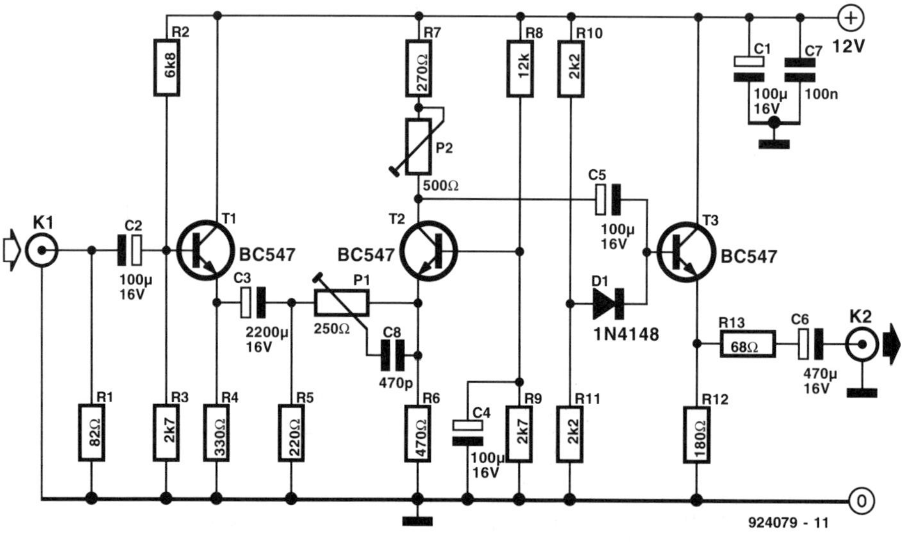

Preamplifier for Kalundborg frequency reference

133 In all cases where reception of Kalundborg is marginal using the ferrite rod at the input of the S042P, the two-stage preamplifier shown here can help you ensure longer 'lock' periods of the frequency standard.

The antenna inductor, L_1, consists of 150 turns of 0.3 mm dia. enamelled copper wire on a 12.3 mm dia plastic, thin cardboard or paper former, positioned centrally on a ferrite rod with a length of 238 mm and a diameter of 9.6 mm. The inductor is resonated at 243 kHz by a fixed capacitor, C_9, and a trimmer, C_{10}. To keep the load on the antenna inductor, as small as possible, the first stage of the preamplifier is formed by a FET Type BF245.

The amplfied 243 kHz signal is fed to one of the inputs of the S042P at the input of the frequency standard (see previous article). Using the preamplifier, the author (who lives in Denmark) measured a signal level of 1.4 V_{pp} on pin 6 of IC_2 in the frequency reference.

The noise level normally measured after Kalundborg switches off at 0.30h CET is about 100 mV_{pp}.

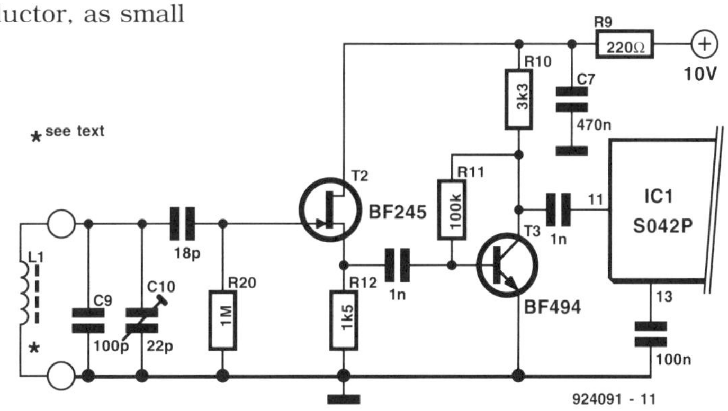

Wideband active telescopic antenna

134 Most VHF/UHF amplifiers for use with a symmetrical antenna (such as an open or closed dipole) have a balun at the input. A balun (short for 'balanced to unbalanced') is an inductive device that converts the symmetrical (balanced) RF signal into an asymmetrical (unbalanced) signal that can be applied to the base of a transistor. Unfortunately, baluns have an inherent loss of 2–3 dB, while most input transistors (typically a BFR91 or similar) have noise figures not better than about 2 dB.This explains the rather poor overall noise figure of 4–5 dB of this type of input stage.

A much lower noise figure is achieved by the circuit shown here, which does not incorporate a balun, and uses the BFG65 low-noise transistor as the amplifying device. The combination of a telescopic rod and a low-noise wideband RF amplifier is referred to as an 'active antenna'. The design shown here has two options: (1) if used with an existing antenna, it ensures a much better S/N (signal-to-noise) ratio in the receiver, or (2) the same S/N ratio can be achieved using a much simpler antenna.

The antenna proper is an open dipole with a total length of 1.6 m, which works as a

0.5 lambda dipole from 60 MHz to about 187 MHz, or as a multi-lambda V-dipole up to about 900 MHz. Unusually, the balanced-to-unbalanced conversion is done at the output of the amplifier, with the aid of a length of coax cable that functions as a kind of choke. The construction of the amplifier on the PCB shown, and its electrical behaviour, enables the antenna to 'see' a balanced load.

The two telescopic rods are connected directly to the solder pads marked 'ANT.1' and 'ANT.2'. Usually, telescopic rods have a kind of 'knee' construction at the base that enables them to be rotated as well as bent up and down. With some dexterity, this mechanism may be retained for use with the present amplifier.

The output inductor, L_1, consists of 10 turns of 2.5 mm dia. 60 Ω coax cable wound on a 10 cm long ferrite rod with a diameter of 10 mm.

The phantom supply for the amplifier is not contained on the printed circuit board, but is simple to install at the input of the receiver, since it consists of a resistor and a capacitor only. If the receiver is located further than about 2 m away from the active antenna, it may be connected to K_1 via a length of ordinary coax cable like RG58.

Provided the antenna is used in an area with relatively low local field strengths, you may lower the noise figure of the BFG65 by increasing R_3 a little. This should not be done, however, if there are strong signals around, in which case the result is an increased risk of cross-modulation.

With R_3 = 560 Ω, the current consumption is about 20 mA. The gain of the dipole signal amounts to about 12 dB, while a noise figure of about 1 dB is achieved. This will ensure results comparable to those of a much larger antenna (yagi), provided signal reflections and multipath reception are not a problem.

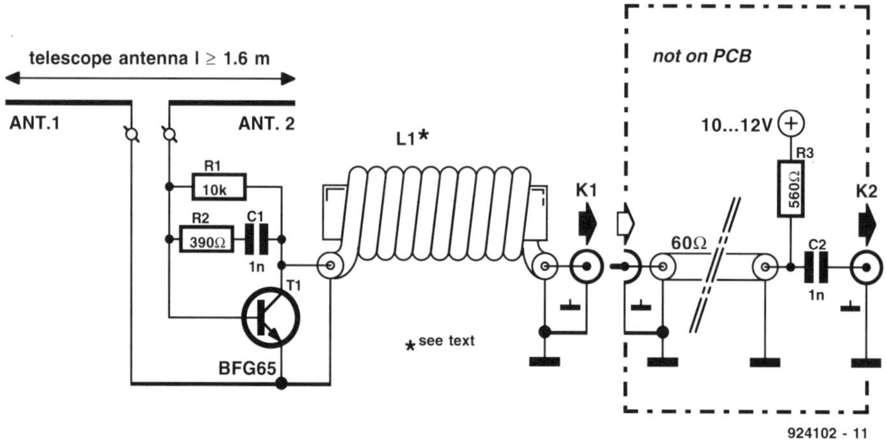

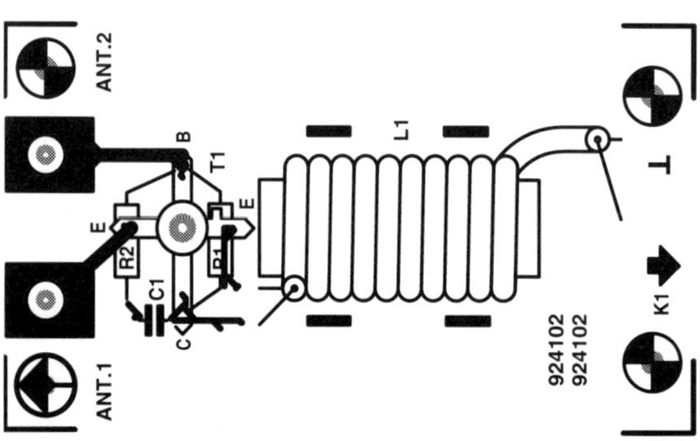

Parts list

Resistors:
R_1 = 10 kΩ
R_2 = 390 Ω
R_3 = 560 Ω (see text)

Capacitors:
C_1 = 1 nF ceramic
C_2 = 1 nF ceramic

Semiconductors:
T_1 = BFG65

Inductor:
L_1 = see text. Materials: 2.5-mm dia. 60-Ω coax cable; 10-mm dia. ferrite rod.

Miscellaneous:
ANT.1;ANT.2 = telescopic rod antenna for PCB connection.
Min. length = 80cm.
K_1 = BNC socket.
PCB Ref. 924102.

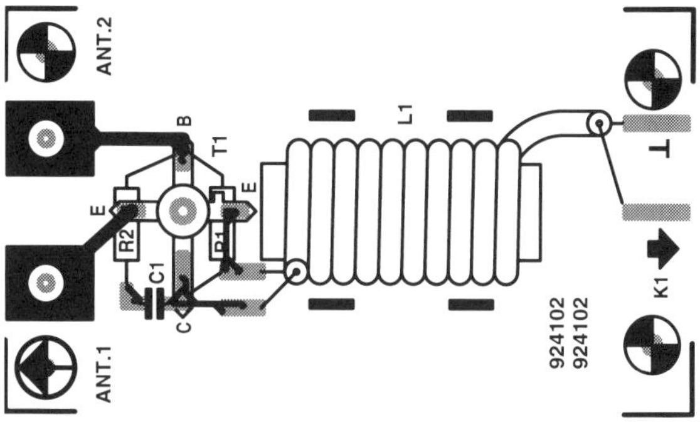

Ferguson BSB receiver as D2MAC decoder

135 After the demise of BSB (British Satellite Broadcasting) in early 1991, the electronics surplus market was flooded with tens of thousands of brand new, but otherwise unusable, satellite TV indoor units plus 55-cm dish antennas or squarials. Initially, unmodified ex-BSB sets went at give-away prices of £20 and less, which is a fraction of the original price. Unfortu-nately, these receivers are designed to handle DMAC only, a TV transmission standard used by BSB and a couple of Scandinavian broadcasters only. Not surprisingly, the 'scrapped' BSB sets were soon modified to handle the more widely used D2MAC format, and then sold as 'upgraded' receivers. One of the companies that has been actively involved in this is Trac Satellite Systems (address below). The modification from DMAC to D2MAC essentially consists of fitting a new EPROM, reprogramming the receiver's built-in EEPROM, and disabling the access control module (ACM). The result is a complete D2MAC satellite receiving system that costs less than a D2MAC decoder alone. The pictures and sound produced by the modified sets are, in a word, superb.

In addition to complete, modified BSB receiver sets, there are also many 'scrapped' receiver boards around. This article shows you how to use such a board, in this case from a Ferguson SRB1 receiver, as a D2MAC decoder for use with an Amstrad SRX200 Astra receiver. The conditions for being able to do so are: (1) your Ferguson board is in working order, and (2) it has been modified for D2MAC.

One transistor does the job! The interface shown here consists of an inverter, which is required to ensure the correct polarity of the baseband signal fed into the D2MAC decoder. The baseband signal is taken from pin 4 on the 15-way sub-D socket on the rear panel of the SRX200, and fed to a simple transistor inverter. Switch S_2 is added to enable the tuner inside the Fergu-son receiver to be used as a second input to the D2MAC decoder. This allows you to keep using the Ferguson set for reception of the D2MAC channels on the German and French high-power DBS channels at 19° West. A second switch, S_1, selects the polarity of the video signal

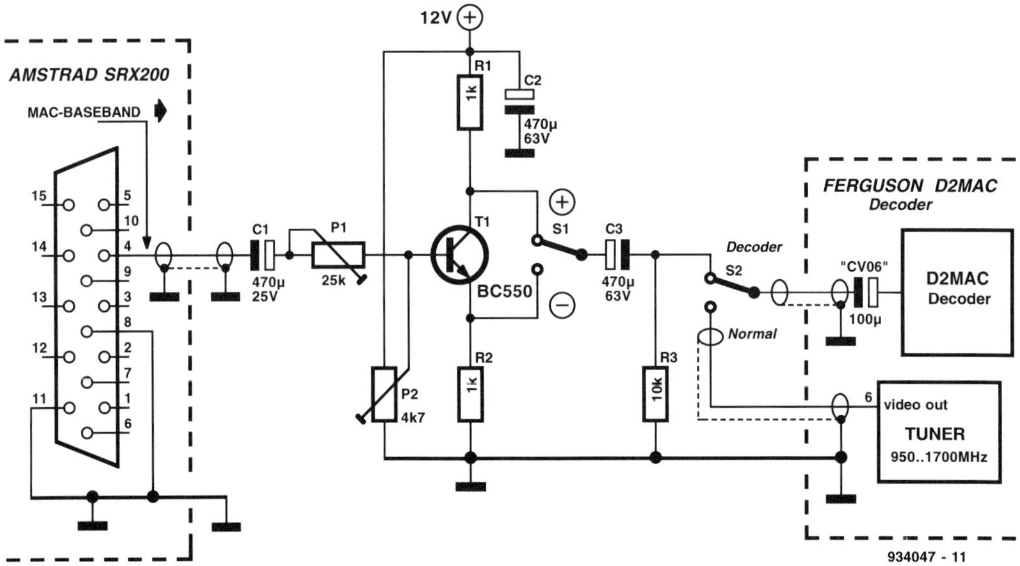

(which can give quite interesting effects on some D2MAC broadcasts). Finally, the interface should be connected with short coax cables. If you cannot avoid the use of a relatively long cable (for instance, a standard length SCART cable), fit a 10 pF speed-up trimmer across the fixed terminals of P_1.

Converted SRB-1 receiver systems are available from Trac Satellite Systems, Commerce Way, Skippers Lane, Middlesbrough, Cleveland TS6 6UR. Telephone: (0642) 468145. Fax: (0642) 440927.

Infra-red a.f. receiver

136 This infra-red receiver is the counterpart of the infra-red transmitter. The output from the infra-red diode D_1 is amplified by IC_1. The diode is a Type BPW41N which works well with the transmitter and is fairly fast (≈ 200 ns). The output pulses of IC_1 are taken to comparator IC_2, which separates them from the ambient light. To that end the d.c. component of the output of IC_1 is stored in C_9 and used as reference for IC_2.

The received signal is decoded by a simple binary scaler, IC_{3a}. Each pulse results in a change of level at the output of this stage, so that the output resembles the original pulse-duration-modulated (PDM) signal.The output is filtered by R_7-C_{11}, decoupled by C_{12}, resulting in an audio signal that is suitable for driving high-impedance headphones.

The binary scaler stage contains a delay network, R_6-C_{10}, which prevents the scaler being clocked by ambient noise. Owing to this network, the previous level is retained at the D input for a microsecond during which the state of the clock input is immaterial.

The output signal may be enhanced by an active 3rd or higher order low-pass (10–12 kHz) filter, which may also serve as a buffer.

In combination with the infra-red

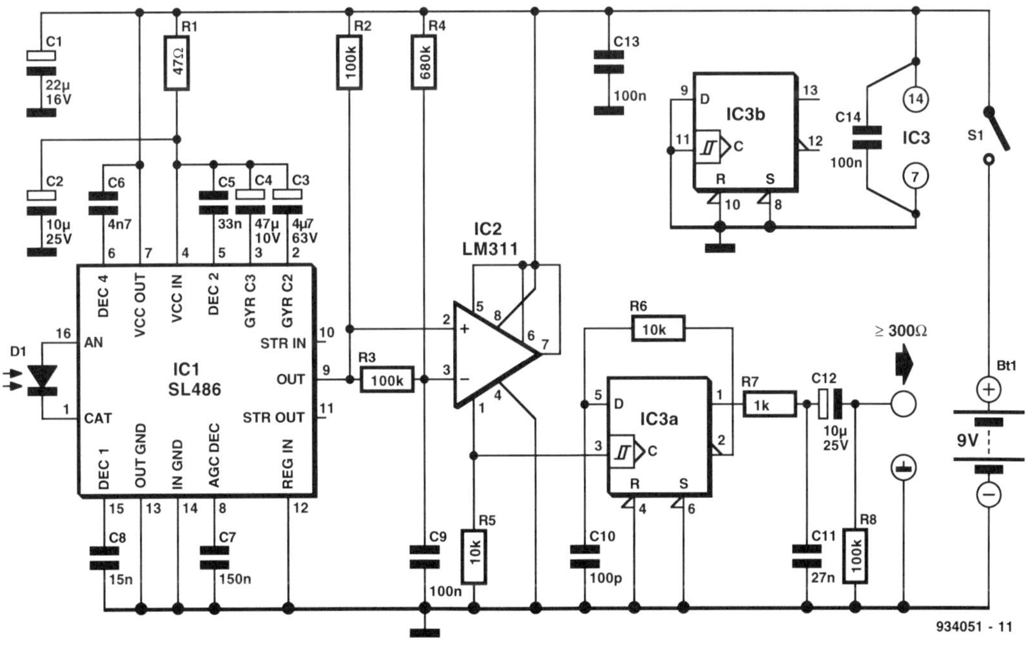

transmitter, and ignoring the residual high-frequency components, the audio signal is virtually undistorted. With the receiver at a distance of 1 m (3.3 ft) from the transmitter and 30% modulation, the signal-to-noise ratio is about –50 dB. The maximum distance the prototypes worked without interference was about 5 m (16 ft). The circuit draws a current of some 15 mA, which makes it advisable to use a rechargeable (NiCd) battery.

Infra-red transmitter

137 It often happens that someone wants to watch television without the sound disturbing other people in the room. To avoid having a long cable trailing over the floor, the present transmitter (and accompanying receiver) make it possible to listen to the TV sound 'wirelessly'.

The transmitter uses pulse-duration (=pulse-width) modulation. The modulated signal is produced in the traditional manner of having the audio signal set against a pure high-frequency triangular signal in a comparator. A suitable triangular signal generator can be found in Chapter 6. If another generator is used, make sure that its off-set is equal to half the supply voltage of 5 V and its peak value is 2.5 v_{pp}.

For a good range, the current through the LEDs must be large. Since, however, LEDs should not draw large continuous currents, the pulses must be short, and this is why PDM is used. (In this type of modulation, the time of occurrence of the first and last transition, that is, the leading and the trailing edge, is varied from its unmodulated position).

The pulses are generated by XOR gate IC_{2d}, which compares the original PDM signal with one delayed by R_5-C_3-IC_2. This results, for every level change, that is transition, into a short pulse whose width is equal to time constant R_5-C_3.

The output signal of IC_{1d} switches T_1. In this way, the current drawn by the LEDs (with a supply voltage of 10 V) is restricted to a peak value of 400 mA with the aid of R_6. The average current drawn by the circuit is then some 90 mA.

Presets P_2 and P_3 serve to remove any differences between the pulses caused by a leading edge and those resulting from a trailing edge. In the absence of an input signal, set P_3, with the aid of an oscilloscope, so that all pulses in the output signal have the same width. Then set P_2 so that the pulse spacings are equal.

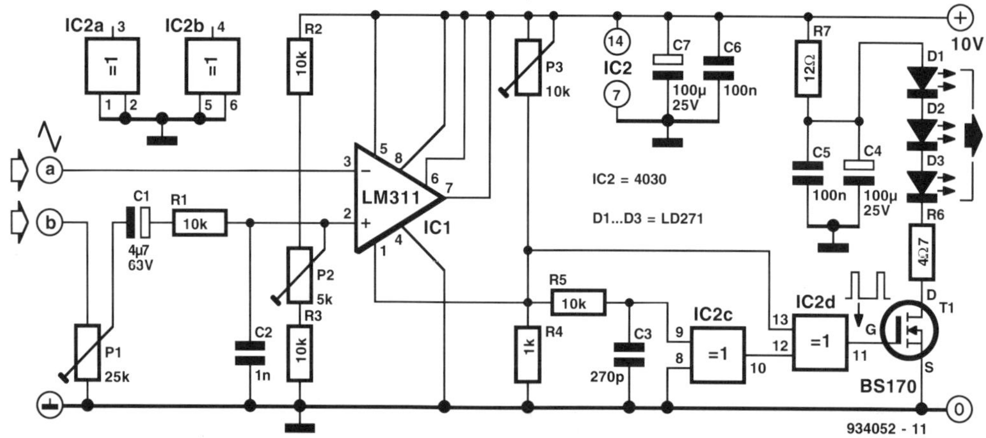

The output of IC_1 should then be a perfect square wave.
With the receiver used in combination with the transmitter (with maximum input), set P_1 to obtain the lowest possible interference on the received signal
Finally, note that IC_{2a} and IC_{2b} may be used for the triangular wave generator.

Optical squelch

138 Many of the less expensive short-wave receivers have no squelch circuit and continue to emanate noise when no usable signal is coming in. Fortunately, they normally have an LED which lights to indicate that the level of an incoming signal is at or above a predetermined limit. This LED can be used to control a light dependent resistor (LDR). The potential across the LDR can serve to actuate a circuit that switches the loudspeaker or headphones off in the absence of a suitable signal. This is possible without any work in the receiver itself by connecting an external loudspeaker to the headphone output. Inserting the plug into this socket automatically switches off the internal loudspeaker. The connection to the external loudspeaker is via the contact of relay Re_1.
The circuit is based on comparator IC_1. P_1, R_1 and R_2 provide the reference voltage at the inverting input of the opamp, which in the quiescent state (when no light falls on to R_4, which then has a high resistance) is higher than the potential at the non-inverting input. When light falls on to R_4, its resistance decreases, whereupon the level at the non-inverting input of the comparator becomes higher than that at the –ve input. The output of the comparator then changes state and switches on T_1, whereupon the relay is energized. The relay contact then connects the loudspeaker to the headphone output.
During construction it should be borne in mind that the less ambient light falls on to the LDR, the better the squelch will perform.
Preset P_1 should be adjusted so that the relay is not energized when the LED is out, but is actuated as soon as the diode lights. If difficulties arise in the setting of P_1, they are probably caused by the LDR having an incorrect value. In the diagram, the values of R_1, R_2 and P_1 are suitable for use with an LDR that has a resistance when light falls on to it of about 6 kΩ. Measure this resistance with a multimeter; the value of R_1 should be made (about) equal to this, and that of P_1 about twice as high. The value of R_2 and R_3 should be roughly the sum of the values of P_1 and R_1.
The power supply may be taken from the receiver; if that is not possible, a separate mains adaptor should be used. The circuit draws a current of about 5 mA plus the relay current.

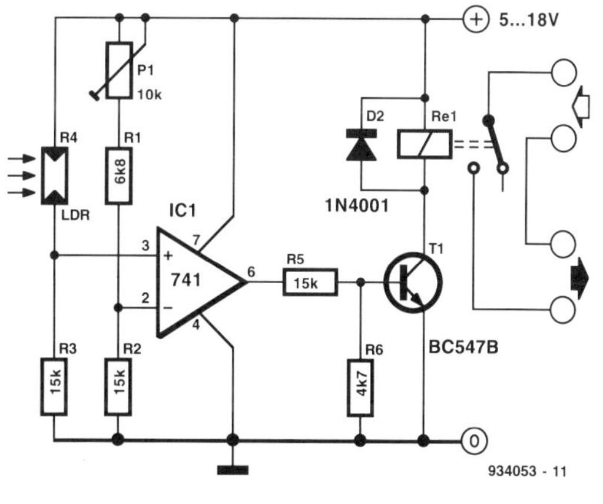

Infra-red controlled remote switch

139 Nowadays, several manufacturers produce ICs that combine an infra-red receive LED with associate amplifier and demodulator. The Type SFH505A from Siemens, used in the present circuit, has in addition a band-pass filter to minimize any interference. It is best used in conjunction with the infra-red transmitter in the next article.

The output of IC_1 is limited to a maximum pulse width. This is used to advantage by the associated transmitter to span as long a distance as possible. Most commercial remote controls use modulated data transfer, in which the pulse width is smaller than IC_1 can handle. The difference is easily detected by an integrator, R_2-C_2, and a Schmitt trigger, formed by the –T input of monostable IC_{2a}. The time constant is just short enough to ensure that at the maximum pulse width of IC_1 the voltage across C_2 is just under the trigger threshold of IC_{2a} (IC_1 has an active low output, which is why it is connected to the negative trigger input of IC_{2a}). Note that remote control units of, for instance, Sony or Philips, have no effect on the present circuit.

The monostable is retriggerable, so that any bounce at the transmitter does not affect the wanted state of the circuit. The mono time has been set at about half a second. The circuit cannot, therefore, switch on and off rapidly. If, nevertheless, several or repeated pulses arrive at the trigger input, they cause a lengthening of the output pulse, but the state of the circuit gets changed only once.

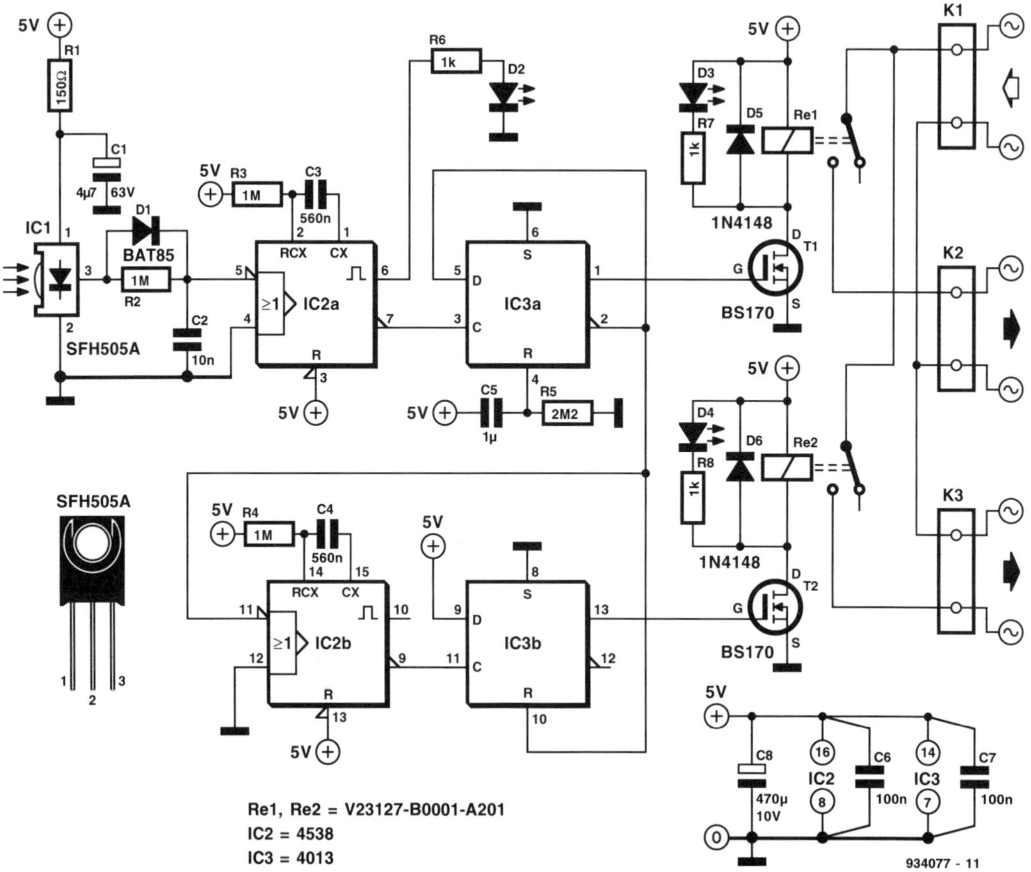

The output at pin 7 of IC_{2a} clocks D-type bistable IC_{3a} at the last trailing edge. The bistable is arranged as a binary scaler, so that the circuit can be switched on and off by repeated transmissions.
Diode D_2 indicates that IC_{2a} has received a trigger pulse. If interference is experienced from other remote controls, this is indicated by D_2.
The output of IC_{3a} energizes one of the relays via T_1. At the same time, it triggers monostable IC_{2b}, which energizes the second relay via IC_{3b} and T_2. In this way, the second relay is energized a good half a second after the first. This enables several apparatuses to be switched on in two steps. Diodes D_3 and D_4 indicate when the relays are actuated. The relays can switch up to 2000 VA (8 A).
The switch-on current of an equipment may be limited by first switching on the mains voltage with one relay via a series resistor and use the second relay to short out the series resistor.
The circuit draws a current of about 0.6 mA when the relays are unenergized and the LEDs are out. When the relays are actuated and the LEDs light, this increases to 125 mA.

Basic infra-red transmitter

140 The transmitter is intended primarily for use with the receiver in the preceding article. It works from two small 1.5 V dry batteries or one 3 V lithium battery. To make the unit compact and yet have reasonable frequency stability, the design uses a phase-shift oscillator, T_1, which provides good frequency stability. To ensure that the transistor has sufficient amplification at the low supply voltage, it is a darlington.
Each of the branches of phase-shift network, R_1-C_1-R_2-C_2-R_3-C_3 has roughly the same time constant. To ensure that the branches do not affect each other unduly, $R_2 \approx 3.8 R_1$, and $R_3 \approx 3.8 R_2$. Because of this mutual influence, the gain of the amplifier is somewhat higher than the theoretical value of 18 dB.
The value of R_5 is a compromise. It needs to be low so as not to affect the phase-shift network (the output impedance of T_1 also affects the oscillator frequency). At the same time, it must not be too low, because that would increase the current and reduce the gain.
To lessen the influence of the base-emitter resistance of buffer T_2, this transistor is driven by T_1 via an independent resistor, R_6. The buffer is necessary to provide sufficient current for the LEDs. Owing to the low supply voltage, these diodes cannot be connected in series, and they are, therefore, driven independently via their own series resistor, R_7 and R_8.
The transmitter is switched on and off by connecting or disconnecting the supply voltage with S_1. A drawback of this is that current continues to flow as long as S_1 is pressed. Briefly pressing of S_1 is, however, sufficient to switch the receiver on or off.
The current drawn by the circuit depends on the supply voltage and on how long S_1 is

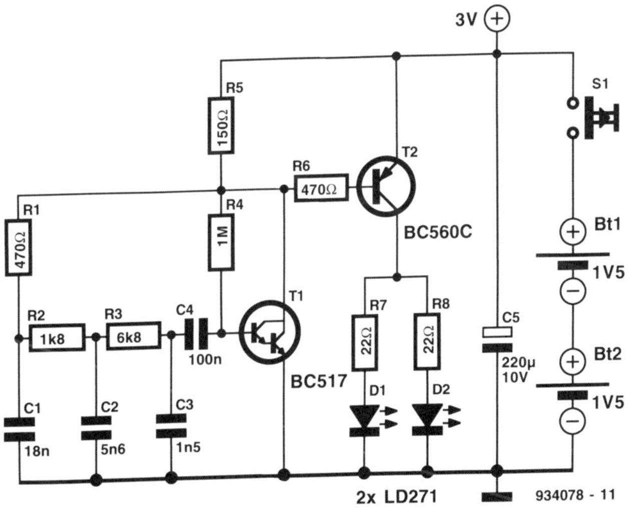

pressed. When U_b = 2 V, the frequency is 29.3 kHz, the peak current through each of the LEDs is 25 mA, and the total current drain is 27 mA. When U_b = 3.2 V, the frequency is 30.4 kHz, the peak current through each of the LEDs is 64 mA, and the total current drain is 63 mA.. At this supply voltage, the range between the prototype transmitter and receiver was 13 m (43 ft).

The current drain may be reduced by connecting in series with S_1 a parallel *RC* network (R = 10 kΩ; C = 1000 µF, 6.3 V). When S_1 is then pressed, only a brief current pulse ensues; even if S_1 is held down, the current does not rise above about 300 µA.

Frequency doubler

141 The circuit of the frequency doubler may be looked at from different angles. With input signals ≥1 V, T_2 and T_3 operate as full-wave rectifiers. This means that the fundamental frequency of the input signal is automatically doubled. With input signals <1 V, the two anti-phase signals produced by T_1 from the input signal are applied to the emitters of T_2 and T_3 and summed. This means that the fundamental frequency will virtually disappear, so that, because of non-linearities, only the harmonics remain: the first harmonic will then become the new fundamental frequency in the output signal. This means, of course, that the signal strength reduces appreciably: of an input of 25 mV, only 6 mV remain.
Assuming that the input is sinusoidal, the suppression of the fundamental frequency is optimized with P_1. The operating point of T_3 should be set with P_2 for as near a sinusoidal output as possible. In the protype, the output signal showed a distortion of 5.5% with an input signal frequency of 1 kHz.
The input frequency range is 80 Hz to well over 100 kHz.

Any tendency of T_2 or T_3 to oscillate may be suppressed by soldering a small ceramic capacitor (about 56 pF) between the base and collector of the transistor.
The circuit draws a current of about 4 mA.

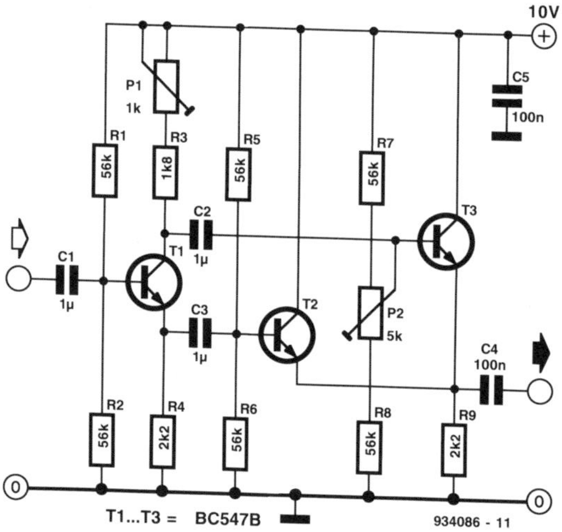

Infra-red signal repeater

142 The repeater is intended to extend the operating range of a standard infra-red (IR) remote control system used with an audio or video system to around 10 metres (33 ft). The original signal is picked up by the receiver (Fig. 1), amplified and then fed to the transmitter (Fig. 2) via a length of coaxial cable. It enables, for instance, control of an audio system located in the lounge from the dining room (where parallel speakers are fitted).
Although in Fig. 1 the IR receive diode is a Type BPW34, good results were also obtained with a BP104. Performance of the receiver

built on the PCB in Fig. 3 was found to be appreciable better than one built on a prototyping (Vero) board.
The receiver draws a current that depends to a large extent on the digital code used by the remote control system and lies between 30 mA and 100 mA. The supply lines to T_1 and T_2 should be kept as short as possible.
The range of the transmitter in Fig. 2 depends to some extent on the room in which it is used, but in normal circumstances should be around 10 metres (33 ft). The frequency of the output signals may be between 5 kHz and 200 kHz. Effective hum suppression is provided.
Care should be taken that optical feedback between the receiver and transmitter is avoided. When they are used in different rooms, that is, of course, no problem.

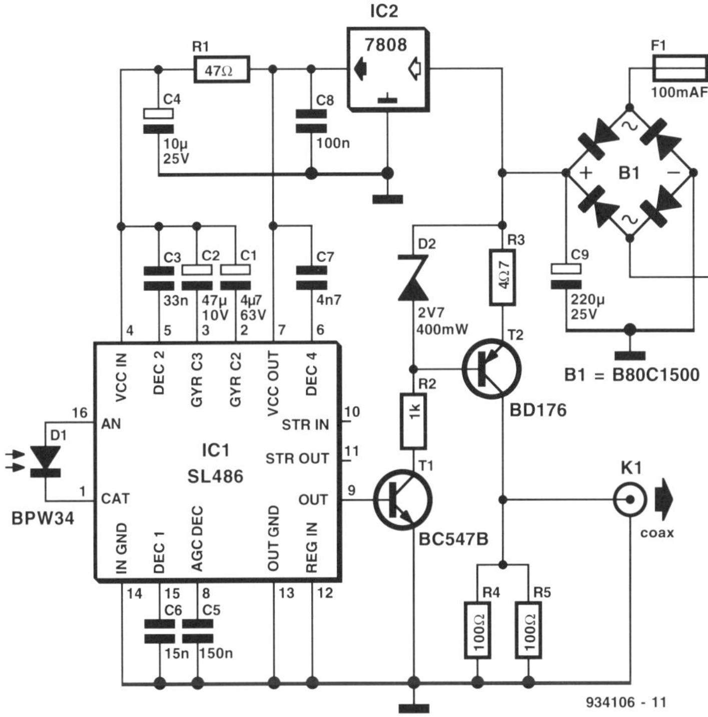

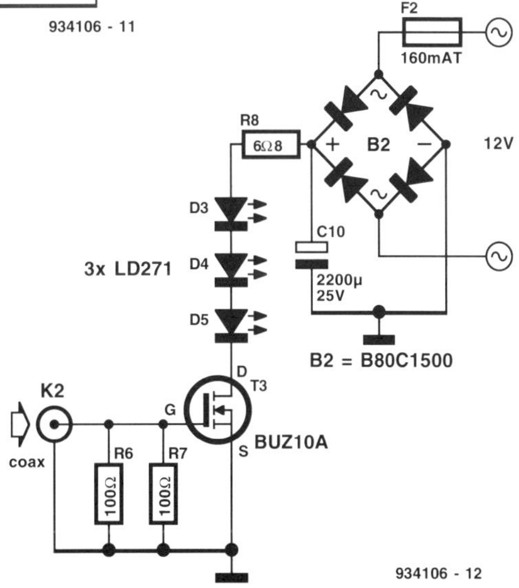

Video data change detector

143 The circuit detects a change in video information and uses this to switch on an alarm in a closed-circuit television guard system or a video recorder.

The circuit is usable only where the same picture is shown on the TV monitor for some time. When the picture changes, the average brightness alters and this is detected by a window comparator.

Capacitor C_1 prevents any d.c. component in the video signal reaching the circuit. Opamp IC_{1a} clamps the signal. On average, the potential at pin 3 of IC_{1a} is always equal to half the supply voltage.

The potential across C_2 is buffered by IC_{1b}. The output signal of this opamp is applied to window comparator IC_{1c}-IC_{1d} via presets P_1 and P_2.

The reference voltage for the comparator is held at half the supply voltage ($^1/_2\ U_b$) by R_8-R_9. Preset P_1 is adjusted to make the potential at its wiper slightly higher than $^1/_2\ U_b$, and P_2 to make its wiper voltage slightly lower than $^1/_2\ U_b$. The potentials at the wipers of P_1 and P_2 are applied to the comparator via fairly long (>10 s) time constants (R_6-C_3 and R_7-C_4). The smaller the set window, the faster the reactor will change state.

When the level of the input signal rises, the potential across C_2 decreases, whereupon the levels across P_1 and P_2 also drop. When the voltage at pin 9 of IC_{1c} drops below $^1/_2\ U_b$, the output of this stage goes high. This change of state, indicated by the lighting of D_2, is used to give an external apparatus a trigger or start pulse of up to 65 V via the open collector output of T_1. Resistor R_{13} serves as a current limiter.

When the level of the input signal decreases, a similar action follows, but in this case IC_{1d} changes state, which is indicated by the lighting of D_3.

When there is no change in the average input signal, T_1 remains off.

The circuit draws a current of about 12 mA.

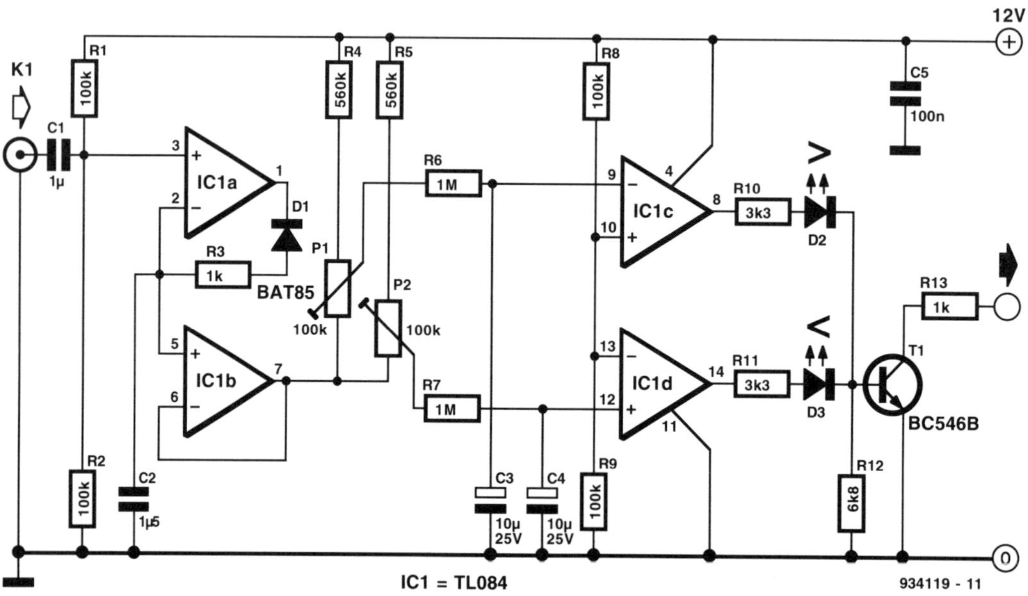

Inexpensive video enhancer

144 Often, a video signal applied to the TV receiver via the SCART or AV input appears rather less well-defined than a signal applied directly from the antenna socket. A considerable improvement may be obtained by connecting an inductor of about 10 μH in series with the input resistor of the SCART or AV socket (this resistor is normally 75 Ω or 82 Ω). The inductor raises the input impedance at higher frequencies, so that these are attenuated less than the lower frequencies. The improvement is particularly noticeable with signals of limited bandwidth (such as from a video cassette recorder): the picture is sharper and the colours are richer.

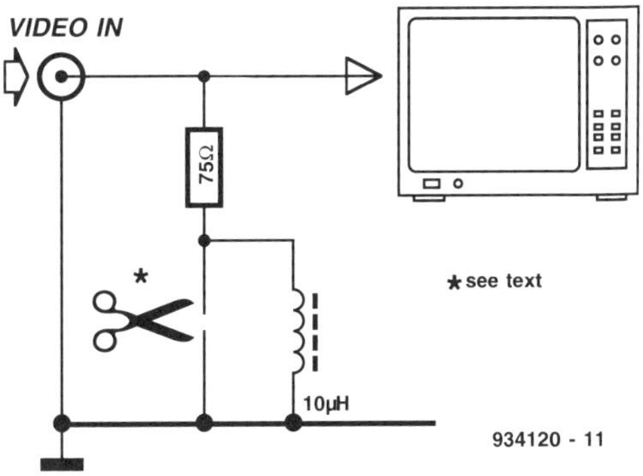

934120 - 11

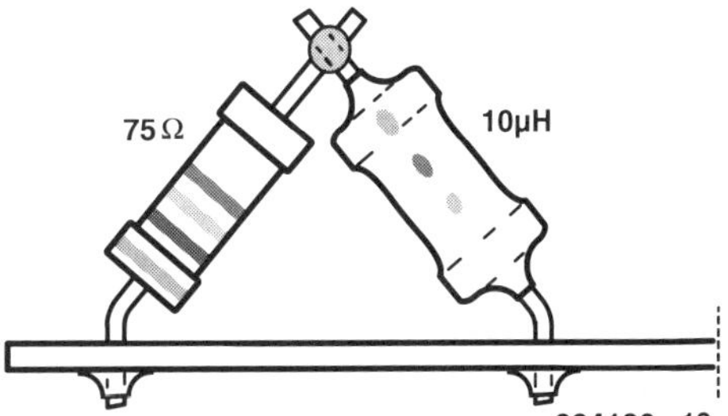

934120 - 12

Digital LED voltmeter

145 In this somewhat unusual digital voltmeter, the measurand (voltage to be measured) is digitized in an analogue-to-digital (A-D) converter and then displayed in three decimal digits. The display is not the usual seven-segment type, but consists of three groups of ten LEDs. Although this type of display is a little unusual, the measured value can be read without problem after only a short familiarization period: even voltage **changes** can be readily interpreted. Note that the meter can only be used for measuring direct voltages.

The A-D converter is based on a CA3162, which can process direct voltage up to 999 mV (1 V full-scale deflection–f.s.d.).The f.s.d. is extended to 10 V with the aid of potential divider R_1-R_2-R_3. Other ranges are possible by altering the values of the resistors.

The measured value is read from three bars of LEDs: the first one of these, D_1-D_{10}, shows units; the second, D_{11}-D_{20} tens; and the third, D_{21}-D_{30}, hundreds. The circuit is nulled with P_1 when the input is open circuit. Zero here means that diodes D_1, D_{11} and D_{21} light. Diodes D_{10}, D_{20} and D_{30} represent the figure 9.

Next, apply a known voltage to the input and adjust P_2 till the LEDs read the correct value.

Some people may find it helpful to use a different colour for each of the three groups of LEDs.

When the input voltage is too high, the display goes out. When the input is negative, the 'unit LEDs' do not light.

Note that variations in the supply voltage affect the measurement adversely; it is, therefore, advisable to use a regulated source, for instance, a 9 V PP3 battery and a Type 7805 regulator. Since only three LEDs can light at any one time, the current drawn by the circuit does not exceed about 30 mA.

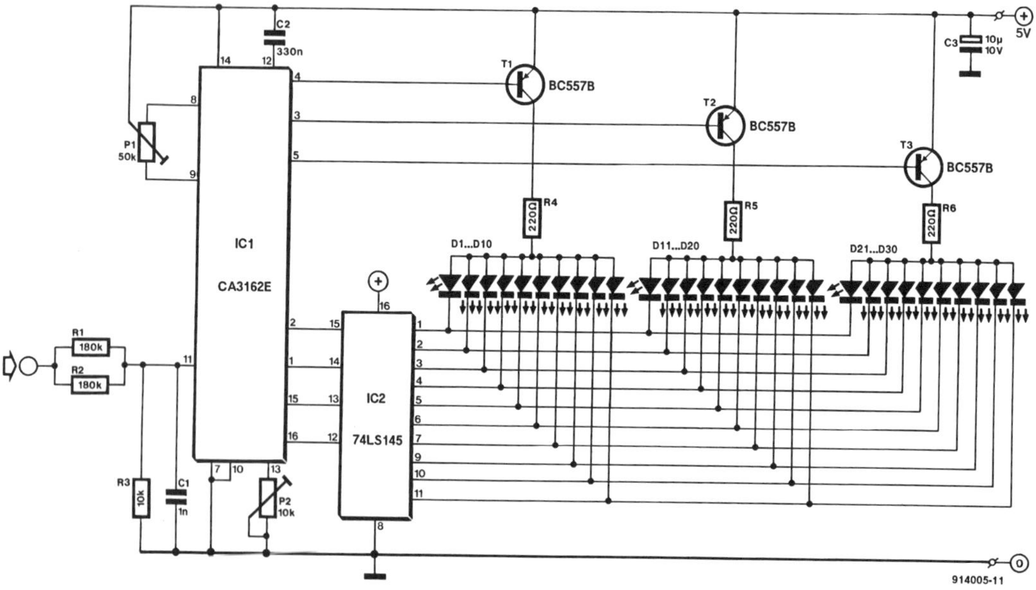

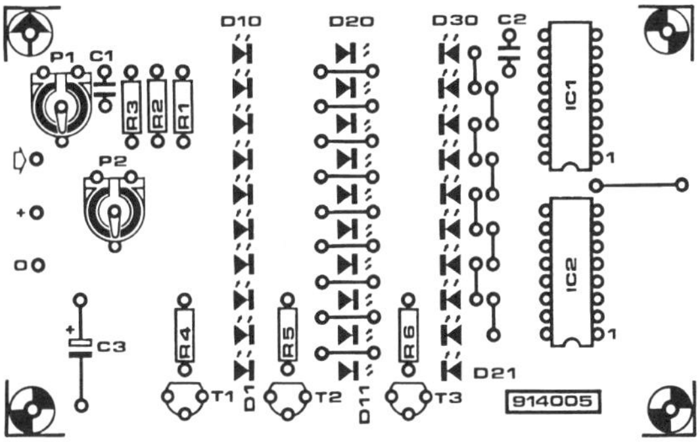

Parts list

Resistors:
R_1, R_2 = 180 kΩ
R_3 = 10 kΩ
R_4, R_5, R_6 = 220 Ω
P_1 = 50 kΩ preset
P_2 = 10 kΩ preset

Capacitors:
C_1 = 1 nF
C_2 = 330 nF
C_3 = 10 μF, 10 V

Semiconductors:
D_1–D_{10} = LED, red
D_{11}–D_{20} = LED, yellow
D_{21}–D_{30} = LED, green

Integrated circuits:
IC_1 = CA3162E
IC_2 = 74LS145

Miscellaneous:
Enclosure 70×125×48 mm

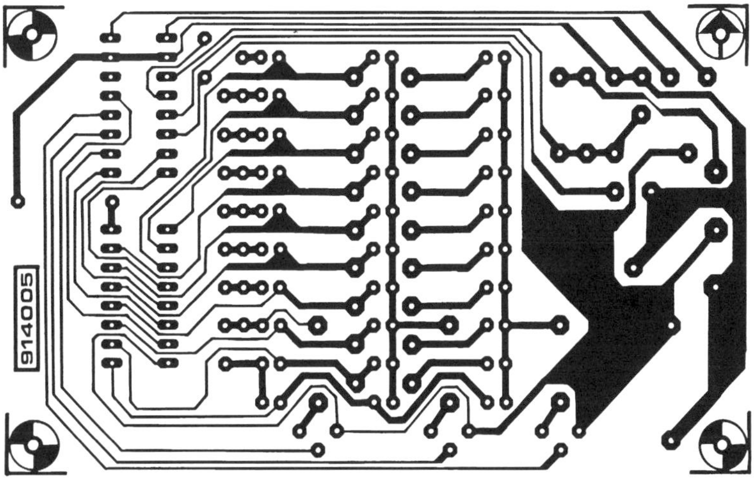

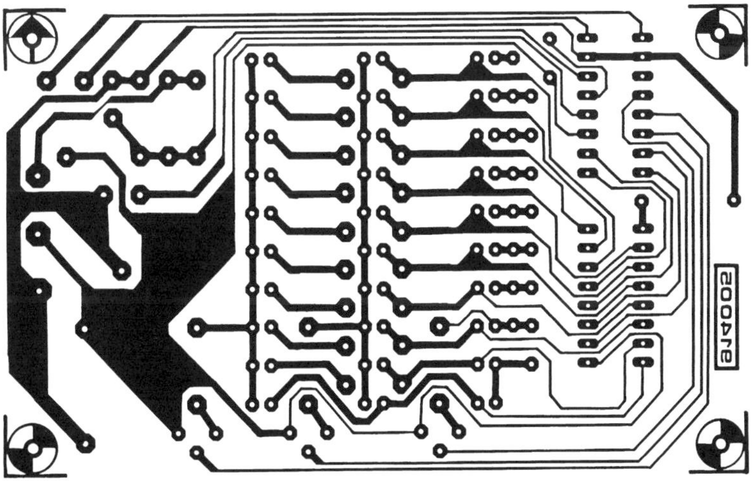

Wien bridge with asymmetrical power supply

146 Normally, a Wien bridge oscillator contains two identical capacitors and two identical (variable) resistors. That being the case, the transfer factor of the bridge in Fig. 1 is 1:3. For example, when a potential of 1 V is applied to the non-inverting input of the opamp, the output voltage of the amplifier will be 3 V.

In many cases, a smaller transfer factor is required. With reference to Fig. 2,

$$U_p/U_o = 1/(1+R_1/R_2+C_2/C_1)$$

from which it follows that the factor becomes smaller if the value of R_1 or C_2 is increased.

The frequency is altered when the value of both capacitors or of both resistors is changed. That makes it possible to vary the frequency by using a dual-gang potentiometer in place of the two resistors. Since the two resistances are then always identical, the ratio U_p/U_o will be 1:12 when C_2 = $10C_1$. To ensure sufficient positive feedback for the oscillator to start, the amplification of the opamp must be >12. With values as shown in Fig. 1, the amplification is

$$\alpha = 1+(R_5+P_1)/R_3 = 13.8.$$

Stability of the output voltage is ensured in the traditional manner by two anti-parallel connected diodes in the feedback loop. Preset P_1 is adjusted so that the sinusoidal output voltage is just not clipped by the supply voltage.

The frequency of the output signal may be set between 150 Hz and 1500 Hz with P_2; higher frequencies may be obtained by altering the values of C_1 and C_2.

The supply voltage, which must be regulated, may lie between 9 V and 12 V. When the oscillator is not loaded, it draws a current of about 6 mA.

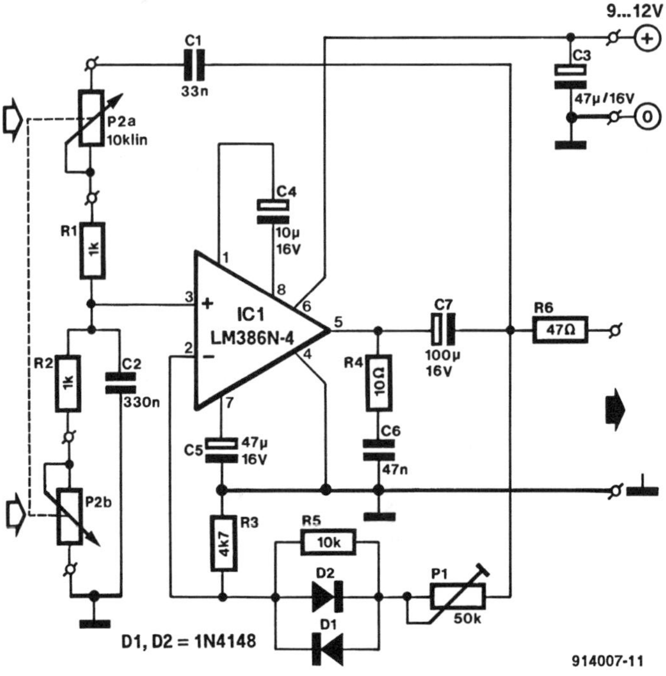

Parts list

Resistors:
R_1, R_2 = 1 kΩ
R_3 = 4.7 kΩ
R_4 = 10 Ω
R_5 = 10 kΩ
R_6 = 47 Ω
P_1 = 47 kΩ preset
P_2 = 10 kΩ, dual-gang
linear preset

Capacitors:
C_1 = 33 nF
C_2 = 330 nF
C_3, C5 = 47 µF, 16 V, radial
C_4 = 10 µF, 16 V, radial
C_6 = 47 nF
C_7 = 100 µF, 16 V, radial

Semiconductors:
D_1, D_2 = 1N4148

Integrated circuits:
IC_1 = LM386N-4

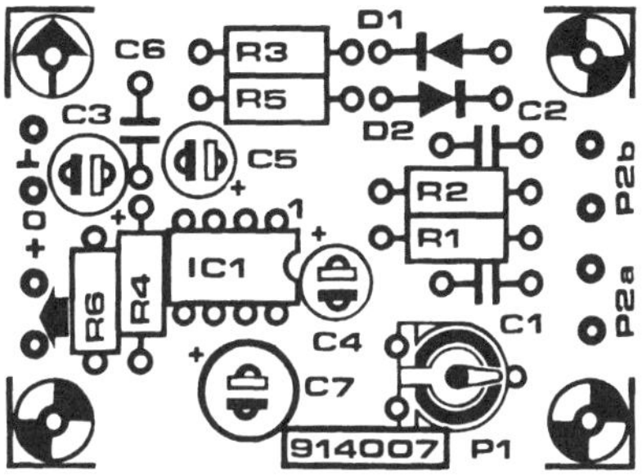

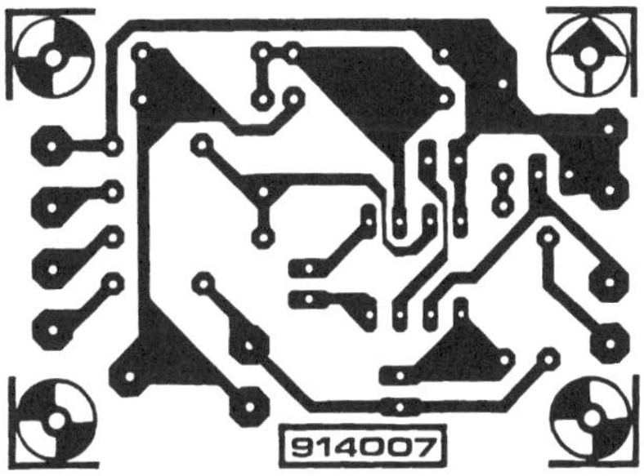

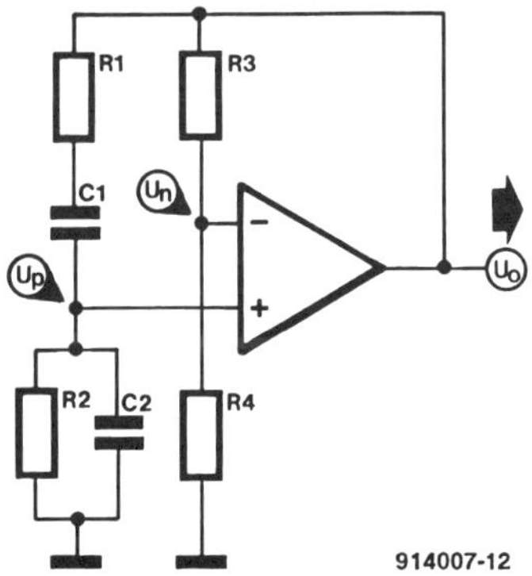

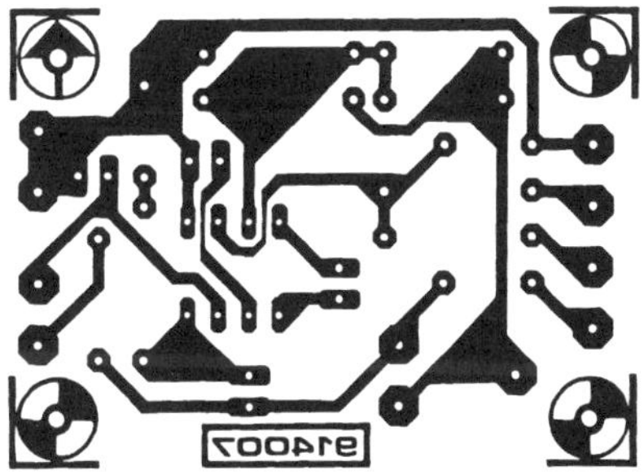

Remote temperature module for digital multimeters

147 National Semiconductor's LM334Z is a temperature-dependent adjustable current source supplied in a plastic TO-92 package. In Fig. 1, a 226 Ω resistor, R_1, is used to set a current gradient of 1 µA K^{-1}.

The remote temperature sensor is formed by IC_1, R_1 and C_1. Since its output is a temperature-dependent current, a simple two-wire connection may be used between the sensor and the DMM (digital multimeter) interface. Constant current

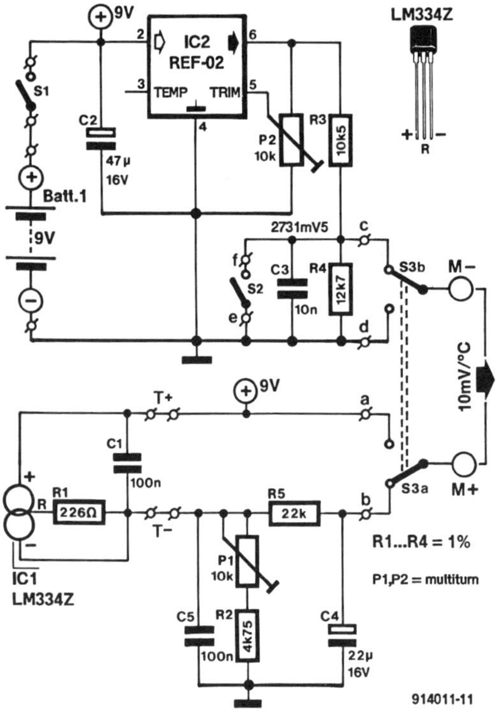

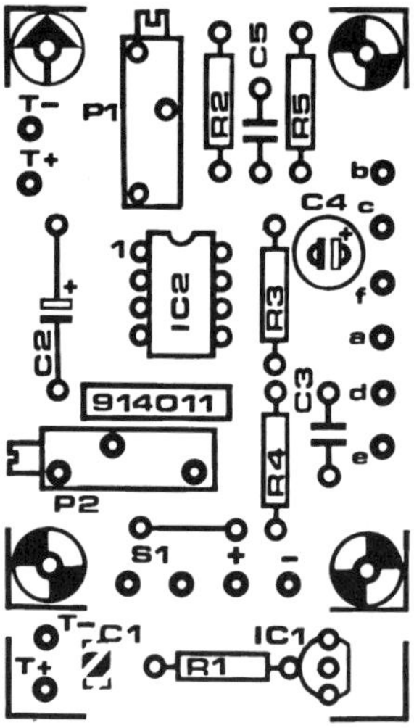

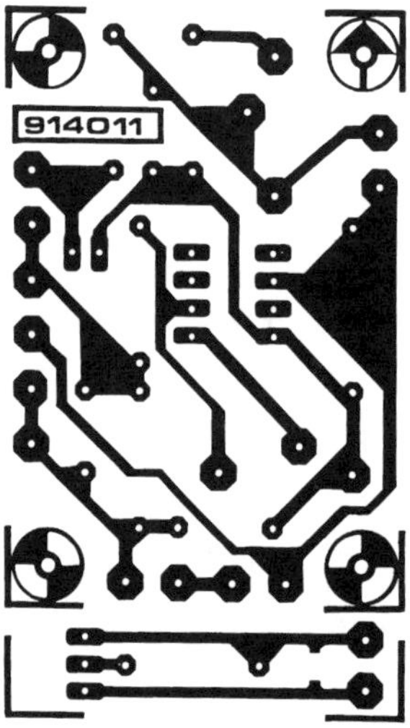

drive as applied here eliminates problems with voltage drop and expensive low-loss wiring associated with voltage drive. Also, the voltage drop across a relatively long cable is temperature dependent, which calls for a fairly complex compensation circuit. By contrast, when the sensor is a constant current source, the length and the total resistance of the wire between it and the interface at the DMM side have virtually no effect on the output signal. This obviates a compensation circuit, and allows the sensor to be fitted at quite some distance (up to 25 m = 80 ft) from the DMM using inexpensive wiring.

Components P_1 and R_2 convert the current supplied by the sensor into a voltage with a gradient of 10 mV K^{-1}. Capacitor C_1 suppresses high-frequency interference which may be picked up by the cable.

To prevent problems with ground levels, the current source must be powered by a separate 9-V battery as shown in the diagram. To allow temperature readings in degrees Celsius (°C), an adjustable high-stability voltage regulator Type REF-02 from Precision Monolithics Inc. is used to subtract a fixed amount of 2731.5 mV from the converter output voltage. This is achieved by 'lifting' the converter ground by 2731.5 mV (the REF-02 output voltage) when switch S_2 is opened. When S_2 is closed, the converter produces a temperature reading in kelvin.

The circuit is switched on and off by S_1. A quick battery condition check is available by switching S_{3a} to position 'a', and S_{3b} to position 'd'. Replace the battery if the DMM indicates less than 7.1 V.

Calibration of the converter is fairly simple. First, adjust multiturn preset P_2 until a voltage of 2731.5 mV is obtained across R_4 (open S_2). Next, adjust the temperature gradient (preset P_1) by comparing the DMM reading to that produced by a calibrated thermometer. Set the DMM to the 2 V range for degrees Celsius (°C) readings. An indication of, say, 0.217 V (on a $3^1/_2$ digit instrument) then corresponds to a measured temperature of 21.7 °C. Properly adjusted, the temperature sensor achieves a resolution of 0.1 kelvin. Finally, the current drain is about 2 mA.

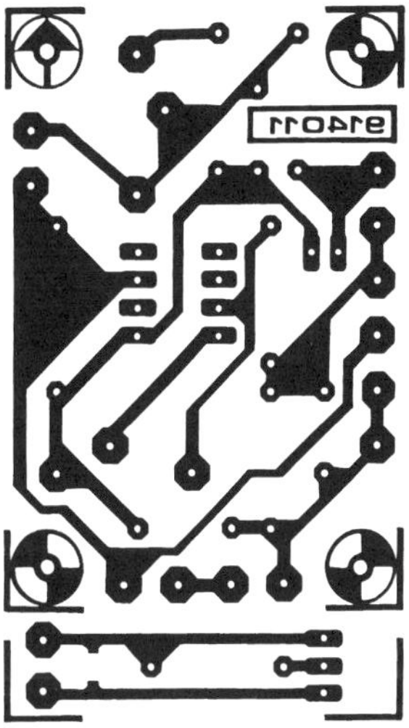

Audible tester

148 The tester, which is very useful for testing parts of electronic circuits, consists of an oscillator that generates a 1 kHz test signal and a detector that amplifies the detected signal which is then made audible by a small loudspeaker or buzzer. The tester draws only a small current so that it can be powered by a 9 V (PP3 or 6F22) battery.

Circuit IC_{1a} functions as a rectangular-wave generator whose frequency is determined by the time constant R_4–C_2. With values as shown, the frequency is about 1 kHz and this is hardly affected by variations in the supply voltage.

The oscillator signal is fed to the circuit on test via C_3, R_5, potentiometer P_1 and C_4. With a 9 V supply, the maximum voltage at the wiper of P1 is about 3.5 V_{pp} When S_1 is closed, the voltage at the output terminals is reduced to $^1/_{14}$th.

The measurand is input to the detector via sensitivity control P_2. The circuit is protected against too high input voltages by R_9, D_1 and D_2. After the signal has been buffered by IC_{1b}, it is applied to power amplifier IC_2 via C_6 and P_3. The signal is raised to a level that enables its driving a small loudspeaker or buzzer.

The prototype drew a current of 7 mA in the absence of a signal, and this is increased to nearly 200 mA with a strong input signal. The maximum drive level to the power amplifier, and thus the maximum current drain, is determined with P_3.

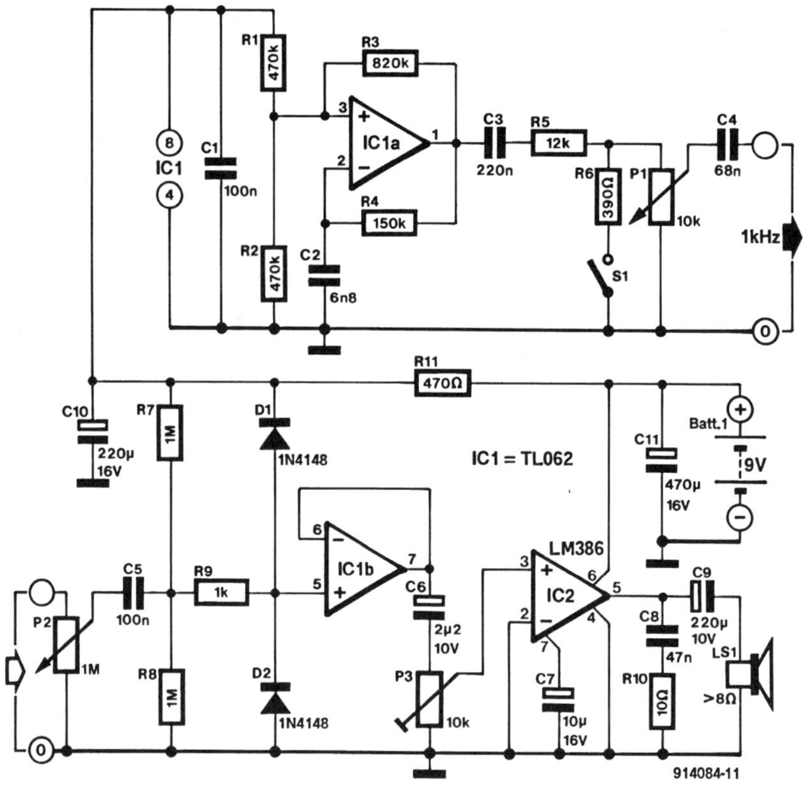

Pulse generator with one 4066

149 The diagram shows how the inexpensive and widely available 4066 quad bilateral analogue switch can be used to build a pulse generator with adjustable 'high' and 'low' times of the output waveform.

Assuming that switch IC_{1a} is open, the control input of IC_{1b} is logic high, and this switch is therefore closed. This results in low levels at the control inputs of IC_{1c} and IC_{1d}. Capacitor C_2 is charged via preset P_1, and C_3 via preset P_2. When the voltage across C_2 reaches a certain level, IC_{1a} is closed so that the control input of IC_{1b} is pulled low. The outputs of the circuit, ① and ②, are then logic high. ① has a swing of 5 V, and ② a swing virtually equal to the supply voltage (max. 15 V).

Meanwhile, switch IC_{1c} is closed, so that C_2 is discharged. Switch IC_{1a} is opened, and C_3 is charged via P_2. When the voltage across C_3 has reached a certain level, IC_{1b} is closed, and the outputs of the circuit change to logic low.

The'low' and 'high' times of the output waveform are adjusted with P_1 and P_2 respectively. With the component values specified, the 'low' time can be set between 136 μs and 3.75 ms, and the 'high' time between 15 μs and 330 μs. Other on and off times can be created by changing C_2 and C_3.

The circuit draws a current of about 8 mA at a supply voltage of 10 V. Note that ① produces a less than perfect waveform, and has a low fan-out. The other output, ②, is buffered and should be used for most applications.

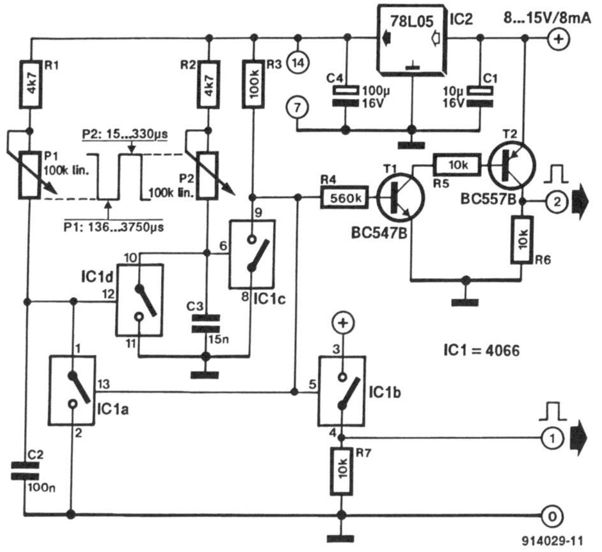

Battery tester

150 Various types of dry and rechargeable battery (with an e.m.f. of ≤2.7 V) can be tested with the circuit shown in the diagram. It is based on the well-known Type LM3914 LED-display driver from National Semiconductor. The circuit compares the e.m.f. of the battery with a reference voltage that is derived from an internal source. The reference potential (pin 8) can be set between 1.5 V and 2.7 V by R_1-R_2-P_1. The voltage at pin 8 refers to the maximum scale value of the LED series. That is, if that voltage is 1.5 V, each LED represents 150 mV. It is recommended to set it to 1.5 V for NiCd batteries, and to 2.0 V for dry batteries. Resistor R_1 arranges a current of 12.5 mA for each LED.

It is advisable to test dry batteries on load, since the terminal voltage depends on the residual capacity. And, of course, the e.m.f. of even an almost flat battery is still close to its specified value. Rechargeable

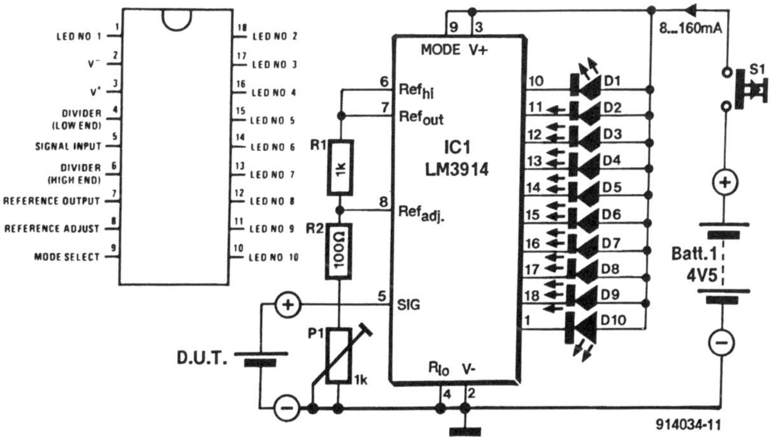

batteries retain their specified e.m.f. until they are virtually discharged, when it drops fairly rapidly. It is, therefore, of not much use to check the residual capacity of these batteries on the basis of their e.m.f.: the test is limited to an indication of whether the battery is fully charged or (nearly) flat.

Semiconductor tester

151 The tester can be used to test virtually any kind of semiconductor device, ranging from switching diodes to power transistors. In addition, it provides a rough indication of the gain ofbipolar transistors, and, more generally, can be a useful aid in finding functional, short-circuited and internally open devices in semiconductor batches.

The tester is based on a single CMOS IC and a bi-colour LED as a visual indication. Gate IC_{1a} forms an *RC* oscillator. The oscillator signal is buffered and made available in true and inverted form by the three remaining gates in the IC.

The bi-colour (red/green) LED indicates the direction of the current that is allowed to pass through the test probes or the device under test. Resistor R_1 functions as a current limiter.

The signals at the input and the output of gate IC_{1c} are applied to a pair of test probes, a two-terminal test socket for diodes, and a three-terminal transistor socket. The base current for the transistor on test (TUT) can be set with preset P_1. The preset may be calibrated with the aid of known, functional transistors to give an approximate gain scale.

Only one LED lights when a semiconductor is functional. The LED colour then indicates the polarity (n-p-n or p-n-p, or cathode/anode). When the component is internally open, no LED lights. A semiconductor with an internal short-circuit is easily recognized by the green and red LEDs lighting simultaneously at about equal intensity. Transistors must be connected with the base, collector and emitter pins to the indicated socket terminals, so check the pinout before running the test!

The circuit may also be used as a simple continuity tester. It draws a current of about 300 μA without a DUT (diode on test) or TUT (transistor on test) connected, and about 7.5 mA with the probes short-circuited.

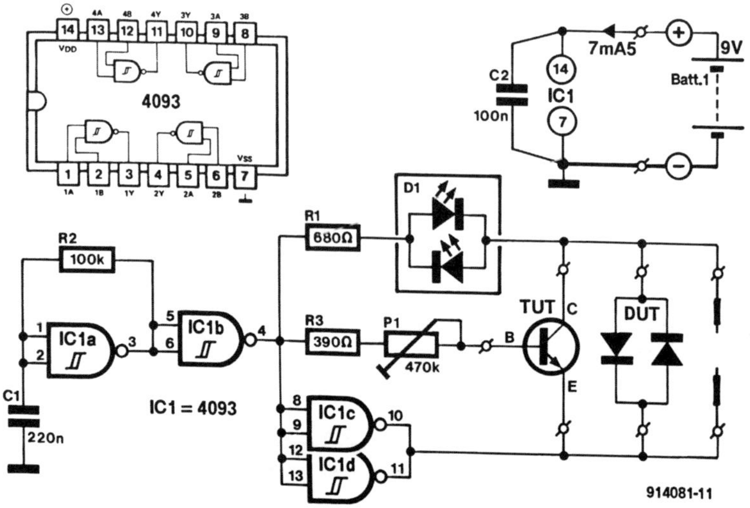

Triggered sawtooth generator

152 The sawtooth generator shown in the diagram may be used with an oscilloscope. It is linear, retriggerable, and is enabled automatically in the absence of a trigger signal.

When a positive pulse is applied to its input B, monstable IC_{3a} generates a sawtooth, which, owing to the 'standard' *RC* network, is not linear. If, however, the resistor in that network is replaced by a current source, consisting here of T_1, R_4, P_2, linearity returns. The period of the sawtooth depends on the settings of S_1 and P_2.

The sawtooth signal is buffered by T_2 to prevent its quality being affected adversely. The voltage level at the gate of T_2 is 0–3.5 V, that at the source is slightly lower. The rectangular signal at the Q output of IC_{3a} is used to suppress the electron beam during the flyback.

The trigger signal is provided by Schmitt trigger IC_{1a}, which has the small drawback that the input signal must be at least 1 V r.m.s. Preset P_1 provides compensation for the d.c. level at the input. Some experimentation with the values of R_1 and R_2 is advisable since these resistors determine the operating range of P_1.

As long as trigger signals are input, IC_{3a} generates sawtooth signals. Since IC_{2a} receives the same trigger signals, it is enabled continuously. There is, however, a fundamental difference between the two circuits: the 74HCT123 is retriggerable, whereas the 74HCT221 is not; it is enabled only after the current period has been processed; all intermediate trigger pulses are ignored.

In practice, the Q output of IC_{2a} is high as long as trigger pulses are input. When these pulses cease, the Q output goes low, which results in IC_{3b} being started via the Q output. After its mono time has elapsed, IC_{3b} enables IC_{3a} and this in its turn reactuates IC_{3b}, so that, even in the absence of input trigger signals, the

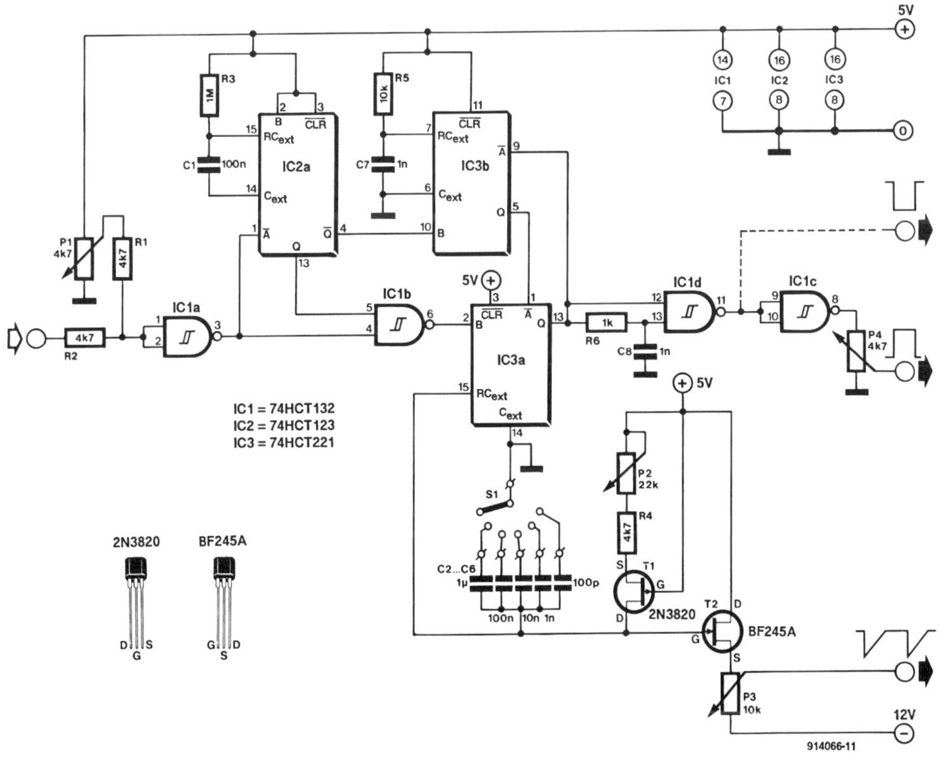

sawtooth generator continues to operate.
The flyback is suppressed by IC_{3a}; network R_6-C_8 provides a slight delay to ensure that the beam is suppressed at the right moment. Since these components have a fixed value, it may happen at very low frequencies that small portions of the sawtooth are not suppressed. This may be obviated by switching capacitors of different values in the C_8 position via a second wafer on S1.

The range of the generator may be refined by adding capacitors in the time base whose ratio is 1:2:5; the capacitors in the diagram, C_2–C_6, have a ratio of 1:10. With values as shown, the period of the sawtooth in the upper position of the switch can be set between 1 μs and 6 μs; periods below 1 μs can not be obtained.
The circuit as shown draws a current of about 7.5 mA from the 5 V supply. The negative supply is not critical and may be between 5 V and 12 V.

General purpose test probe

153 THE COMPACT test probe is made from five transistors, three LEDs, a zener diode, and three resistors. It enables rapid 'measurement' of voltage levels at digital gates, fuses, diodes, batteries, and others. Of course, it does not provide absolute values, but rather a good indication of correct operation or otherwise.
Measurements are carried out with pins A and B. If the potential difference between A (the reference pin) and B is 1.9–2.0 V, D_2 will light. If the voltage at B is ≤1.4 V higher than that at A, D_3 will light. Finally, if the potential at B is ≥11 V with respect to that at A, D_1 will light.
Transistor T_5 is used as a zener diode, which keeps the total current drawn down (since transistors 'break down' in a stable manner at a lower current than zener diodes).
The probe allows measurement of alternating voltage.
The maximum input voltage is highly dependent on the dissipation allowed in R_1. For example, when this resistor is a 0.5 W type, the input voltage may be as high as 200 V r.m.s.
The current drawn by the circuit depends on the number of lighting LEDs: it is not more than 10 mA at a supply voltage of 3 V. In quiescent operation, the current is so low (about 5 μA) that an on/off switch is not necessary.

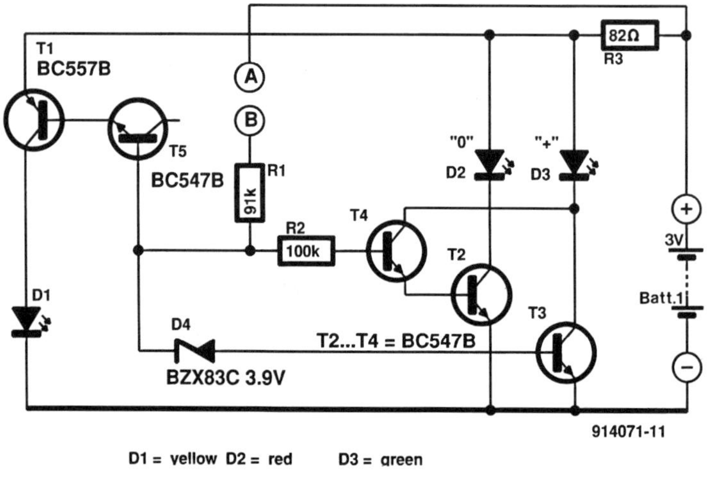

Sine wave converter

154 The converter, based on a Type CA3080 operational transconductance amplifier (OTA) converts a triangular voltage into a sinusoidal one with very low distortion. An OTA is a device whose amplification can be varied by a control current.
The circuit needs a stable, precisely symmetrical supply voltage, since the control is derived directly from this via resistor R_2. It also needs a symmetrical triangular voltage at a level of 350 mV p-p at pin 3 of IC_1. The output in those conditions is a sinusoidal voltage at a level of 2.85 V r.m.s. Provided the circuit is constructed carefully and P_1 is set correctly, the distortion does not exceed 1.2%. This figure deteriorates rapidly when the supply voltage is not accurate or precisely symmetrical.

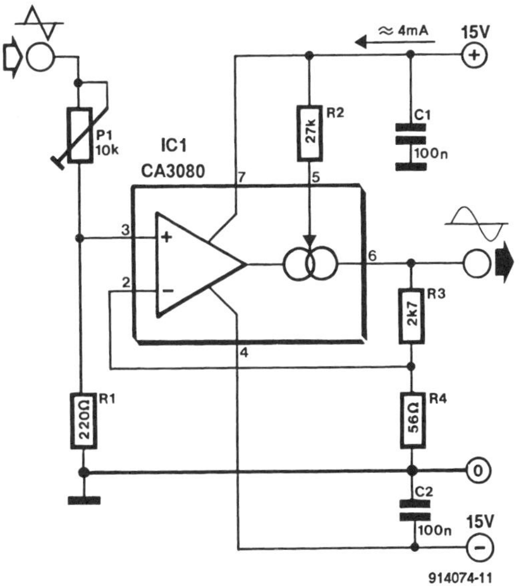

Binary logarithmic ladder network

155 The network provides an attenuation between 0 dB and 78.75 dB that is presettable in 64 stepswith the aid of a 6-bit code.
Six independent, relay switched attenuator sections may be connected in series in accordance with the input code. Since the sections have identical 1 kΩ impedances, they do not affect each other and can, therefore, be interchanged without any problem. The only requirement is that the network is terminated correctly, here by R_{20}. The fairly low characteristic resistance of 1 kΩ was chosen to keep the noise generated by the attenuator low.
Buffering of the attenuator is effected by a Type NE5532, an IC that operates effortlessly with a 1 kΩ input impedance.
The use of relays in the various sections ensures that there are no linearity problems with switching elements. Furthermore, relays make it possible for the control circuits and the attenuator to be electrically isolated.
The relays are driven by a transistor, so that the control inputs require relatively little energy: even a simple logic circuit (TTL or CMOS) can switch the relays in this manner.
The signal-to-noise ratio of the attenuator, with component values as shown in the diagram, is 92 dB (A-weighted even 107 dB), provided the input signal is not smaller than 1 V r.m.s. The maximum input voltage is 7 V r.m.s. Total harmonic distortion at frequencies up to 20 kHz is not greater than 0.003%. The current drawn by the circuit depends primarily on the relays: in the prototype it was about 120 mA.

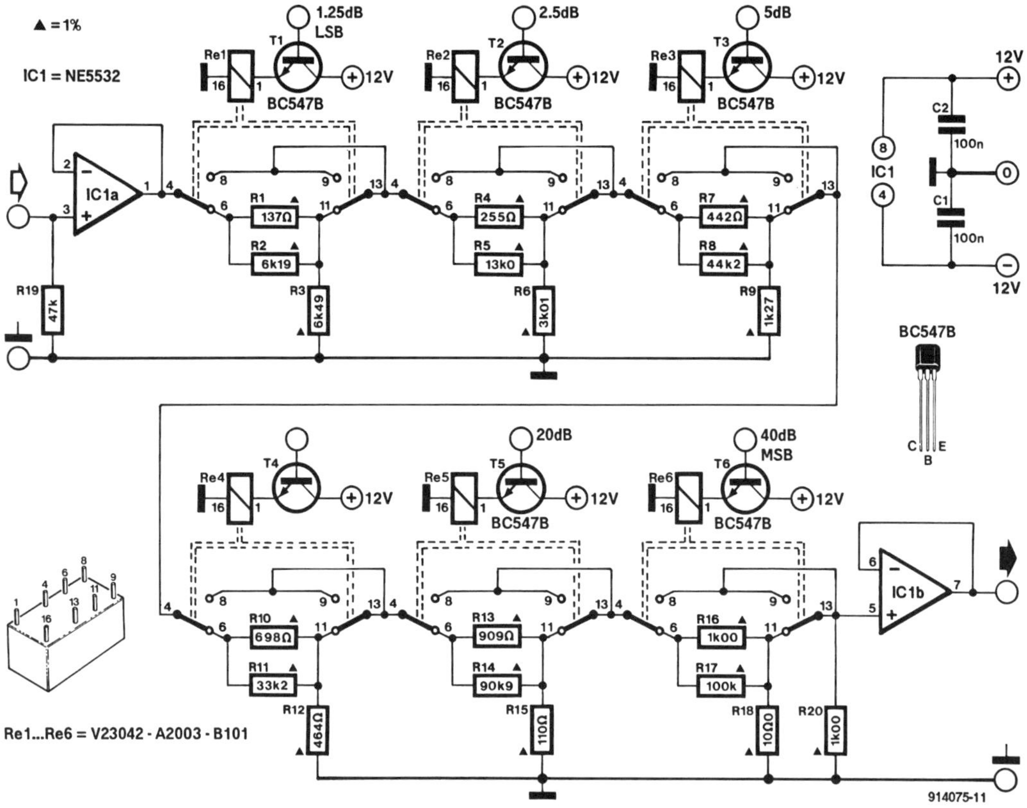

LED VU meter

156 Although there are several ICs on the market with which it is quite simple to construct an LED VU meter, there is, no doubt, still interest in building such a meter from discrete components.

The meter shown in the diagram is based on eight opamps, contained in two Type LM324 chips, that function as comparators.

The inverting input of each of the opamps is provided with a reference voltage derived from potential divider R3–R10. The values of these resistors are chosen to give a 5 dB step between adjacent resistors. Resistors R1 and R12 ensure that the reference voltages are higher than half the supply voltage.

The non-inverting input of each opamp is supplied with the rectified input signal (D1 and D2), which is also superimposed on to half the supply voltage.

When the voltage level at the positive input of an opamp rises above that at the inverting input, the output of that opamp goes high and the associated LED lights. The higher the input signal, the more LEDs will light. If D3 and D4 are red, the circuit may be used as a simple peak indicator, showing when the input signal exceeds a certain value.

The supply may be between 10 V and 15 V. The current drawn from it depends largely on the number of LEDs that light; it is not more than 160 mA at 10 V and 110 mA at 15 V.

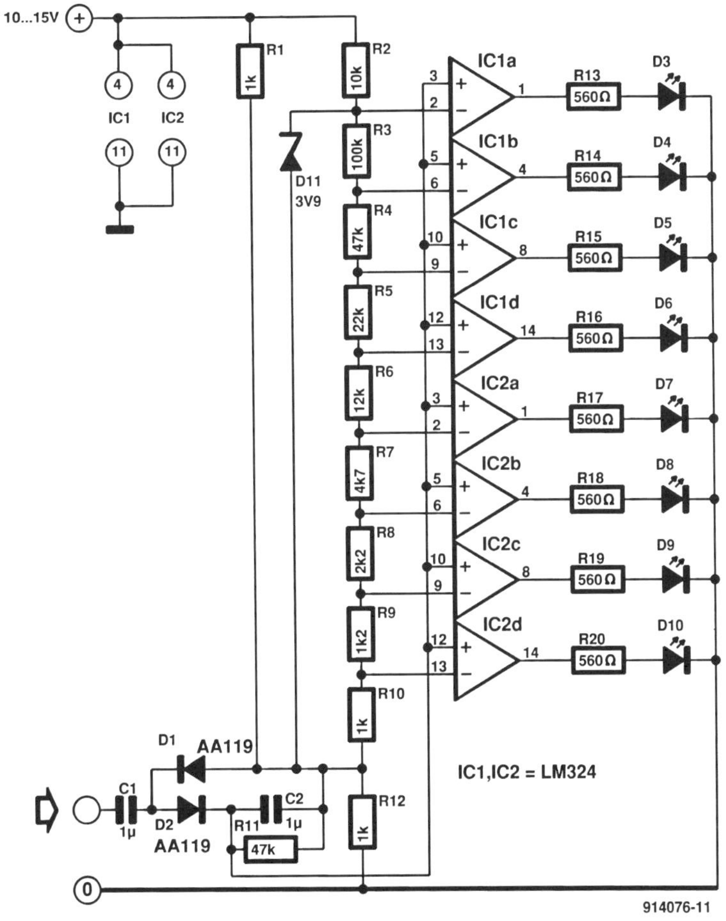

Pulse shaper

157 The diagram shows how the four gates contained in a 4077 may be used to build a circuit that doubles the frequency of a signal applied to it. In other words, it generates a pulse for every edge of the input signal; the pulse width is determined by the internal delay of gates IC_{1a}–IC_{1c}. To that end, the original signal at pin 13 is compared with the delayed signal at pin 12. Because of the XNOR function of IC_{1d}, any level difference between the two is translated into a level change at the output of IC_{1d}.
The quiescent output level at pin 11 is set by connecting the ± input of IC_{1a} to ground or to the +ve supply rail. If it is taken to ground, IC_{1d} delivers a zero level followed b y positive pulses for each of the edges of the input signal. If it is linked to the +ve supply rail, IC_{1d} outputs a high level followed by negative pulses for each of the edges of the input signal.
Instead of a 4077, a 4030 or a 4070 may be used, in spite of the fact that these have XOR instead of XNOR gates. Only the pulse width of the output signal will be slightly different owing to the changed transfer times in the gates.
The current drawn by the circuit depends on the signal frequency: at very low frequencies, it is virtually nil.

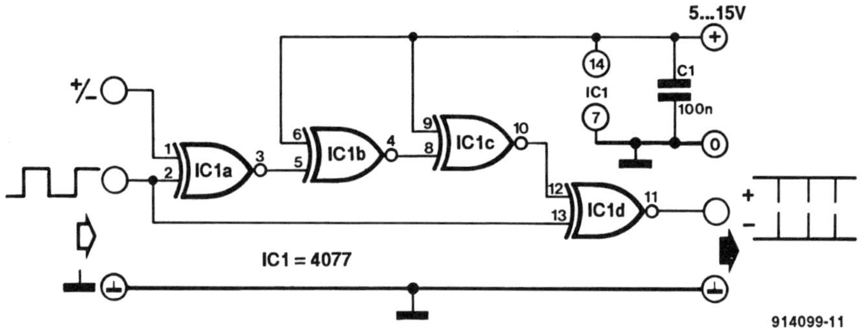

Audible level indicator

158 A 600 Hz signal at a level of 2.4 V_{pp}, generated by the oscillator on board an LM1830 (National Semiconductor) is applied to a probe via C_2. The probe is immersed in the liquid whose level is to be monitored. Because of C_2, there is no direct voltage at the probe, so that there are no electrolysis problems.

As long as the probe makes no contact with the (conductive) liquid, the signal level at the input of the detector is equal to the level of the oscillator signal. When the liquid touches the probe, the detector input is connected to ground (or nearly so). This causes the level at pin 10 to drop. When it becomes more than 0.6 V lower than the oscillator signal, the detector switches on the internal output transistor in the rhythm of the oscillator frequency, since that is not suppressed by the detector.

The consequent signal at pin 12 is used to drive a simple output stage, T_1, which drives a small loudspeaker, LS_1.

The supply for the circuit is best taken from a 9 V PP3 battery. In quiescent operation, the current drain is 3 mA; when the alarm sounds, the current rises to about 80 mA.

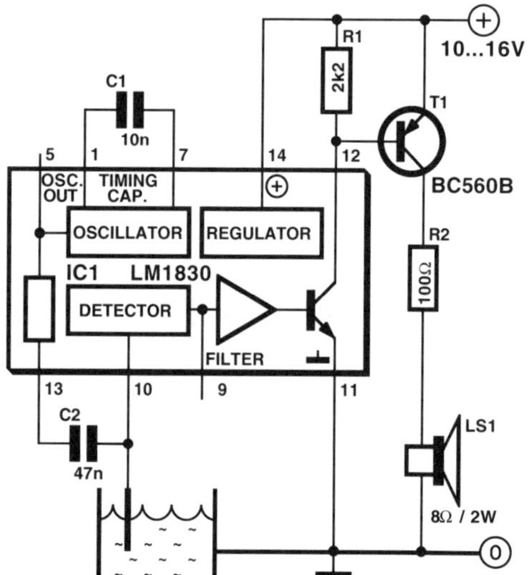

Measuring electrolytic capacitors

159 Most capacitance meters have no facility for measuring large electrolytic capacitors. The circuit described here makes it possible for such capacitors to be measured with some degree of accuracy, in spite of the large tolerances these components normally have.

Opamp IC_{1a} is arranged as an astable. Capacitor C_2 is charged via R_2; as soon as the potential across it reaches the level of that at the non-inverting input of IC_{1a}, which is determined by voltage divider R_1-R_3-R_4, the opamp toggles and C_2 is discharged till the voltage across it reaches the new level at the + input of IC_{1a}.

The measuring circuit consists of switched resistors R_6–R_9, and S_1, R_{10}, P_1 and P_2. The capacitor on test is charged via T_2 and discharged rapidly via T_1.

Comparator IC_{1b} compares the level (0.65 V) at its non-inverting input with that at its inverting input. When the capacitor on test is connected across the input terminals, P_2 is adjusted till the LED just lights. The potentiometer must be given a scale to enable the value of the electrolytic capacitor to be read directly. The scale can be calibrated with the aid of a capacitor of known value for each range (1–4.7 μF; 4.7–47 μF; 47–470 μF; and 470–4700 μF). Basically, it is linear, but it may be necessary to make a scale for range 1 empirically.

To ensure the best possible accuracy, it is advisable to use a regulated power supply. The circuit draws a current of about 20 mA (almost all of it through the LED).

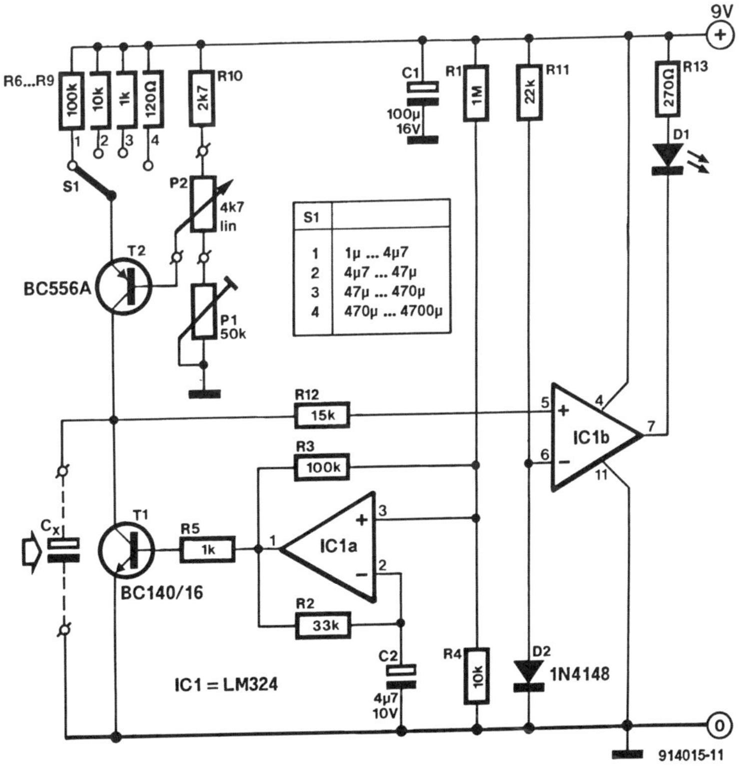

Precision rectifier for digital voltmeters

160 This simple circuit, based on a single opamp in non-inverting mode, is a precision rectifier extension for digital voltmeters. The circuit can be connected to a high-impedance voltage divider without the need of an additional buffer stage that increases the cost and, more importantly, the power consumption. Another advantage of the circuit is that the accuracy is not affected by the offset voltage of the opamp. The output of the rectifier is differential to allow ready connection to the IN-LO and IN-HI inputs of digital voltmeter (DVM) ICs like the familiar 7106 and similar types.

Circuit IC_1 is a LinCMOS operational amplifier operating in the high-bias mode. The TLC271 used here achieves a good high frequency response at a low current consumption of about 1 mA. For all practical purposes, the gain of the operational amplifier is $2R_1/R_2$, where $R_1 = R_3$ and $R_2 = R_4$. With the values shown, the gain is nearly equal to 1.1107, which is the r.m.s. (root-mean-square) shape factor for sinusoidal waveforms. Capacitors C_1 and C_2 are optional. They improve the response and stability of the rectifier at high frequencies.

Any d.c. component at the input, as well as the offset voltage of IC_1, appears as a common-mode voltage across C_3 and C_4, and is therefore rejected. The low-frequency response of the rectifier is determined by the time constant R_2C_3 (or R_4C_4). With the component values specified, the 1%-accuracy bandwidth extends from 25 Hz to about 20 kHz.

The circuit is powered by the 9 V battery used for the DVM module. The ground of the rectifier is connected to the COM terminal of the module, which is at a potential of about 2 8 V below the positive supply. The DVM should be set to a full-scale input voltage of 200 mV.

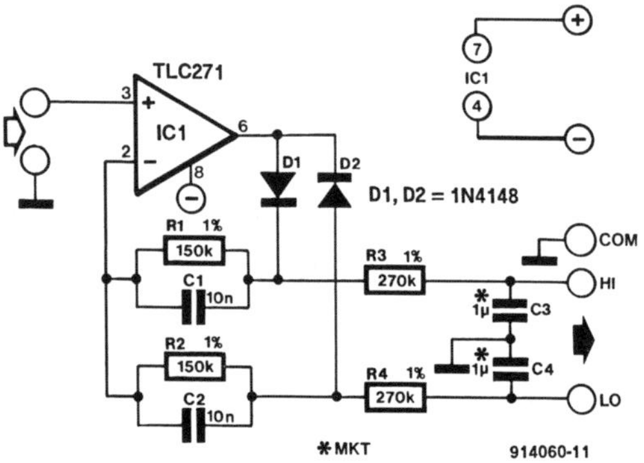

Continuity tester

161 A continuity tester is very useful for checking printed-circuit boards. It indicates a sound connection by a squeak from a buzzer; there is, therefore, no need to continually watch a meter,

The voltage across test terminals TP_1 and TP_2 is only 80 mV. That is not sufficient to test diodes, but it obviates the risk of damage to electronic components.

The design consists of a comparator, IC_1, and an astable consisting of T_1 and T_2, whose frequency is around 1250 Hz for a supply voltage of 3 V. The astable is actuated the moment the output of IC_1 (pin 6) goes logic high. This happens when the resistance between TP_1 and TP_2 is low, so that the voltage at pin 2 of the comparator is lower than that at pin 3, which depends on the setting of P_1.

The circuit is powered by a 3-V battery (two 1.5 V cells in series), but may be maximum 12 V. At such a high supply voltage, it may be that the tone of the buzzer is too high; in that case, the values of C_1 and C_2 should be increased,

If a Type TLC271C is used instead of a TLC251 for IC_1, the supply voltage must be not lower than 3 V. At 3 V, the circuit draws a current of about 1.4 mA.

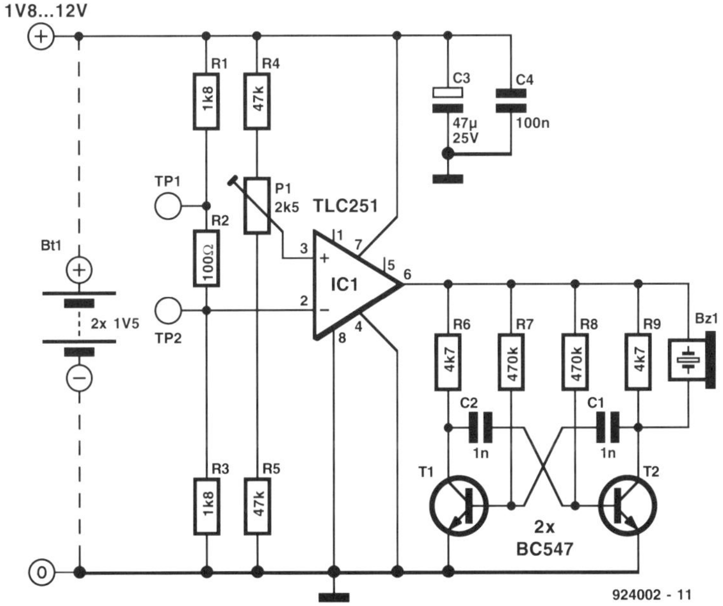

MIDI (cable) tester

162 A tester for MIDI connections is a real asset for many constructors. The one presented here is kept fairly simple as shown in Fig. 1.

The LED shows whether MIDI data are present where the tester is connected. Since these data consist of very short pulses, the LED must be a high-efficiency type.

To test both ends of a cable, the tester must be built in duplicate: one terminated into a socket and the other into a plug. A possible design is shown in Fig. 2.

The connectors should preferably be metal DIN types. The LED can be housed in the rubber support sleeve as shown in Fig. 3. The connecting wires to the LED must, of course, be well insulated.

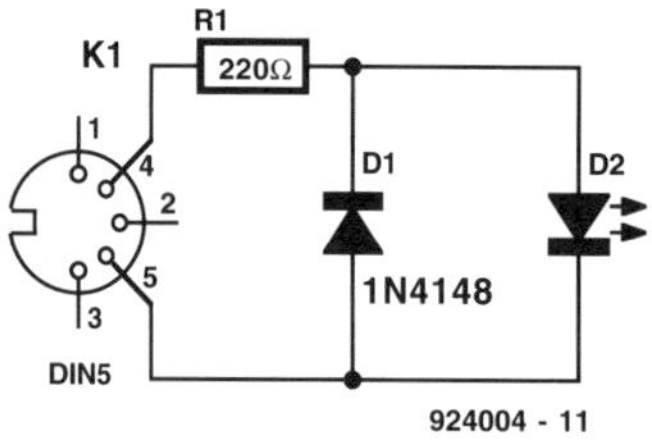

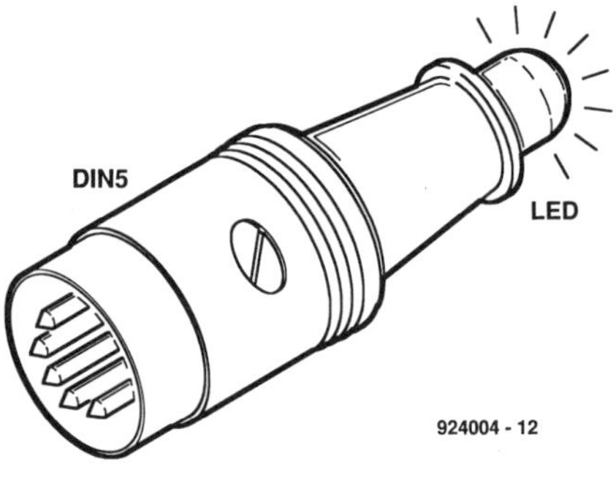

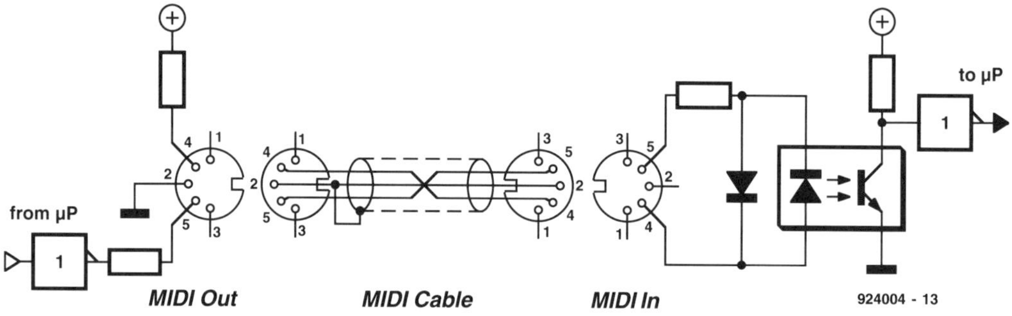

Repeating pulse generator

163 The circuit uses one of the four bilateral switches contained in the CD4066, and is a kind of astable multivibrator. The main limitations of the circuit are the relatively poor shape of the output signal, and the output signal swing, which is not digitally compatible. These drawbacks are easily overcome, however, by a buffer at the output. When the switch closes, the current in the chain produces a voltage across R_3, which raises the switching level at the trigger point by about one volt. Likewise, R_1 reduces the voltage applied to R_4 to a level below the switch-off voltage.

The mark-to-space ratio (duty factor) of the output signal may be varied by using parallel charging resistors, one with a series diode (preferably a germanium type), as shown in the circuit diagram.

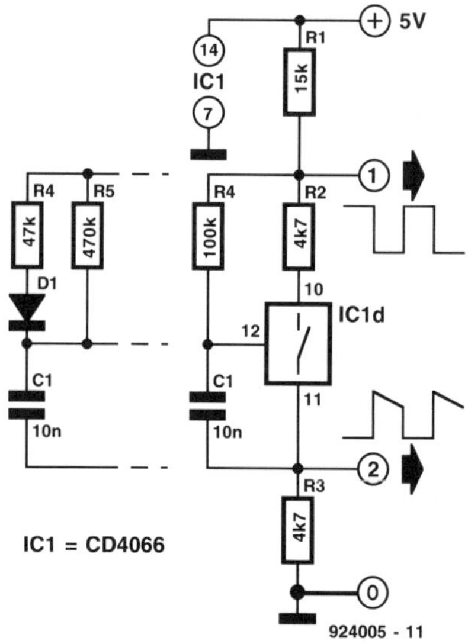

In the prototype, the component values shown resulted in an output frequency of 957 Hz with a supply voltage of 5 V exactly. The duty factor was about 0.4 without the parallel *R-D* network. With the network fitted, the output frequency and duty factor were 317 Hz and 0.06 respectively.
The current drain varied between 0.12 mA and 0.39 mA.

Fuse monitor

164 When an appliance ceases to operate, there may be various causes for this, one of which is the blowing of the mains fuse. The monitor proposed here contains an LED that lights when that is the case. It is suitable for use with fuses rated from milliamperes to amperes.
As long as the fuse is intact, the full mains voltage exists across C_2-R_3-D_3. Capacitor C_2 and resistor R_2 serve to limit the base current of T_1. Diode D_3 prevents C_2 from being charged, which would cause the base current to drop to zero rapidly.
Capacitor C_1 and resistor R_2 limit the current through D_1, while D_2 ensures that the voltage across the LED does not exceed 2.7 V. At the same time, D_2 prevents C_1 from being charged.
As long as the mains voltage exists at junction F_1-C_2, transistor T_1 conducts and short-circuits D_1 and D_2. When F_1 blows, T_1 is switched off, whereupon current flows through D_1 and D_2: the LED then lights.
Resistor R_1 must conform to relevant safety regulations. Furthermore, capacitors that still carry the mains voltage after the appliance is switched off, must become discharged (via R_1) within a stipulated time. When the monitor is in use, remember at all times that certain of its parts are at, potentially lethal, mains voltage.

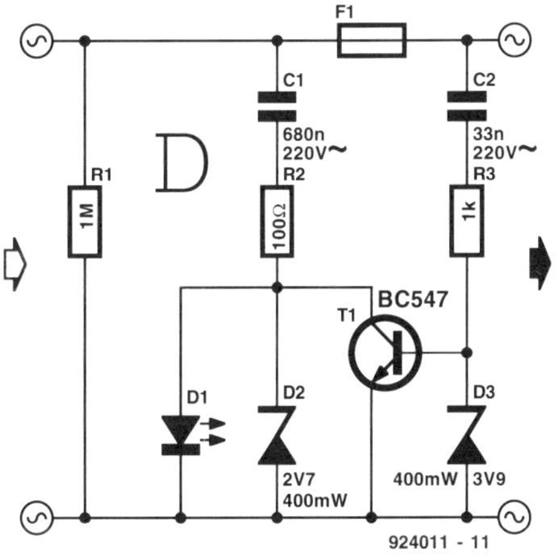

Power supply tester

165 This little circuit enables you to measure the so-called dynamic response of a d.c. power supply. A power MOSFET, T_1, is used to switch the supply load on and off at a user selectable rate. The response of the supply to these fast load variations is displayed on an oscilloscope.
The switching rate is selected with the aid of a rotary switch, S_1, which also serves as the on/off switch. The available switching frequencies are: 10 Hz, 100 Hz, 1 kHz and 10 kHz. The well-known 555 timer IC is used to supply the switching signal. Diodes D_3 and D_4 cause the astable multivibrator to supply an output signal with a duty factor of about 0.5. The switching transistor, T_1, is protected against too high currents by a fast 10-A fuse inserted in the drain line. The tester may be powered by any regulated d.c. supply with an output voltage between 6 V and 15 V. However, this must not be the supply under test! Given the low current consumption of the tester (40 mA max.), a 9-V battery is an excellent power source.

The tester is extremely simple to use. First, select the load resistance of the supply you wish to test; say, 12 Ω/15 W for a 12-V, 1-A PSU. This resistor is connected between output 'R' of the tester, and the '+' output of the PSU. The '0' output of the tester goes to the '-' (or '0') terminal of the PSU. Next, connect the scope input to the PSU outputs, and the trigger input to K_3 of the tester. Switch on the scope, the PSU and the tester. The scope will now display the dynamic regulation characteristic of the PSU at the given output current (1 A) and the selected switching rate (initially, 10 Hz).

Construction of the tester is straightforward on the small printed circuit board shown here. The power MOSFET is bolted on to a small PCB-mount heat sink, and will not run very hot even when the maximum permissible drain current (about 10 A) is approached.

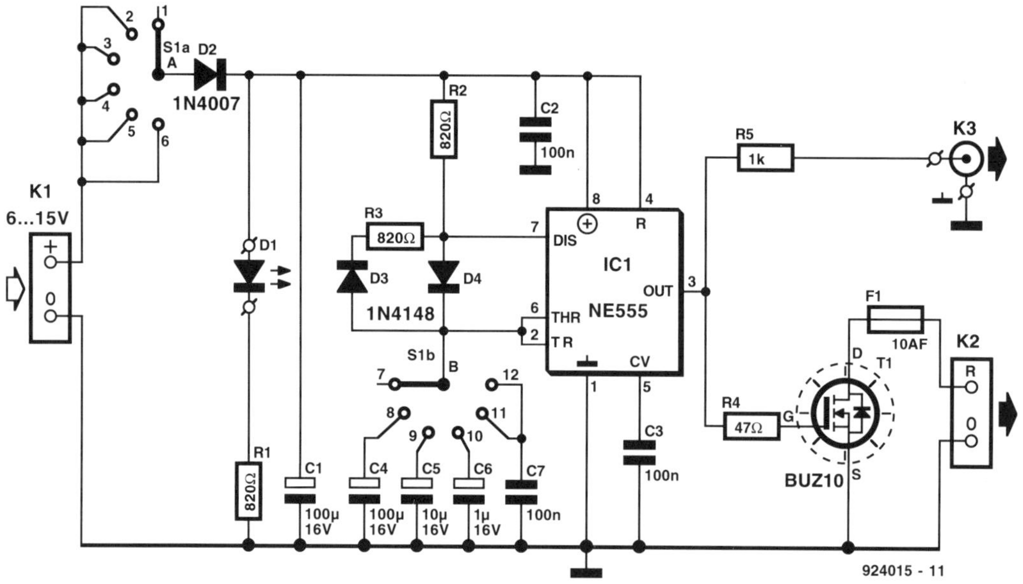

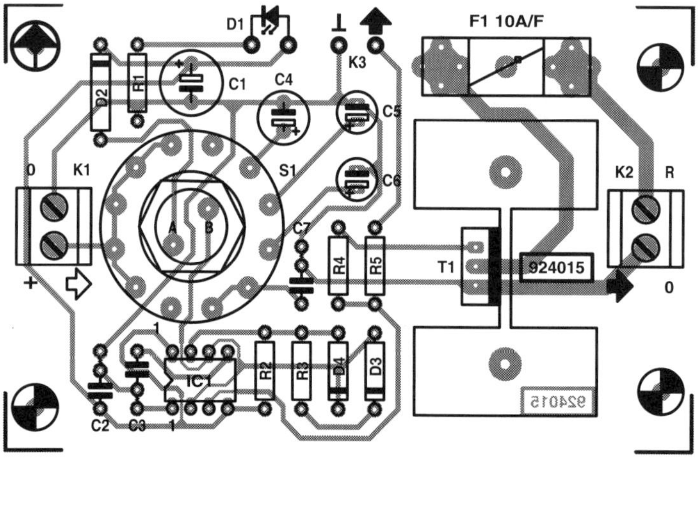

Parts list

Resistors:
R_1; R_2; R_3 = 820 Ω
R_4 = 47 Ω
R_5 = 1 kΩ

Capacitors:
C_1 = 100 µF, 16 V radial
C_2; C_3; C_7 = 100 nF
C_4 = 100 µF, 16 V radial
C_5 = 10 µF, 16 V radial
C_6 = 1 µF, 16 V radial

Semiconductors:
D_1 = LED, red, 5mm
D_2 = 1N4007
D_3; D_4 = 1N4148
T_1 = BUZ10

Integrated circuits:
IC_1= NE555

Miscellaneous:
K_1; K_2 = 2-way PCB terminal block; pitch 5 mm.
K_3 = panel mount BNC socket.
S_1 = 2-pole 6-way PCB mount rotary switch.
F_1 = 10A fast fuse plus PCB mount holder and cap.
Heat sink 5 K W^{-1}, e.g. SK129/38.1mm (Siemens)

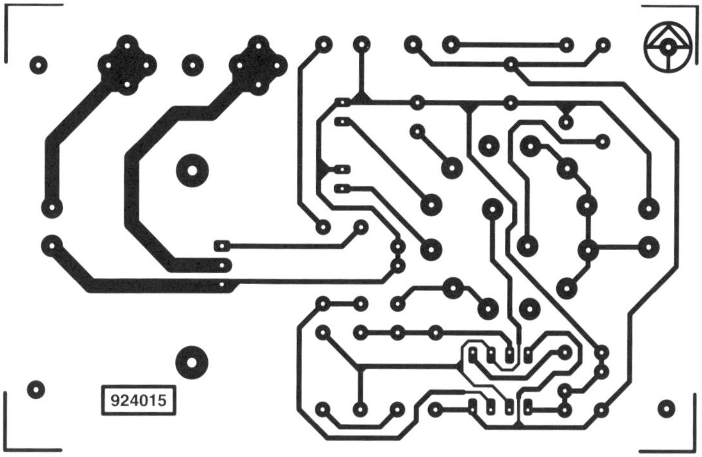

Digital temperature converter

166 This little circuit converts temperature measured by an NTC (negative temperature co-efficient) resistor into a digital signal. The resistance of the NTC, which is an inverse function of the ambient temperature, determines the frequency of an oscillator built around the familiar TLC555 timer. The astable circuit is designed to give a pulse repetition frequency (p.r.f.) of about 250 Hz at 25 °C that rises with temperature. The non-linear relationship between temperature and oscillator frequency is not a problem here, because it is relatively simple to 'straighten' by arithmetic operations performed by a computer or a microcontroller system.

Basically, three temperatures are measured, and the corresponding oscillator frequencies stored as reference points, which serve to interpolate other values in between.

The components used are all surface mount devices (SMDs), but the printed-circuit board allows the use of a traditional NTC in position R_1.

To ensure fast response to temperature changes, the completed PCB is fitted into a small metal tube with a diameter of 13 mm or larger. Great care should be taken to isolate the board and the components from the metal tube; if necessary, use heat-shrink sleeving! The tube is sealed hermetically with potting compound or two-component resin glue, through which three wires are passed: + supply; output signal; ground.

Current consumption of the converter is smaller than 1 mA.

The multifunction measurement card for PCs (Ref. 1) is perfect for processing the converter output signal. Use is made of the frequency meter input and the program modules found in the Turbo Pascal library 'PMEASURE.PAS' on diskette ESS1751, and in Borland's 'Numerical Toolbox', which provides some handy interpolation routines.

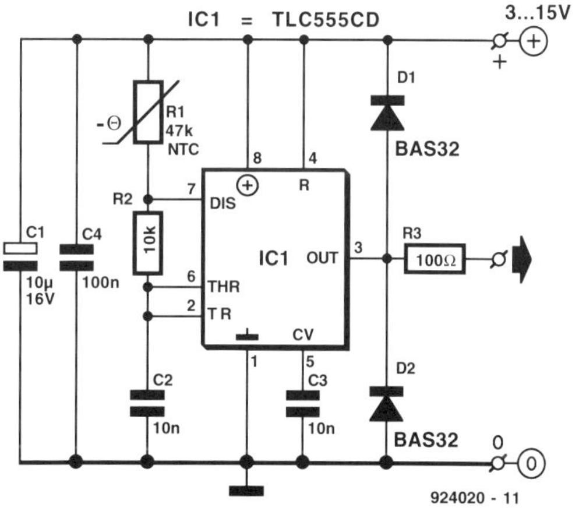

Parts list

(all components surface-mount devices)

Resistors:
R_1 = 47 kΩ NTC
R_2 = 10 kΩ
R_3 = 100 Ω

Capacitors:
C_1 = 10 µF 16 V
C_2;C_3 = 10 nF
C_4 = 100 nF

Semiconductors:
D_1;D_2 = BAS32

Integrated circuits:
IC_1 = TLC555CD

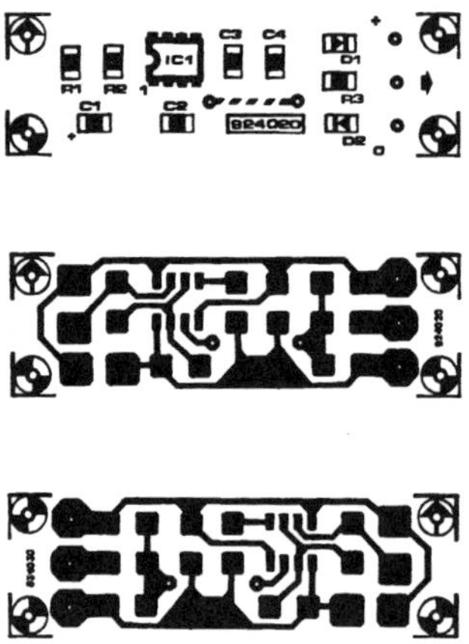

***Reference*:**
"Multifunction measurement card for PCs", *Elektor Electronics* January, February 1991.

Simple signal generator

167 This signal generator provides a 440 Hz sine wave output at two levels. The power supply may be 1.5–16 V, so that even a single 1.5 V battery can do.

Opamp IC_{1a} operates as a rectangular-wave generator; the values of R_4 and C_1 determine the frequency at which the output of the device toggles. Preset P_1 enables the output to be set as a square wave (it may be adjusted by listening for minimum distortion).

Network R_5-R_6-C_2 reduces the output of IC_{1a} by 3 dB (50%), after which the signal is superimposed on half the supply voltage (derived from the average level of the rectangular signal by C_2). This voltage is needed for setting the d.c. operating point of IC_{1b}.

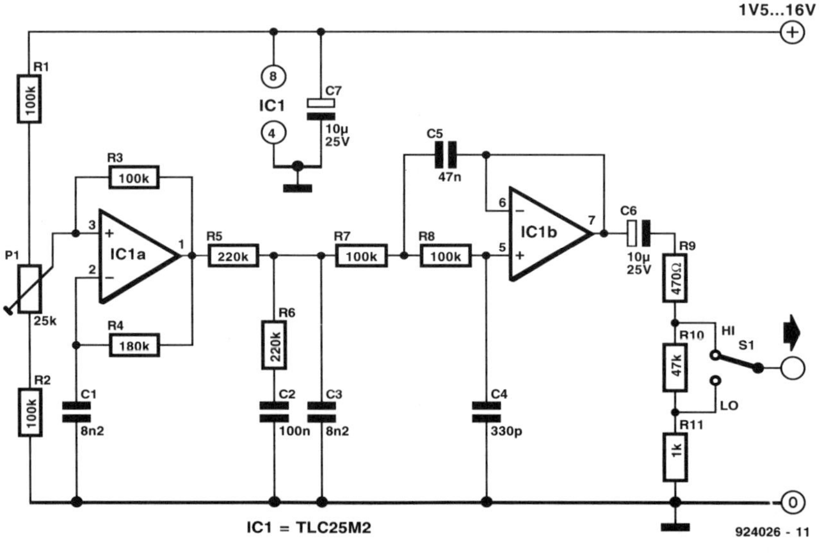

Opamp IC_{1b} forms a third-order Chebishev filter with a cut-off frequency of 400 Hz. This filter removes the majority of harmonics from the rectangular signal, so that the output is a reasonably clean sine wave.
The level of the output signal is selected with S_1 from potential divider R_9-R_{10}-R_{11}, depending on the requirements of the circuit on test. With a power supply of 16 V, the output level is 1.5 V r.m.s. or 30 mV r.m.s.; with a 1.5 V supply, the output levels are 150 mV and 3 mV. The output frequency is somewhat dependent on the supply voltage and varies from about 440 Hz at 16 V to around 370 Hz at 1.5 V.
The circuit draws a current of 300 µA at 16 V and 80 µA at 1.5 V.

Discrete frequency–voltage converter

168 There are a number of ICs available which convert frequency to voltage, but that can also be done with discrete components as shown in this circuit. Such a simple design has, of course, its limitations: the input signal must be a square wave, must have a constant amplitude, and be provided by a low impedance source (≤50 Ω).
The circuit is called a transistor pump and is related to the charge pump that is used, for instance, in a voltage doubler or voltage inverter. Here, the added transistor arranges for C_1 to be charged very rapidly (U_i=low) when the left-hand side of that capacitor is pulled toward zero because of u_i. The right-hand side of C_1 is thus clamped to a potential of $U_{C2}-U_{BE}$. This means that the circuit is not dependent on the pulse width of the input signal. During the leading edge of U_i the charge on C_1 is transferred to C_2, which isdischarged through R_1.
The transfer function of the circuit is derived from the balance in which the increase of U_o during a leading edge must be equal to the voltage reduction caused by R_1 in each period. This yields the following formula

$$U_{o(fi)} = (\hat{u}_i-0.7)R_1C_1f_i.$$

With component values as shown in the diagram, the output voltage increases by about 4.3 mV Hz^{-1} if the level of the input signal is 5 V. The speed with which the output voltage reacts depends on the time constant of the circuit:

$$\tau = R_1(C_1+C_2).$$

Since the value of both R_1 and C_1 is fairly large, the reaction is not very fast. There are a number of applications where that does not matter, however.
The ripple on the output voltage may be calculated from

$$\hat{u}_r = (\hat{u}_i-0.7)C_1/(C_1+C_2).$$

With values as before, the ripple is 400 mV.
The supply voltage for the circuit must be at least 2 V higher than the output voltage

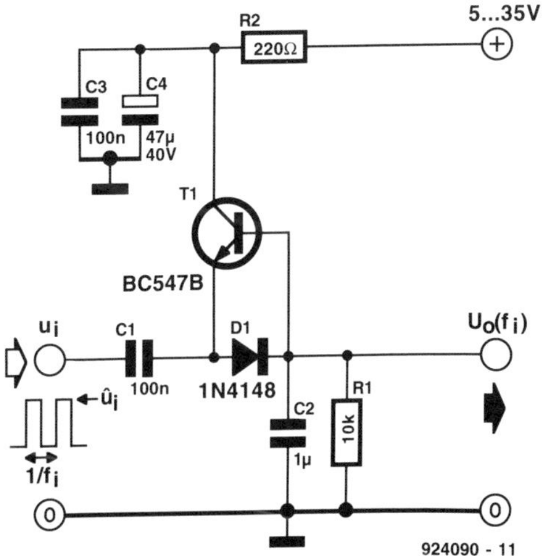

at the highest input frequency to be measured. The circuit draws a current of not more than a few mA.

Filter R_2-C_3-C_4 smooths current peaks during the charging of C_1 and thus limits the RF interference the circuit produces.

The output must be terminated into a high impedance (digital voltmeter or buffer). If that is not possible, a high-value terminating resistor ($\geq 10R_1$) should be used.

Current tongs for PCB tracks

169 Current tongs are instruments for measuring alternating currents in cables. Their great advantage over most other methods is that the circuit need not be broken. Their action depends on the same principle as that of a transformer: after the tongs (transformer core) have been placed around one or more current-carrying conductors (primary winding), magnetic fields arise in the core material. Because of the alternating magnetic field components a voltage is induced in the coil wound around the core (secondary winding). In accordance with Faraday's law of induction, this voltage is a measure of the vector sum of the currents through the wires in the cable.

A frequent requirement is the measurement of an alternating current in the supply lines on a printed-circuit board. In that case, technicians are often more interested in the waveform than in the r.m.s. value. Traditional current tongs can, of course, not be used then. It is, however, not too difficult to construct current tongs. or rather a current probe, for this purpose. All that is needed is a short length of ferrite material,a few metres (yards) of thin enamelled copper wire,a BNC plug, and a metre (yard) of 50-Ω coaxial cable, for instance, Type RG58/U.

Ferrite has a low magnetic resistance and draws the fields around the track towards itself. To ensure optimum coupling, the ferrite must be placed as close as possible to the track. The most convenient shape of the ferrite is a bead cut in half lengthwise: the hollow inside can then envelop the track at an angle of 180°—see diagram.

About 40 turns of the copper wire are wound around the half bead as shown. The terminals of the winding are soldered to the coax cable. Solder one terminal of the coil to the conductor of the cable, insulate the connection with tape and then fold the earth braid back over the conductor. Then solder the second terminal of the winding to the braid,

To give the probe some mechanical rigidity, push a narrow tube (such as an empty ballpen) over the cable on to the winding. When that is done, fit the tube to the sensor with superglue or epoxy resin.

Terminate the other end of the coaxial cable into the BNC plug.

The self-inductance of the sensor winding is some 20–30 µH, which gives a reactance ($= 2\pi fL$) in the wanted frequency range that is much greater than 50 Ω. The voltage measured with a spectrum analyser or oscilloscope is thus virtually independent of the signal frequency.

The core of the prototype shown in the photograph is 4 mm thick and 15 mm long. The self-inductance of the sensor winding is 80 µH, which makes the probe suitable for frequencies >500 kHz. As long as the core material is not saturated, the sensitivity of the probe is about 0.2 V A^{-1}.

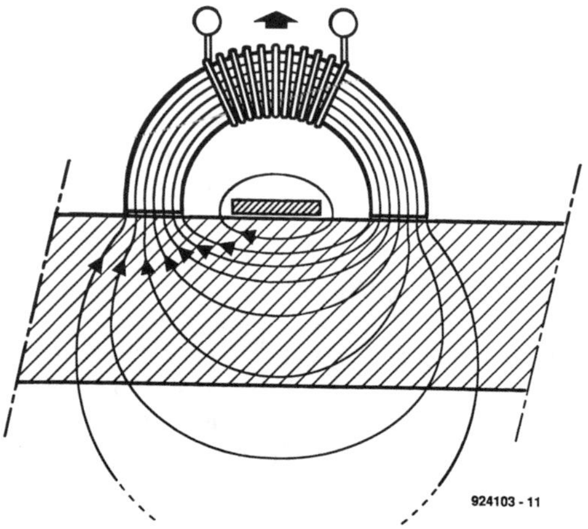

Smartec temperature sensor

170 The Smartec Type SMT160-30-18 is a temperature sensor with a digital output, and housed in a TO-18 case. Pins 1 and 3 are the supply input pins. With a nominal 5 V supply, the current drain is not more than 200 µA. At pin 2, a short-circuit-proof and TTL compatible rectangular output voltage with a pulse repetition frequency (p.r.f.) of 3 kHz is available. The value of the frequency is not all that important, because the actual temperature data are stored in the duty factor (that is, the ratio of the pulse width to the pulse spacing). There is a linear relation between the temperature, *T* and the duty factor:

$$\text{duty factor} = 0.32 + 0.0047 \times T.$$

Thus, at a temperature of –45 °C, the lowest at which the sensor can be used, the duty factor is 0.109. The maximum temperature at which the sensor can be used is 130 °C. The sensors are calibrated to an accuracy of ±0.25 °C during production.

In principle, it would be feasible to apply the rectangular signal to a moving coil meter. This would indicate a value that is directly proportional to the average voltage level of the rectangular signal and thus with the duty factor and the temperature. In practice, it is, however, much more sensible to connect the sensor to a digital input port of a peripheral interface or a microcontroller. Sampling the rectangular signal enables the computer system to carry out temperature measurements with a minimum of external components as shown in the diagram. A suitable PC measurement card is published Ref. 1. Connector K_1 is a 26-way box header that is linked to connector K_6 of the measurement card via a short length of flatcable. The 5 V supply is taken directly from the computer. On the the prototype, L_1-C_1-C_2 proved essential to prevent jitter of the signal, which caused the first digit after the decimal point to move to and fro.

The program for controlling shown here is available, with other basic routines for the measurement card, from Elektor Electronics (Publishing) under reference ESS 1753.

Ref. 1. 'Multifunction measurement card for PCs', *Elektor Electronics*, February 1992, p. 14.

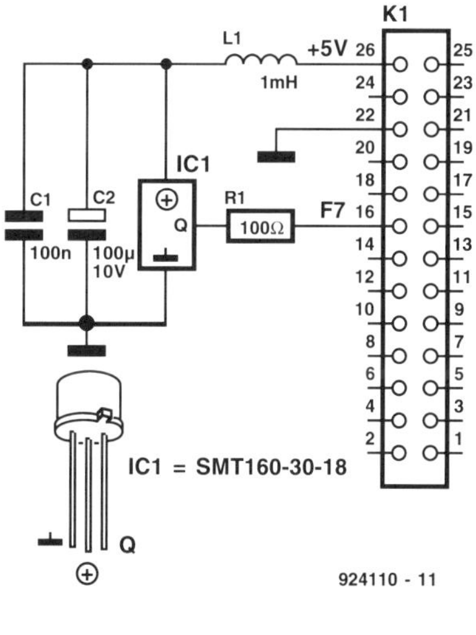

Capacitance meter

171 The simple capacitance meter described here is able to measure capacitances between 100 pF and 1 µF over five ranges.

The circuit consists of a variable oscillator, a scaler and a measuring stage. The oscillator is based on an inverter contained in a Type 74HC14 and generates a frequency, *f*, that is inversely proportional to the capacitance between terminals C_x. Roughly,

$$f = 1.2RC_x,$$

where *R* depends on the position of S_1. With values as shown, the frequency lies between 240 Hz (C_x = 1 µF) and 12 kHz (C_x = 100 pF).

Scaler IC_3 divides the output frequency of the oscillator in a manner which ensures that the maximum output frequency in each range is 120 Hz.

The measuring stage is driven by current source T_1. During one half of each period of the output signal of IC_3 capacitor C_2 is charged via T_1. During the other half of the period, T_2 is switched on by the signal, so that C_2 is short-circuited. In this way, the maximum potential across C_2 depends on the frequency of the signal. The potential is buffered by opamp IC_{1b} and integrated by R_1-C_1. The resulting direct voltage is used to deflect the meter via opamp IC_{1a}.

The circuit is calibrated by connecting a known capacitance of about 100 nF (0.1 µF) across the measurement terminals and adjusting P_1 so that the meter reading corresponds to the capacitance.

Since normally the meter is used only occasionally, it may be powered simply by a 9 V PP3 (6F22) battery.

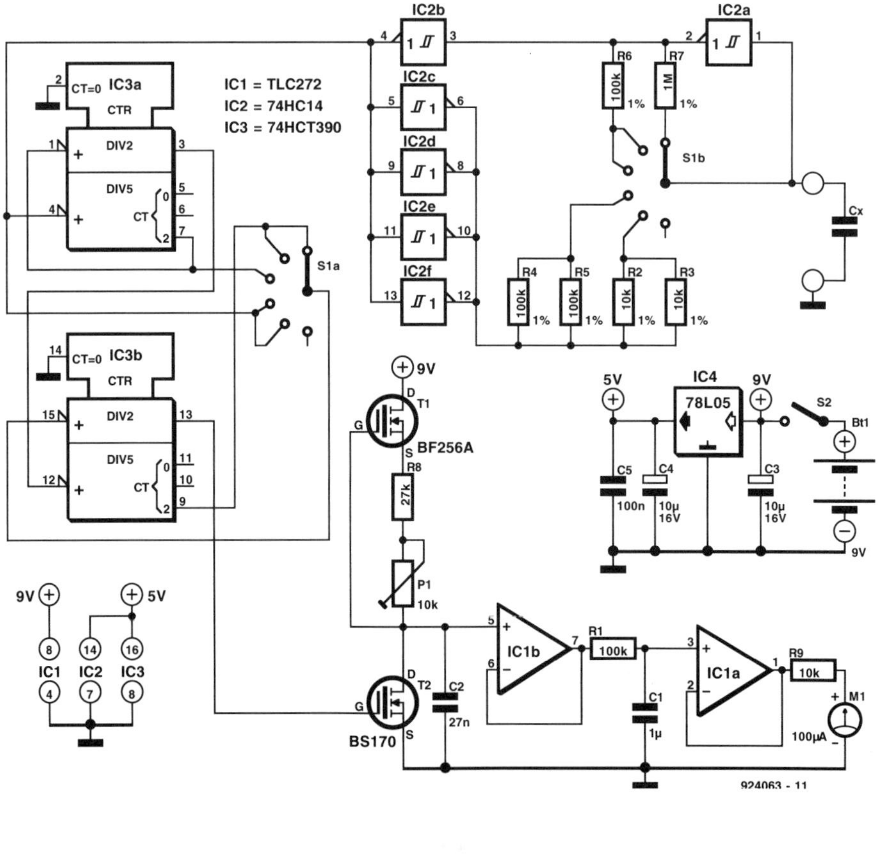

Frequency probe

172 The frequency probe enables 'listening in' to the speed of a computer. It is, however, also suitable for use with other digital circuits, because it makes high frequencies audible, so that signals can be conveniently monitored.

A 12-bit counter serves as 'frequency detector'. The signal measured in a computer or digital circuit is divided by 1024 and is output at pin 14. It is then used to control a transistor, T_1, which in turn drives piezo buzzer Bz_1.

The scale factor has been chosen to convert MHz into kHz, so that the clock frequency of, say, an XT computer will be heard as a shrill 8 kHz tone. If higher frequencies need to be monitored, scale factors of 2048 and 4096 can be obtained by connecting R_2 to pin 15 or pin 1 respectively. If an HCT circuit is used, the measuring limit is some tens of MHz. For frequencies <4 MHz, a standard Type 4040 may be used: that has the advantage that the supply voltage need not be exactly 5 V. The supply connections and the probe are best made from flexible wire terminated into crocodile clips.

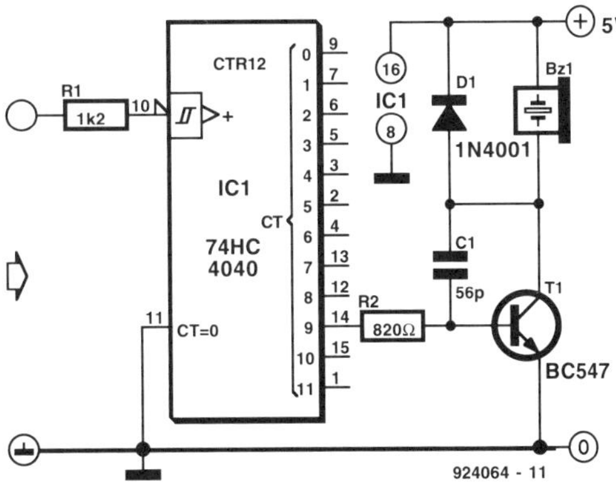

Serial data generator

173 The data generator offers a simple means of testing a circuit.It does not need software and all the components are readily available. The design is based on the timing of a Type SDA3002 circuit. In contrast to the two-wire I²C format, a three-wire connection is used. If only one fixed data word needs to be sent, DIP switches S_2 and S_3, and the eight-fold pull-down resistor arrays, may be omitted; the inputs of IC_3 and IC_4 are then held at fixed levels.

The generator is based on two cascaded shift registers, IC_3 and IC_4, that are used as 16-bit parallel-to-serial converters.

The clock is provided by IC_2, which is started by setting bistable $IC_{1a,b}$ with press button switch S_1. When this switch is pressed, a 100 μs pulse is applied via differentiating network R_1-C_1 to the set input of the bistable. The capacitor is discharged via R_2. The pulse duration is not changed by keeping S_1 depressed; this is necessary because if it were to exceed a half clock period, the timing would go astray. The output of IC_{1b} is the ENABLE signal for the circuit to be tested.

The start pulse is used also to load the logic levels set with the DIP switches on to shift registers. The registers accept the parallel data as long as the signal at their pin 1 is low. This is another reason that the start pulse has to be short.

The start pulse begins the entire test cycle—see the timing diagram in Fig. 2. Output 2^3 (Q3, pin 7) of IC_2 is inverted by

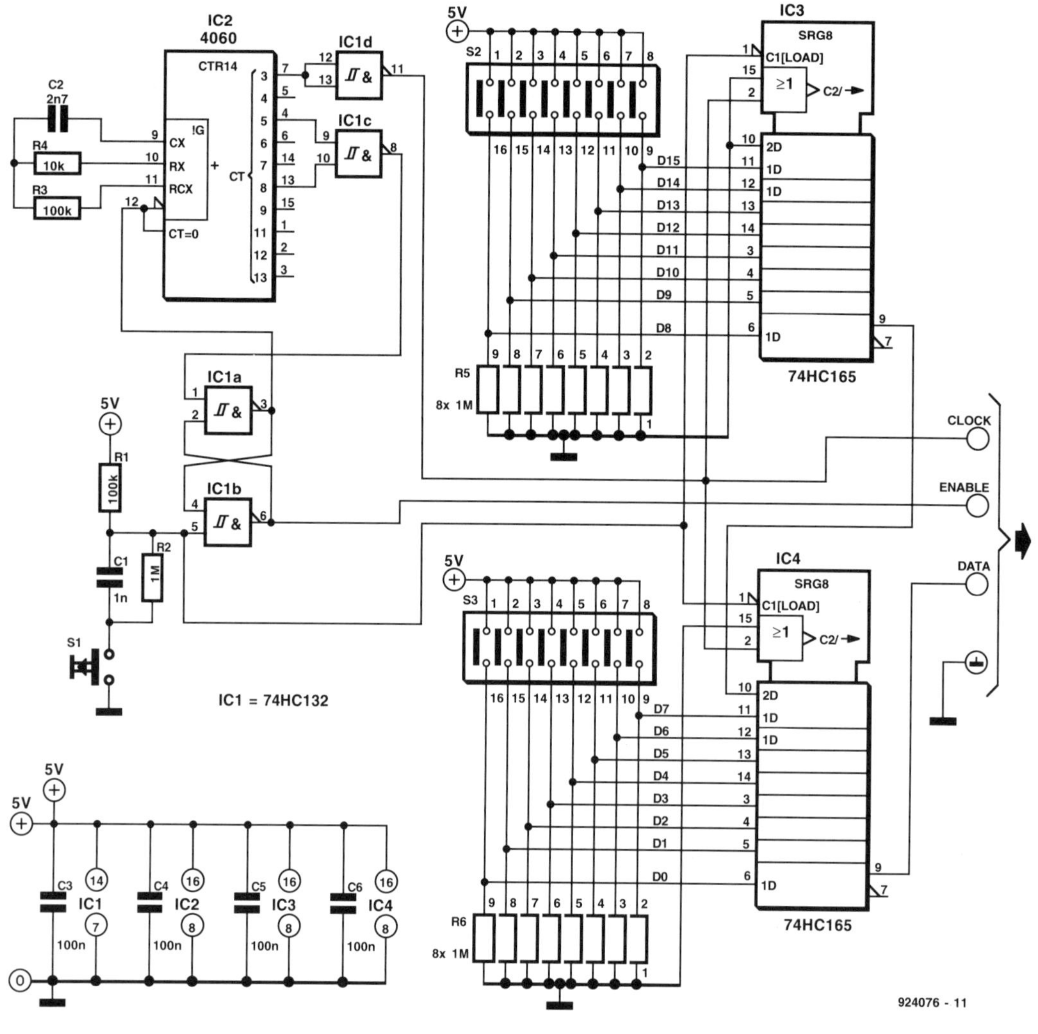

IC_{1d} to provide the clock for the shift registers. This ensures that D0 is present at the output for an entire clock period (1.25 ms).

After output 2^3 of IC_2 has been high 18 times, the bistable is reset and the enable line is pulled low by IC_{1c}, which is connected to outputs 2^5 (Q5, pin 4) and 2^8 (Q8, pin 13) of IC_2. This completes the cycle.

In this example, 18 bits are shifted. The two MSBs are determined by the serial input (pin 10) of IC_3. If this input is high, D16 and D17 are '1'. If these bits are required to be variable, one or two additional D-type bistables, connected as shift registers, must be added.

The current drawn by the generator, determined primarily by the pull-down resistors, is very low: if all bits are high, it is about 80 μA. After S_1 has been pressed, the oscillator is started briefly, which doubles the current.

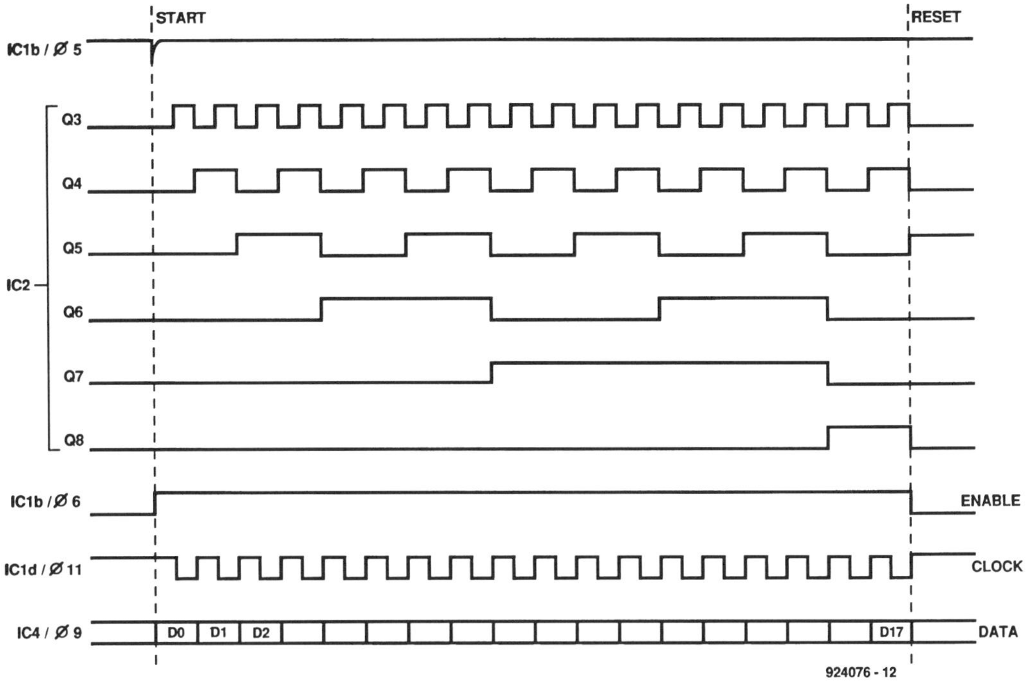

Relative humidity sensor

174 The Type NH-02 humidity sensor from Figaro consists of a capacitive humidity sensor, Z_s, in series with a thermistor, Z_t on an aluminium substrate. The parameters of the thermistor ensure that it compensates the temperature-dependence of the dielectric of the sensor almost completely. Over the temperature range 15–35 °C, the temperature dependence of the NH-02 is about 0.3% K^{-1}.

Before the sensor can operate, it needs to be energized by an alternating voltage with a frequency of 50–1000 Hz. In the present circuit, that signal is generated by a Wien oscillator based on IC_{1a}, whose frequency is set to about 1 kHz by bridge R_3-R_4-C_3-C_4. The two diodes in the feedback loop ensure amplitude stability. The level of the alternating signal, and thus the maximum output voltage, can be set over a wide range with P_2. This preset is coupled to the sensor by buffer IC_{1b}, which prevents the impedance of the sensor influencing the set voltage.

The output voltage of the sensor is buffered by IC_{1c} and then applied to peak rectifier IC_{1d}-D_3-D_4. The output voltage of the rectifier is smoothed by R_{10}-C_6, after which the direct voltage is amplified ×5 by IC_{2b} ($\alpha = 1+R_{12}/R_{11}$).

The characteristic of the relative humidity vs the output voltage is not linear as shown in the graph. For simple on-off switching applications that does not matter, but for others it may be necessary to make the voltage linear. This is best done with a computer and suitable software.

The power supply may be an asymmetric stabilized 8–15 V type. Opamp IC_{2a} produces an artificial earth at half the supply voltage, so that the circuit is powered by a symmetric 4–7.5 V supply. Note that the supply voltage must be larger than the wanted output voltage.

The circuit is best aligned with the aid of

an oscilloscope. With P_1, set the amplitude of the oscillator to a peak value of 2 V (measured between pin 1 of IC_1 and earth). Then, with P_2, set the wanted peak value of the supply to the sensor. In many cases, a peak value of 1 V will be found suitable; the output voltage then varies between 0 V and 5 V.

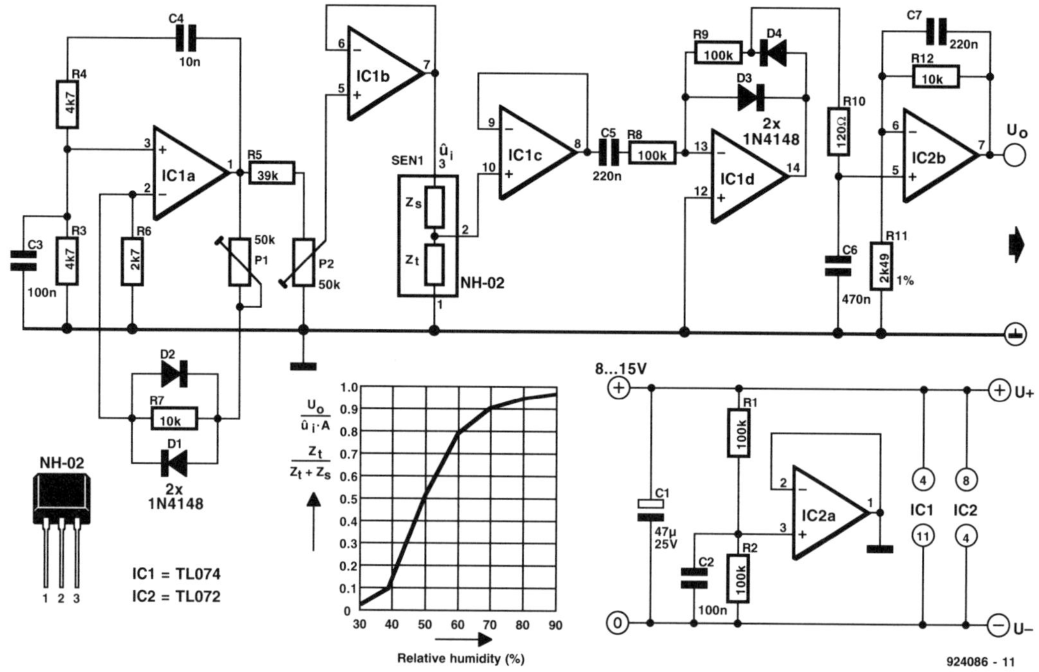

Differential temperature indicator

175 The circuit in the diagram enables the monitoring of two temperatures, t_1 and t_2. The sensors are negative temperature coefficient (NTC) resistors, R_3 and R_4, which may be connected to the circuit via lengths of circuit wire that may be up to a few metres long.

Diodes D_1 and D_2 indicate whether t_1 and t_2 are close to each other or not. 'Close' means that the difference between the two voltages from the sensors is smaller than the level set with P_2. If the temperatures are close, both LEDs light; if they are not, one LED will go out: D_1 if t_1 is higher than t_2, and D_2 if t_1 is lower than t_2. Apart from this optical indication, it is possible to obtain an acoustic one by connecting a d.c. buzzer to terminal S.

It is also possible to connect a 7–12 V relay with a maximum energizing current of 400 mA, to this terminal. Free-wheeling diode D_5 protects T_1 against a possible, destructive back-e.m.f.

The circuit draws a maximum supply current of 35 mA, largely on account of the LEDs.

A certain temperature off-set may be preset with P_1. Normally, this preset will be at the centre of its travel: when $t_1 = t_2$, the potential at the wiper will then be half the supply voltage.

The 'window' within which temperatures are monitored may be preset with P_2 to a value of 1–25 °C at a regulated supply voltage of 8 V and t_1 = 25 °C.

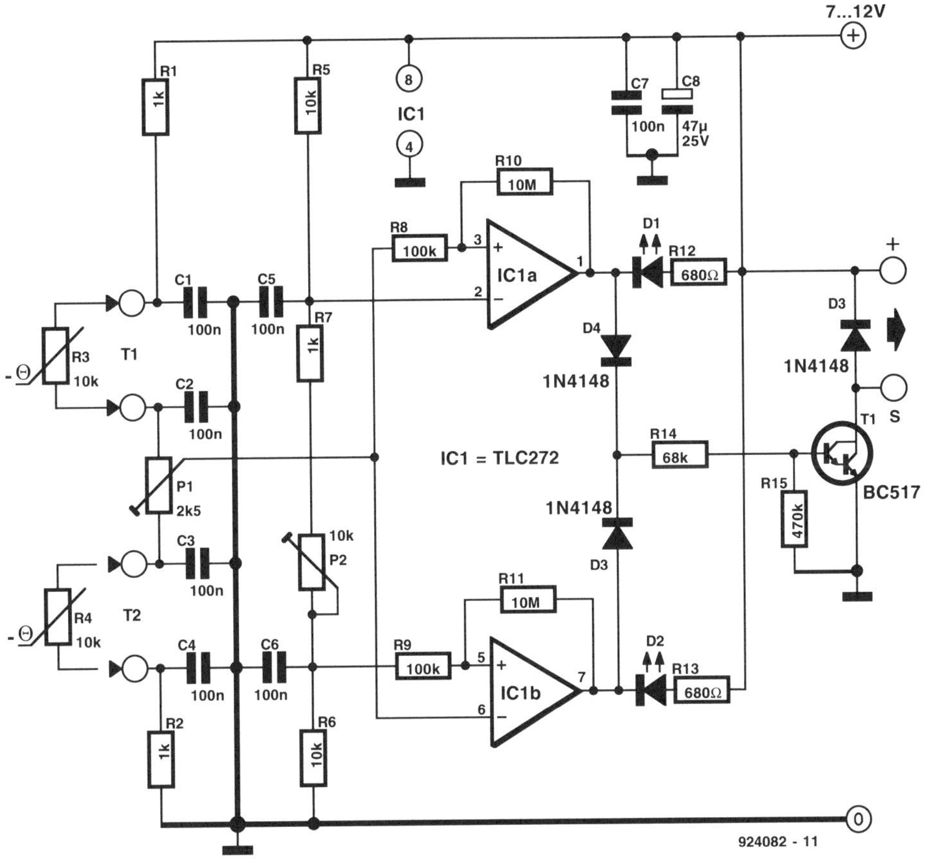

Thermocouple-to-DMM interface

176 Thermocouples are economical and rugged devices for temperature measurements. Because of their small size, they respond quickly, and are good choices where fast response to temperature changes is important. Type K thermocouples have a wide temperature range, and are used from cryogenics to jet engine exhaust analysis. The present circuit converts thermocouple outputs into a direct voltage with a gradient of 10 mV °C^{-1}, which is just about all that is needed to enable a digital multimeter (DMM) to be used as a readout. Alternatively, the interface output signal may be fed to a computer system for more advanced temperature recording applications.

The AD595A from Analog Devices is a complete instrumentation amplifier and thermocouple cold junction compensator on a monolithic chip. It combines an ice point reference with a precalibrated amplifier to produce an output of 10 mV °C^{-1} directly from a thermocouple signal. The AD595A is laser trimmed for type K (chromel-alumel) thermocouples.

The application of the AD595A shown here could not be simpler. The Type K thermocouple is connected to a special socket, K_1. Resistor R_1 is included to ensure that common-mode voltages induced in the thermocouple loop are not converted to normal mode. The AD595A features an alarm output, +ALM (pin12), which is used here to drive a low-current LED. The +ALM output goes low when one or both of the thermocouple leads are interrupted. It should be noted that the cold junction compensation provided by the AD595A will be affected whenever the alarm circuit is actuated. This means that readings taken when the alarm output is actuated are invalid.

Because a thermocouple output voltage is non-linear with respect to temperature (see Table), and the AD595A linearly amplifies the compensated signal, the following transfer function must be used to determine the actual output voltage:

$$U_0=247.3(U_{th}+0.011)$$

where U_0 is the output of the AD595A and U_{th} is the thermocouple output in mV.

Since ANSI Type K and DIN NiCr-Ni thermocouples are composed of identical alloys, both may be used with the present interface.

Construction of the interface is straightforward since very few parts are involved. Note the copper area under the converter chip to improve the thermal contact between the thermocouple socket and the IC. A low thermal resistance is important here to ensure that the on-chip ice point reference operates correctly.

The interface may be powered from a symmetical or an asymmetrical power supply. In the first case, the '0' and 'ground' terminals are interconnected, and taken to the '–' terminal of the battery. Note, however, that temperatures below 0 °C cannot be measured if a single-ended supply is used. When a symmetrical supply is used, the full temperature range becomes available. A symmetrical supply is best provided by two 9-V batteries.

The current drain of the circuit is not greater than 1 mA with the thermocouple connected, and not greater than 10 mA with the thermocouple disconnected (alarm LED lights).

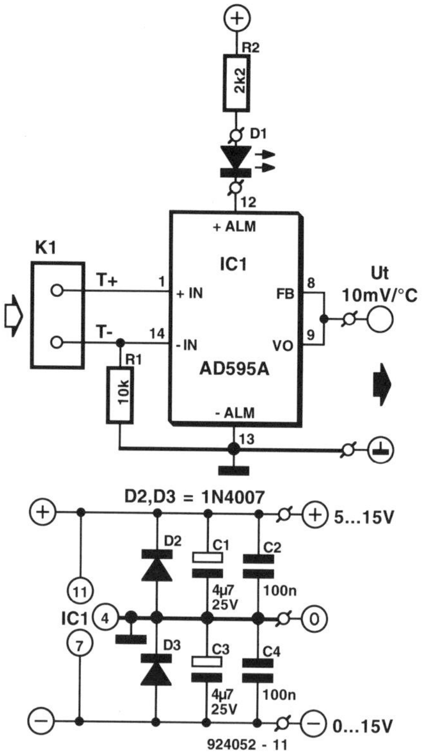

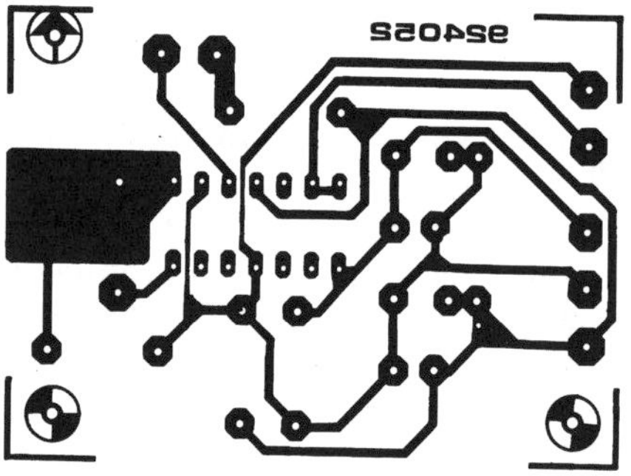

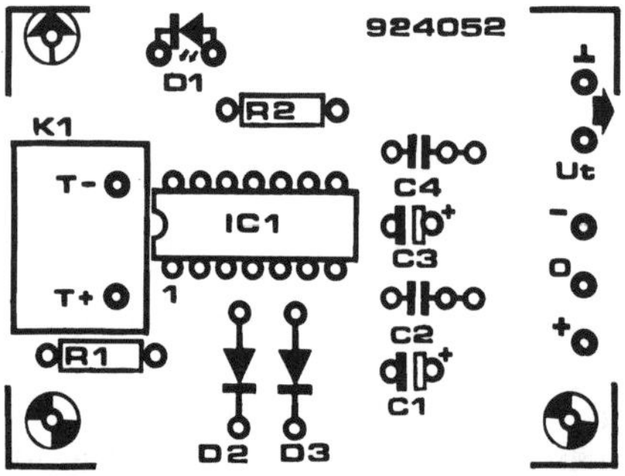

Parts list

Resistors:
R_1 = 10 kΩ
R_2 = 2.2 kΩ

Capacitors:
C_1; C_3= 4.7 µF, 25 V, radial
C_2;C_4 = 100 nF

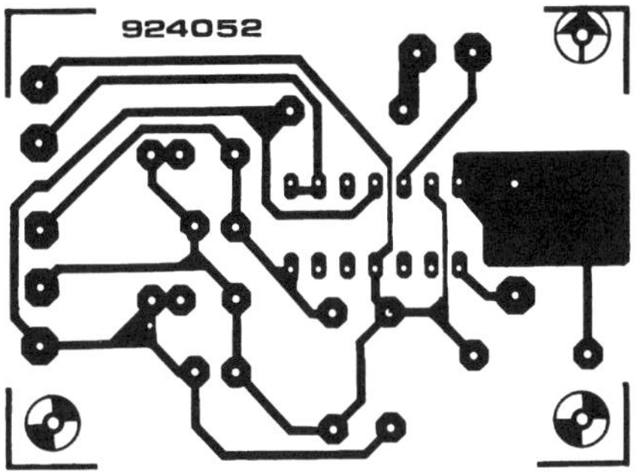

Semiconductors:
D1 = LS3369EH (low-current LED; red)
D2; D3 = 1N400

Integrated circuits:
IC_1 = AD595A (Analog Devices)

Miscellaneous:
K_1 = special thermocouple type-K socket, e.g, RS Electronics 473-127.
Plastic enclosure with battery compartment; approx. size 80x60x20mm

***Further reading*:**
"Temperature measurement techniques", *Elektor Electronics*, December 1991.

"A fast, precise thermometer", *Elektor Electronics*, January 1992.

Audible crystal tester

177 A crystal cannot be tested audibly unless its output is scaled down to the audio frequency range by a circuit as shown in the diagram.

A divider that is particularly suitable for this purpose is the Type 4060 CMOS IC. This circuit contains not only a 14-stage binary scaler, but also a complete oscillator.

The crystal to be tested is connected across the input terminals and S_2 set as indicated in the table. The crystal frequency is scaled down in IC_1 and, depending on the setting of S_2, one of the outputs of the 4060 drives transistor T_1 via R_2. The transistor, in turn, drives a small loudspeaker, LS_1. The power delivered to the speaker is limited by R_5 to prevent damaged eardrums.

It is, of course, not possible to use one scale factor for all sorts of crystal, and that is why switch S_2 enables selection of one of three different factors. For crystals <1 Mhz, the scale factor is 128; for crystals in the range 1–10 MHz, the scale factor is 4096; and for crystals >10 Mhz, the scale factor is 8192.

Also, crystals operating above 10 MHz oscillate readily at somewhat higher voltages than low-frequency ones. That is why S_{2a} and D_2 lower the supply voltage to 4.7 V when crystals below 10 MHz are tested.

If a mains-operated power supply is preferred over a 9-V battery, a 12 V/50 mA one is recommended. Inn that case, D_2 must be a 6.8 V type.

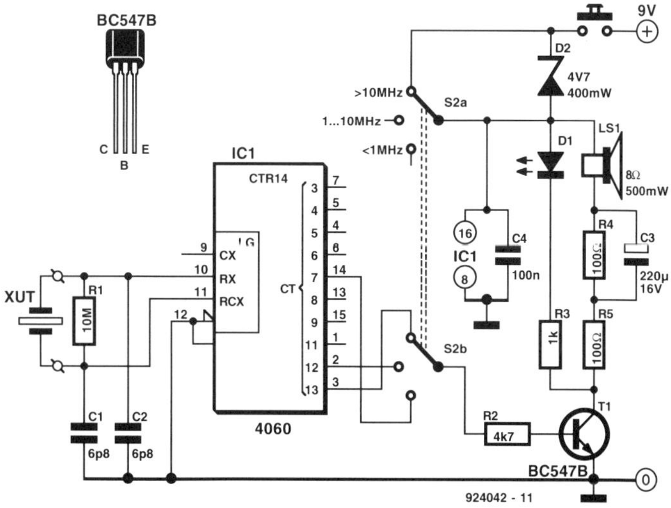

Inductive proximity switch

178 Inductive proximity switches are used, for instance, for measuring motor speeds or determining the position of metal objects. They do not suffer from mechanical wear or contacts that spark. The latter is particularly important in spaces where explosive materials are stored.

Virtually all commercial proximity switches are constructed as shown in Fig. 1. An inductor in the resonant circuit of an oscillator serves as the sensor. If a conducting object enters the magnetic field of the coil, eddy currents are set up in the inductor. This damps the resonant circuit so that the voltage across it drops. This voltage drop is monitored with a Schmitt trigger. When the object gets very close to the inductor, and the voltage across the circuit drops sufficiently, the Schmitt trigger changes state. The trigger is followed by an output stage.

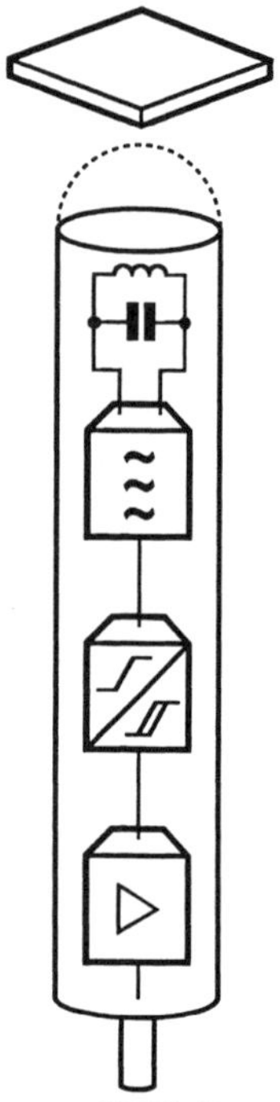

924073 - 12

The sensor, IC_1, used in the present circuit translates the approach of an object into a falling current through the sensor. In the absence of an object, the current is about 4 mA; when an object is at a distance of 4 mm from the sensor, the current is only 1 mA.

The sensor current is converted into a voltage by R_1. This voltage is applied to the non-inverting input of Schmitt trigger IC_2. Actually, since the hysteresis is small, IC_2 functions more as a comparator that likens the voltage across R_1 to that across R_3. When an object is within the proximity limit of 5 mm of the sensor, the potential across R_3 is larger than that across R_1, whereupon the output of IC_2 becomes logic high.

The potential across R_3 is dependent on the supply voltage, of course, but over the range of supply voltages stated in the diagram, it is always greater than the smallest drop across R_1 and always smaller then the largest drop across R_1. This ensures correct switching of IC_2 in all situations.

The only quantity affected by the supply voltage is the output potential of the circuit.

The sensor is a Type IFR10-82-01 from Baumer Electric. It has a diameter of 10 mm and is 5 mm long.

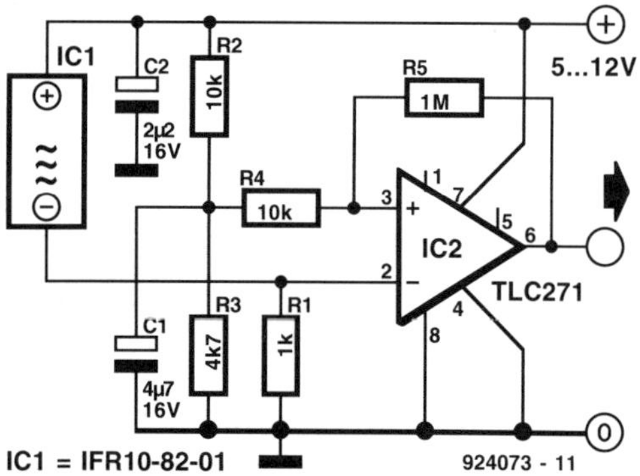

924073 - 11

Pulse generator

179 This pulse generator can provide up to to 12 pulses at its output, depending on the setting of hexadecimal switches S_2–S_{13}.
The (rectangular) pulses are generated by IC_{3c}; with circuit values as shown, their p.r.r. (pulse repetition rate) is a few kHz.
The output of the generator is taken to one input of NOR gate IC_{2c} via buffer/inverter IC_{3b}.
When switch S_1 is operated, bistable IC_{1a} receives a clock pulse, whereupon a logic 1 is passed to the D input (pin 12) of IC_{1b} via IC_{2a}. The other input of IC_{2a} is low because counter IC_3 is disabled via its reset input. At the next leading clock edge, the bistable is set, so that IC_3 and, via IC_{2b}, the pulse output are enabled.
When the relevant number of pulses has been output, junction R_3-S_2 goes high, whereupon the output of IC_{2a} goes low and IC_{1b} passes a logic 0 from its D input to pin 9 (Q\ output).
IC_{1a} is reset via C_1, so that S_1 can be operated again, and the counter is reset via pin 8 of IC_{1b} (Q output). Also, pin 8 disables gate IC_{2b}, so that no more pulses can appear at the output.
The generator draws a current of only 2.5 mA.

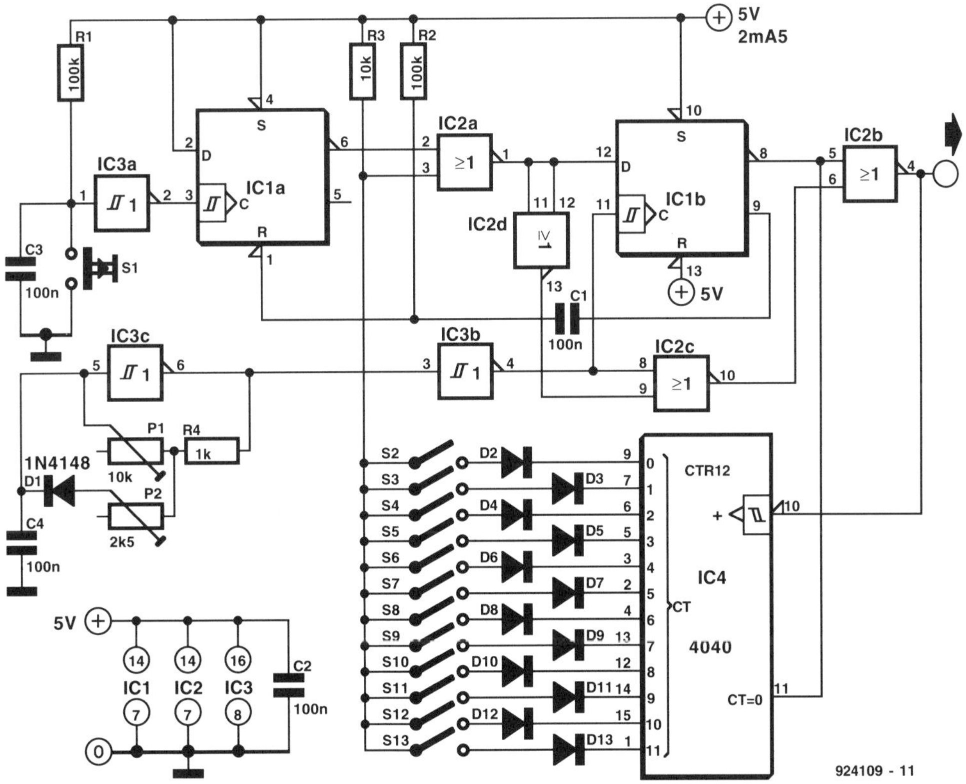

Noise generator

180 Noise generators are used for measuring the self-noise of amplifiers and receivers and for some acoustic measurements. The noise of traditional low-frequency noise generators is based on the stochastic properties of an ion current resulting from a gas discharge. A simple noise generator can, however, be designed without a special gas discharge tube: the reverse-biased base-emitter junction of a bipolar transistor is a compact and inexpensive alternative.

In the circuit diagram, the noise voltage is taken from the emitter of T_1. The base-emitter junction of this p-n-p transistor begins to behave like a break-down diode at a reverse bias of about 9 V, but it **really** starts to generate noise at 10 V.

To ensure that the circuit works satisfactorily with a battery supply of 9 V, a step-up generator, based on IC_1, is used. This stage provides a rectangular output voltage at a frequency of around 2750 Hz. The diode pump, consisting of C_1, C_3, D_1, and D_2, doubles the battery voltage, which results in a stable direct voltage of 10 V across D_3.

Low-pass filter R_{17}–C_6 prevents frequency components of the rectangular-waveform generator appearing in the noise spectrum. Each of the opamps in IC_2 raises the noise voltage in the frequency range from 10 Hz to 300 kHz tenfold. The amplitude of the output voltage can be preset with P_1. The noise signal can be tested with the a.c. buzzer by closing S_2.

Powered by a fresh 9-V battery, the circuit draws a current of 5–6 mA.

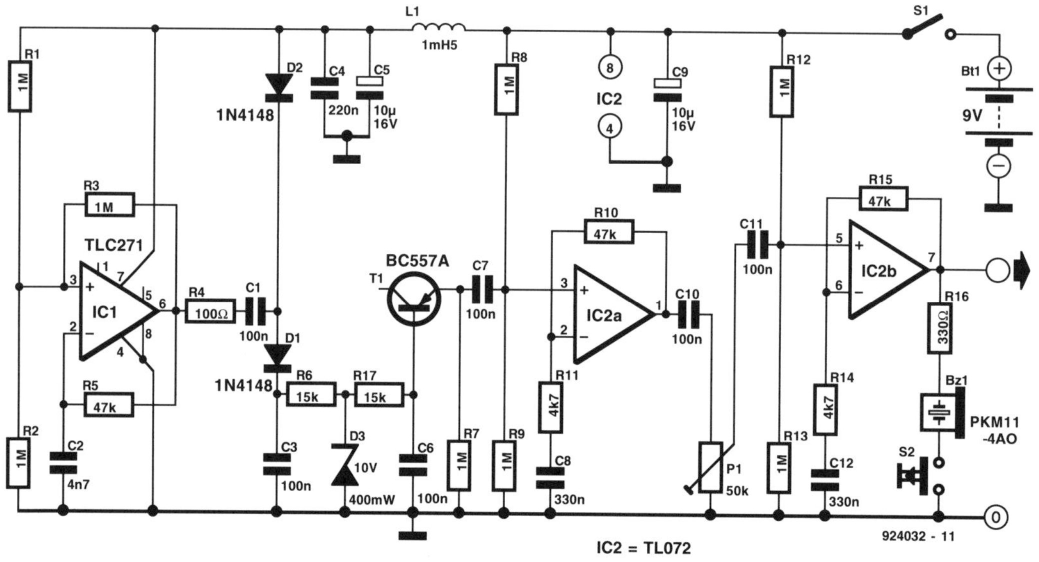

Digital pattern generator

181 During the construction and testing of digital circuits, there is often a need of a accurately defined bit pattern. The generator described here uses an EPROM to store the bit information for the desired pattern. When switch S_1 is pressed, a given pattern is generated once; seven different patterns can be generated simultaneously. The eighth data output of the EPROM is used to mark the end of a pattern.

After S_1 has been pressed, gate IC_{1d} arranges the resetting of address counter IC_2. Gate IC_{1c}, which is connected to the input of the address counter, functions as a start-stop oscillator that is switched by two NAND gates. These gates are designed to keep the oscillator disabled until the switch is released. Only then are pulses fed to the address counter. The frequency range of the clock can be varied to individual requirements by altering the value of C_3. With values as shown, the frequency can be set with P_1 between 15 Hz and 150 Hz.

When a bit pattern is generated, data line D7 must be logic high. At the end of the pattern, this line goes low, whereupon the clock is stopped.

The maximum length of a bit pattern in the circuit as shown is 8191 clock pulses (8×1024–1 stop pulse). If link JP_1 is set to position B, an external stop pulse can be used.

The circuit draws a current of about 8 mA.

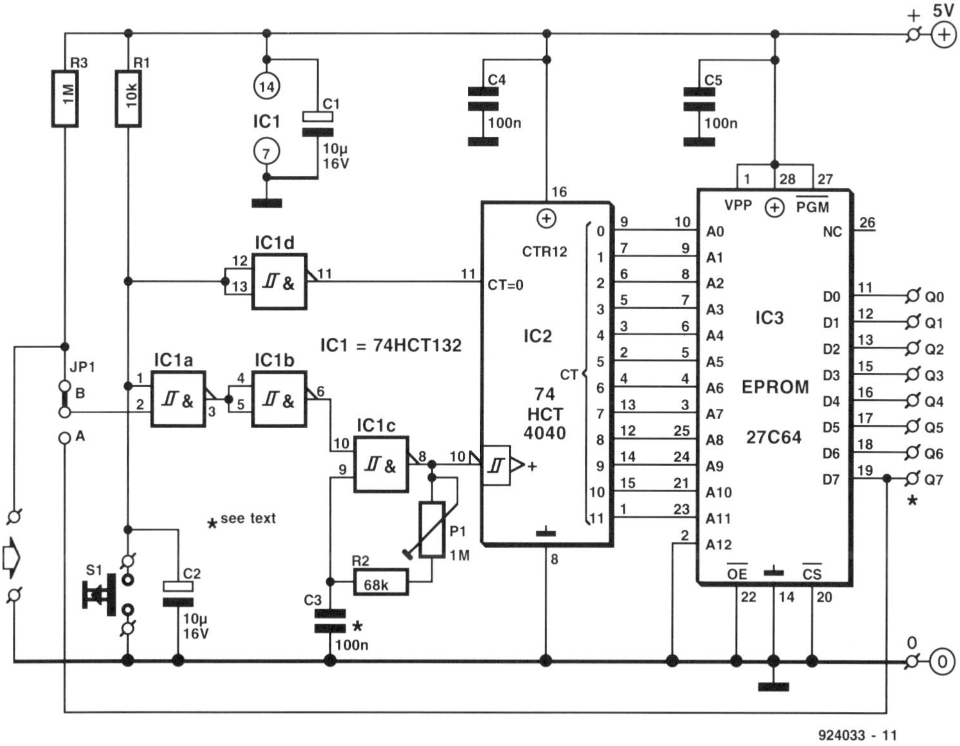

Audible transistor tester

182 The tester has no operating controls and draws current only when the loudspeaker sounds.

It comprises three oscillators that drive a common loudspeaker via buffering OR gates. Each oscillator consists of a permanently fitted transistor and the transistor on test (TOT). The oscillator frequency depends on the time constant of the appropriate *RC* network (R_5C_2, $R_{11}C_4$, or $R_{15}C_5$) and the current amplification of the TOT. The n-p-n and p-n-p sections are virtually identical, but, of course, the supply lines have been reversed. The buffer stages differ slightly to ensure that the loudspeaker remains connected correctly.

The stage for testing n-channel FETs is basically derived from the design of the n-p-n stage. In operation, the gate must be negative with respect to the source. Transistors T_1, T_3 and T_6 must be high-gain types to prevent TOTs with a low β being assessed as non-functional.

The tester has been designed to enable unknown transistors to be tested without any danger of damage. The pinout of bipolar transistors may be EBC or BCE. Test leads with terminals EBCE or even CEBCE (both n-p-n and p-n-p) are, therefore, recommended. The transistor to be tested is connected with its pins to different terminals until the loudspeaker sounds. This does not only give the pinout, but also makes it clear whether the TOT is an n-p-n or a p-n-p type. The frequency of the emitted tone indicates whether the emitter and collector have been connected correctly. When they have not, the current gain is only about 1, so that the oscillator frequency is high. If the two connections are interchanged, the current gain is higher and the oscillator frequency lower.

Owing to the symmetry of n-channel FETs, it is not possible to distinguish between source and drain. When the loudspeaker sounds, it means only that the gate has been connected correctly.

The test leads may be terminated into crocodile clips or into quick-release, spring-loaded terminals.

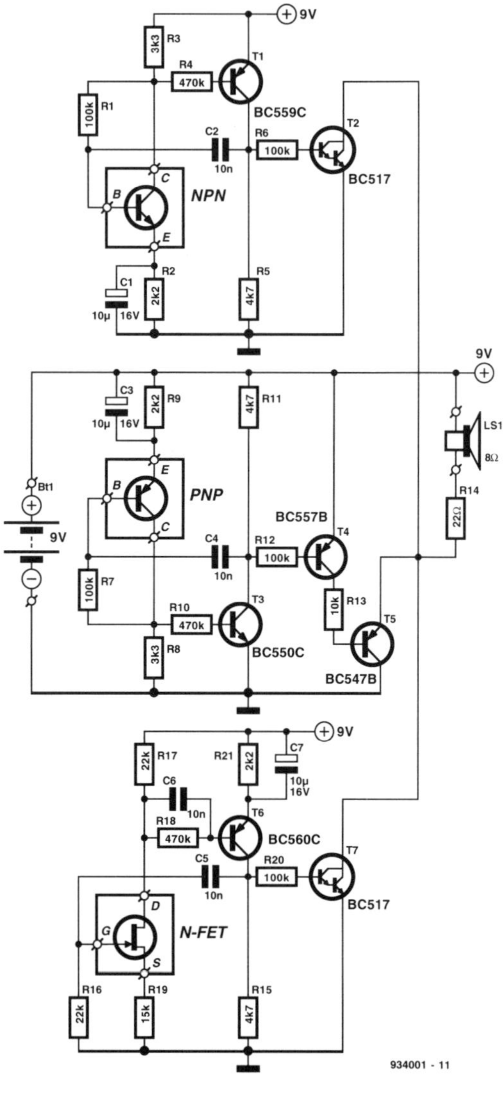

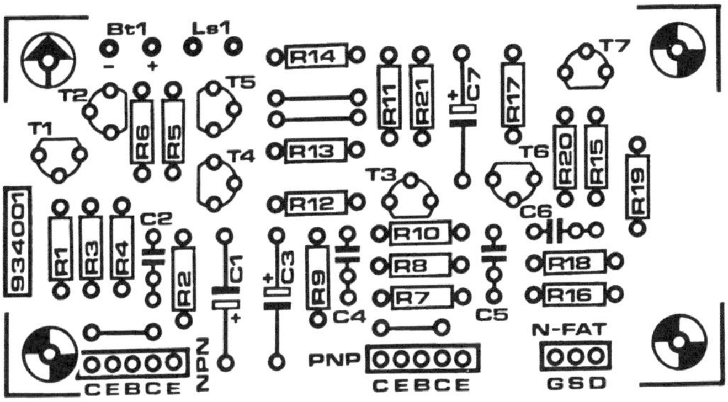

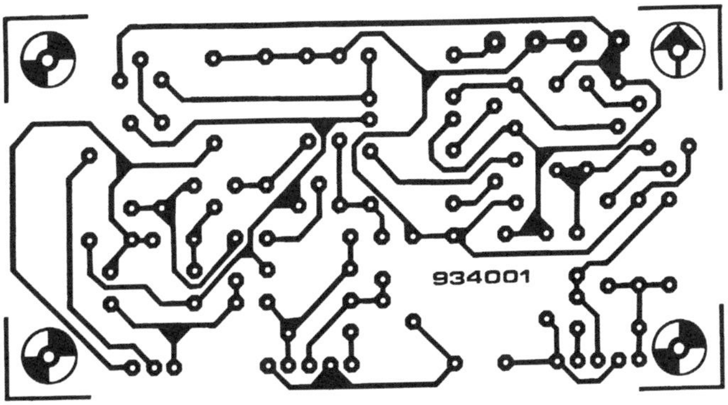

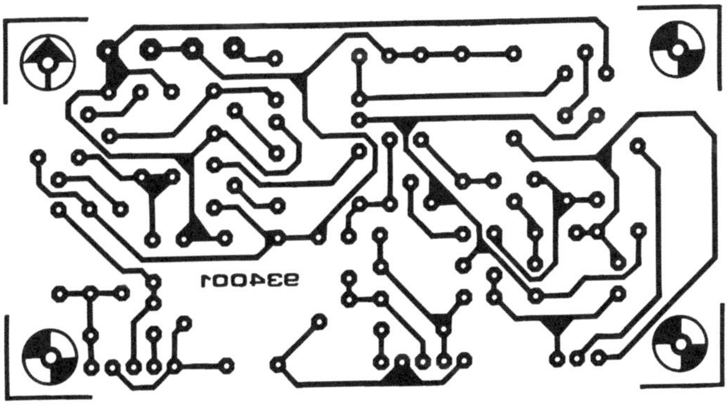

Audible tester

183 A simple unit for straightforward testing an audio, TTL or CMOS circuit is always useful.

A simple oscillator, based on IC_{1a}, generates a frequency that can be preset between about 500 Hz and 1.5 kHz with P_1. Before the signal is ready for use, it is buffered by IC_{1b}.

Switch S_2 serves to match the signal to the equipment on test. With the switch in the upper position, the tester can be used with audio circuits; in the centre position, with CMOS circuits, and in the lowest position, with TTL circuits (only HC or HCT types, not standard TTL logoc or LS2 types). If testing of LS2 type circuits is

required, the value of R_2 must be reduced to 1 kΩ. The 4093 cannot provide sufficient current for testing standard TTL logic circuits.
The output of the probe that is used to check various points in the equipment on test is applied to the input of IC_{1c}. If there is a signal present, the LED lights and the buzzer sounds.
The tester may be powered by a 9-V battery or a 9-V mains adaptor. It may also be possible to obtain the power from the equipment on test. The tester draws a current of about 11 mA.

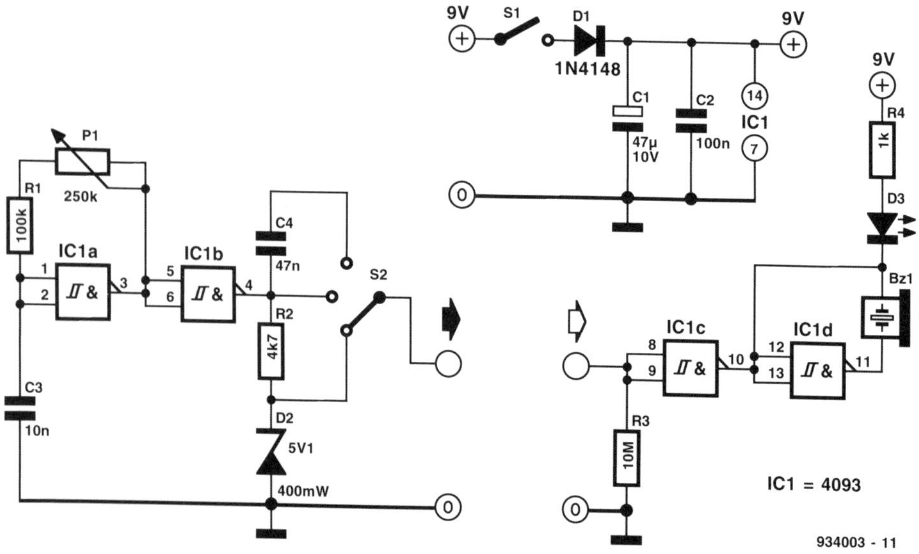

General purpose digital tester

184 The tester can not only detect rectangular-wave signals, it can also generate them. The test signals may be generated continuously or for a preset period as required.
The detector part is based on IC_2. A signal input to K_3 is applied to IC_2, where it is divided by 1024 and then fed to a buzzer. The buzzer will sound at a frequency between 1 Hz and 10 kHz if a signal is present at a frequency between 1 kHz and 10 MHz respectively.
The generator is formed by monostable IC_{1a} and variable oscillator IC_{1b}. The pulse width of the monostable may be set with P_1 to 1s maximum. The monostable is started when push-button switch S_1 is pressed. Rotary switch S_2 serves to select a continuous or a burst signal; in its third position there is no output.
The oscillator operates over three ranges selected by S_3: 1 Hz, 100 Hz, 10 kHz. The wanted frequency is selected with P_2. The required signal is available at K_2.
Although the tester may be powered by a 9-V battery, provision has been made for an external 5–15 V source. Diode D_1 serves as power on indicator. Switch S_4 is the on-off switch. The tester draws a current of about 50 mA.

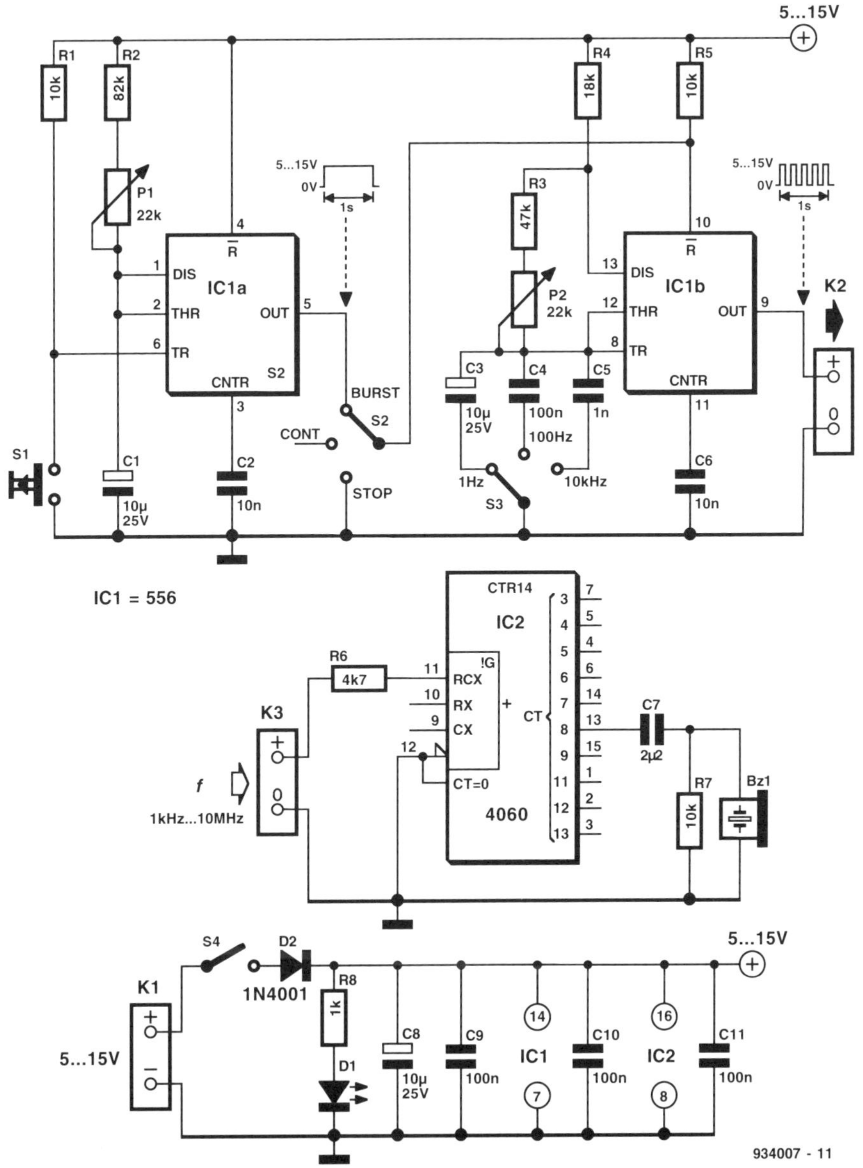

934007 - 11

Multi-phase rectangular-wave generator

185 The generator consists essentially of two distinct parts. A rectangular-wave oscillator, based on IC$_1$, generates two signals which may be phase-shifted between 0° and 360° with respect to one another. The frequency of the signal at a supply voltage of 5 V is around 1 kHz and at a supply voltage of 12 V about 1.4 kHz.

The stage based on by-eight-counter/divider IC$_2$ provides eight signals, each of which is 45° out of phase with the preceding and the following signal.

The output of IC$_{1a}$ may be shifted between

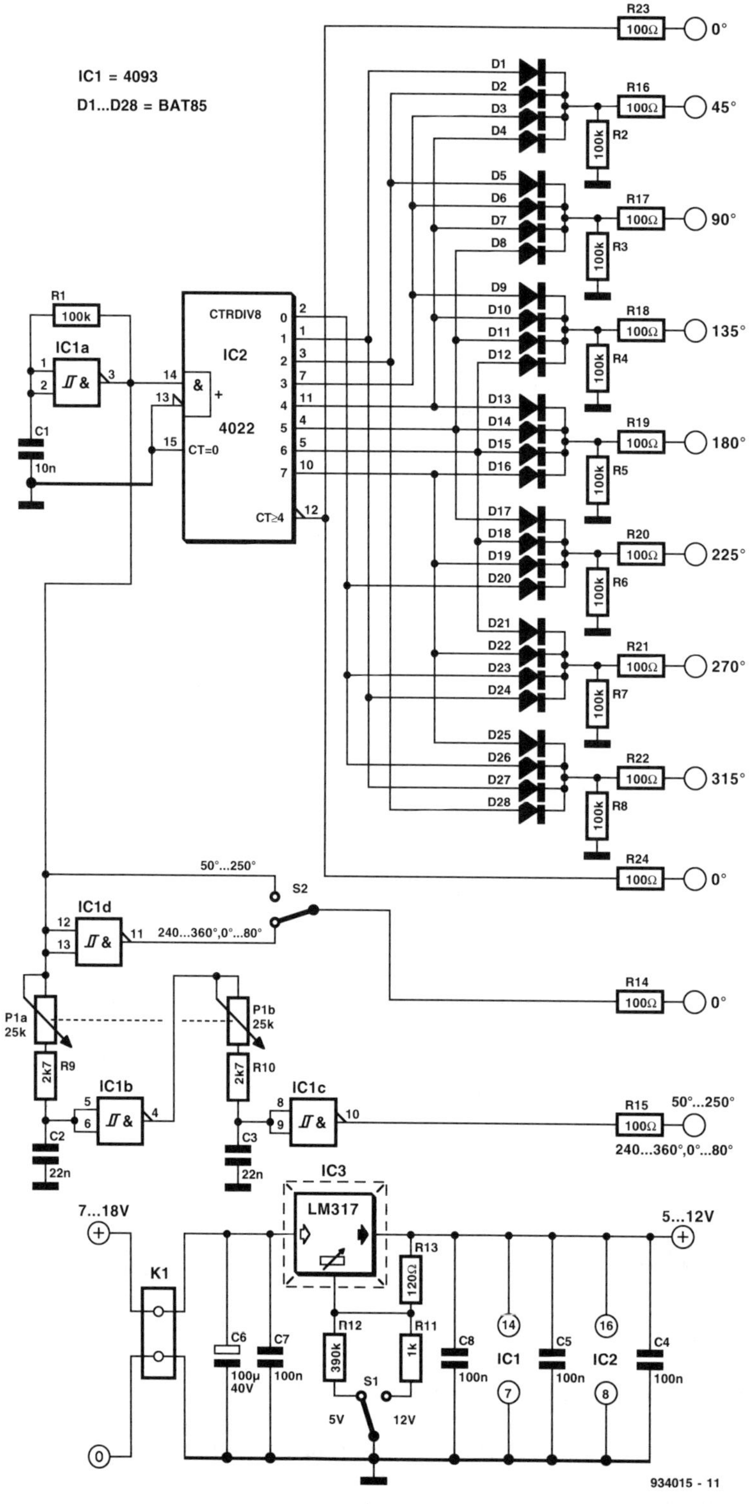
IC1 = 4093
D1...D28 = BAT85
R1
100k
IC1a
CTRDIV8
IC2
4022
CT=0
CT≥4
R23
100Ω
0°
R16
45°
R17
90°
R18
135°
R19
180°
R20
225°
R21
270°
R22
315°
R24
0°
R2
R3
R4
R5
R6
R7
R8
100k
C1
10n
IC1d
S2
50°...250°
240...360°,0°...80°
R14
0°
P1a
25k
P1b
25k
R9
2k7
R10
2k7
IC1b
IC1c
C2
22n
C3
22n
R15
50°...250°
240...360°,0°...80°
IC3
LM317
7...18V
5...12V
K1
R13
120Ω
R12
390k
R11
1k
S1
5V
12V
C6
100µ
40V
C7
100n
C8
100n
C5
100n
C4
100n
IC1
IC2
934015 - 11

50° and 250° with networks P_{1a}-R_9-C_2 and P_{1b}-R_{10}-C_3. Since the 0° signals may be inverted (which appears like a 180° shift) with IC_{1d} and S_2, the range 240°–180° is also covered with P_1 (via 360° = 0°). The 0° output at R_{14} must not be confused with those at R_{23} and R_{24}, because (since IC_2 is a by-eight divider), the signals at the R_{14} output and the other two differ in frequency by a factor eight.
Each counter state of IC_2 is reached when the associated output (of eight) goes high. If the counter cycle is thought of as a circle (360°), each counter state corresponds to a phase angle of 45°. The eight output signals are obtained by connecting four of the IC_2 outputs to the circuit outputs via an OR gate for every subsequent step in the count cycle. In this way, the circuit output is a square-wave (duty factor 1:1).The OR gates are formed by D_1–D_{28}. No OR gate is required for producing a 0° signals, since that is already arranged by the carry output of IC_2 (CT≥4) which is high only when the counter state is ≥4.
The use of the circuit is enhanced by S_1 which allows the supply voltage to be switched between 5 V and 12 V.
The current drawn by the generator is negligible.

Continuity tester

186 The continuity tester is based on the well-tried μA741, which is used as a comparator that operates a buzzer at its output. The inverting (–) input is at half the supply voltage via R_3 and R_4. When the potential at the non-inverting (+) input is lower than that at the –ve input, which is the case when the resistance between the two terminals is small enough, the buzzer will be actuated.
With values of R_1, R_2 and P_1 as specified, the sensitivity of the tester is about 1 kΩ. The circuit is set by connecting a 1 kΩ resistor between the input terminals and adjusting P_1 so that the buzzer just does not sound. The sensitivity may be increased to 100 Ω by making the values of R_1, R_2 and P_1 ten times smaller. The circuit then draws a slightly larger current (2.5 mA instead of about 2 mA).

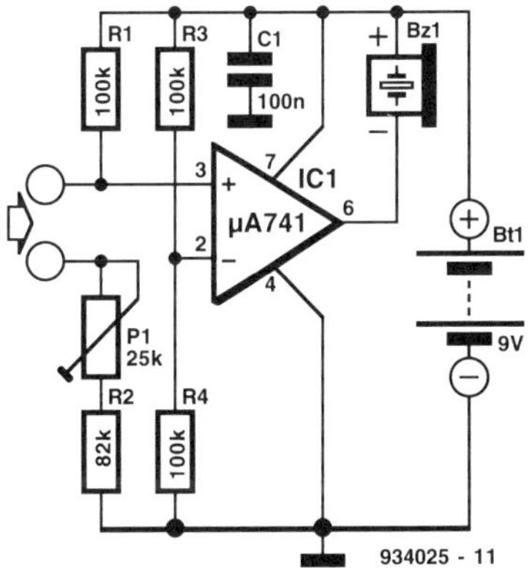

Audible continuity tester

187 Whatever the latest trend in digital multimeters, no electronics workshop is complete without a continuity tester for fast and reliable testing of connections, components and PCB tracks. The little instrument discussed here is an audible continuity tester whose output frequency is determined by the resistance measured. Handheld and extremely simple to use, the tester allows you to concentrate on the connections you wish to examine rather than on a LED, a display or a moving-coil meter. So, using the continuity tester is a matter of listening as you 'probe through' the circuit under investigation.

The circuit diagram shows a two-transistor oscillator whose output frequency is a function of the resistance measured between the input terminals, A and B. The feedback element in the oscillator is formed by capacitor C_1. Diode D_1 and the base-emitter diode in T_1 ensure smooth charge reversal in C_1, without creating a potential divider with base resistor R_1. The collector resistance of T_1 is formed by a diode, D_2, which ensures stable amplification in spite of battery voltage reduction. The distortion caused by this arrangement is of little consequence since the output is not a clean sine wave anyway. The transistors draw the greater part of their supply current through a passive piezo buzzer, and so generate an audible tone.

A combination of one n-p-n and one p-n-p transistor is used to make sure that no battery power is consumed when the test terminals are not connected. This obviates an on/off switch. Depending on the setting of preset P_1, the current drain of the actuated circuit is a modest 3–5 mA.

The frequency of the tone produced by the tester drops with increasing resistance measured. It is practically impossible to damage components or connections with the tester, since the voltage difference between the input terminals is about 8 V, and the maximum current about 50 μA.

Preset P_1 is adjusted such that an agreeable tone is produced with the input terminals short circuited. Next, connect a resistor of 22 kΩ, whereupon the tone frequency should drop considerably.

The completed printed circuit board is fitted into a probe-type enclosure, together with the 9-V (PP3) battery. Input 'A' is connected to a solid pin (the actual probe tip) that protrudes from the front side of the enclosure, while 'B' is connected to a flexible, red wire fitted with an insulated croc clip.

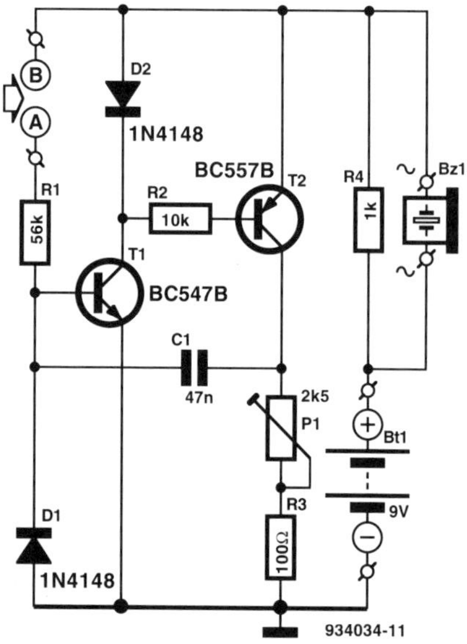

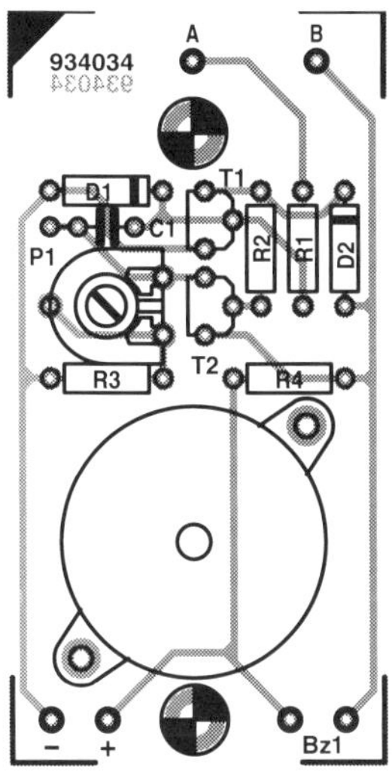

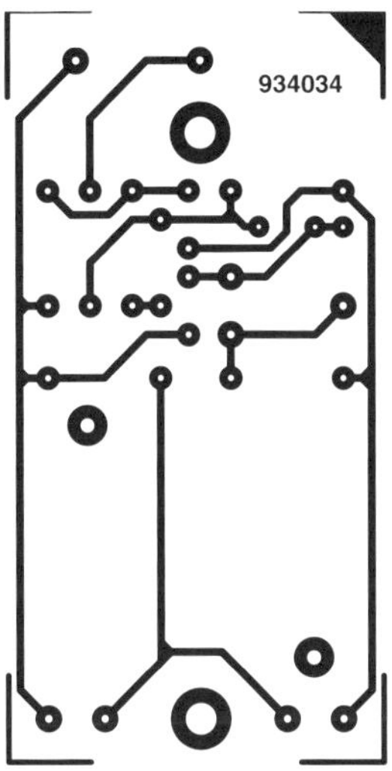

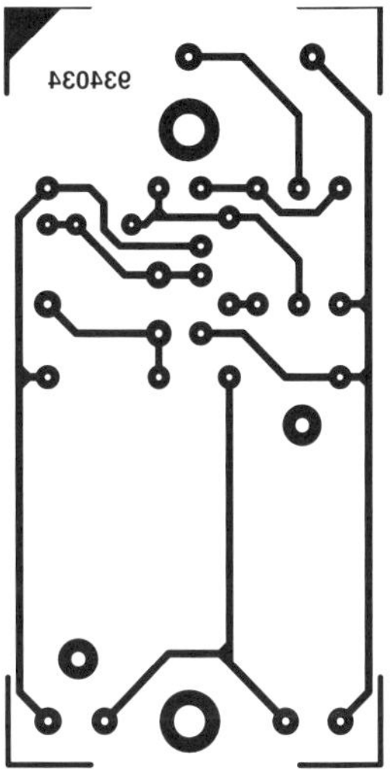

Parts list

Resistors:
R_1 = 56 kΩ
R_2 = 10 kΩ
R_3 = 100 Ω
R_4 = 1 kΩ
P_1 = 2.5 kΩ preset, H

Capacitors:
C_1 = 47 nF

Semiconductors:
D_1, D_2 = 1N4148
T_1 = BC547B
T_2 = BC557B

Miscellaneous:
Bz_1 = passive, a.c. type piezo-ceramic buzzer
Bt_1 = 9-V (PP3) battery with clip-on leads
Probe enclosure, e.g. Conrad Type 52.68.86-66

Logic level tester

188 The tester shows on a seven-segment, common-cathode display whether its input is logic 1 (H on the display) or logic 0 (L on the display). An undefined level is indicated as 'n'.
When the input is low, T_1 is switched off and T_2 and T_3 conduct. This results in a high output at IC_{1a} and a low one at IC_{1b}, whereupon the segment 'd' lights. Together with the always actuated segments 'e' and 'f', this causes an 'L' to be displayed.
When the input is high, T_1 is on and T_2 and T_3 are off. The output of IC_{1a} is then low and that of IC_{1b} is high. Apart from segments 'e' and 'f', 'b', 'c' and 'g' also light, which causes an H to be displayed.
When the input is undefined, or open, all transistors are on (because of R_1, R_2 and R_3). The outputs of IC_{1a} and IC_{1b} are then both high, which causes segments 'a', 'b' and 'c' to light together with 'e' and 'f; an 'n' (no level) is then displayed.
The buzzer is not essential, merely useful; it may thus be omitted. If it is used, it should be a type with integral oscillator.
The type of the components used is not critical. The display may be any seven-segment, common-cathode type. IC_1 may be a standard 7400 or an LS version of this or the open-collector version 7401; HC and HCT types are less suitable because these cannot provide the required output current.
The switching thresholds of the tester are at 1 V and 3 V, which is not standard, but in practice they do well. The 3 V threshold

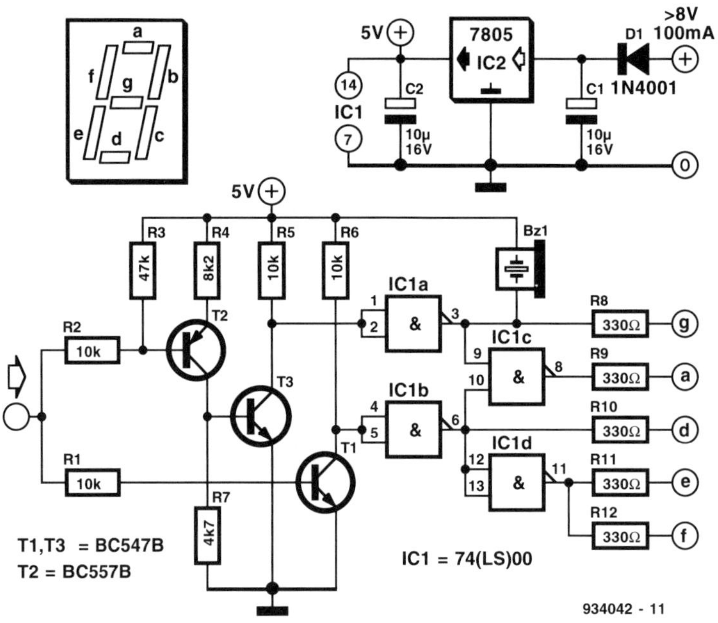

may be lowered slightly by increasing the value of R_4.

The input impedance of the tester is 5 kΩ, so that it does not affect the circuit on test.

The current drawn by the tester is determined by the type of display; in the prototype, it was about 60 mA.

Multi-function test probe

189 Based on only two ICs, this circuit can perform three test functions as selected by switch S_1.

The circuit is based on a clock generator, IC_{1a-1c}, that provides a rectangular signal at a frequency of 3 Hz.

With S_1 in position A, T_1 applies the clock signal to the probe. This buffer stage can deliver a current of up to 100 mA via limiting resistor R_7. In practice, this means that the probe can inject the 3 Hz signal into any digital circuit, thereby overruling all logic outputs. Only short circuits can cause the signal to disappear and the probe can, therefore, quickly detect them. This is, however, a rough and ready test, because it means that the tester short-circuits all outputs connected to the track that is being examined, whereas manufacturers allow this to happen to only one output at a time.

With S_1 in position B, the tester can detect logic levels. When the probe is not connected to anything, the clock is present at the input of IC_{1d}, so that D_1 flickers in a 3 Hz rhythm. When a logic level is applied to the probe, it suppresses the clock; if the level is low, the LED goes out; if the level is high, the LED lights continuously.

The switching thresholds of IC_1 lie at about $^1/_3$ and $^2/_3$ of the supply voltage, which is suitable for the CMOS 4000 series and the HCT family. Standard TTL, LS, and HCT ICs may be used, but the thresholds are then not optimum.

The tester can work with input signals at frequencies of up to 30–40 MHz.

With S_1 in position C, the input signal is applied to the clock input of shift register IC_2. After eight clock pulses at the probe, the high level that is constantly present at inputs A and B of IC_2 will have arrived at the Q_H output, so that D_2 lights. Since IC_2 is constantly reset by the 3 Hz signal, the LED will flicker as long as there is a clock at the probe, provided that its frequency is higher than 25 MHz.

The circuit draws a current of 10 mA when both LEDs flicker. To ensure that the pulse generator works satisfactorily, the power supply should be able to provide a current of some 100 mA.

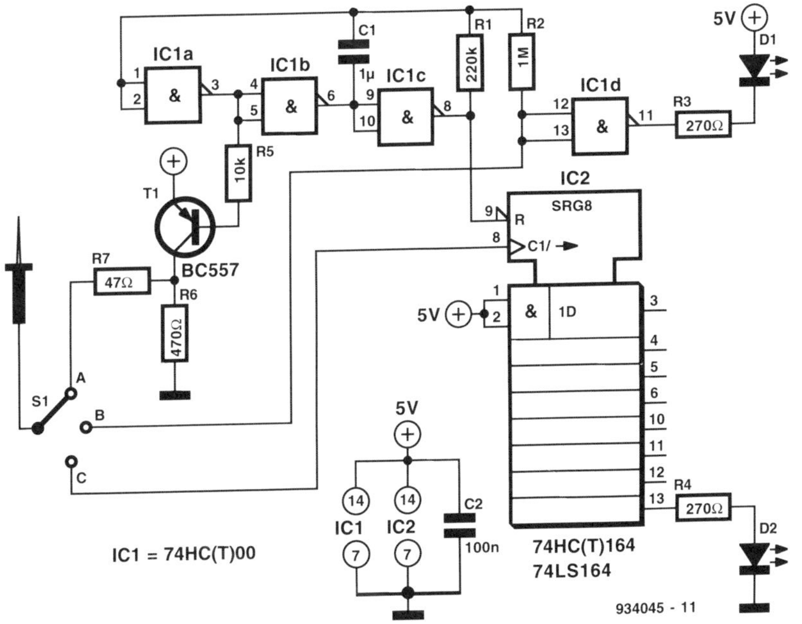

Triangular-signal generator

190 Pulse duration modulators require a perfect triangular signal, that is, one that is absolutely symmetrical with straight sides. Any deviation from this results in distortion.

The design is the usual rectangular-triangular oscillator. A Schmitt trigger converts a triangular wave into a rectangular one, which is used to continuously charge and discharge a capacitor via current sources. Positive current source T_1 arranges the charging and negative current source T_2 the discharging. The Schmitt trigger is formed by XOR gates IC_{1a} and IC_{1b}, which are connected as inverters. The remaining gates in the IC are not used here, but are available, for example, for a pulse duration modulator. Resistors R_1 and R_2 determine the hysteresis of the convertor.

The output of the trigger switches on T_1 and T_2 alternately. If the output of IC_{1b} is high, T_2 conducts; if it is low, T_1 is switched on. Resistors R_3 and R_4 ensure that the base voltage of T_1, and thus the potential across emitter resistor R_7, is constant if the output of IC_{1b} is low. This ensures that C_1 is charged via T_1 with a constant current. Diode D_1 arranges T_1 to be switched off rapidly. Tran-sistor T_2 operates similarly, but when the output of IC_{1b} is high.

The amplitude of the triangular signal must not be greater than the base voltage of T_1 and T_2. Assuming a supply voltage of 10 V, this means that the amplitude of the triangular signal is 2.5 V_{pp}.
Opamp IC_2 is a good-quality, fast type (200 V μs^{-1}) that functions as a buffer between the output and the input of the Schmitt trigger. Since it is a good-quality type, it has virtually no effect on the shape and quality of the triangular signal. If required, the symmetry of the triangular signal may be corrected to a small degree by connecting a low-value (1 kΩ) preset in series with R_3 or R_6. The value of the chosen resistor must be reduced by half the value of the preset.
The circuit can generate signals at frequencies up to 300 kHz. With values as shown, the output frequency is 38 kHz. The frequency may be altered by changing the value of C_1.
The circuit draws a current of about 8 mA, of which 5 mA flow through IC_2.

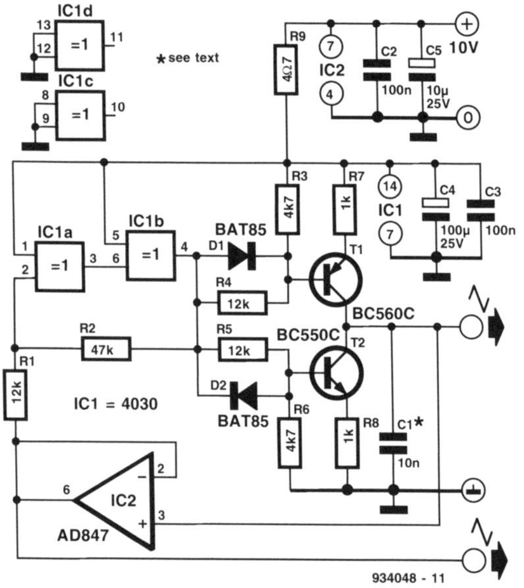

Manually controlled I²C output

191 During experiments with I²C circuits, it sometimes happens that the bus fails for no apparent reason. Since not everybody has a logic analyser available, the circuit presented here may in a number of cases prove to be useful. It is no more, and no less, than a manually operated I²C output. The clock and data lines are made high or low with the aid of two switches. Two LEDs indicate the levels as they are **really** on the bus. Note that if the I²C circuit has a built-in time-out (normally of about 1 ms), the present circuit is of no use, since it is impossible to react fast enough manually. The circuit on test will 'see' a circuit that does not react and will cease its I²C routines.
The present circuit consists of two identical halves: one for the data line (SDA) and the other for the clock (SCL). A change-over switch, S_1 or S_2, and a bistable, IC_{1a} or IC_{1b}, produce a logic 1 or 0. The bistable is essential to suppress the switch bounce, which, without fail, would be interpreted as a a series of pulses. The position of the switch indicates whether a 1 or a 0 is applied to the bus. The real level on the bus is indicated by D_1 and D_2. This does not necessarily coincide with the level that is applied by the present circuit, because the bus has an open collector structure. It is for that reason that the outputs of the bistables are not connected directly to the bus but via T_2 and T_4 respectively.
Pull-up resistors R_4 and R_{11} are needed only once on the bus, so that they may be omitted if they are already present elsewhere.
The present circuit may also be used as a simple I²C monitor. With both switches in the '1' position, the bus is not affected and the LEDs then show whether the bus is active. Much activity is needed, because a single transmission cannot be indicated by an LED.
The current is determined primarily by the LEDs and amounts to about 20 mA.

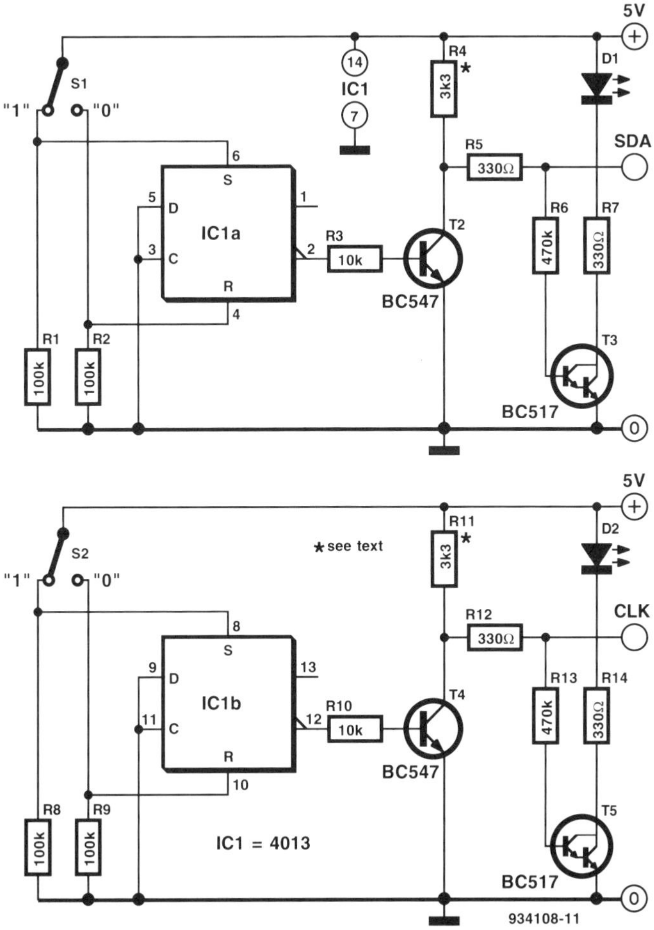
5V
S1
"1"
"0"
14
IC1
7
R4
3k3
*
D1
SDA
R5
330Ω
6
S
5
D
IC1a
1
3
C
2
R3
10k
T2
R
4
BC547
R6
470k
R7
330Ω
R1
100k
R2
100k
T3
BC517
0
5V
S2
R11
*see text
3k3
D2
CLK
R12
330Ω
8
9
13
IC1b
11
12
R10
10k
T4
BC547
10
R13
470k
R14
330Ω
R8
100k
R9
100k
IC1 = 4013
T5
BC517
0
934108-11

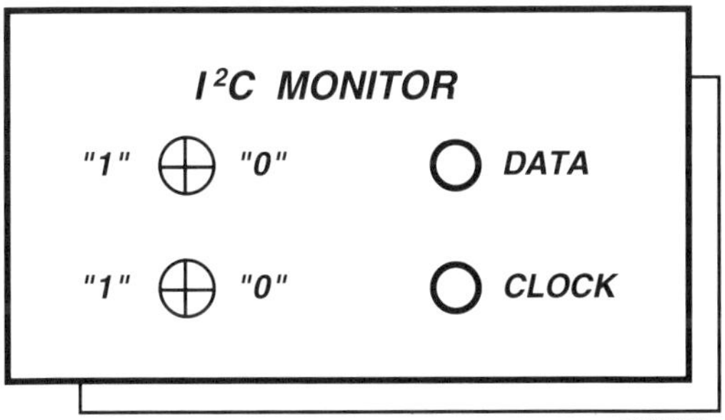
I²C MONITOR
"1"
"0"
DATA
"1"
"0"
CLOCK

934108-F

Storage oscilloscope tester

192 The tester generates two different signals for checking a digital storage oscilloscope. The first is a stepped voltage on to which glitches are superimposed; the second is a rectangular 2 kHz signal on to which a 15 Hz analogue signal is superimposed.

The first signal quickly shows whether the oscilloscope eliminates the glitches during signal processing. If it does, the measurements cannot be fully trusted. It can also be used to check the trigger function; if this is poor, it will not be able to cope with the signal properly.

The second signal shows whether at a given setting low-frequency signals remain clearly visible (they should).

The stepped voltage is generated by oscillator IC_{1a} and a digital-to-analogue converter (DAC)—see diagram. The converter consists of IC_2, D_1–D_4, R_3–R_6, and R_9. The signal is buffered by T_1. Diode D_5 compensates the forward bias of the base-emitter junction of T_1. The spikes are introduced by C_4.

The second signal is produced by switching T_2 in rhythm with the 2 kHz signal and changing the collector voltage with the 15 Hz signal generated by oscillator IC_{1b}.

The circuit needs a supply voltage of 9–15 V and draws a current of about 50 mA.

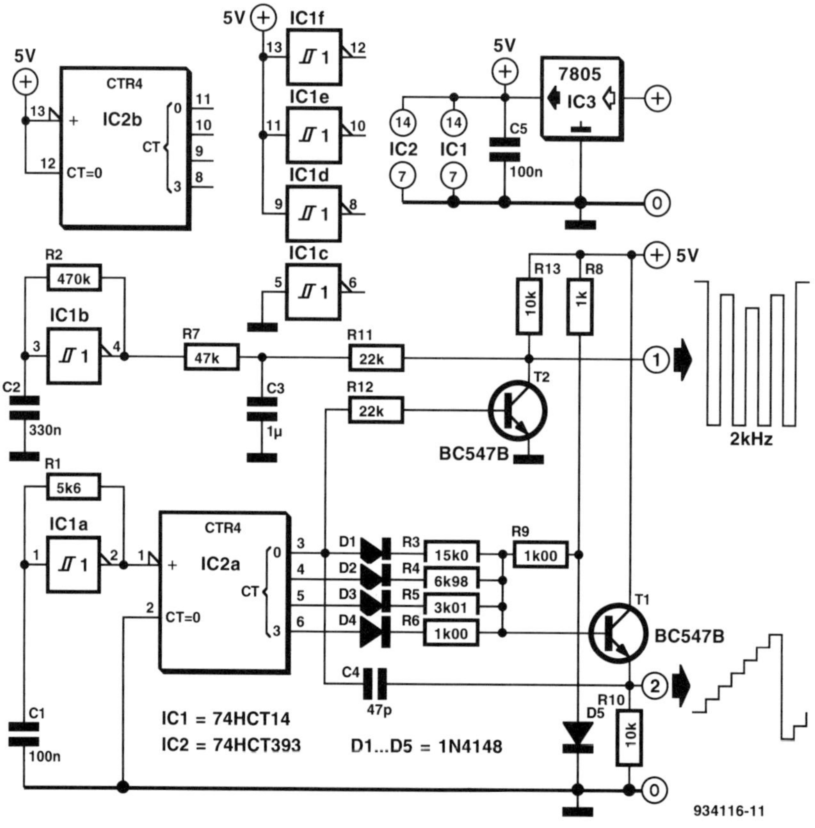

Electronic reversing circuit for model trains

193 Many model train enthusiasts find the mechanical reversing system for trains in the H0 series from Märklin and other manufacturers primitive and unreliable. The system is based on a.c. motors and a mechanical reversing assembly operated by a small electromagnet. The motor speed is determined by the track voltage, which can lie between 4 V and about 16 V. When the knob on the speed controller is turned fully anti-clockwise, the a.c. voltage on the track is briefly increased to 24 V. Ideally, this causes the electromagnet in the loco to be actuated and overcome the counterforce of a small spring. In practice, this way of changing the direction of a model train is fraught with difficulties as the tension of the spring is a very critical factor. In not a few cases, the voltage pulse fails to actuate the reversing mechanism, and instead cause the loco to hurl itself at a turnout where it is derailed. When the spring is too loose, it may happen that a loco, running at full speed, reverses suddenly with 'disastrous' results.

Some ten years ago Märklin recognized the disadvantages of the voltage-operated reversing system, and came up with an electronic alternative in the form of a zener diode and two transistors. Unfortunately, this upgrade proved expensive and difficult to fit in existing locos, which many modellers would be loathe to give up.

In all-electronic reversing systems developed a few years ago, the direction of the loco is 'stored' in a small button cell. This is necessary to prevent the information being lost as there is no supply voltage when the locostands still. The present circuit uses a 100 µF tantalum capacitor to keep the control circuit powered for up to 8 hours. The capacitor, in the author's opinion, is more elegant and environmentally safer than the battery. The circuit described below is based partly on SMDs (surface-mount devices), and is designed to be as economical as possible as regards power consumption.

When the circuit is not actuated, transistors T_1 and T_2 are off, and the inputs of

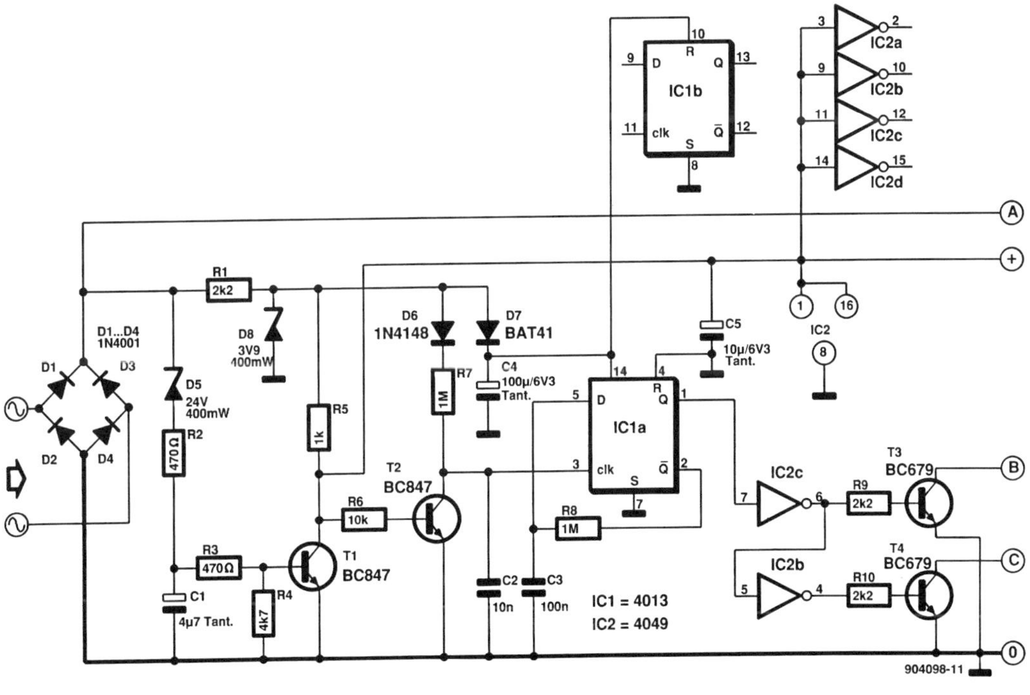

IC_1, a bistable Type 4013, are effectively not connected. The last direction of the loco is stored in the bistable. When the loco runs, D_5 blocks and keeps T_1 off. The 4049 (IC_2) is supplied with about 3.5 V via R_5, so that the motor driver transistors, T_3 and T_4, can be controlled. Transistor T_2 conducts and supplies IC_{1a} with a clock pulse. When the track voltage rises to 24 V, T_1 is turned on and removes the supply voltage from IC_2. T_2 is switched off and supplies the bistable with another clock pulse via D_6 and R_7. The active transistor, T_3 or T_4, is changed, and the motor changes direction in a reliable manner. Since the loco motor is powered with d.c. after installing the circuit, you may avail yourself of the opportunity to isolate the loco lights from the chassis and fit diodes to couple the lighting to the direction control.

The construction of the circuit is illustrated in the photographs. The dimensions of the circuit board are such that it can take the place of the relay removed (carefully) from the loco. No part of the circuit may touch the metal chassis.

The points marked **B** and **C** on the PCB are connected to the field terminals of the motor, and point **A** to the terminal previously connected to the slide contact. The slide contact and the loco chassis are connected to the bridge rectifier inputs. Finally, note that the printed-circuit board allows you to build five reversing circuits.

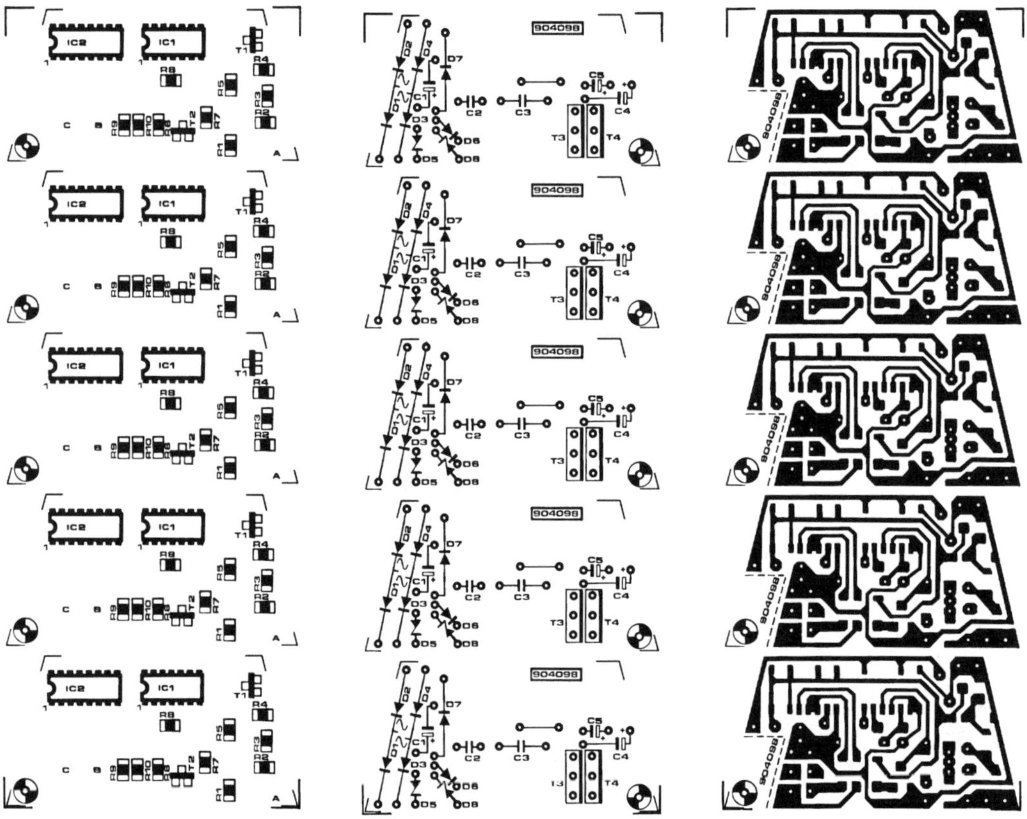

Parts list

Resistors (all SMD):
R_1, R_9, R_{10} = 2.2 kΩ
R_2, R_3 = 470 Ω
R_4 = 4.7 kΩ
R_5 = 1 kΩ
R_6 = 10 kΩ
R_7, R_8 = 1 MΩ

Capacitors:
C_1 = 4.7 µF, 16 V
C_2 = 10 nF, tantalum
C_3 = 100 nF, ceramic
C_4 = 100 µF, 6.3 V, tantalum

Semiconductors:
T_1, T_2 = BC846B (SMD)
T_3, T_4 = BD679
D_1–D_4 = 1N4001
D_5 = zener, 24 V, 400 mW
D_6 = 1N4148
D_7 = BAT41
D_8 = zener, 3.9 V, 400 mW

Integrated circuits
IC_1 = 4013 (SMD)
IC_2 = 4049 (SMD)

Miscellaneous:
PCB 904098 (for 5 reversing circuits)

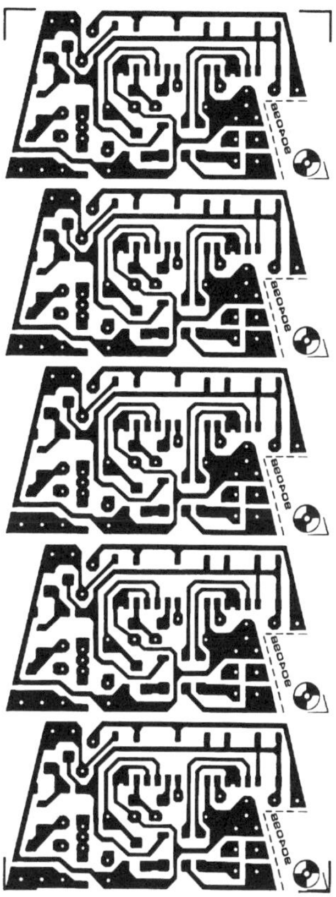

HCT crystal oscillator

194 The wide frequency range, low power consumption and well-defined switching levels of HCMOS inverters make these devices eminently suited to building quartz crystal oscillators with TTL compatible outputs. Here, the six gates in a 74HCT04 package are used to build three crystal oscillators. The only difference between the 2 MHz, 16 MHz and 24 MHz oscillators is the capacitance around the quartz crystal, which in all cases must be a type that resonates at the fundamental frequency; overtone crystals cannot be used here.

For output frequencies, f_o, other than the ones used here, use the following design data:

$$C_2 = 723 / f_o;$$
$$C_1 = C_2 / 4.$$

where f_o is in MHz and C_1, C_2 in pF. For 1-MHz crystals with a high impedance:

$$C_1 = C_2 / 10.$$

When one of the oscillators is not built, make sure to fit a wire link in PCB position C1, C3 or C5. This ensures a low level at the input of the first oscillator gate, preventing high current consumption and spurious oscillation of the HCT04.

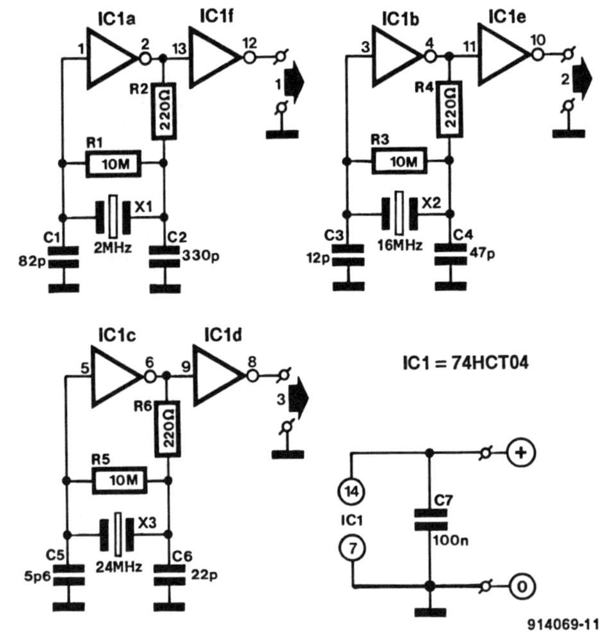

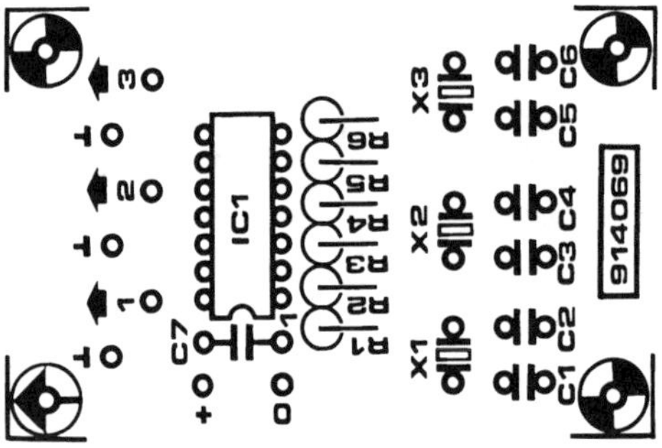

Parts list

Resistors:
R_1, R_3, R_5 = 10 MΩ
R_2, R_4, R_6 = 220 Ω

Capacitors:
C_1 = 82 pF
C_2 = 330 pF
C_3 = 12 pF
C_4 = 47 pF
C_5 = 5.6 pF
C_6 = 22 pF
C_7 = 100 nF

Integrated circuits:
IC_1 = 74HCT04

Miscellaneous:
X_1 = crystal 2 MHz
X_2 = crystal 16 MHz
X_3 = crystal 24 MHz

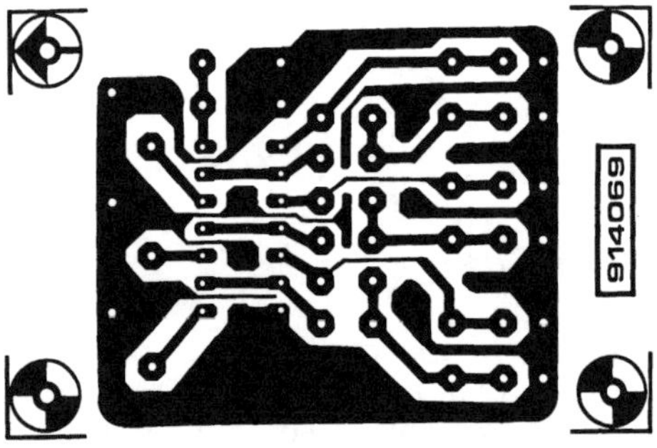

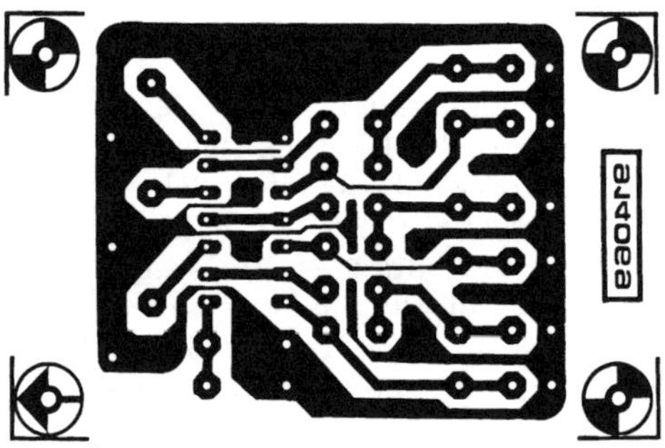

Safe solid-state relay

195 Although the S202DS2 solid-state relay (SSR) from Sharp is a useful and interesting electronic component, it fails to meet the minimum requirements for electrical safety in many countries where the mains voltage is 220 V or 240 V. This is mainly because the breakdown voltage of the optocoupler in the S202DS is too low, and the pin spacing of the device is too small.

For the many applications where electrical safety is a primary concern, Sharp have developed another SSR, the S201S04. The small SIL enclosure (shown in the photograph) contains an optocoupler complete with a series resistor, a zero-crossing switch and a power triac. The presence of the zero-crossing switch implies that the SSR is suitable for non-reactive loads only. Furthermore, since the value of the series resistor is only 130 Ω, an additional, external, resistor will often be required to prevent too high a current through the LED in the internal optocoupler.

For reasons of safety, the solid-state switch is best built on the printed-circuit board shown here (this board is not available ready-made). The value of the external series resistor, R_1, depends on the control voltage and the trigger current. The trigger current, in turn, depends to some degree on the current to be switched, and will typically lie between 5 mA and 20 mA. The optimum value is best determined empirically, observing a maximum current of 40 mA. The minimum value of the series resistor, R_1min, in Ω, is calculated from

$$R_1\text{min} = 25\,(U_s - 2.4) - 130$$

where U_s is the control voltage applied to connector K_1.

Diode D_1 protects the SSR against reverse control voltages, and D_2 indicates whether the SSR is supplied with a control current. Network R_2-C_1 is connected across the SSR output to protect the device against voltage surges on the mains.

When connected to a mains supply of 220 V or 240 V, the circuit may be used with non-reactive loads up to 330 W, which corresponds roughly to the maximum permissible effective load current of 1.5 A.

WARNING. Since the circuit carries dangerous voltages at a number of points, it is essential that proper electrical insulation is applied. Never work on the circuit when the mains is connected to it. Make sure that no part of the circuit can be touched when it is being set, adjusted or used. Good guidance on safety precautions may be obtained from the IEEE Wiring Regulations, a copy of which is available in most Public Libraries in the UK.

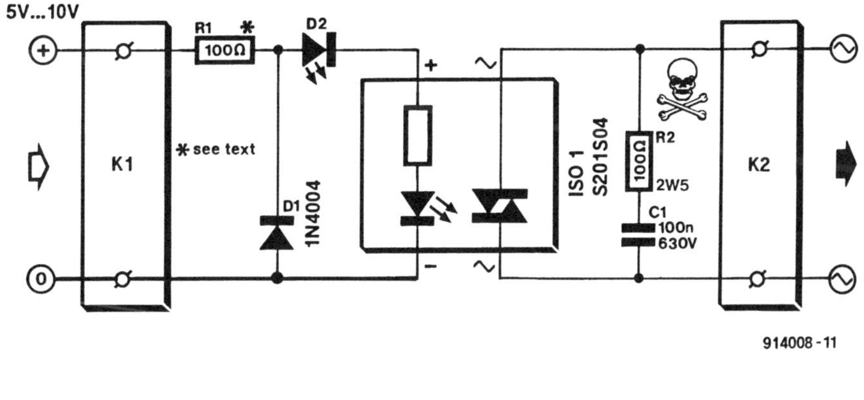

914008-12

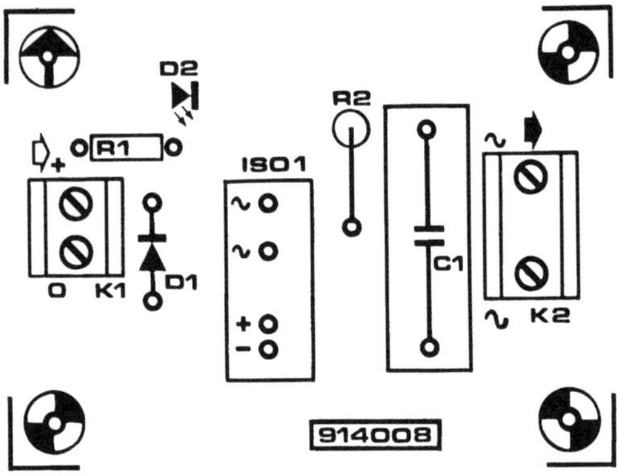

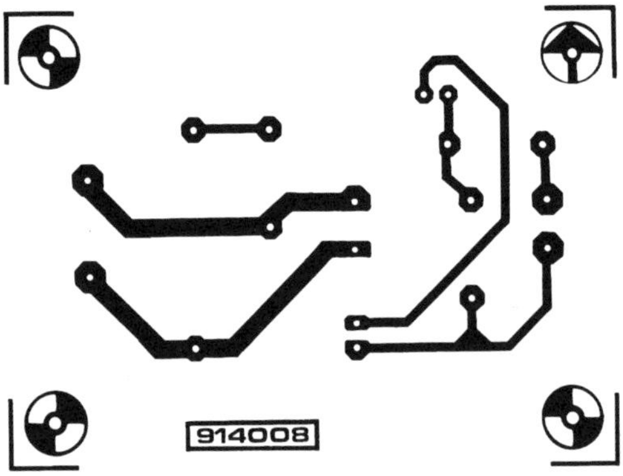

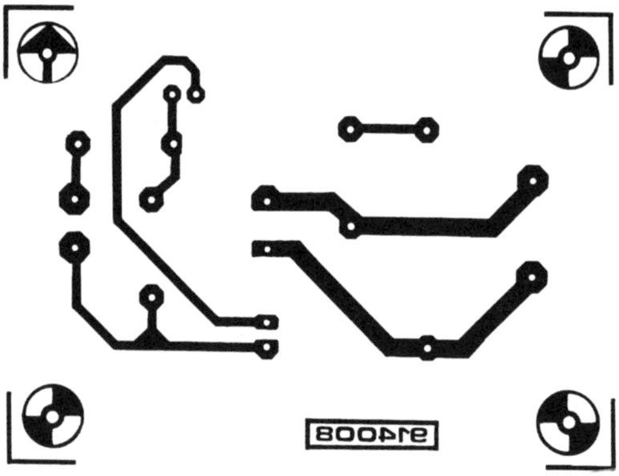

Parts list

Resistors:
R_1 = 100 Ω
R_2 = 100 Ω, 2.5 W

Capacitors:
C_1 = 100 nF, 630 V

Semiconductors:
D_1 = 1N4004
D_2 = LED, red
ISO_1 = S201S04 (Sharp)

Miscellaneous:
K_1 = PCB terminal block, 5 mm pitch
K_2 = PCB terminal block, 10 mm pitch

Windscreen wash-wipe circuit

196 In many older cars the windscreen wash pump is not coupled to the windscreen wipe function. This circuit switches on the wiper motor for a predetermined time each time the pump is actuated. The wipers start to work when the pump switch is pressed, and keep on working a while after the switch is released.

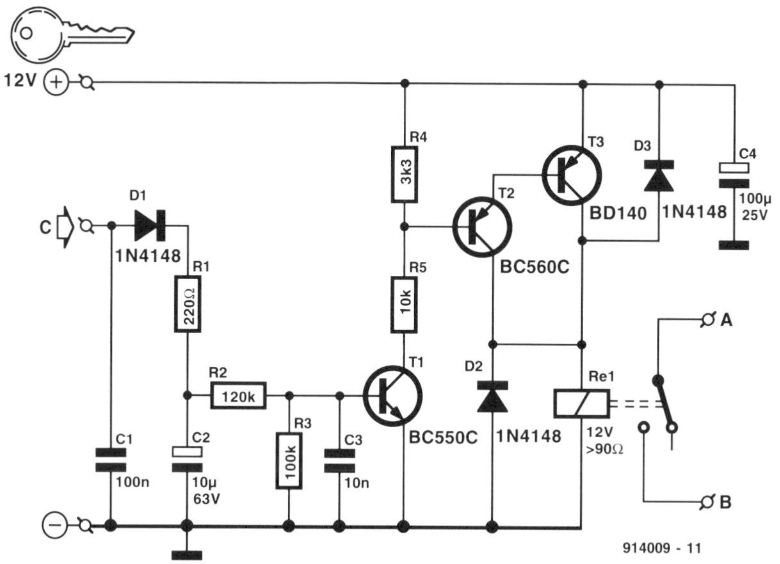

The anode of diode D_1 is normally taken to ground via the wash pump motor. When the pump is powered, C_1 charges rapidly via D_1 and R_1. Consequently, T_1, T_2 and T_3 are switched on, and relay Re_1 is energized. C_1 is charged as long as the pump motor and the wipers operate. When the pump switch is released, the washing stops, but the wiping continues for a time determined by R_2-C_2. Diode D_1 prevents C_2 being discharged via the pump motor. Diodes D_2 and D_3 protect the circuit against back e.m.f. from the relay coil.

Installing the circuit in a car should be easy as there are only three connecting points (apart from the supply voltage). Note, though, that the circuit is designed to work with a pump motor that has a fixed contact to ground, while its positive contact is taken to the wash switch.

The relay on the board is capable of switching 10–20 A. Its contacts are connected across the wiper switch via terminals 'A' and 'B' by means of heavy-duty wiring. The connections to the board are made via spade terminals and mating sockets as used in car electrical systems. The spade terminals are screwed direct to the board, and soldered to minimize the contact resistance. Finally, the relay may be an Omron Type G2L-113P-4S-SV or a Bosch Type 0 332 016 101. The Omron type fits on the PCB, the Bosch type does not.

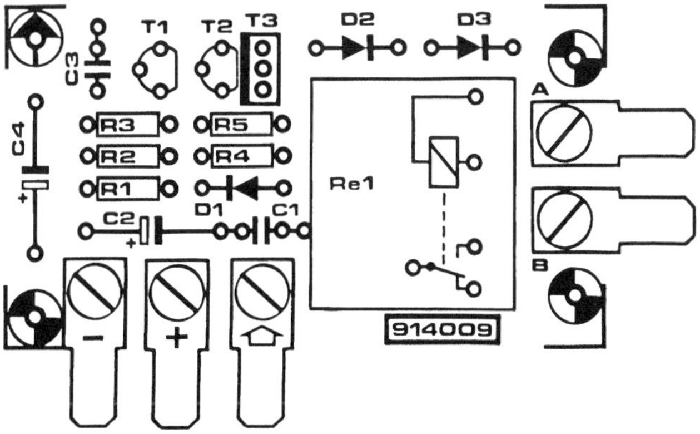

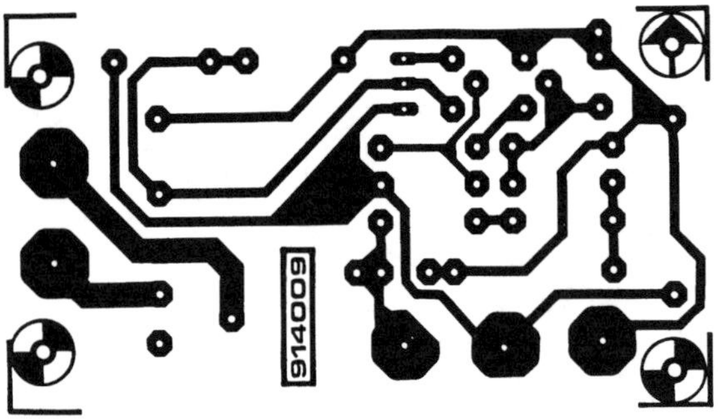

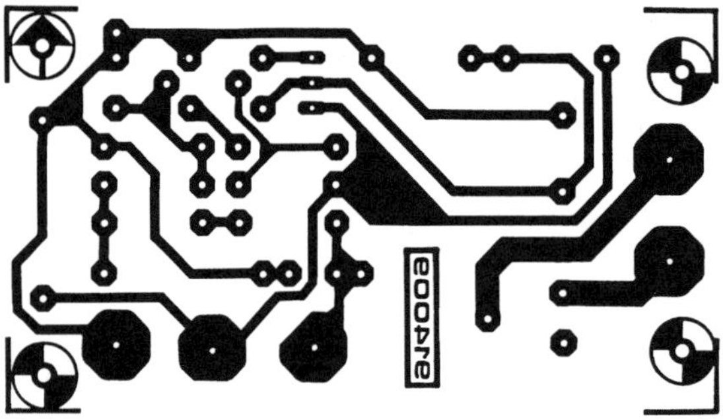

Parts list

Resistors:
R_1 = 220 Ω
R_2 = 120 kΩ
R_3 = 100 kΩ
R_4 = 3.3 kΩ
R_5 = 10 kΩ

Capacitors:
C_1 = 100 nF
C_2 = 19 µF, 63 V
C_3 = 10 nF
C_4 = 100 µF, 25 V

Semiconductors:
D_1, D_2, D_3 = 1N4148
T_1 = BC550C
T_2 = BC560C
T_3 = BD140

Miscellaneous:
Re_1 = 12 V relay for PCB mounting, Z>90 Ω,contact rating >10 A

Solid-state light-sensitive switch

197 This electronic switch is designed to be connected direct to the mains, which obviates a low-voltage supply and so keeps the cost and space requirement to a minimum. The circuit switches a lamp on when it gets dark, and switches it off again when it gets light. The switching is done without a relay, avoiding problems with sparks and mains pollution caused by the contacts and the coil inductance.

The switch is powered by the mains via R_{10}, C_4, D_3, D_2 and C_3. A voltage reference, D_1, supplies 8.2 V to a light measuring network, R_2-P_1. As the light intensity drops, the resistance of the LDR (light-dependent resistor), R_2, increases. Consequently, the voltage across P_1 drops, so that the gate-source voltage of FET T_1 drops also. When switch S_1 is closed, time constant R_3-C_2 causes the gate voltage of T_1 to change more slowly than the resistance of R_2. This is necessary to prevent the circuit responding to quick changes in the ambient light intensity.

Components T_1, T_2, R_4, R_5, R_6, and R_8 form a Schmitt trigger. Normally, T_1 conducts so that T_2 is off. When the gate voltage of the FET drops below a certain level, T_2 is switched on. Consequently, T_3 starts to conduct, and supplies the gate current necessary to trigger triac Tri_1. The load, lamp La_1, is then switched on. When the light intensity increases above the level set with P_1, T_1 is switched on, so that the load is switched off.

Switch S_1 is included to disable the time constant during adjustment. Resistor R_9 serves to discharge C_4 after the circuit has been disconnected from the mains.

WARNING. Since the circuit carries dangerous voltages at a number of points, it is essential that proper electrical insulation is applied. Never work on the circuit when the mains is connected to it. Make sure that no part of the circuit can be touched when it is being set, adjusted or used.

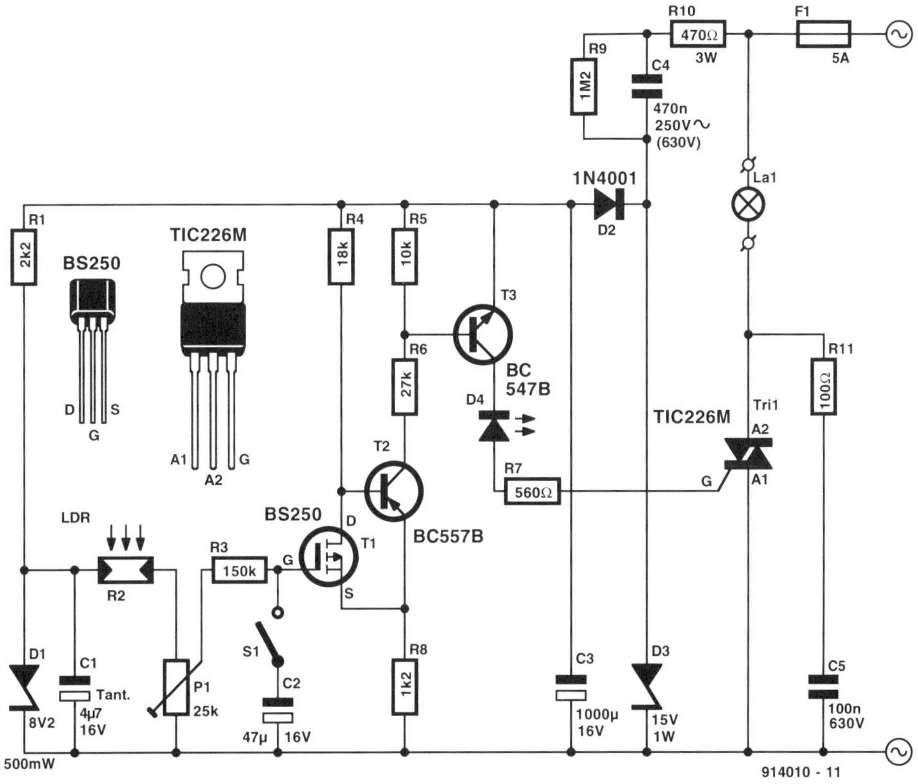

Voltage-controlled current source

198 The source, based on a Type TL084 quadruple opamp, is intended to convert an input signal of 0–5 V into a current of 0–20 mA. This type of circuit is used, for instance, to transfer measurands (quantities being measured) over long leads. Since the resistance of the leads is part of the current loop, it is of no consequence and can not affect the measurement.

Opamp IC_{1a} is a straightforward input amplifier. Opamp IC_{1b} adjusts the direct voltage component of the amplified input signal: the operating point may be shifted with P_2. It is, for example, possible to arrange an output current of 4 mA for an input voltage of 0 V. The output current range is then 4–20 mA.

Opamp IC_{1c} and T_1 convert the output of IC_{1b} to a signal of 15 V. This makes it possible for IC_{1d} and T_2 to function as a voltage-to-current converter. The output current flows to earth via load resistance R_L.Varying the values of R_2 and P_1 allows the amplification to be altered as required.

The circuit may also be used as a temperature-to-current converter by making the potential divider at the input consist of a fixed resistance and one with a negative temperature coefficient.

When the requirements are exacting, the two zener diodes should be temperature compensated.

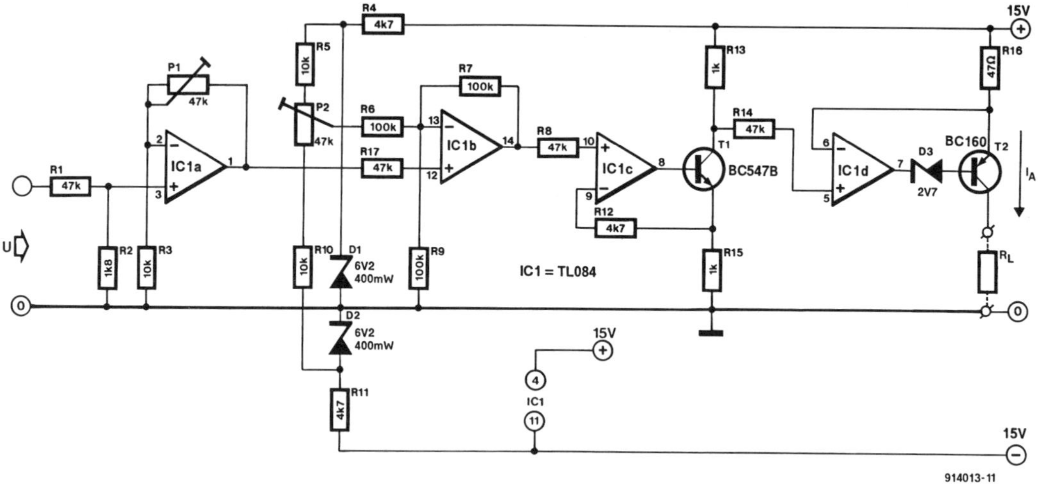

Maintenance of rechargeable batteries

199 It is often stated that NiCd batteries must not be totally discharged. This is not entirely true, but in practice it is difficult to prevent polarity reversal of one or more cells.

The polarity of a cell in a battery that gets discharged first will be reversed by the other cells in series with it. This is usually the end of the cell because its internal structure cannot cope with this condition. Manufacturers try to prevent this by making the positive electrode of battery cells more robust. But even then, the cell will be damaged by regular polarity reversal, particularly at high charging currents. Once a cell is damaged, its capacity is reduced, which causes it to get discharged more quickly, ..., and a vicious circle is set up.

A possible solution to this problem is to shunt each cell with a Schottky diode that can handle a current equal to, or larger than, the charging current.

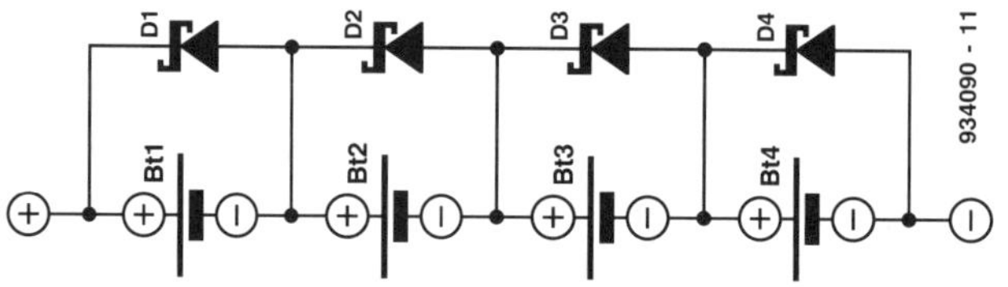

Fault signalling circuit

200 The present circuit was developed to make it possible for different sensors to be added to an existing alarm installation. These sensors may be gas or smoke detectors, door switches, infra-red detectors, and others.

In quiescent operation, the level at all inputs must be zero. When the relevant sensor is actuated, pin 3 of IC_{1a} is made high via R_5. Since this opamp inverts, its output, and thus the cathode of D_9, is at 0 V. Since the anode of D_9 is at +12 V, the diode lights to indicate an alarm condition. Across both D_2 and D_3 a potential drop of about 1.2 V then occurs and this results in T_1 being switched on and T_2 being switched off. Relay Re_1 is then deenergized and opens the contact via which the alarm installation is controlled. Since that contact is closed in quiescent operation, a supply failure is also signalled.

When the cause of the alarm has been removed and the installation has been reset, all inputs return to zero volts, T_1 is switched off, T_2 is switched on, and the relay is re-energized. This condition is indicated by the lighting of D_5. Since D_5 and R_4 are in series with the relay coil, the load on the relay is then slightly less, so that the relay draws a smaller current. Capacitor C_2 ensures that at switch-on R_4 and D_5 are short-circuited, guaranteeing that the relay is energized.

Where the current must be kept low, the standard LEDs may be replaced by low-current types. The value of the relevant bias resistors (R_7, R_{10}, R_{40}) must then be increased to 8.2 kΩ.

Networks C_3-R_5, C_4-R_8, and C_{14}-R_{38}, form low-pass filters that prevent noise voltages actuating the alarm. That is important, because the cables between sensors and inputs may be very long.

The circuit is protected against voltage peaks by zener diodes D_6, D_8 and D_{28}. This makes it possible for the control voltage to be higher than 12 V, although regulations prohibit the use of voltages above 42 V.

Diode D_1 protects the circuit against polarity reversal.

Capacitor C_1 decouples the supply voltage. The current drawn, dependent on the relay, is about 200 mA.

Type 4050 ICs may be used in the IC_1 and IC_2 positions, but it should be noted that these are non-inverting devices, so that part of the circuit action is then reversed.

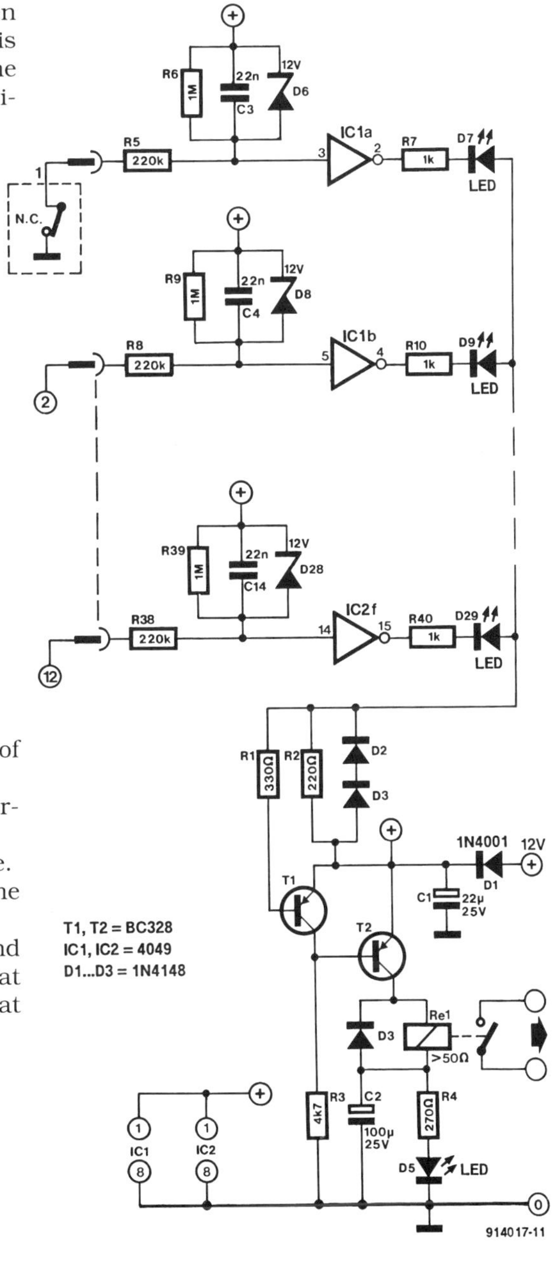

Automatic cycle lights

201 To prevent the dynamo-driven lights of your bicyle going out when you stop in the dark for, say, traffic lights, the simple circuit here may offer help.

The circuit uses four NiCd batteries with a capacity of between 0.25 Ah and 1.25 Ah, which are constantly charged when the dynamo is driven via R_1 and D_1. Since the battery voltage is rather less than the dynamo output, the lights are dimmed to a small extent when the cycle is stopped, but in practice that is hardly noticeable.

Monostable IC_{1a}, which has a mono time of 1 s (R_5-C_2), is used to detect whether the dynamo generates a voltage with the aid of D_3, R_3 and R_4. As long is there is a voltage, the monostable holds IC_{1b}, also a monostable, in the reset state. The relay is not energized and the lights are powered by the dynamo.

When the dynamo voltage drops, IC_{1a} is no longer triggered so that its outputs change level. This causes the reset state of IC_{1b} to be removed, whereupon its T input is actuated and remains so for two minutes, during which time the relay is energized and the cycle lights are powered by the batteries.

Strictly speaking, IC_{1b} is not essential, but it does ensure that the lights are switched and that the battery can not be discharged completely.

The relay should be a type that operates faultlessly when its supply voltage reaches 4.8 V.

It is advisable to build the circuit in a watertight, or at least waterproof, enclosure.

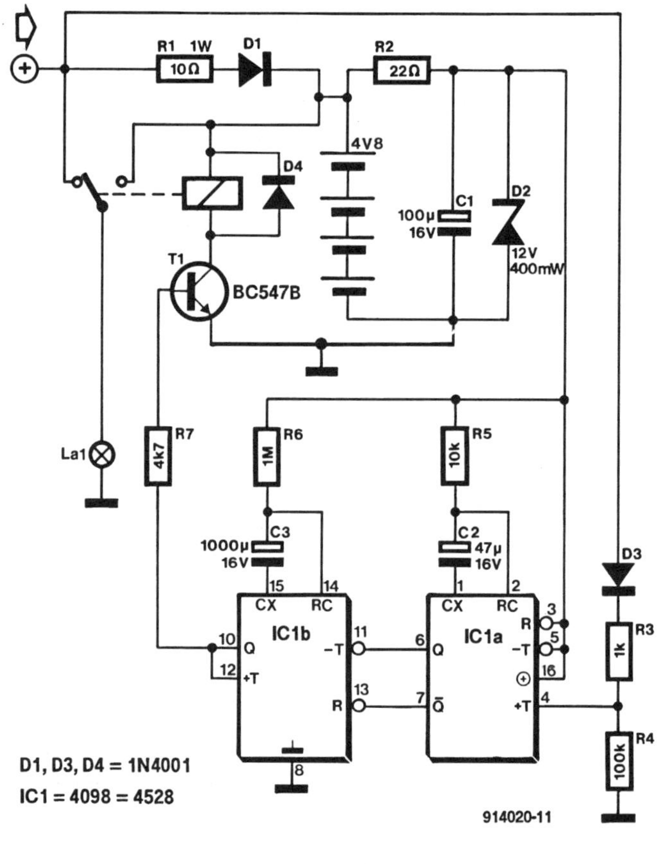

General purpose time switch

202 This circuit meets the requirement for a simple, low-cost timer with a timing range of a less than a second to several tens of hours, and a thyristor-based mains interface capable of controlling inductive loads.

The circuit consists basically of a timer section and a switching section. The timer derives its accuracy from the mains frequency: 50 Hz or 60 Hz. Components R_4, R_3, R_2, C_1 D_1 and D_2 convert the mains voltage into a clock signal for binary ripple counter IC_1, a 4040. Counters IC_1, IC_3, and an 8-input AND gate, IC_4, form a programmable divider. When the time set with the jumper wire(s) at the outputs of IC_2 and IC_3 has elapsed, the output of IC_4 goes low and causes bistable IC_{2b}-IC_{2c} to toggle. As a result, the gate drive of thyristor Th_1 is removed, and the load connected to K_2 is switched off. To prevent mains pollution and switching noises, the load is switched during the zero crossing of the mains voltage. The timer can also be used to switch on a load after the preprogrammed time <197> this only requires wire jumper 'Y' to be fitted instead of 'Z'.

The time is set by fitting a maximum of eight wires between the 8-input AND gate and the counter outputs. The actual time is the sum of all selected times listed in the table.

SWITCHING TIMES

IC_2 output	Time 50 Hz	Time 60 Hz
Q1	0.02 s	0.02 s
Q_2	0.04 s	0.04 s
Q_3	0.08 s	0.07 s
Q_4	0.16 s	0.13 s
Q_5	0.32 s	0.27 s
Q_6	0.64 s	0.53 s
Q_7	1.28 s	1.07 s
Q_8	2.56 s	2.13 s
Q_9	5.12 s	4.27 s
Q_{10}	10.24 s	8.53 s
Q_{11}	20.48 s	17.07 s
Q_{12}	40.96 s	34.13 s
IC_3 output		
Q_1	1 m 22s	1 m 8 s
Q_2	2 m 44 s	2 m 16 s
Q_3	5 m 28 s	4 m 33 s
Q_4	10 m 55 s	9 m 6 s
Q_5	21 m 50 s	18 m 12 s
Q_6	43 m 41 s	36 m 24 s
Q_7	1 h 23 m	1 h 12 m
Q_8	2 h 55 m	2 h 25 m
Q_9	5 h 50 m	4 h 51 m
Q_{10}	11 h 39 m	9 h 43 m
Q_{11}	23 h 18 m	19 h 18 m
Q_{12}	46 h 36 m	38 h 50 m

The 12-V supply voltage for the circuit is obtained from the mains with the aid of a single-phase rectifier, D_3. The supply voltage is stabilized and smoothed by zener diode D_5 and reservoir capacitor C_6.

The two series resistors in the rectifier, R_{16} and R_{17}, must not be replaced by a single 100-kΩ resistor. Remember, the total voltage drop across these resistors is of the order of 220 V (at a mains voltage of 240 V), which is too much for a single resistor rated at 0.5 W. The same applies to resistors R_3 and R_4.

The timer is actuated by pressing push-button S2. Note that actuating the timer means that the load is switched on or off, depending on whether jumper Y or Z is fitted. During the timing cycle, the circuit can be reset by pressing S2 again. This starts a full timing cycle, irrespective of when the switch was pressed. The timing cycle can be stopped by pressing S_1. Provision is made for electronic control of the start and stop functions. This is achieved with the aid of two optocouplers, IC_5 and IC_6. The electronic control inputs, K_3 and K_4, are electrically isolated from the timer, and can be driven with control voltages of 5 V.

The construction of the timer is straightforward. A minimum of wiring is required since all parts are contained on a single printed-circuit board. The start and stop switched must be rated at 250 V because they are at mains potential. The circuit must be fitted into an ABS enclosure with non-metallic screws. Take care to provide adequate strain reliefs and insulation of the mains input and output cables.

WARNING. Since the circuit carries dangerous voltages at a number of points, never work on the circuit when the mains is connected to it. Make sure that no part of the circuit can be touched when it is being set, adjusted or used.

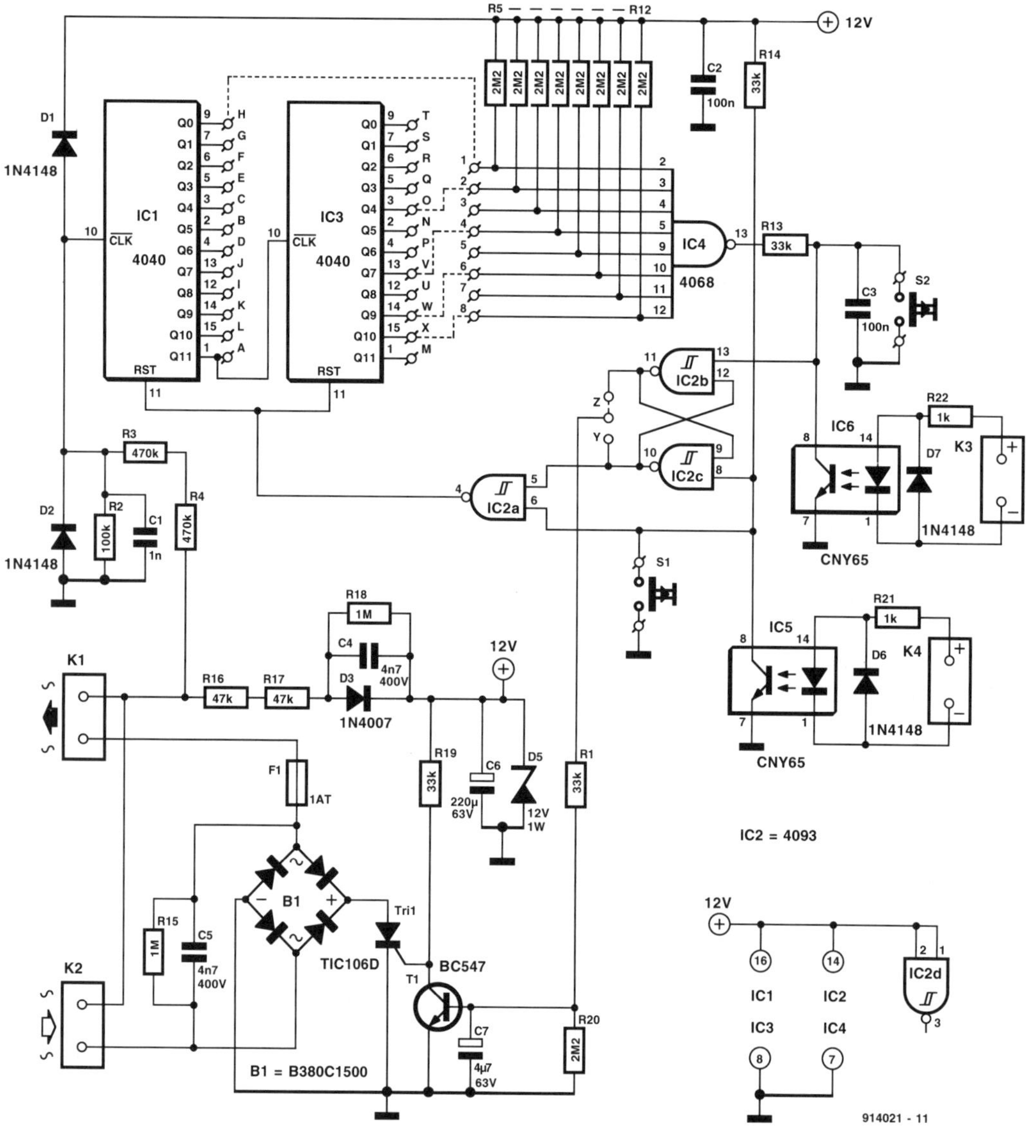

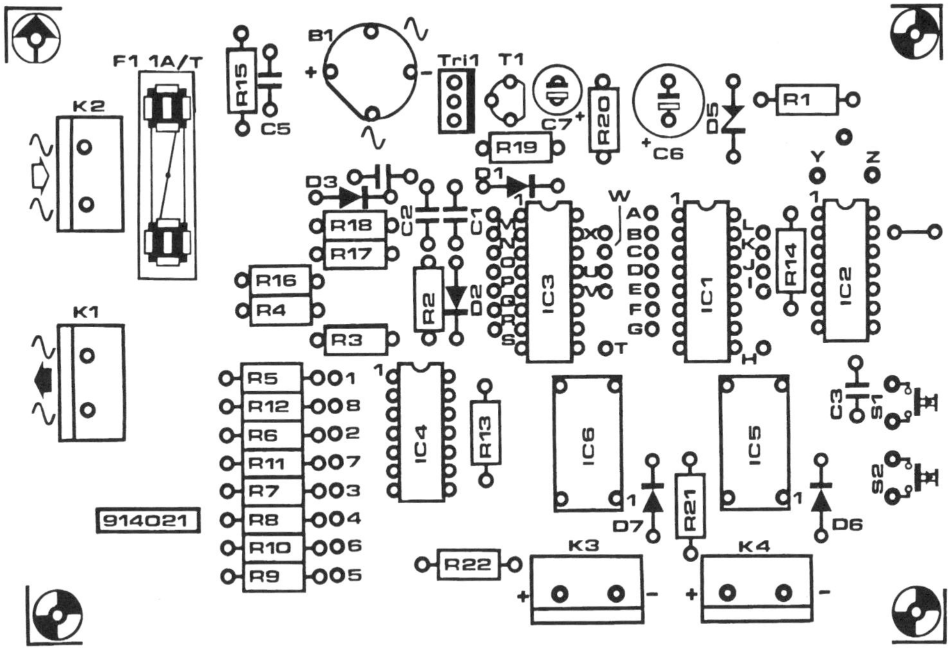
F1 1A/T
K2
K1
B1
R15
C5
Tri1
T1
C7
R20
C6
D5
R1
Y
N
R19
D3
D1
W
C2
C1
R18
R17
R16
R4
R3
R2
D2
IC3
IC1
IC2
R14
A
B
C
D
E
F
G
T
H
R5
R12
R6
R11
R7
R8
R10
R9
IC4
R13
IC6
IC5
C3
S1
S2
D7
D6
R21
914021
R22
K3
K4

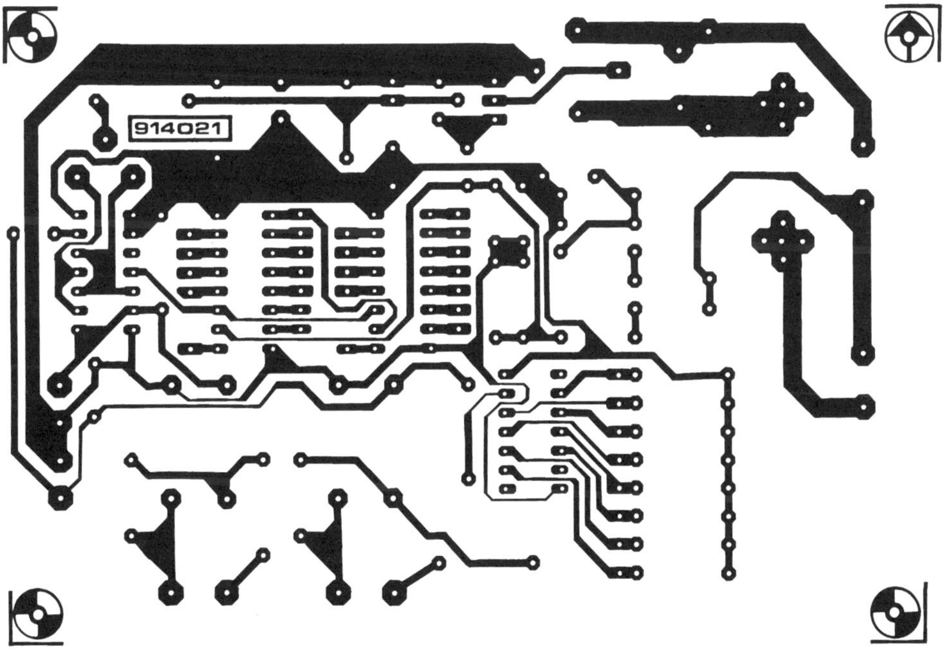
914021

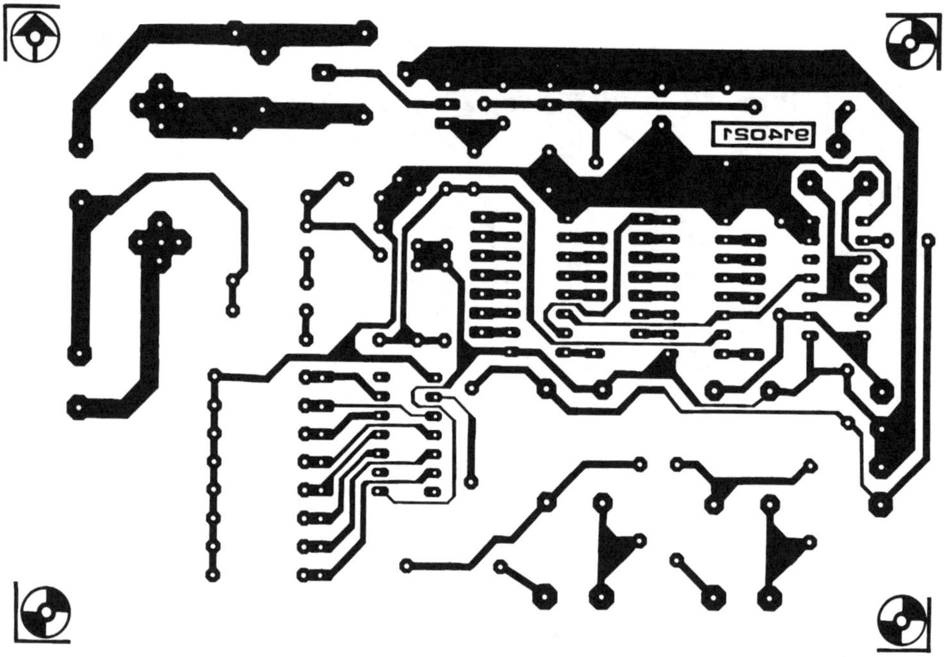

Parts list

Resistors:
R_1, R_{13}, R_{14}, R_{19} = 33 kΩ
R_2 = 100 kΩ
R_3, R_4 = 470 kΩ
R_5–R_{12}, R_{20} = 2.2 MΩ
R_{15}, R_{18} = 1 MΩ
R_{16}, R_{17} = 47 kΩ
R_{21}, R_{22} = 1 kΩ

Capacitors:
C_1 = 1 nF
C_2, C_3 = 100 nF
C_4, C_5 = 4.7 nF, 400 V
C_6 = 220 µF, 63 V, radial
C_7 = 4.7 µF, 63 V, radial

Semiconductors:
D_1, D_2, D_6, D_7 = 1N4148
D_3 = 1N4007
D_4 = not used
D_5 = zener diode, 12 V, 1.3 W
B_1 = B380C1500 (round type)
T_1 = BC547B
Tri_1 = TIC106 D

Integrated circuits:
IC_1, IC_3 = 4040
IC_2 = 4093
IC_4 = 4068
IC_5, IC_6 = CNY65

Miscellaneous:
K_1–K_4 = 2-way PCB terminal block, 7.5 mm pitch
S_1, S_2 = push-button switch for 250 V a.c.
F_1 = fuse, 1 A, delayed action, with holder for PCB mounting
Enclosure, e.g., Bopla E430

Bounce-free switch

203 Switches with change-over contacts with traditional debouncing circuits are not always usable or economical. Keyboard switches, for instance, seldom have change-over contacts. Furthermore, change-over switches have one extra connection, which can not always be accommodated in the construction.

The small circuit here operates with one make contact or one break contact. Which one does not matter in practice, because the Q or $\overline{Q}$ output may be chosen to invert the switch action.

The logic level at the input of the circuit is determined by pull-up resistor R_1 and the position of S_1. The input signal goes straight to the data input of bistable (US: flip-flop) IC_{1a}, where it is clocked as soon as the contact bounce has disappeared (after 0.5–10 ms).

The clock is generated by IC_{2a}, an XOR gate. Every time its input level alters, this gate generates a pulse, whose width is determined by R_2 and C_1. That pulse is, however, not devoid of contact bounce, which is, therefore, filtered out by C_2 and the output resistance of IC_{2a}. The potential across C_2 is smoothed and inverted by IC_{2b} before the pulse is applied to the clock input of the bistable. The result of all this is that the output signal is clean, albeit delayed by a few milliseconds.

Since the bounce filter uses the output resistance of IC_{2a}, this circuit can not be replaced by just any other type. When replacing is unavoidable, the value of C_2 must be adapted to the new circumstances, or a resistor connected in series with the output of IC_{2a}.

The current drain of the circuit is 3 mA.

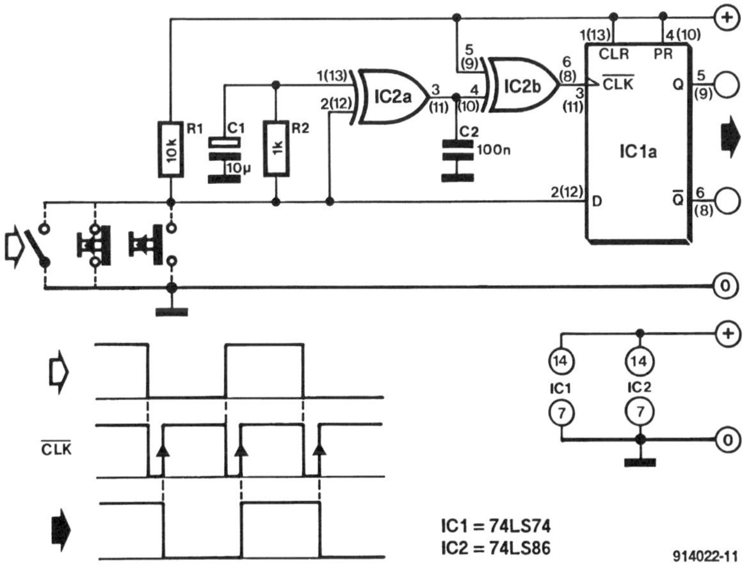

Switch for central-heating pump

204 The pump in some central-heating systems has two or even three speeds. At the lowest speed, not much hot water is pumped around the circuit and this may result in the boiler overheating. The switch circuit proposed here prevents that happening.

The electrics of the pump are as shown in the diagram. The main winding is normally 175 Ω and the auxiliary windings 135 Ω: these values may, however, be different in certain pumps and this should, of course, be checked. The capacitor in series with the windings provides the necessary phase shift that enables the motor of the pump to rotate. In positions *a* and *b*, the impedance is increased and this results in a weaker field so that the motor runs more slowly and the pump displaces less water.

A simple circuit enables the automatic switching between *a* and *b* or between *b* and *c*. Its 24-V input is parallel with the drive of the gas valve of the main burner. When this valve closes, the speed of the pump increases and the boiler can not overheat.

If the installation uses a 240 V drive for the gas valve, a small transformer may be used to obtain the 24 V.

On a safety note, bear in mind that the 24 V supply, which reaches the thermostats in various rooms, is insulated from the mains only by the relay. Therefore, this relay should be a heavy-duty type that provides adequate insulation

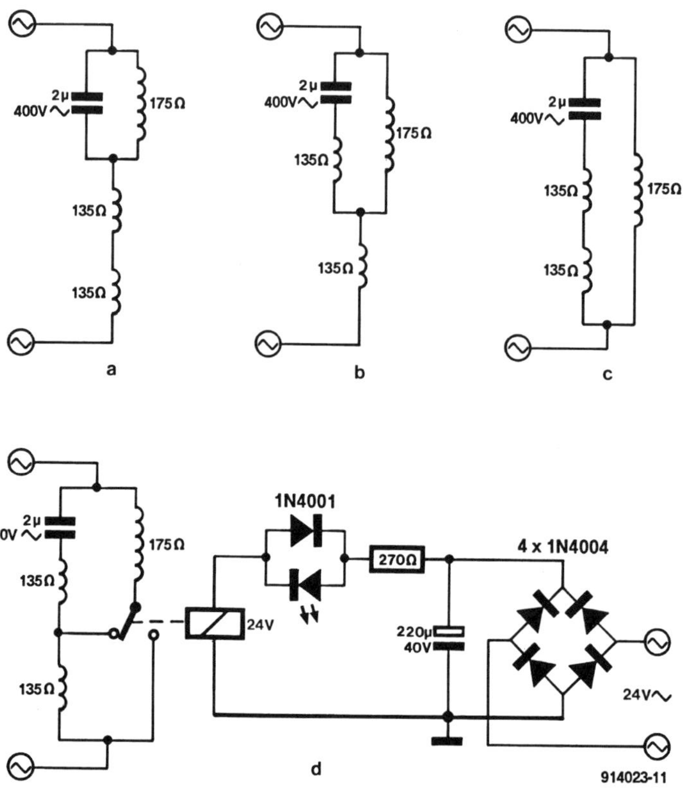

Time delay with one 555

205 Many electronic circuits frequently require the brief delay of a pulse. Such a delay, here between 100 µs and 100 s, is easily provided by a simple circuit based on the popular 555. That is more than adequate for most applications.
The output of the 555 can go high only if the potential at pin 2 drops below a third of the level of the supply voltage, provided that the level at pin 4 is high. In quiescent operation, the level at pin 4 is low and C_1 is charged via T_1, so that the output is low. When the input goes high, T_1 is switched off and C_1 is discharged via R_1. In that condition, the reset state is cancelled, and after a time delay that depends on the state of discharge of C_1 the output of the 555 goes high. The time delay in seconds is calculated from $t = 0.69R_1C_1$, where R_1 must be greater than or equal to 10 kΩ.

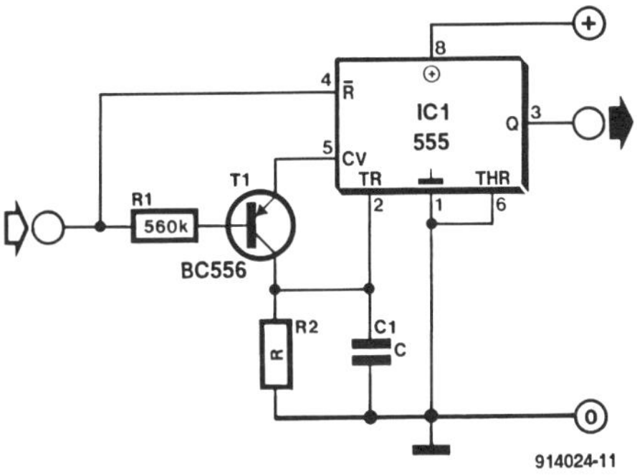

Precedence detector

206 The detector was designed primarily for use with quiz games. It indicates who has first pressed a push button by the sounding of a buzzer and the lighting of an LED. The quiz master can then reset the circuit.
In the diagram, the four push buttons, S_1–S_4, are connected to the D inputs of four bistables contained in IC_2. They are also connected to the clock input of IC_2 via OR gates IC_{1a}, IC_{1b} and IC_{1c}. The Q outputs of the bistables drive D_1–D_4.
After the bistables have been reset with S_5 (which briefly makes the CLR input low), all Q outputs are high, so that the LEDs are out. If one of the push-buttons, say S_1, is pressed, a high level ensues at the associated D input. This high level is applied to the clock input of all bistables, whereupon the existing levels at the D inputs of the bistables are stored and applied to the outputs; D_1 will then light
Since the switches and the LEDs are connected to the supply line via a common resistor, R_4, the voltage at points after R_4 will drop to about 2 V owing to the lighting of D_1. Because of potential divider R_9-R_{10}, the voltage across the switches will be only about 1 V. If then one of the switches, other than S_1, is pressed, the voltage at the associated D input(s), as well as at the clock input, will be too low for the IC to react. In this way, the circuit is disabled after one of the switches has been pressed.
Darlington T_1 drives the d.c. buzzer. The base of T_1 is connected to the anodes of the LEDs via R_3 and C_4. At the instant one of the LEDs lights, C_1 passes this low level to T_1, which then energizes the buzzer. After about 0.5 s, C_4 is recharged via R_2 and R_3, so that T_1 is switched off.
The circuit draws a current of about 5 mA when the LEDs are out. When one of the LEDs lights and the buzzer sounds, the current rises to some 50 mA.

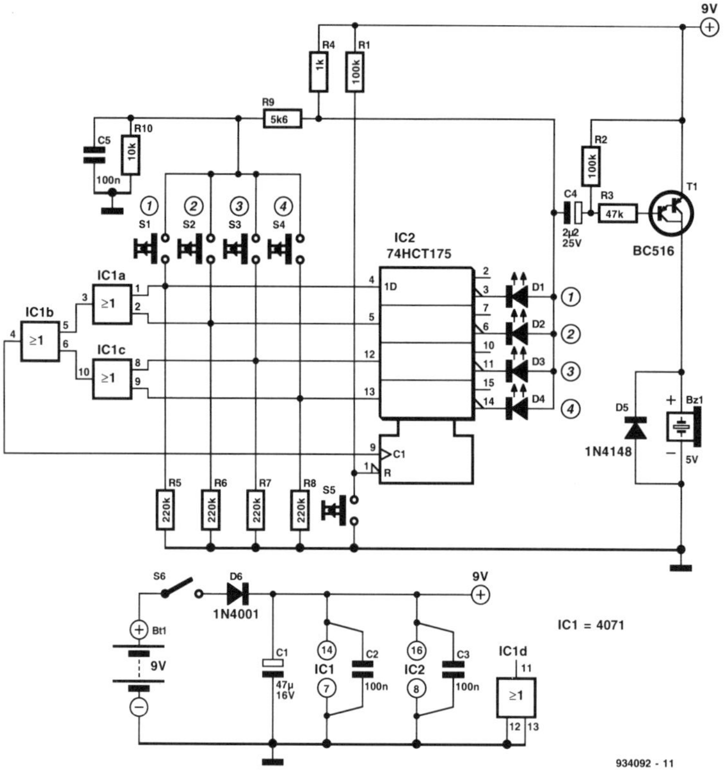

BCD rotary switch

207 Binary-coded-decimal (thumb-wheel) switches are not only relatively expensive, but are also often not available to the wanted specification. A good alternative is shown in the present circuit that uses generally easily available components and a 12-position rotary switch (the most popular type) of which two positions are disabled. The terminals of the switch are connected to the inputs of a prority encoder, IC_1. When an input goes low, the IC puts the number of that input as an inverted BCD code at the output. The four XOR gates enable the inverted code to be inverted again into a standard BCD code. This operation is effected with link JP_1: in position P, the output carries the standard BCD code, and in position N, the inverted BCD code.

The circuit is powered by 3–18 V (because CMOS ICs are used). The use of a 5 V source enables LS, TTL, HC and HCT inputs also to be connected to the switch. The circuit draws a current of only about 200 μA.

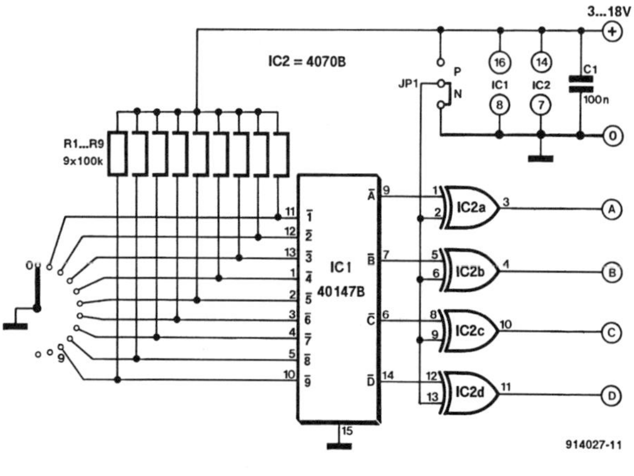

House telephone

208 Telephones are now readily and cheaply available: two identical ones and a handful of components enable a simple house telephone to be set up.
Since the two telephones are connected in series as shown in the diagram, half the supply voltage exists across either of them. Neither buzzer will sound, since the potential across zener diodes D_1 and D_3 is below their breakdown voltage.
If, say, the handset of telephone 2 is lifted, a virtual short circuit ensues across this telephone. The potential across telephone 1 then rises to almost the supply voltage. Breakdown then occurs in D_1, which causes a sharp increase in the reverse current through the diode. The buzzer will then sound and the LED light. If then the handset of telephone 1 is lifted, the supply voltage is again divided symmetrically across the two telephones, which is sufficient for carrying out a conversation.
The buzzers may be continuous-tone or intermittent-tone types to personal preference. Similarly, the LEDs may be standard or flashing types.
The power supply may be a standard 12-V mains adapter. When the supply voltage is too high, the buzzers will sound even when both handsets are lifted. If dissimilar telephones are used, one or both zener diodes need to be replaced by different types to ensure that during quiescent operation the voltage drops across the telephones are identical.

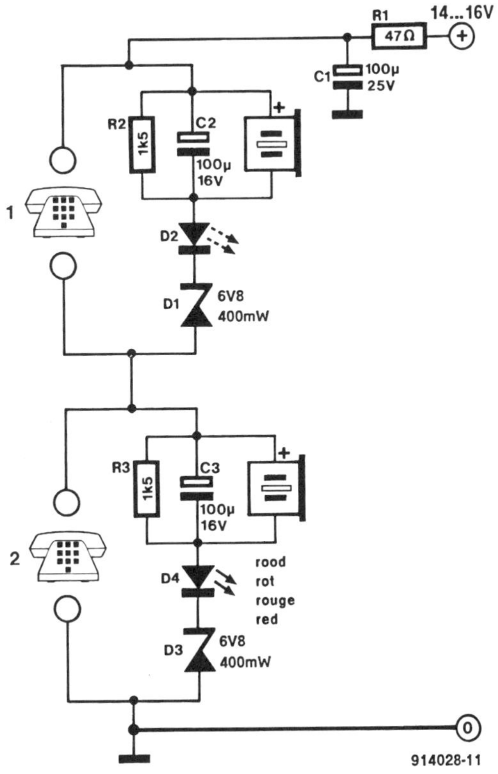

Darkness-sensitive switch

209 The circuit presented here enables the automatic switching on of outside lighting when its gets dark and, what's more, it does so for a predetermined period. A new period can be begun only when it has been light again.
The switch is a solid-state relay. From the instant that T_4 and T_5 are on, the LED in the relay lights and lamp L_1 is powered. As soon as one of the transistors switches off, the lamp will go out.
Whether T_5 is on depends on phototransistor T_3. If light falls on to this, it is switched on and removes the base current from T_5. In other words, T_5 can be on only when it is dark.
The base-emitter junction of transistor T_2 is also connected in parallel with T_3 and it, too, will therefore be off when it is light. This causes a constant reset on IC_1, all of whose counter outputs are then low.

As soon as it gets dark, base current for T_2 is provided by R_7 and the transistor switches on. The counter can then count the pulses of its internal oscillator, while the lamp remains on. When, after a short time, output Q13 goes high, transistor T_4 switches off. This causes the LED in the solid-state relay to quench and the lamp goes out. Since at the same time the oscillator is stopped via T_1, Q13 remains high. This state is maintained until it gets light again and IC_1 is reset, whereupon a new cycle can be started.

The period the lamp is on may be set to between 1 and 5 hours with P_1.
No special transformer is needed for the power supply, which may be derived direct from the mains. Diodes D_1–D_5 rectify the mains voltage and the result is smoothed by C_4. Capacitor C_5 operates as a resistance, and it should, therefore, have a working voltage rating of not less than 400 V, although 630 V is preferred.
Note that mains voltage exists at several points in the circuit: great care should be taken to insulate the switch unit adequately

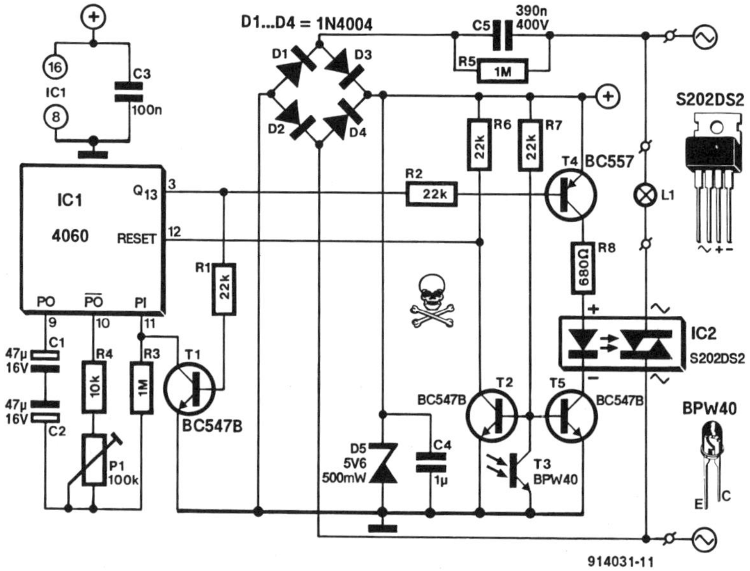

Basic timer [I]

210 The time-determining element in the present circuit is capacitor C_1, which is charged via a current source based on T_1. The voltage across the capacitor is, therefore, a ramp. A comparison of this ramp with a preset voltage gives a fairly accurate time indication. With values shown in the present circuit, that time lies between 1 and 10 minutes.

The ramp and reference voltage are compared by IC_{1a} and IC_{1b}, both open-collector output types. Ignoring IC_{1a} for a moment, the output transistor of IC_{1b} will be off as long as the set voltage is higher than the potential across C_1. If that potential becomes higher than the reference voltage, the output transistor in IC_{1b} switches on and T_2 goes off. In practice, this would mean that the buzzer sounds until the set voltage level is exceeded: not exactly ideal. Since the intention is that the buzzer sounds only briefly when the set level is exceeded, IC_{1a} is needed. This comparator evaluates the reference voltage with a potential that is slightly higher than that across C_1. This off-set is provided by R_4-P_1-D_3. The result is that IC_{1a} reacts in a different way from IC_{1b}: the output transistor in IC_{1a} is on when the potential across C_1 plus that at the wiper of P_1 is lower than the reference voltage, and is switched off when the reference voltage is exceeded. Thus, this comparator starts the buzzer sounding for an infinitely long period just before the set time has elapsed.

Summing up the action: one output actuates the buzzer after a certain period of time has elapsed, while the other output switches the buzzer off again after a slightly longer period of time has elapsed. Thus, the buzzer sounds only during the short overlap of these two periods. The length of this overlap is set with P_1. The time period before the buzzer is actuated is set with P_2. The timer is started when S_1 is pressed.

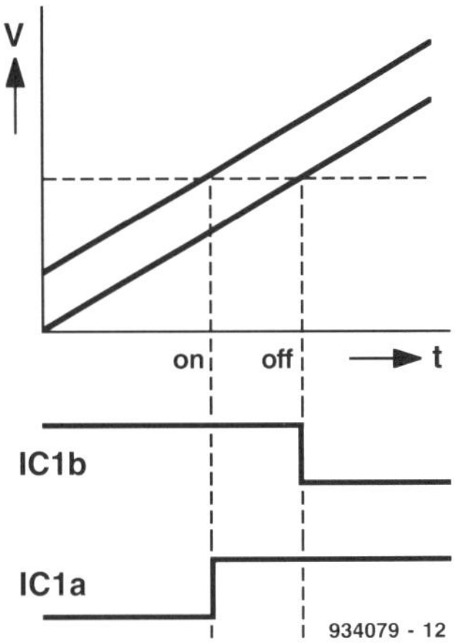

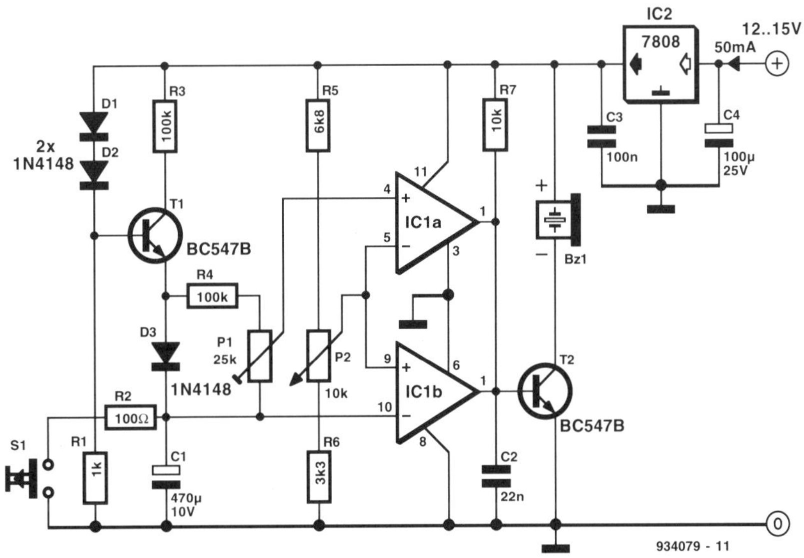

Switch-off delay for battery supply

211 A frequent annoyance with battery-operated equipment is that just after you switch it on you notice that the battery is flat. Quite probably, the last user (you?) has forgotten to switch it off. The circuit described here makes sure that this will never happen again. A touch on the button, S_1, is sufficient to let the equipment work for a predetermined period only.

An interesting feature of the circuit is that its quiescent current is 0.00 mA, because T_1 switches the timer off completely at the end of the cycle. Switching on is effected by the energy contained in the power-on pulse. When S_1 is pressed, the supply voltage is available immediately across C_2. Because of the differentiating action of R_2-C_2, the supply voltage is briefly connected to the V+ input of IC_1 via D_1. This energy is sufficient

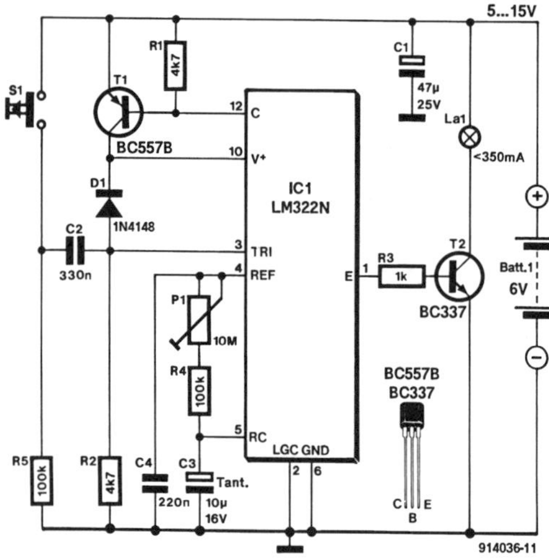

to enable the IC and start the timer, whereupon T_1 is switched on. This transistor provides energy to the IC for the remainder of the cycle. At the end of the cycle, T_1 is switched off and provides no more energy.
The on-time, *t*, is determined by:

$$t = (P_1+R_4)\ C_3 \text{ seconds.}$$

The maximum current switched by T_2 must not exceed 350 mA.
The supply voltage may lie between 5 V and 15 V.
The minimum trigger amplitude is 5 V.
The switch-off time with component values shown in the diagram is 1–100 seconds.
The current drawn by the circuit during the switching interval is about 4 mA when the supply voltage is 6 V.

Telephone buzzer as switch

212 Rectifying the buzzer signal on a telephone line results in a voltage that may be used to switch one or more loads, for instance, a light to show the deaf or hard-of-hearing that the telephone is ringing.
In the present circuit, the buzzer signal is rectified by D_1–D_4; the rectified voltage is smoothed by C_2 and held at 15 V by zener diode D_5. The voltage is used to drive the LED in optoisolator IC_1 via R_3. The optoisolator provides safe isolation between the telephone line and the present circuit and its load. Note, however, that in spite in this, the telephone authorities in some countries may not permit the use of the present circuit: it is always best to seek the advice of your local telephone manager.
As soon as a call signal appears on the telephone line, the LED in the optoisolator ensures that the integral transistor is switched on. This in turn switches on T_1, whereupon relay Re_1 is energized.
Capacitor C_3 is charged as long as the transistor in the optoisolator conducts, but discharges via R_4-R_5-T_1 when the call signal fails: this prevents T_1 being switched off during the intervals between the various pulses

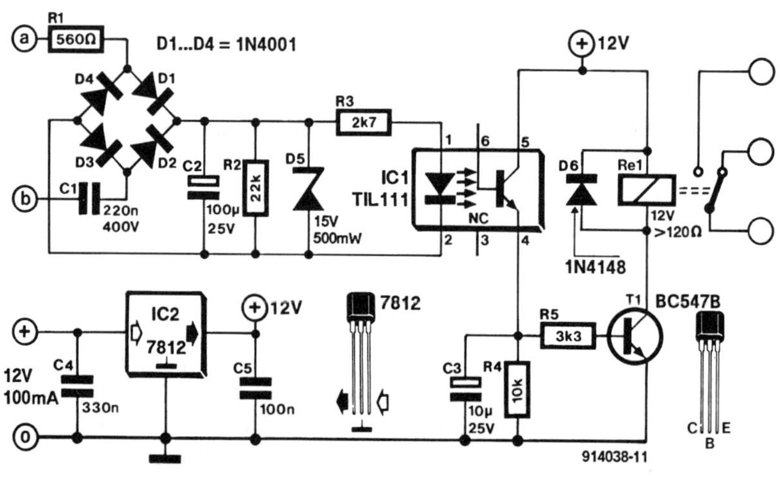

of the call signal. When that signal fails, the relay will be denergized after a short delay. The circuit can be supplied by a 12 V mains adapter. The current it draws depends on the type of relay used, but should not exceed 100 mA. Make sure that the relay can handle the switched voltages and currents.

Telephone extension

213 In many countries, unlike most of the UK, it is still not possible to parallel two or more telephones easily. The circuit described here may either be connected in parallel to an existing telephone or be used as a stand-alone unit, when it will energize a relay on receipt of an incoming call. The relay may be used to operate an optical or an aural indicator, or cause a trigger pulse to be generated that actuates an interface, which in turn operates a suitable telephone.

The incoming a.c. signal is applied to terminals *a* and *b* and from there fed to optocoupler IC_1 via C_1 and R_1. The negative half-waves are returned via D_1, while the positive half-waves are returned via the LED in the optocoupler. The resulting pulsating d.c.

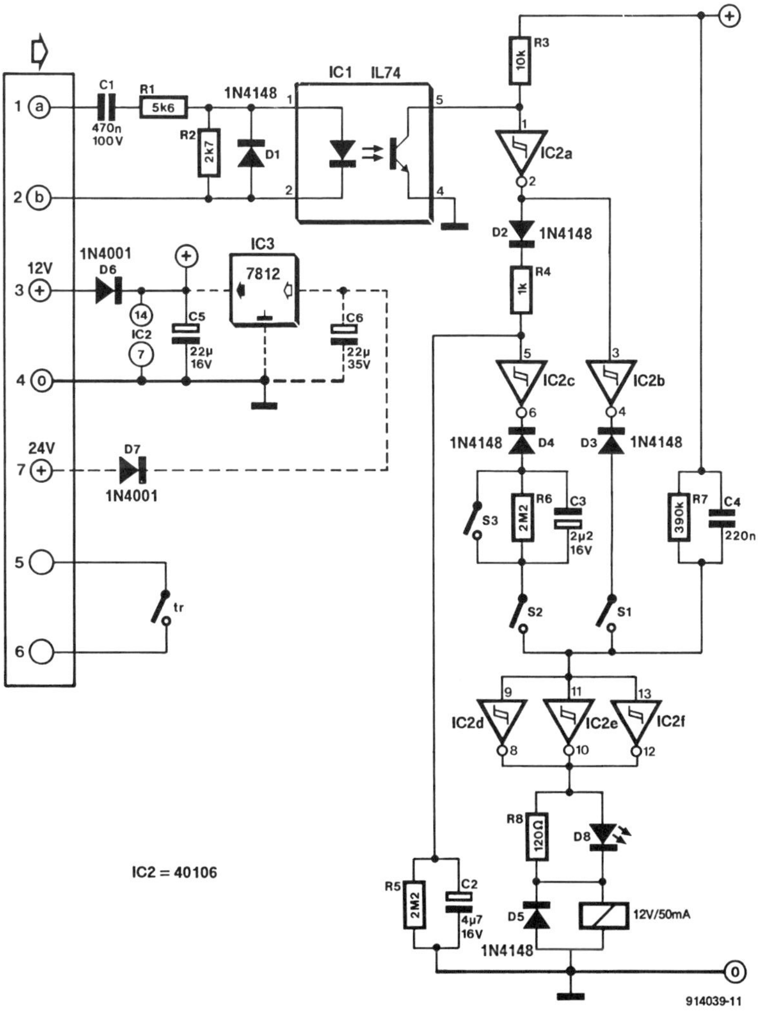

output of the phototransistor in the opto-coupler is applied to inverter IC_{2a}. This (call) signal is smoothed by D_2, R_4, C_2 and R_5, and this results in a direct voltage at the input of IC_{2c} during the pulse spacing. The consequent low- level, short-duration pulses at pin 6 of IC_2 are passed to inverters IC_{2d}–IC_{2f} that serve as relay drivers, and which energize the relay (preferably a high-resistance type). Diode D_8 indicates the state of the relay. The circuit thus functions as a monostable whose time constant is formed by R_7-C_3.
When both S_2 and S_3 are closed, C_3 does not produce a pulse from the direct voltage output of IC_{2c}. The low-level potential then remains at the inputs of the relay drivers to ensure that the relay stays energized in this condition. When S_1 is closed instead of S_2, the relay is energized during the pulse widths and de-energized during the pulse spacing. This mode of operation is best for optical call indicators.
The power supply may be a simple 12-V mains adapter. Diode D_6 protects the circuit against incorrect polarity. Higher supply voltages make a 12-V regulator (IC_3) and an additional electrolytic capacitor (C_6) necessary. The current drawn by the circuit is only a few mA.
It is advisable to check with your telecommunications authorities whether the circuit is allowed in your locality before building it.

Sequential control

214 Sequential controls are used where continuous remote control of mechanical installations, such as rotating antennas or valves, is required.

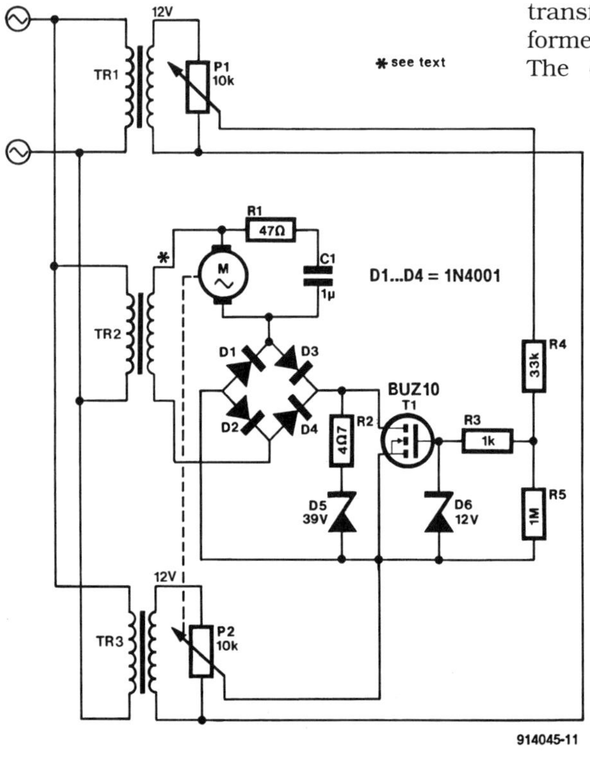

That shown in the diagram offers a setting accuracy of 2.5%, although it has only a few components.
The control motor is in series with a bridge rectifier across the secondary of mains transformer Tr_2. The rating of the transformer must accord with that of the motor. The other two mains transformers are light-duty, 12-V types from across whose secondary a small alternating voltage may be taken by means of P_1 and P_2.
The wiper of P_1 is connected to the gate of T_1 via resistive network R_3-R_5. The wiper of P_2 is connected to the source of T_1. The source-gate junction of the transistor serves as a null-point detector. When the circuit is balanced, the potential difference between the two wipers is zero, so that T_1 is switched off. No current can flow through the motor since the current loop through the rectifier bridge is broken for each half cycle. When one of the potentiometers is adjusted, the circuit is no longer balanced and T_1 is switched on during either the positive or the negative half cycles, depending on which of the potentiometers was adjusted.

Current then flows through the motor, D_4, T_1 and D_1, or through the motor, D_3, T_1 and D_2. In other words, the motor can rotate in either direction. If the motor is coupled mechanically with P_2, P_1 may be used for remote control of the motor.

The circuit as shown is intended for 12 V motors; if different motors are to be used, bear in mind that they are operated from half-wave rectified voltages, which means that the transformer must be rated at 1.52 times the motor voltage.

Variegating LED

215 When the control voltage at the input of the circuit is varied from 0 V to +12 V, the LED will first light up green and then gradually, via orange and yellow, turn to red.

The two sections of the bi-colour (red and green) LED are driven separately: the green one by IC_{1a} via R_7, and the red one by IC_{1b} via R_8.

Opamp IC_{1b} has an amplification of ×2, which results in the red LED lighting from input voltages of about 0.5 V. This section lights at maximum brightness when $U_{in}>U_b/2$.

Opamp IC_{1a} is an inverting amplifier with an amplification of ×2. Moreover, its non-inverting input is at a level of $U_b/2$. When the input voltage $<U_b/2$, its output is high. When the input rises above $U_b/2$, the green section of the LED will gradually become less bright until it goes out completely when $U_{in}=U_b$.

The supply voltage should not exceed 30 V: when it is higher than 12 V, the value of resistors R_7 and R_8 should be altered accordingly.

The current drawn by the circuit depends mainly on the LED: with a 34 mcd type and a supply voltage of 12 V, it is about 35 mA.

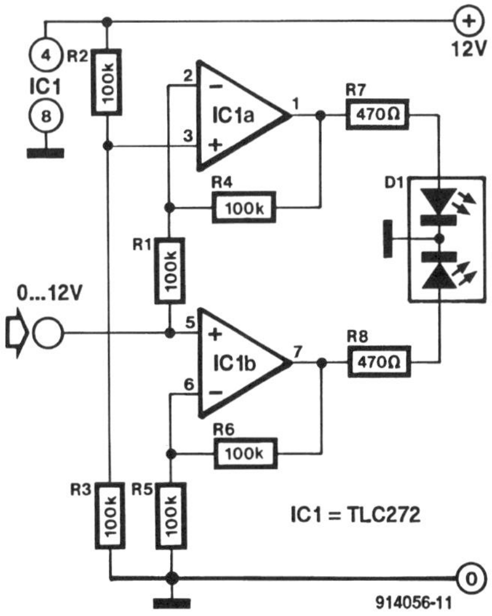

Variable time switch

216 The switch described has two time ranges that are selected with a push-button. In the prototype, the ranges were 5 minutes and 20 minutes, but these can be altered easily. Pressing the switch three times in succession switches off the load (here, a lamp).

When S_1 is pressed, IC_3 is enabled via D_1 and NAND gate IC_{2b}; at the same time, capacitors C_3 and C_4 are charged. When the switch is released, IC_3 gets a clock pulse from IC_{2a} via R_2, so that its QA output goes high. This results in the relay being energized by T_2, so that its contact closes and the lamp comes on.

At the same time, the QA output switches on transistor T_1, so that R_4 is short-circuited. Capacitor C_4 is then discharged via R_3. After the potential across C_4 has dropped to the lower trigger level of IC_{2b}, which takes about 5 minutes, IC_3 is reset and the relay switches off the lamp.
When S_1 is pressed twice in succession, the counter gets two clock pulses and its QB output goes high. Transistor T_1 then remains off and C_4 is discharged via R_3+R_4, which takes about 20 minutes. As an indication that the longer time has been selected, immediately after the second press on S_1, the lamp goes out briefly. This is effected by IC_{2d}, which is given a pulse by differentiating network R_9-C_5 when QB of IC_3 goes high.
When S_1 is pressed three times in succession, both QA and QB of IC_3 go high, which results in the output of IC_{2b} becoming 0 and C_4 being discharged rapidly. The lamp is then switched off instantly.
Rotary switch S_2 enables switching the lamp on or off permanently. When this switch is set to its centre position, the relay is controlled by the timing circuit.

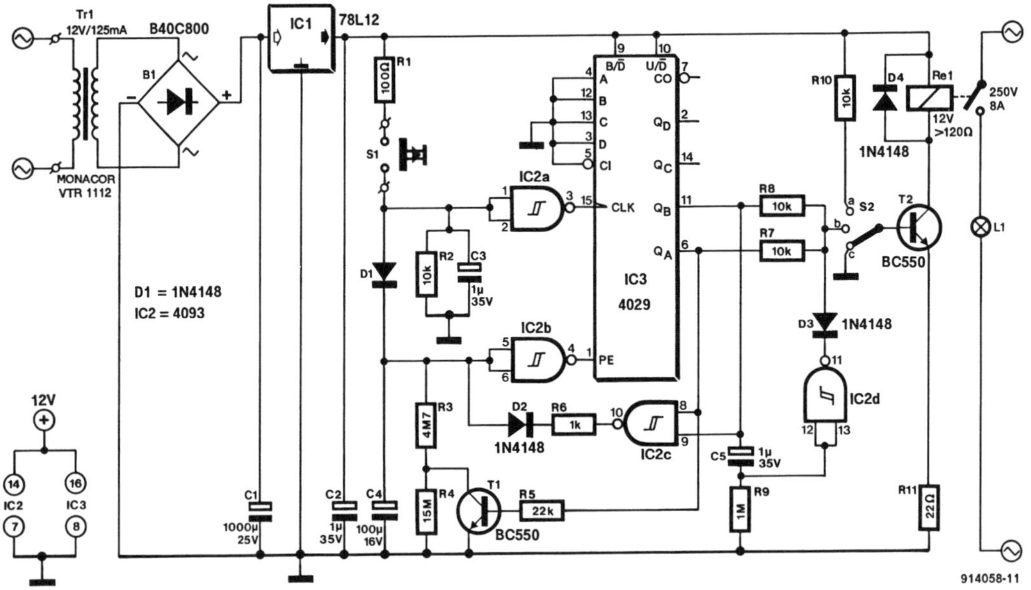

Momentary-action push-button

217 The circuit described here is a kind of remote control for all sorts of equipment that must be started or switched on with the aid of a trigger pulse. The authoress actuates it with a flash of the headlights to switch on the garage lights
When light falls on to the six series-connected solar cells, T_1 is switched on, which results in T_2 being switched off. Capacitor C_2 is then charged.
When the light is removed from the solar cells, T_1 is switched off and T_2 is switched on. Capacitor C_2 will then be discharged through the LED of the electronic relay, which is consequently energized for an instant. In that way a trigger pulse is generated for electrically isolated equipment without the need of an additional power supply.
In the diagram, C_1, R_1 and R_2 ensure a

stable switching operation.

The solar cells used in the prototype gave an effective range of 2–3 m (7–10 ft).

The circuit is intended as a momentary action push button, not as a locking switch. The latter function may be obtained by adding, for example, a latching relay

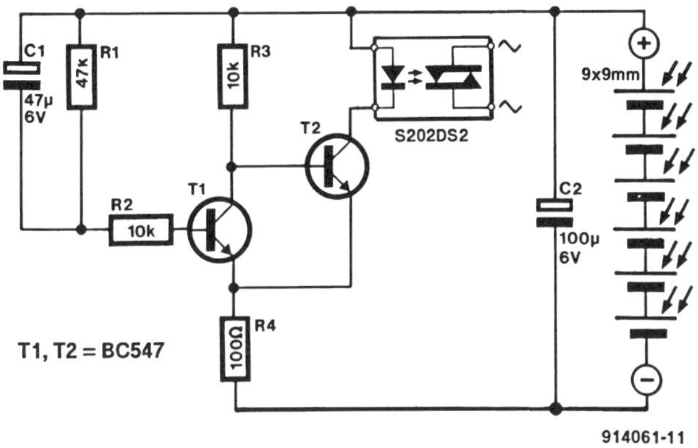

Long-period mains timers

218 Siemens Type SAB0529 IC may be programmed for periods from 1 second to 31 hours 30 minutes. It may be used for switching staircase lights, battery chargers, and many others.

The chip is programmed via pins E–I with the aid of switches S_1–S_4. When S_1 is closed, the IC is enabled for a period of 1 h; S_2: 4 h; S_3: 10 h; and S_4: 16 h. All sorts of combination are also possible: for instance, if both S_2 and S_3 are closed, and the other two switches are open, the chip is enabled for a period of 14 h. The IC controls a 4-A triac, which can switch fairly large loads. The timer is started with S_5; in an emergency, it may be stopped prematurely with S_6.

Great care should be given to the construction, since dangerous mains voltage is present at several points in the circuit.

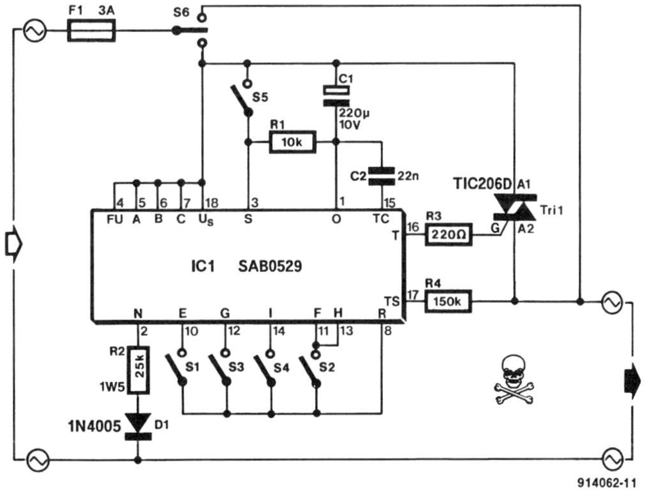

Complementary doorbell

219 It is often not possible to put a second bell in parallel to the existing one, because the bell transformer is normally already fully loaded. The circuit proposed here offers a solution to this problem.

The buzzer sounds **after** the bell-push has been released, The energy for this is stored in capacitor C_2 while the bell-push is pressed. This causes only a small (and brief) additional load on the bell transformer.At the same time, T_1 ensures that T_2 is off. When the bell-push is released, T_1 is switched off and T_2 begins to conduct. As long as the charge on C_2 lasts, the buzzer will sound. The length of the sound depends on the value of C_2 (the higher the value, the longer the sound). Since all the energy is supplied by C_2, the buzzer cannot be replaced by a normal bell or gong, since these draw far too large a current.

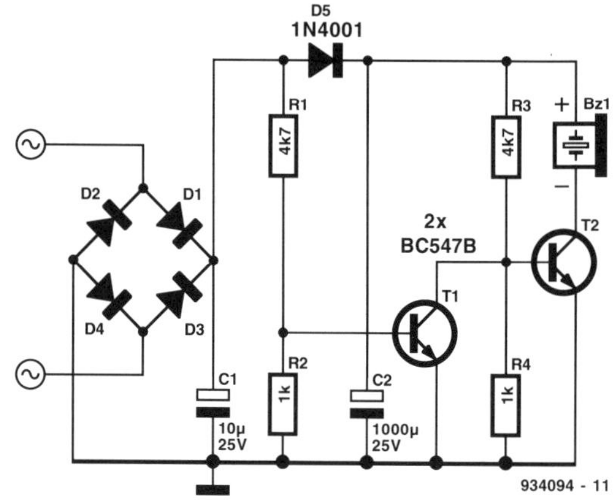

LED indicator for temperature logger

220 The measurement card for the indoor/outdoor thermometer* may be provided with the additional LED indicator shown in the diagram to check the operation of the software.

Every fifteen seconds, a pulse is placed via the background program TLOGGER on to line PBO, which is taken outside via connector K_6. The level on this line is, of course, not switched irrespective, but only at the instant that TLOGGER checks whether a temperature measurement is being carried out. If that is so, the level on the line remains high for about 1 s (depending on the speed of the computer). When no measurement is being carried out, this time is appreciably shorter: of the order of 80 ms. The line is also uscd by the software to signal an error condition outwards. If, for instance, TTRANS.CFG or TLOGGER.CFG is not found, or the path is not correct, or the disc is full, pulses at a frequency of 10 Hz are placed on it.

The diagram shows how the LED is connected to K_6 via a short length of flatcable. Power is supplied by the computer.

* *Elektor Electronics*, March 1991

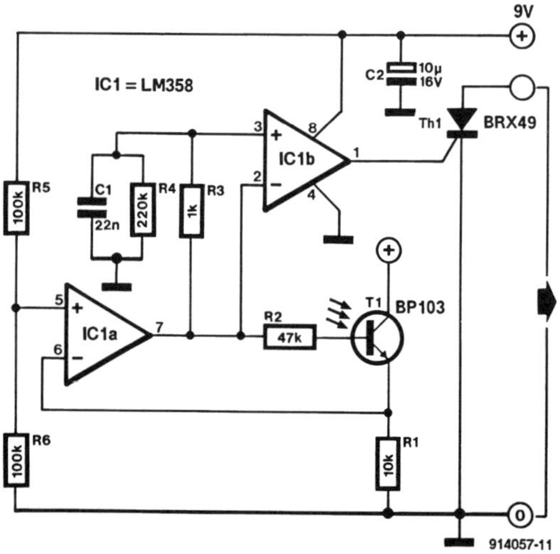

Automatic blower fan control for cars

221 If you were not aware of it, the rush hour can affect your health. Stuck in a traffic jam or caught in the dense traffic in a big city, car drivers are often forced to switch off the fan to prevent being choked by the exhaust fumes produced by the vehicle in front of them. Switching the fan on and off every minute or so to keep the fumes out is a nuisance, and calls for an automatic switch controlled by the engine speed. Such switches exist, but unfortunately for most of you they are only found in top-of-the-range cars equipped with an airconditioning system.

The circuit shown here will not set you back too much, yet puts you at a par with certain BMW 7xx drivers—well, at least as far as the fan control is concerned. The control automatically switches off the fan when the engine runs at a relatively low speed. After a short delay, the fan comes on again automatically when you are clear to drive after passing a congested spot or start moving again in a traffic jam.

The circuit consists essentially of (1) an engine speed monitor circuit set to a switching level of 1,800 rev/min; (2) an integrator to prevent fast switching around the engine speed; and (3) a time constant that introduces a delay before the fan is switched on.

The engine speed monitor serves to sense whether you are moving or not. This is achieved by detecting when the car engine idles, i.e., runs at a relatively low speed. The monitor consists of two monostable multivibrators, IC_{1a} and IC_{1b}. The first is supplied with pulses from the contact breaker. Resistors R_1-R_2 and diodes D_1-D_2 serve to reduce the pulse level to the maximum supply voltage of the circuit. This protection is necessary because in some cars the contact breaker pulses can have a peak value of up to 200 V. Monostable IC_{1a} supplies pulses of a fixed length as long as the period of the input signal is greater than the time constant defined by network P_1–R_3–C_1. If the input pulses are shorter, the Q output of IC_{1a} remains at 1. The time constant, t_1, depends on the number of cylinders, N, in the engine:

$$t_1 = 120 \,/\, (\text{ engine speed} \times N)$$

For example, for a four-cylinder engine, and a switch-off speed of 1 800 rev/min:

$$t_1 = 120 \,/\, (\,1\,800 \times 4\,) = 16.67 \text{ ms.}$$

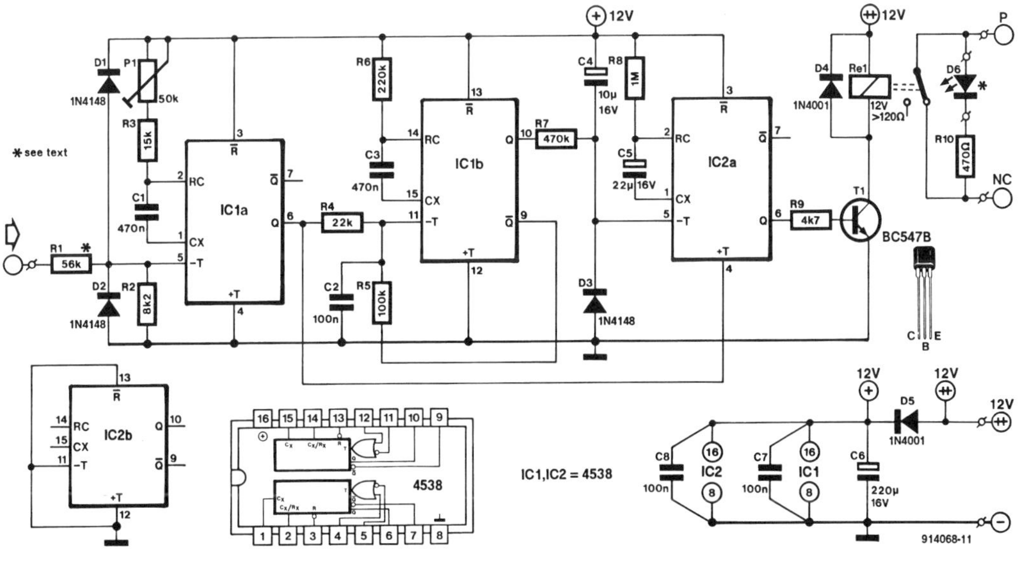

Similarly, for a 5-cylinder engine, t_1=13.32 ms; for a 6-cylinder engine, t_1 = 11.11 ms; and for an 8-cylinder engine, t_1 = 8.32 ms.
The second monostable, IC_{1b}, is triggered as long as it receives input pulses from IC_{1a}. This results in pin 10 of IC_1 going high when the engine speed drops below 1 800 rev/min. To delay the response of the engine speed monitor, a small hysteresis is provided by R_5 and C_2. The monotime of IC_{1b}, t_2, must be greater than the greatest value of t_1. Here, t_2 is set to about 100 ms.
The integrator is formed by network R_7-C_4, whose time constant, t_3, is set to about 3 s. Then follows a third monostable, IC_{2a}, which determines the on time, t_4, set to about 20 s. A transistor driver, T_1, interfaces the control to a relay. Note that the fan is switched off when the relay is energized. This is done to enable the fan to be used when the control is not powered for whatever reason. Finally, the third monostable is connected to the first via the positive trigger input. This is done to prevent IC_2 being re-triggered every time the engine speed drops below 1 800 rev/min.
The circuit is best constructed on the printed-circuit board shown here. Keep an eye on the maximum fan current that flows via the board and the relay contacts—the connections between the fan and the board, 'P' and 'NC' (for normally closed) must be made in heavy-duty terminals and sockets as used in cars.
Test the circuit with the aid of a function generator connected to pin 5 of IC_1. Set a generator frequency that corresponds to the desired engine speed at which the fan control must operate. The generator frequency is $1/t_1$. Set the length of t_1 by adjusting preset P_1 until the signal at pin 10 just toggles. Lower the frequency. The relay must be energized. Increase the frequency, and the relay must be de-energized after a delay of about 20 s. The circuit is now ready to install in the car. The current consumption is about 1 mA in the off state, and about 38 mA when the relay is energized.

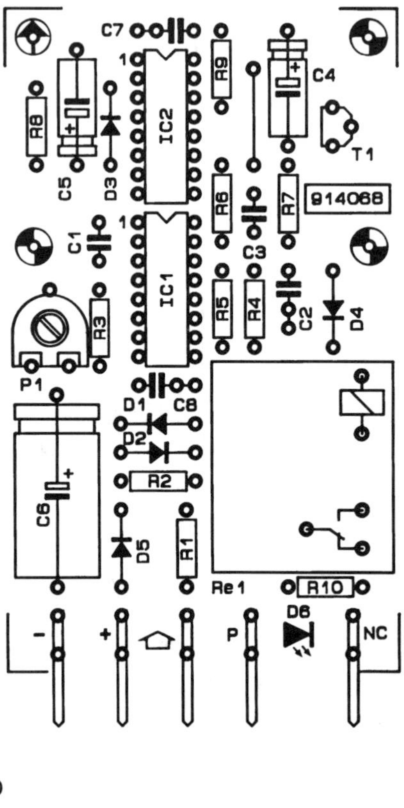

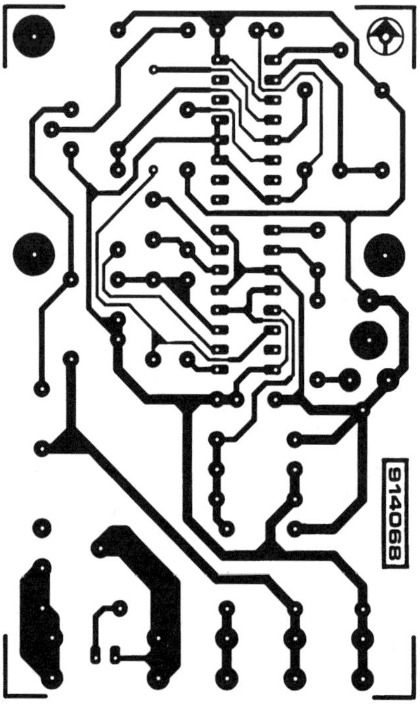

Parts list

Resistors:
R_1 = 56 kΩ
R_2 = 8.2 kΩ
R_3 = 15 kΩ
R_4 = 22 kΩ
R_5 = 100 kΩ
R_6 = 220 kΩ
R_7 = 470 kΩ
R_8 = 1 MΩ
R_9 = 4.7 kΩ
R_{10} = 470 Ω
P_1 = 47 kΩ preset

Capacitors:
C_1, C_3 = 470 nF
C_2, C_7, C_8 = 100 nF
C_4 = 10 μF, 16 V
C_5 = 22 μF, 16 V
C_6 = 220 μF, 16 V

Semiconductors:
D_1, D_2, D_3 = 1N4148
D_4, D_5 = 1N4001
D_6 = LED
T_1 = BC547B

Integrated circuits:
IC_1, IC_2 = 4538

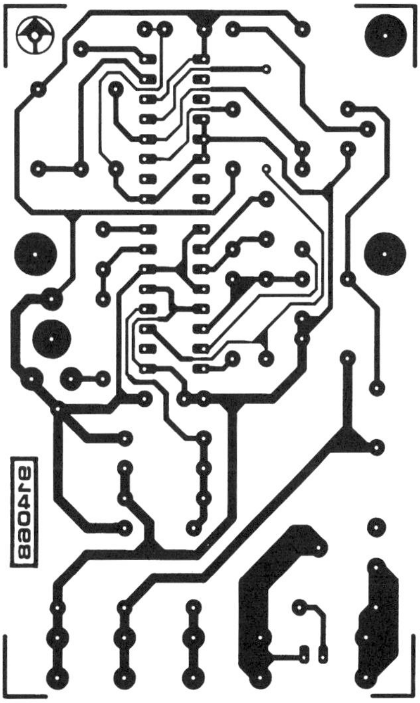

Miscellaneous:
Re_1 = 12 V/330 Ω PCB mount relay, contact rating about 8 A, e.g. Siemens V23127-A0002-A201
5 off angled 'fast-on' pin for PCB mounting

General purpose timer

222 The timer is intended to actuate an apparatus for a predetermined period of time, for instance, an ultra-violet exposure unit or a photographic enlarger. Periods may be set between 0.1 s and 999.9 s.

The periods are set with the aid of thumb wheel switches and stored in four Type 74HCT190 counters, IC_4–IC_7. Setting is facilitated by the set times being displayed instantly on the LED display. After the start button has been pressed, the inputs of the counters are disabled and the count down starts. At the same time, relay Re_1 is energized.

In quiescent operation, the contacts of K_2 are linked to those of K_1. Since K_1 is connected to the mains, the load connected across K_2 is supplied with mains voltage; there is no voltage supplied to the load connected across K_3. When Re_1 is energized, that is, during the switch-on period, the mains voltage is switched from K_2 to K_3. At the end of the switch-on period, a buzzer in the collector circuit of T_1 sounds. In an emergency, the switch-on period may be prematurely terminated with the aid of the stop button.

When the switch-on period is over, oscillator IC_2 is disabled and the relay is de-energized. The preset period is then shown on the LED display again.

Power for the clock may be provided by a simple 300 mA mains adapter combined with a Type 7805 regulator.

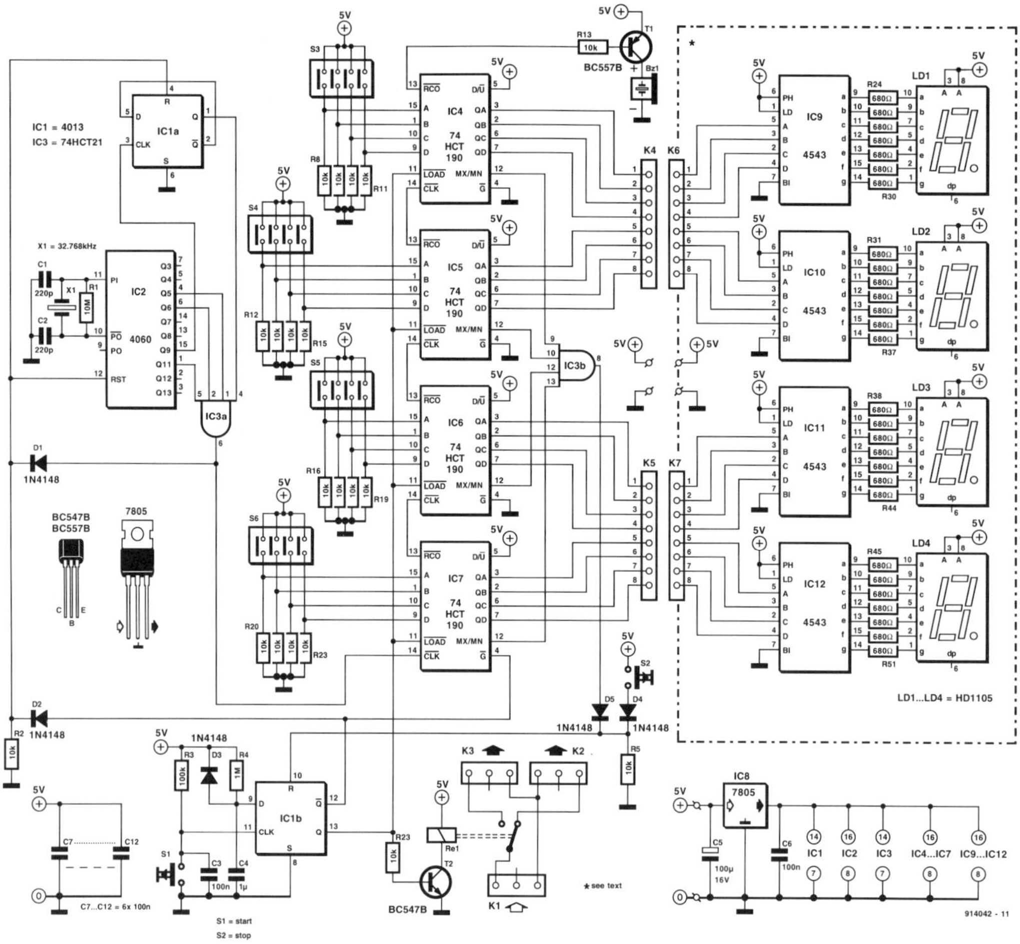

Bedside light timer

223 Many young children will insist on keeping the bedside light on for a couple of minutes after the storybook has been closed and father or mother has gone downstairs. They are also prone to fall asleep with the light on, which is a waste of energy, and a problem for the parent because the light has to be switched off without waking the child.

The timer shown here is an elegant solution to this little domestic problem. Simple to build from a handful of inexpensive components, it lets you determine how long the bedside light remains on after you have said goodnight and actuated the timer.

Pressing switch S_1 causes bistable IC_{1b} to toggle, and produces a debounced clock pulse at the input of the second bistable, IC_{1a}, whose Q output goes high, triggering a low-power thyristor, Th_1. The complementary bistable output, $\overline{Q}$, goes low and enables timer IC_2. The load, a small bulb (max. 60 W) is switched on, and remains on until counter IC_2 resets IC_{1a}.

The counter, a Type CD4541, has an on-board oscillator that operates at a frequency, f, given in Hz by

$$f = 1/2.3\ R_{TC}\ C_{TC}\ ,$$

where R_{TC} and C_{TC} are the resistor and capacitor connected to pin 1 and 2 respectively. The resistor connected to the RS input, pin 3, has a value of about $2R_{TC}$. The scale factor of the 4541 is set to 65 536 (2^{16}) here by tying its A and B control pins to the positive supply rail. This means that the OUT pin changes state after 32 768 clock pulses. The logic levels defined at pins 5, 10 and 9 select a logic low level at the OUT pin when the RESET pin is logic high. Hence, the delay, t, in seconds, introduced by the circuit can be calculated from

$$t = 2.3 \times 32\,768 \times R_5 \times C_5.$$

The circuit is powered direct by the mains. Transistor T_1 forms a 10-V zener diode. A LED, D_2, is used as an orientation aid for the child. The LED forms part of the ITW (Digitast) push-button, S_1.

It should be noted that the actual supply voltage of the circuit may lie between 6 V and 12 V, depending on the characteristics of T_1. The actual value is of little importance, however, as long as IC_{1a} is capable of supplying a trigger current of about 200 μA to the thyristor.

The circuit is constructed on the printed-circuit board shown in Fig. 2, and fitted in a suitable ABS enclosure. In the interest of safety, make sure the input and output cable are properly insulated and secured with strain reliefs. The clearance for the keytop in the top panel of the enclosure must be made as small as possible to prevent any risk of the circuit being touched.

WARNING. Since the circuit carries dangerous voltages at a number of points, it is essential that proper electrical insulation is applied. Never work on the circuit when the mains is connected to it. Make sure that no part of the circuit can be touched when it is being set, adjusted or used.

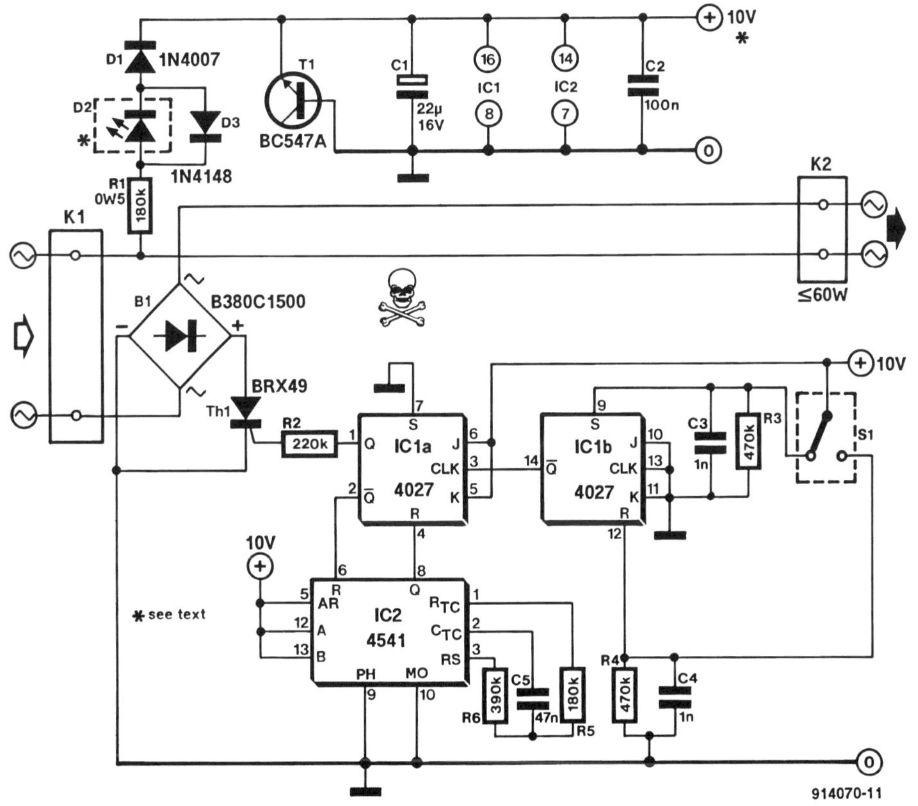

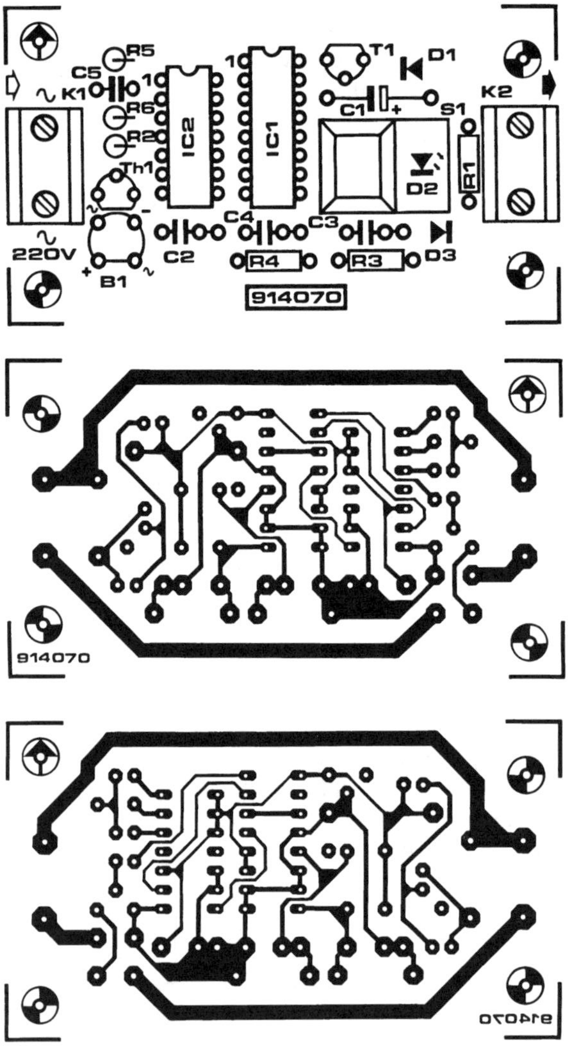

PARTS LIST

Resistors:
R1 = 180 kΩ, 0.5 W
R2 = 220 kΩ
R3, R4 = 470 kΩ
R5 = 180 kΩ
R6 = 390 kΩ

Capacitors:
C1 = 22 μF, 16 V
C2 = 100 nF
C3, C4 = 1 nF
C5 = 47 nF

Semiconductors:
D1 = 1N4007
D2 = LED, high-efficiency (see S1)
D3 = 1N4148
B1 = B380C1500
T1 = BC547A
Th1 = BRX49

Integrated circuits
IC1 = 4027
IC2 = 4541

Miscellaneous:
S1 = digitast push-button switch with integral LED
K1, K2 = PCB-mount terminal block, pin spacing 10 mm
ABS enclosure about 100¥50¥25 mm

Slave mains on-off control

224 The control is intended to switch mains-powered equipment in synchronism with a master control unit. One particularly useful application of it is in audio racks where the signal sources (cassette deck, CD player, tape recorder, tuner, and so on) are switched on and off together with the power amplifier.

The circuit monitors the current drain of the master unit (via K_2) with the aid of optoisolator IC_1. When P_1 is set to maximum sensitivity (corresponding to the highest resistance value), a few milliamperes are sufficient for the control to switch on the slave(s) connected to K_3. The maximum sensitivity should be used only rarely, since allowance must be made for leakage and quiescent currents of the master unit.

When the current drain of the master unit exceeds the preset trigger level, the transistor in IC_1 starts to conduct, causing the output of IC_2 to go low. Consequently, T_1 conducts and actuates the load (the slave) via relay Re_1. After the master unit has been switched off,

C_2 is charged via R_5. At a certain voltage on C_2, the comparator toggles and switches off the slave via T_1 and Re_1. This happens after 500 ms or so.

The state of the control is indicated by two LEDs: one, D_5, is the on/off indicator for the control proper and the other, D_6, shows the on/off state of the slave unit. The maximum loads at the master and slave outputs are 500 W and 750 W respectively.

The control is best constructed on the PCB shown in Fig. 2. The mains connections to the board are made with three 3-way PCB terminal blocks. For safety reasons, the earth track on the board should be strenghtened by a piece of copper wire with a cross-sectional area of 2.5 mm² or larger.

WARNING. Since the board carries dangerous voltages at a number of points, it is essential that proper electrical insulation is applied. Never work on the board when the mains is connected to it. Make sure that no part can be touched when it is being adjusted or used

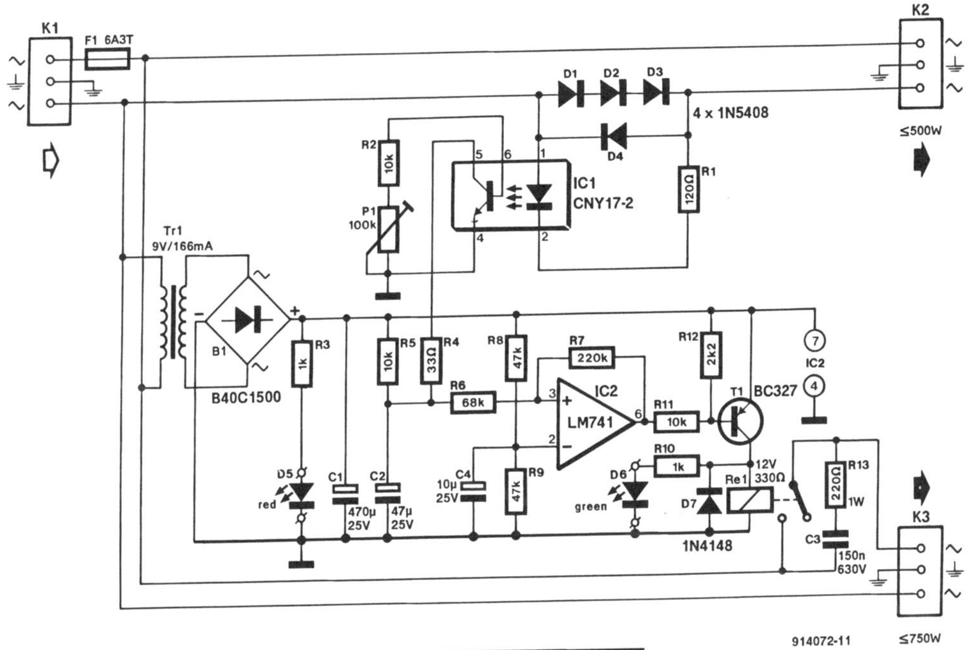

Parts list

Resistors:
R_1 = 120 Ω
R_2, R_5, R_{11} = 10 kΩ
R_3, R_{10} = 1 kΩ
R_4 = 33 Ω
R_6 = 68 Ω
R_7 = 270 kΩ
R_8, R_9 = 47 kΩ
R_{12} = 2.2 kΩ
R_{13} = 220 Ω, 1 W

Capacitors:
C_1 = 470 μF, 25 V
C_2 = 47 μF, 25 V
C_3 = 150 nF, 630 V
C_4 = 10 μF, 25 V

Semiconductors:
D_1–D_4 = 1N5408
D_5 = LED, red
D_6 = LED, green
D_7 = 1N4148
B_1 = B40C1500
T_1 = BC327

Integrated circuits:
IC_1 = CNY17-2
IC_2 = LM741

Miscellaneous:
K_1–K_3 = 3-way PCB block, pitch 7.5 mm
F_1 = 6.3 A fuse with holder
Tr_1 = mains transformer, 9 V, 160 mA sec.
Re_1 = SPST relay, 12 V, 330 Ω
ABS enclosure 190×110×74 mm
($7^1/_2$×$4^5/_{16}$×$2^7/_8$ in)

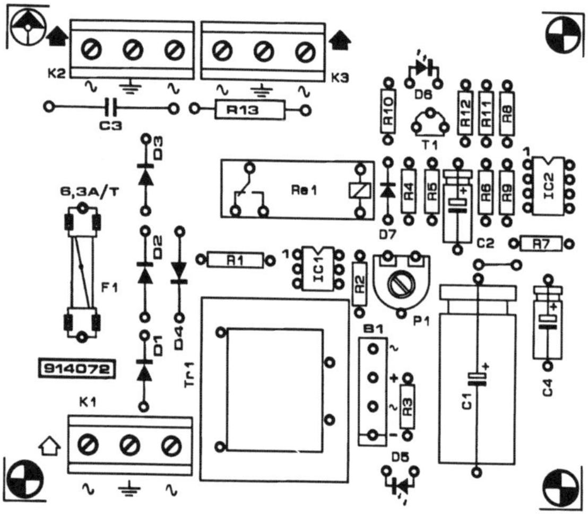

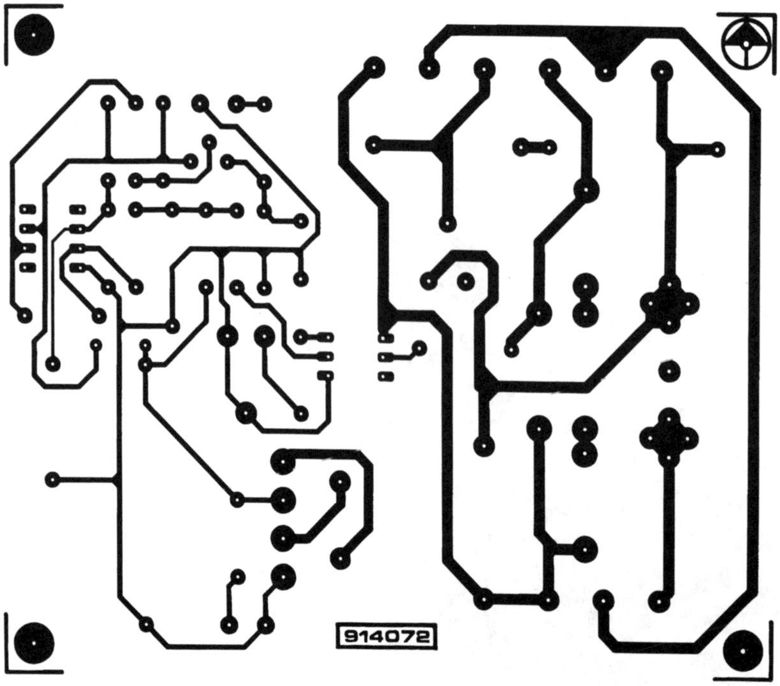

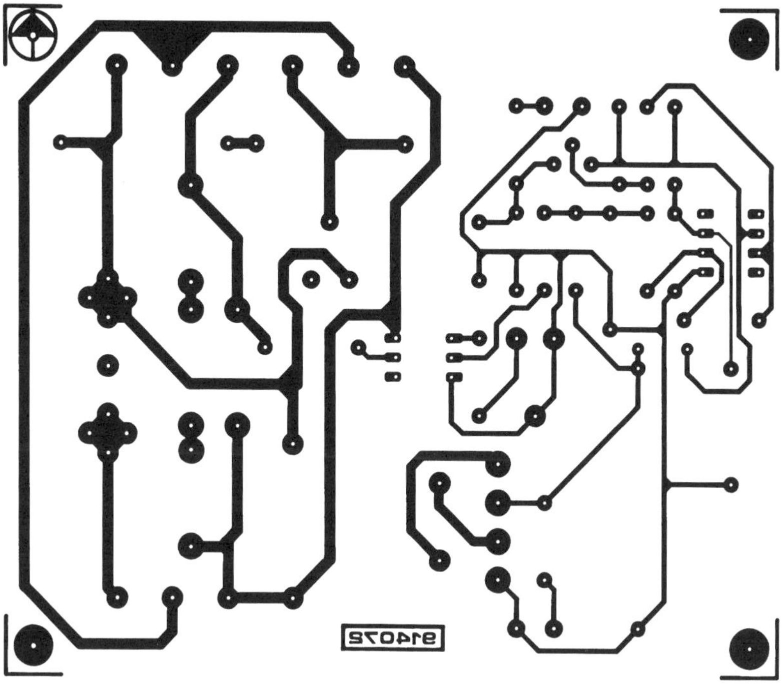

Disco running lights

225 Pop music and light effects are inseparable. The circuits that make the effects possible vary from simple to complex. The circuit presented here is a simple one and is a sort of running light whose rate of change depends on the frequency and intensity of the sound.

The signal is applied to the clock input of counter IC_1 via a single-transistor amplifier, T_1. Its (amplified) level must be high enough to overcome the switching threshold of the counter, while its frequency determines how often the counter is clocked.

The input may be fed with the signal of a preamplifier, but it is also possible, as shown in the diagram, to connect an electret microphone across it, when it functions entirely contact-less. After amplification, the signal is applied to IC_1 via P_1, which controls the sensitivity of the circuit.

Since audio frequencies are too high for making a good visual effect, the signal frequency is scaled down by IC_1 when S_1 connects pin 11 with pin 15. When the switch is in the other position, pin 2 is connected with pin 15: the counter then divides by 1 and the effect assumes a completely different character that no longer resembles a running light.

The actual running light is provided by IC_2, a counter with integral 1-from-10 decoder, which is clocked by the Q0 output of IC_1. Of the ten outputs of IC_2, each of which is connected to an LED, there is always one 'high'. The ten LEDs have a common bias resistor, R_5, an arrangement that is perfectly feasible, since only one LED lights at a time (although it often seems as if more do so).

The circuit may be expanded by adding an LED at pin 12 (carry out) of IC_1. This LED must have its own bias resistor (560 Ω).

The power supply must be able to provide a current of up to 100 mA: at low frequencies the current is appreciably lower.

The colours of the LEDs can be chosen to individual taste.

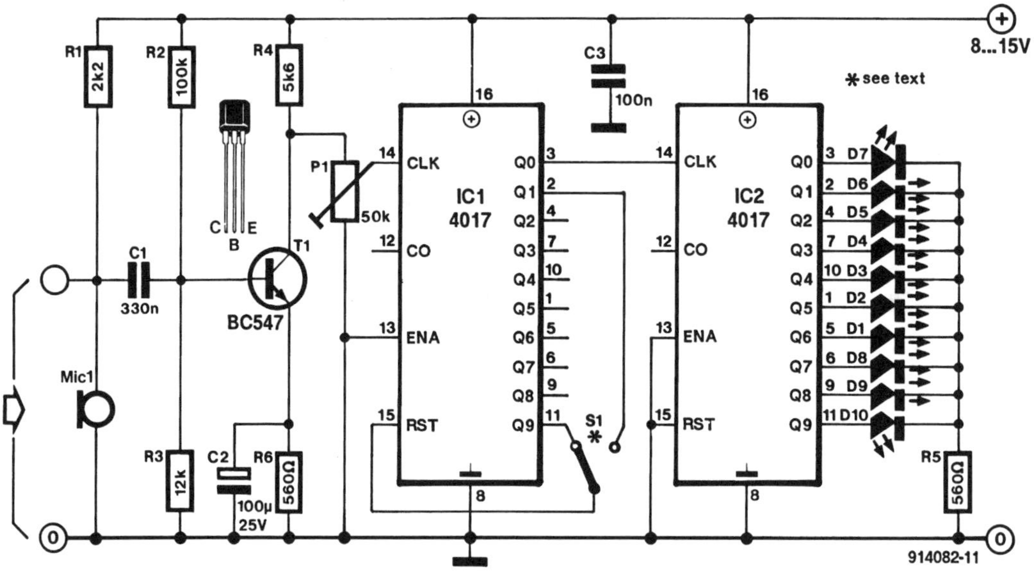

Halogen light switch

226 The switch circuit, intended for low-voltage halogen lights, extends the life of the lamps, because it ensures that the filament current is increased gradually, thus obviating the high peak currents that flow through the lamp with normal switches. The addition of a timer would enable the circuit to switch the lamp off again after a preset period of time.

The lamp is switched via T_1, a MOSFET that has a channel resistance of only 0.08 Ω, which ensures that losses are low (in the prototype ≤250 mW). Control is by means of pulse-width modulation, which also tends to minimize losses.

The circuit is switched on and off with S_1. Bistable IC_{2b} is a debounce circuit that clocks binary scaler IC_{2a}. When the Q output of the scaler is high, the lamp is on or is coming on; when the Q output is low, the lamp is out or going out. The lamp may be switched off automatically by IC_3 after a preset time. The time, *t*, in seconds, is calculated from

$$t=32768x2.3xR_4xC_7,$$

where R_4 is in ohms and C_7 in farads. With values as shown, the time is 700 s (11 min).

When the reset input of IC_{2a} is earthed, the lamp can be controlled only via S_1; timer IC_3 and the associated *RC* network can then be omitted.

When the lamp is switched on with S_1, the voltage across C_6 rises slowly. Because of D_3, even at standby there is a potential across C_6 at a level just below that necessary to toggle comparator IC_{1b}. As soon as C_6 is being charged, the comparator will, therefore, toggle almost immediately. This starts rectangular-wave generator IC_{1c}. However, it is not the rectangular signal that is used here, but the triangular signal across C_8. That signal is compared with the voltage across C_6. This results in a 25 kHz rectangular signal at the output of IC_{1d}, whose pulse width increases slowly. That signal is used to drive T_1, and thus the lamp, which will gradually begin to light.

The voltage across C_6 continues to rise until the toggle level of comparator IC_{1a} is reached. That circuit then toggles, which stops generator IC_{1c}, but T_1 is kept conducting by IC_{1d}. The potential across C_6 is kept just above the toggle level of IC_{1a} by D_2. This arrangement makes it possible, if required, for the lamp to be turned down almost immediately after S_1 is pressed or the time set for IC_3 has elapsed.

When IC_{1a} toggles again, the triangular voltage is compared with the falling potential across C_6, so that the pulse width of the output signal from IC_{1d} decreases. When the voltage across C_6 has reached the level at which IC_{1b} toggles, the generator is switched off again, but this time T_1, and thus the lamp, remains off.

Finding switch S_1 in the dark is facilitated by using a switch with integral LED (D_4).

The power supply consists of a suitable mains transformer (which is probably already present for the lamp) and a bridge rectifier rated at 3 A. The current is drawn primarily by the lamp: with a 20 W lamp, it amounts to 1.6 A.

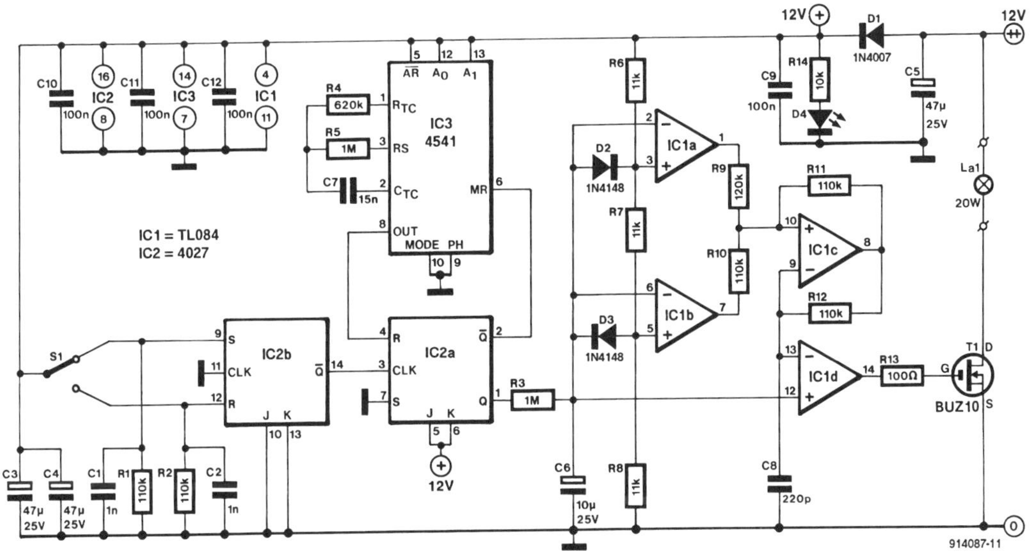

Active two-wire temperature sensor

227 When a physical quantity (measurand) has to be measured at a distance, it is normally necessary to provide the sensor proper with an amplifier. This amplifier needs a power supply, which means that a third wire has to be taken to the sensor. In the present circuit, the measurand and the power supply use the same two lines. To that end, the measurand, buffered by IC_{1a}, is converted into an additional supply current drawn from current source IC_{1b}. In other words, the magnitude of the input voltage may be gauged from the level of the supply current (which, of course, also consists of the current drawn by the opamp and, possibly, the sensor).

When the specified sensor and opamp are used, the current is about 0.7 mA (0.65 mA with a supply voltage of 5 V; 0.7 mA at 10 V; and 0.77 mA at 15 V). With a ful-scale deflection (FSD) of 10 mA, this is a deviation of about 7%, which is, however, easily compensated.

The circuit design is such that when the temperature at the sensor varies from 0 °C to 100 °C, the supply current changes from 0.7 mA to 10.7 mA. Thus, a moving coil ammeter in the supply line enables the temperature to be read quite simply; the supply current to the opamp (0.7 mA) is nullified with the adjustment screw on the meter.

Note that the voltage drop across long lines must be compensated by a higher supply voltage to prevent IC_1 and IC_2 being under-powered (they must get at least 5 V).

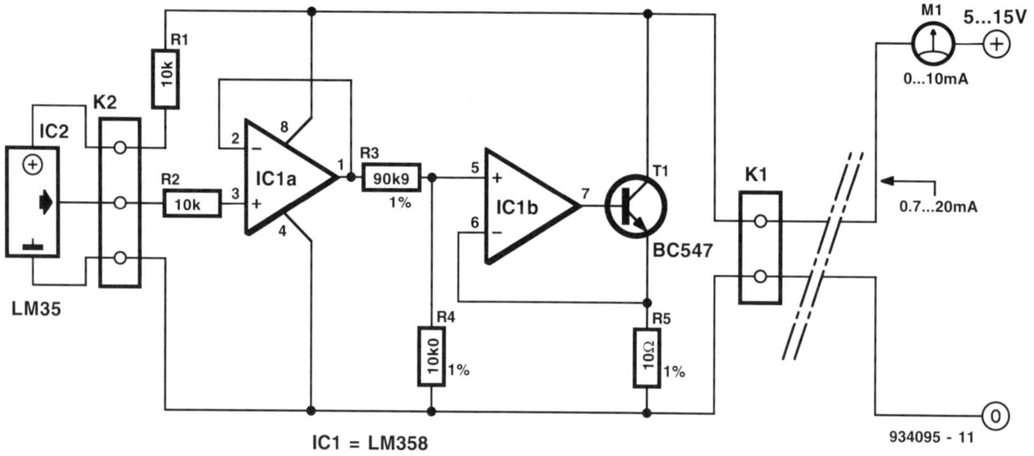

Video camera timer

228 Some video cameras have a socket for a remote control unit. It appears, however, that it is not always easy to connect and interval-control to this socket. For instance, the Blau-punkt 8010 camcorder is not switched on or off in the traditional manner, but by 40–60 ms long pulses. One pulse switches the camera on andf the next one switches it off. Manual operation with a switch is virtually impossible to achieve. However, the timer described here offers a solution.

The timer generates the pulses automatically; the interval between two pulses can be set between about 1 s and 10 s. It operates from a 9 V (PP3 or 6F22) battery: the current drain is only 330 μA.

When switch S_1 is closed (signal A – see Fig. 2), differentiating network R_2-C_1 ensures that IC_{2a} receives only a short pulse (signal B), even if S_1 remains closed for some time. Assuming that the circuit was quiescent before S_1 was closed, pin 1 of IC_{2a} is high, so that the output of the IC (signal C) goes low as a result of signal B.

The output signal of IC_{2a} triggers monostable IC_{1a} via AND gate D_2-D_3-R_3, whereupon the output of IC_{1a} (signal E) switches on transistor T_1. This transistor serves as the stop/start switch of the camcorder; its drain and source are connected to the camera.

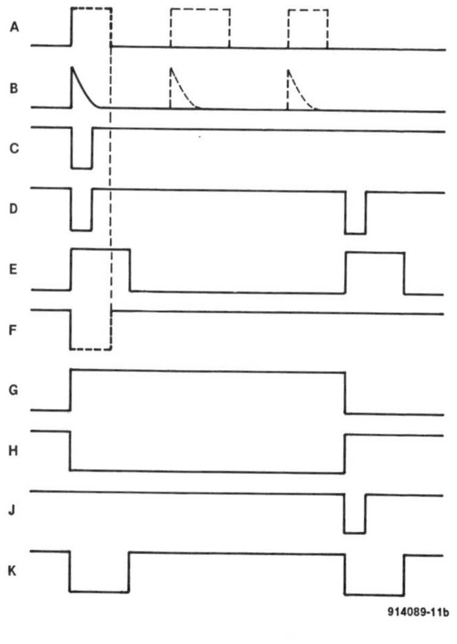

At the same time that IC_{1a} generates the start pulse, IC_{1b} is triggered to commence measuring the time interval between the start and stop pulses. This interval can be preset with P_1. During the interval, signal H is kept low and this disables switch S_1. Once the mono time of IC_{1b} has elapsed, signal H goes high via IC_{2d}, so that IC_{1a} is triggered anew and sends a stop pulse to the camera. The circuit then returns to the quiescent state until S_1 is closed again.

As drawn, the circuit is particularly suitable for adding titles to the filmed material. One touch on S_1 and the title in front of the camera lens is recorded within a few seconds.

Switch S_1 may be replaced by an interval timer for making speeded-up recordings.

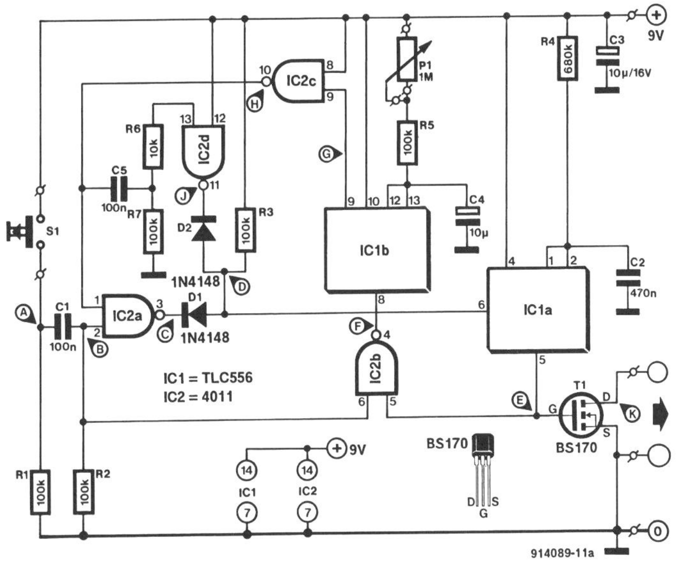

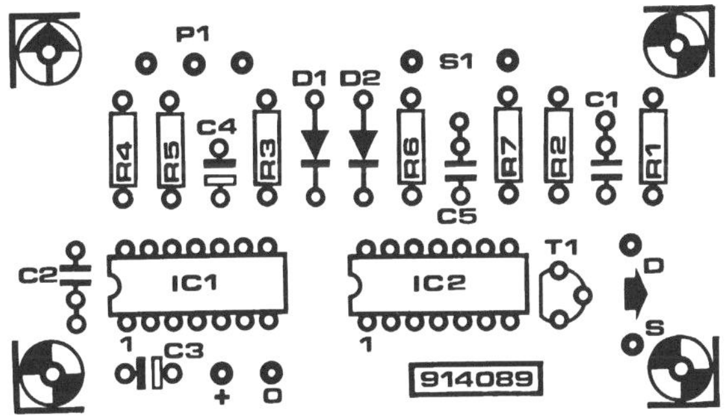

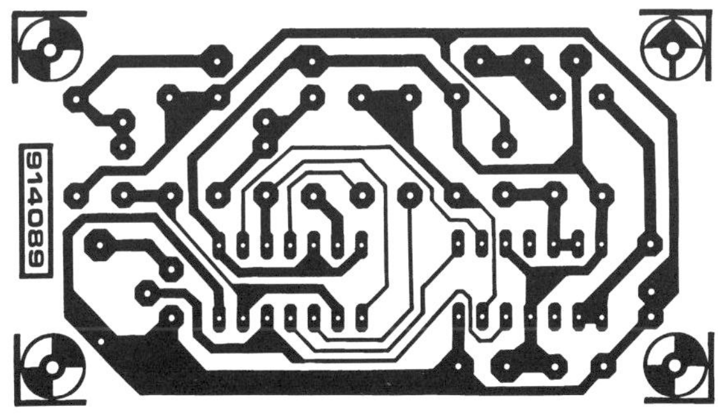

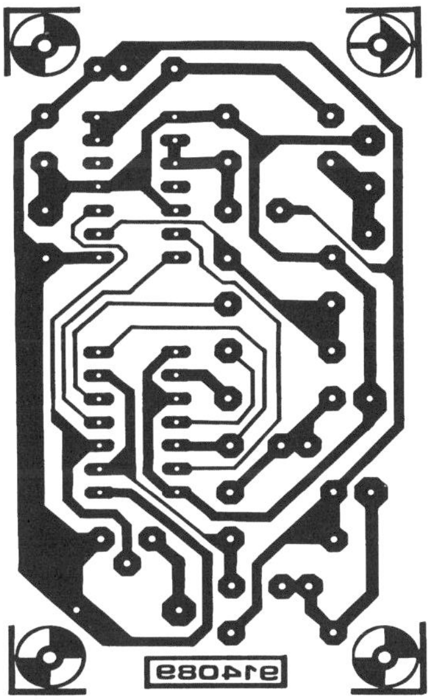

Intelligent power switch

229 Texas Instruments' TPIC2404 is a monolithic high-voltage, high-current, quadruple switch especially designed for driving peripheral loads, such as relays, solenoids, motors, lamps, and other high-voltage, high-current loads from low-level logic. It can switch currents of up to 1 A without any problem; the current may be increased to 4 A by connecting four outputs in parallel. As shown in Fig. 2, the outputs are of the open-collector type.

Connected in a circuit as shown in Fig. 1, the chip may conveniently be used to increase the power handling of the Centronics analogue-to-digital and digital-to-analogue converter published in the May 1990 issue of *Elektor Electronics*. Note that the FAULT output of the IC is connected to input 11 of the converter. This enables the TPIC2404 to indicate the following fault conditions.

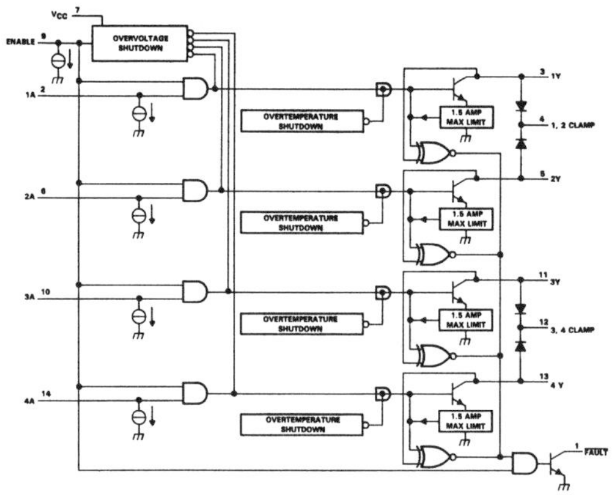

914091-12

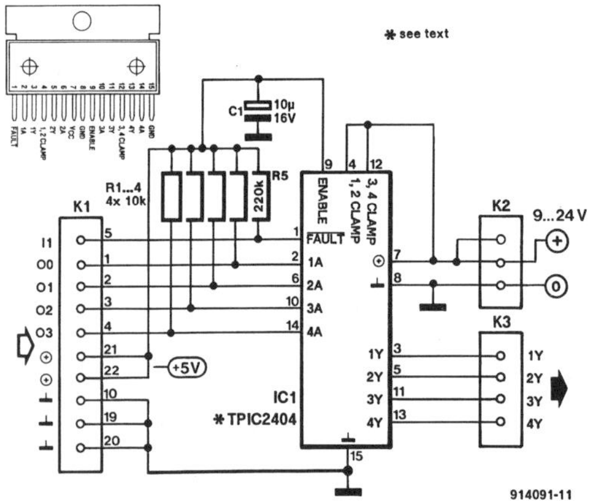

914091-11

- Too high supply voltage (>25.5 V).
- Thermal overload.
- Output short circuit.
- Loads not connected (only if outputs are inactive).

The software regularly monitors the level on line I1 and will indicate a fault condition.

The TPIC2404 must be fitted on to a small heat sink: 8 cm² (1.5 in²). When the chip is used with the converter, R_{31} on the converter board must be disconnected.

The switch draws a current of about 100 mA when all outputs are active.

Fast switching gate

230 Switching transistors are usually driven into saturation, which has an adverse effect on the switching speed. This effect is eliminated, or nearly so, by the use of Schottky diodes at the inputs. It is equally possible to add a diode to a transistor (across its base-collector junction) as shown in the diagram to increase its switching speed.

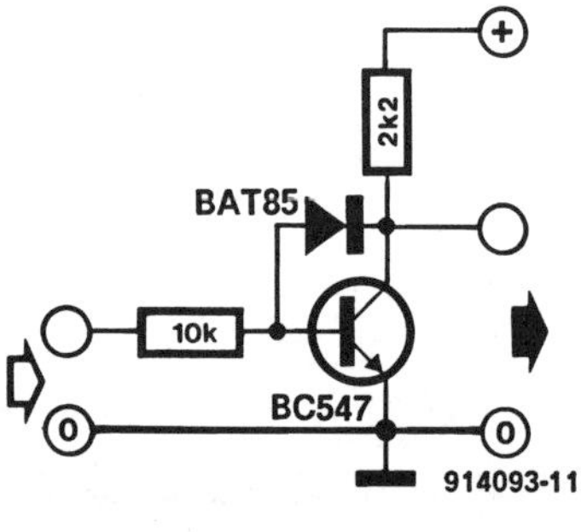

914093-11

When the transistor is driven into conduction, its base current will soon be limited because the diode has a lower transfer potential than the base-collector junction, so that part of the current will flow through the diode. When the transistor is switched off, it will therefore require less time to reach the non-conducting state. The effect is seen clearly in the photograph. Signal 1 is the input signal at a frequency of 166 kHz. Signal 2 is the (inverted) collector signal without diode, and signal 3 is the collector signal with the diode added. It is evident that, owing to the diode, the collector returns to the high-level state much more rapidly.

CMOS dimmer

231 Siemens' Type SLB0586A IC enables the simple construction of a dimmer with touch control. Used in conjunction with a Type TIC206D triac, it allows the dimming of light bulbs of 10–400 without any problem. A 100 µH, 5 A inductor is needed to suppress the switching noise.

Synchronizing pulses are derived from the mains voltage via R_{11}, C_4 and D_4 and these are applied to pin 4 of the IC.

The supply voltage is obtained by means of R_2, C_2, D_1, D_2 and C_3, and lies about 5.3 V below the mains voltage. The touch key used to operate the IC is connected to pin 5 via two 4.7 MΩ resistors, R_5 and R_6, to guarantee the safety of the user.

Since dimmers are often built into an existing circuit, there is frequently a need for operation from two different locations. Conse-quently, the diagram shows an additional push-button switch that may be situated well away from the touch key.

The diagram also shows three jump leads, which are intended for selecting one of three modes in which the IC can work. When jump lead B is used, the light is always switched on at the last used level, whereas when A or C are used, the light comes on at minimum brightness. When B or C are used, the dimming direction reverses every time dimming is used; this is not the case when jump lead A is used.

When the touch key is touched briefly, that is, for 50–400 ms, the light is switched on or off; if it is touched for a longer period, the dimming cycle is started.

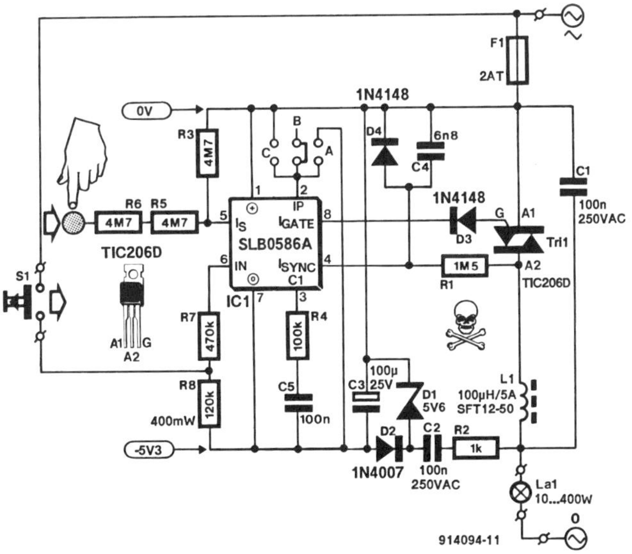

Water level control

232 In some countries, the water supply is irregular at most times; in many other countries only during a drought. A means of making this less inconvenient is offered by the present circuit. It needs two water tanks: one, T_2, at ground level or even underground, and the other, T_1 in the loft or at least appreciably higher than T_2. Tank 2 gets filled by a pump from tank 1 to ensure that there is sufficient water pressure. The circuit shown ensures that the water in tank 1 is kept at a given level; if the water drops below that level, the pump will be switched on. There is protection in case tank 2 is empty. The circuit is operated by a number of sensors mounted in the tanks. Each tank contains a non-corrosive or insulated pin or straight piece of stout wire, *R*. Tank 1 has two sensors, *P* and *Q*, each consisting of a small, non-corrosive disk. Tank 2 has one sensor, *S*, which is identical to those in tank 1. Sensor *P* indicates when tank 1 is full; *Q* signals when tank 1 is empty; *S* shows when tank 2 is empty (no water at all). Pins *R* are connected via a resistor to the positive supply line, while the sensors are linked to the inverting input of three opamps, IC_1–IC_3. The non-inverting inputs of these amplifiers are supplied with a reference voltage derived from potential dividers. Sufficient water between a pin and a sensor causes a virtual short-circuit that results in a high level at the inverting input of the associated opamp. A relay is used to switch the pump on and off: its normally open contact operates the motor and its normally closed contact is linked to the output of IC_2.

When tank 1 is full, the inverting input of all three opamps is at a high level: the outputs of all three are then low and the relay is not energized. When the water level in tank 1 drops, the output of IC_1 goes high, but, since the output of IC_2 is low, the relay will remain inoperative. When, however, the water level drops below sensor *Q*, the output of IC_2 will go high. Transistor T_1 is switched on and the relay is actuated, so that the pump is switched on and the output of IC_2 is open-circuited. The high level at pin 6 of IC_1 will ensure that T_1 remains switched on.

When tank 1 is filled to the level of sensor *P*, the resulting low level at the output of IC_1 will cause the relay to be deenergized, so that the pump is switched off.

When tank 2 is empty, the output of IC_3 becomes high, which switches onT_2 so that T_1 does not get any base current. Conse- quently, the relay can not be energized. The reference voltage for each of the opamps can be preset with P_1–P_3 to obtain the required switching pattern, which, of course, depends also on the sensors used and the composition of the water.

Since there is a direct voltage at the sensors, these musty be inspected regularly. Some advice: the carbon electrodes of a discarded dry battery do not resolve in water and are non-corrosive.

The current drawn by the circuit is determined primarily by the relay coil: the BC517 can switch up to 400 mA. The opamps draw only a few mA.

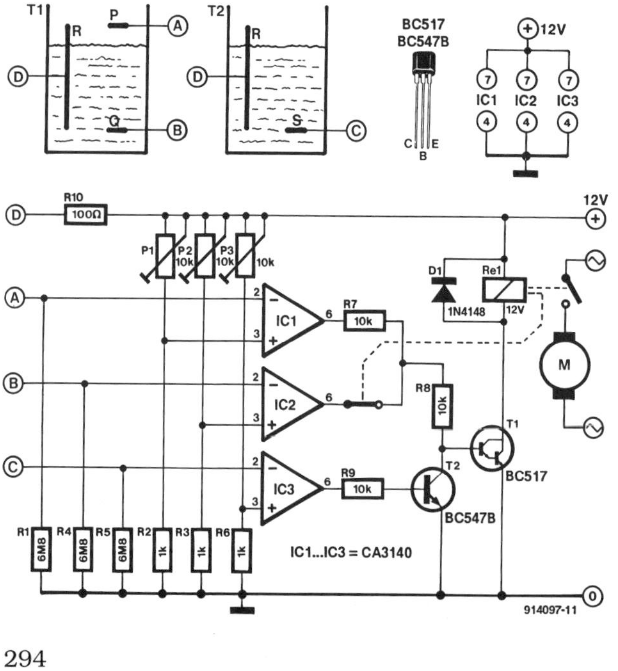

SMT prototyping board

233 Surface mount technology (SMT) components have no wires, and are, therefore, not generally suitable for fitting on Veroboard or stripboard, the traditional cards used to build prototypes of electronic circuits.

The printed circuit board shown here is ideal for starting experiments with SMT components. It has four identical areas with isolated copper spots to accommodate the size of most SMDS (surface mount devices), including ICs. A border of larger copper spots is provided around the prototyping area to facilitate the connection of wires and external components to the board.

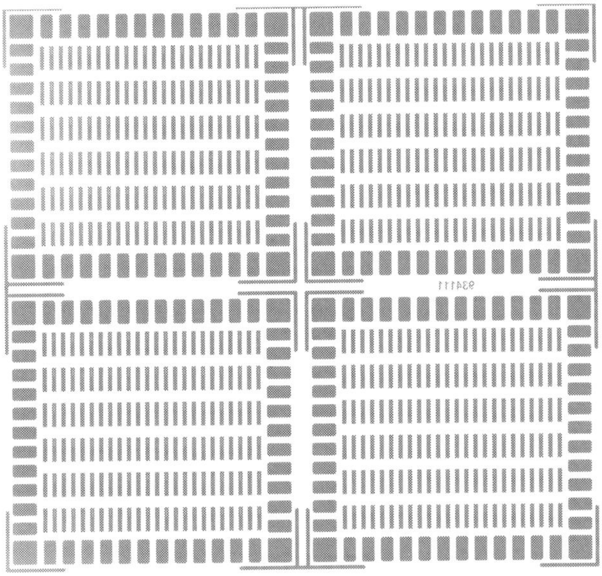

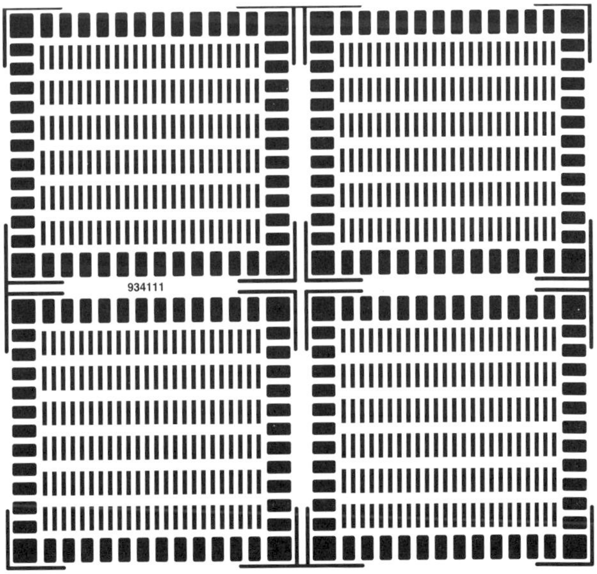

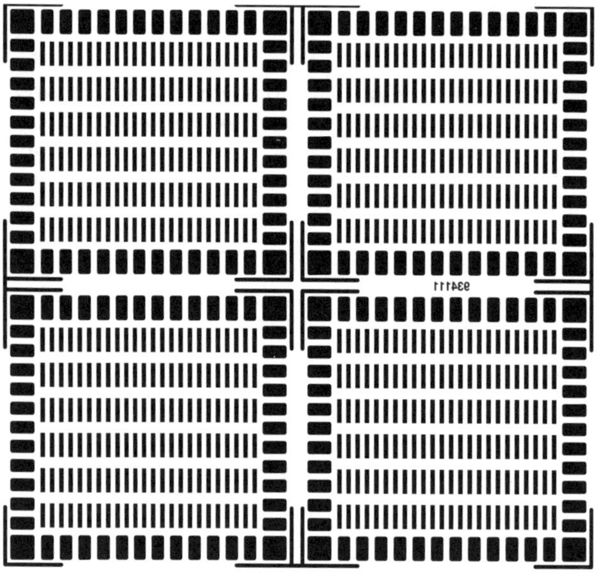

Voltage regulator for cars

234 In older cars, the battery charging voltage is controlled mechanically. The regulator consists of a relay that switches the stator windings of the alternator on and off. This arrangement is prone to breakdown, inaccurate regulation and sensitivity to load variations.

An electronic alternative as shown has the advantages of not containing any moving parts and of providing much more accurate regulation. Moreover, the voltage is measured at the battery terminals, so that losses in the wiring are not included in the measurement.

The regulator, IC_1, is essentially a device that continuously compares the battery voltage with a reference potential. This comparator drives a power transistor that switches the excitation of the alternator.

Terminal Z1 is connected to the + terminal of the battery; Z2 to the ignition switch; and Z3 to the stator winding of the alternator.

The battery voltage is reduced to about 5 V by potential divider R_1–R_2–P_1 and applied to the non-inverting input of Schmitt trigger IC_1. The inverting input of this opamp is at a reference potential of 5 V provided by regulator IC_2. Power transistor T_1 is switched by the output of IC_1 and transistors T_2 andT_3. Diode D_2 functions as an indicator, while D_1 is a free-wheeling diode. Capacitor C_6 attenuates the pulse generated when T_1 is switched on, so that far fewer harmonics are generated and interference on medium-wave radio is suppressed.

The regulator is calibrated by connecting a 12 V lamp or 15 Ω, 10 W resistor between Z3 and earth and a variable power supply and multimeter, set to 15 V, between Z1 and earth. Set the power supply output to 14.3 V and adjust P_1 until the lamp just goes out. When the power supply output is reduced slowly, the lamp should come on again at 13.9 V.

The regulator is best built in a small aluminium case that also serves as heat sink for T_1. The case can be made water-tight with a suitable (hardening) silicone paste.

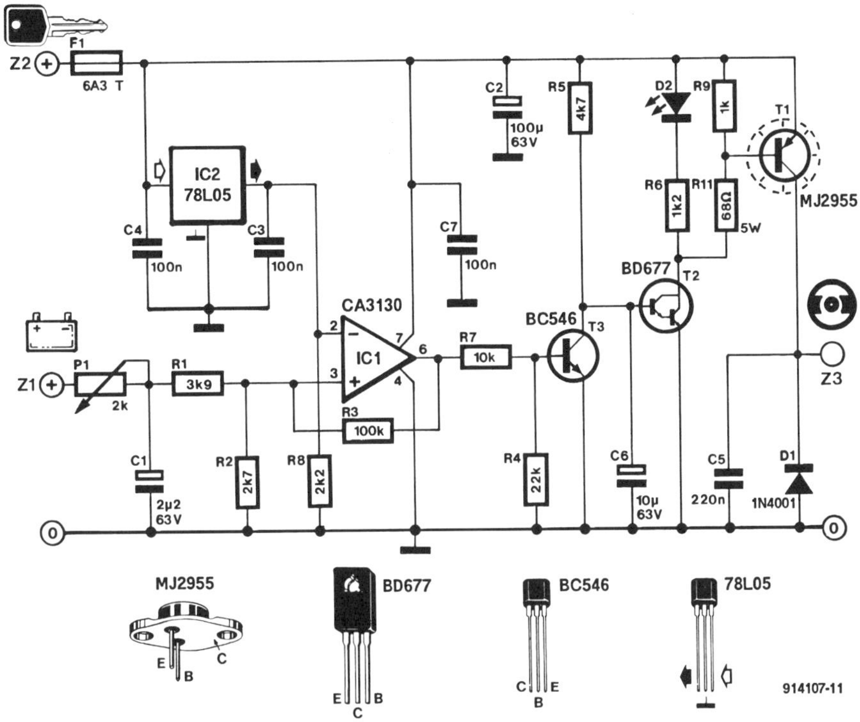

Digital 555

235 There is a digital version of the well-known Type 555 timer IC: the HCT5555. The traditional 555 can be used as astable or monostable, with the timing determined by an *RC* network. Crystal control is not possible and when the time constant is long, accuracy goes by the way.

The HCT5555 is intended purely as a monostable; if operation as an astable is required, a number of external components must be added—see diagram. The timing is arranged, however, by a separate integral oscillator and programmable divider. Owing to the greater number of facilities, the new device is not housed in an 8-pin case like the 555, but in a 16-pin package.

The chip has two trigger inputs: one for first transitions (A) and one for last transitions (B). These inputs can be interlinked: the mono time, determined by the oscillator and the set scaling factor, then starts at each and every transition of the input signal. When one of the inputs is actuated, output Q goes high. There is a complementary output Q.

Some of the new terminals are:

- MR (pin 15)is the Master Reset;
- retriggerable (pin 16), which determines whether the IC reacts to trigger pulses when the mono time has not yet lapsed;
- OSC(illator) CON(trol) (pin 14) via which the integral oscillatgor can be stopped.
- pins 1, 2, and 3, are used for the oscillator, while pins 10-13 determine the scaling factor—see table.

A possible application of the new device is the timer as shown in the diagram that indicates when, for instance, NiCd batteries have been on charge for 14 hours. The LED then goes out (or just comes on if the Q output is used). The oscillator is set to 333 Hz with P_1 (R=1.84 kΩ). Leave all switches open or omit them altogether.

The supply voltage depends on the type of IC used: the HCT version needs 4.5–5.5 V, whereas the HC version can operate from 2–6 V. The IC draws about 0.5 mA.

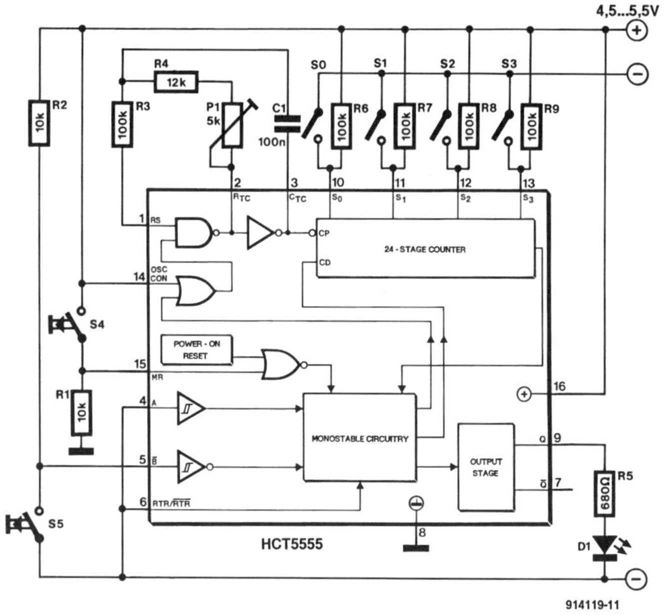

select inputs				divisors	
S3	S2	S1	S0	binary	decimal
L	L	L	L	2^{1}	2
L	L	L	H	2^{2}	4
L	L	H	L	2^{3}	8
L	L	H	H	2^{4}	16
L	H	L	L	2^{5}	32
L	H	L	H	2^{6}	64
L	H	H	L	2^{7}	128
L	H	H	H	2^{8}	256
.	.	.	.	.	.
H	L	L	L	2^{17}	131 072
H	L	L	H	2^{18}	262144
H	L	H	L	2^{19}	524 288
H	L	H	H	2^{20}	1 048 576
H	H	L	L	2^{21}	2 097 152
H	H	L	H	2^{22}	4 194 304
H	H	H	L	2^{23}	8 388 608
H	H	H	H	2^{24}	16 777 216

REF200 from Burr-Brown

236 Burr-Brown's REF200 provides in one 8-pin DIL package two independent current sources, either of which can supply 100 μA, and a 1:1 current mirror. The pinout of the device is shown in Fig. 1. Note that there are no pins for connecting a supply voltage. The correct operation of the current sources is, however, guaranteed when the potential across the source is not less than 2.5 V. The average drift with temperature is only 25 p.p.m. K^{-1}. The output impedance is not less than 100 MΩ. The operating voltage must not exceed 40 V.
The two sources and the mirror make possible a number of configurations, some of which are shown in Fig. 2. That in Fig. 2a is a 50 μA current source whose output voltage may vary between 0 and the positive supply voltage minus 2.5 V. In Fig. 2b, all parts of the IC form a floating 300 μA current source. Finally, Fig. 2c shows a bidirectional 100 μA source.

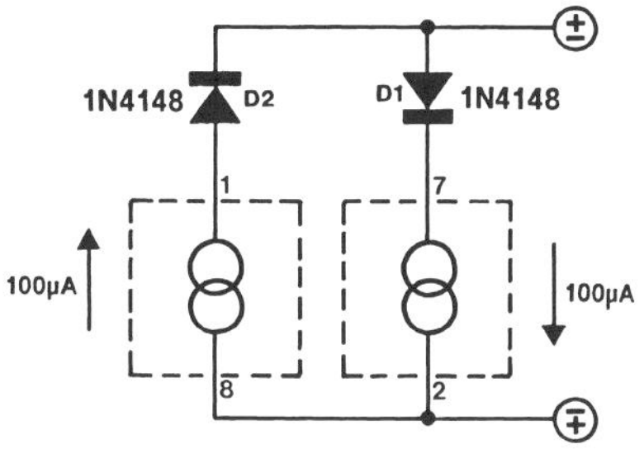

914120-12c

REF 200

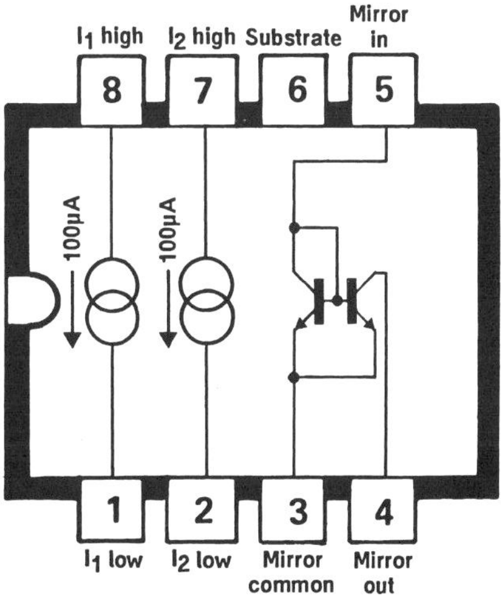

914120-11

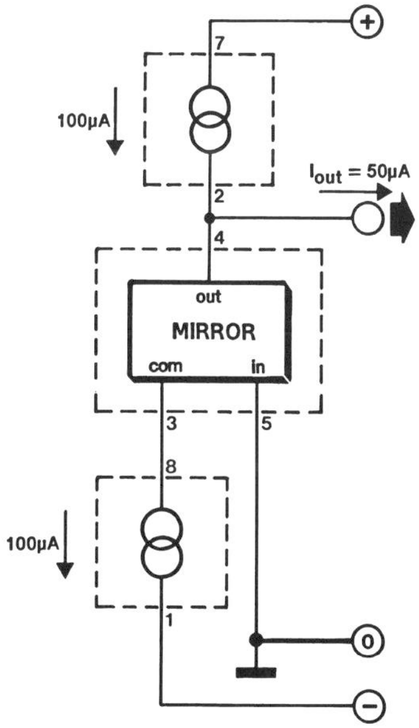

914120-12a

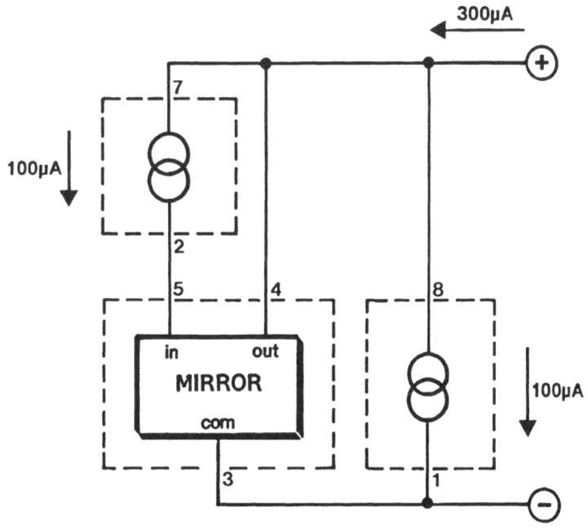

914120-12b

General purpose power-on delay

237 Output Q1 of this delay circuit becomes active after the switch-on period has elapsed and remains so until the next cycle is begun. Output Q2, on the other hand, also functions as a monostable; after its mono period has elapsed, Q2 automatically becomes inactive. The mono period can be set between 1 second and 4 seconds with P_2. The power-on delay can be set accurately between 1 second and 1 minute with P_1 and DIP switch S_1.

The circuit is also provided with a count-down indicator: the digit display on LD1 gives continuous information on the remaining switch-on time. The 7-segment display is driven by IC_4, a decoder chip that makes only the normal digits (1–9) visible.

When the end of the delay period has been reached, counter IC_1 has come to

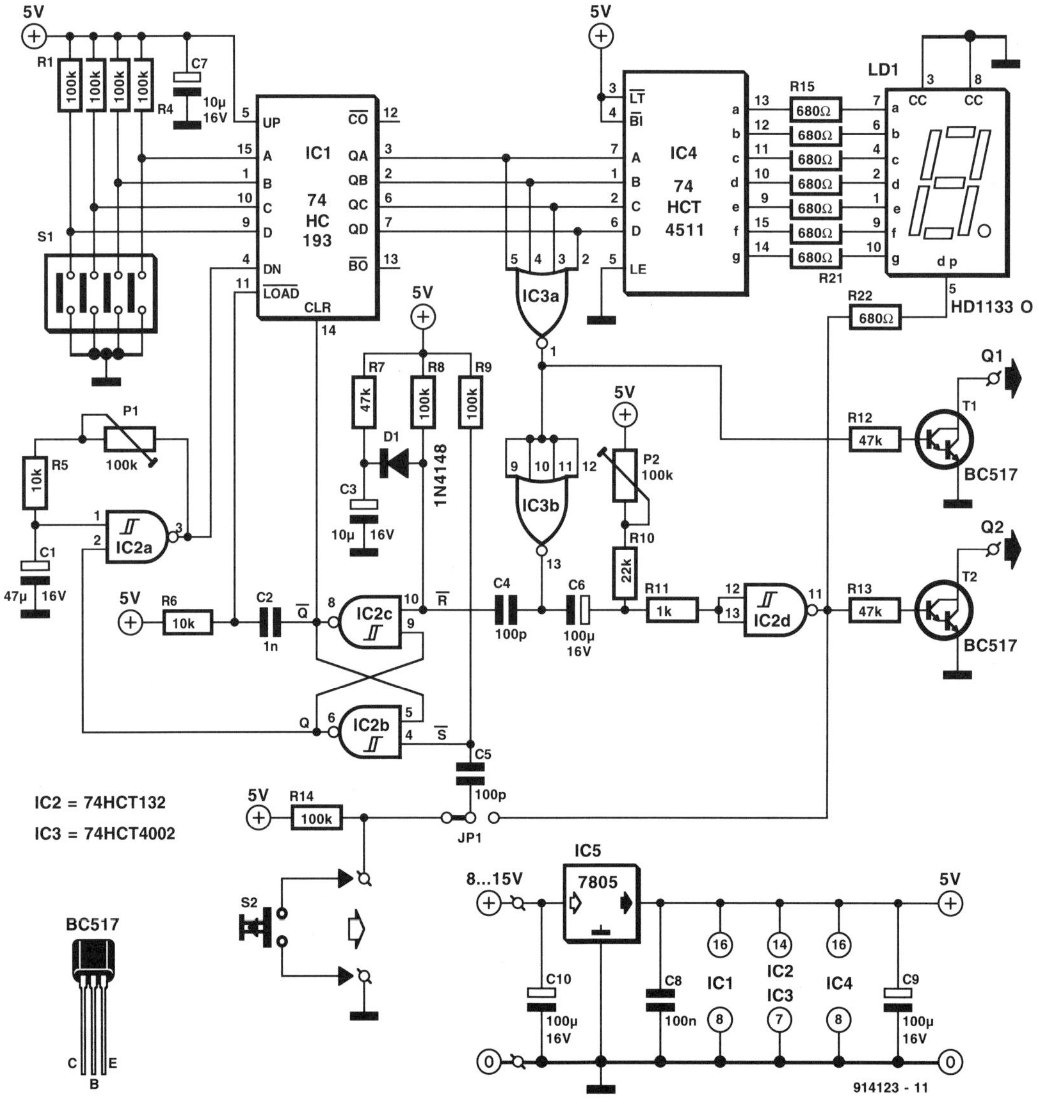

position 0. As a result, the output of IC_{3a} goes high and T_1 is switched on. Since at the same time the output of IC_{3b} toggles from high to low, monostable IC_{2d} is started. The output of the monostable then goes high and switches on T_2, indicated by the lighting of the decimal point of the display. Also, start/stop bistable IC_{2b}–IC_{2c} is reset, which disables the counter and the oscillator. To start the next time-out cycle, the bistable must be set afresh, for which, with link JP_1 in the position shown in the diagram, a last transition (trailing edge) is required at the input. This is arranged most simply by pressing S_2, but the transition may also come from another circuit or from a sensor.

When JP_1 is in the other position, the output of the monostable is linked to the input of the start/stop bistable (flip-flop). This results in the time-out cycle being started anew at the end of the mono time. The circuit then operates as an oscillator.

The circuit requires a power supply of 8–15 V. It draws a current of about 40 mA, most of which by the display. Transistors T_1 and T_2 can switch up to 400 mA.

Basic alarm indicator

238 The most obvious alarm indication is a flashing light. This requires relatively complex electronics, however, to achieve correct flashing. Moreover, in certain circumstances, such as when the lamp is in bright light, the indicator cannot be seen properly.

The present circuit is simplicity itself; it consists of only three components and provides not only a visual, but also an audible indication. It is based on a special LED that contains the electronics to ensure that it lights periodically. When the LED is on, it draws a much larger current than in its quiescent state. This property is used to good effect. When the LED is out, hardly any current flows, so that there is no voltage across the buzzer. When the LED is on, the relatively large current causes a drop across the buzzer, so that this sounds. The capacitor across the LED ensures sufficient buffering (decoupling) of the voltage across the diode. The circuit is actuated when a direct voltage of 7–9 V is applied to it.

The circuit draws a current of about 10 mA.

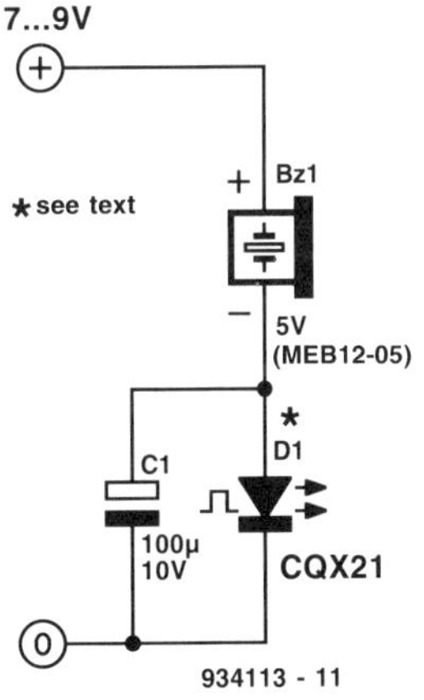

Car battery monitor

239 THE charge-discharge-idle (C-D-I)monitor described here is suitable for all vehicles with a 12-V or 24-V battery of which the negative terminal is connected to the chassis (point B in the drawing)

The current drawn from the battery is measured by monitoring the voltage drop across the heavy cable between the negative terminal of the battery, and the chassis. Usually, this cable is extended to the engine (point C). The positive terminal is usually connected to two cables, a solid one to the starter motor (point A) and a thinner one to the ignition switch.

Two LEDs fitted in the car interior indicate whether the battery is charged or discharged with a significant current, providing a reassuring check on the generator function. A third LED is provided to indicate a kind of neutral area in which the battery is only lightly charged or discharged.

The circuit is essentially a window comparator based on opamps. The voltage drop that exists across the chassis cable of the battery is fed to resistor R_1, which forms part of a measuring bridge consisting of R_1-R_2-R_3-R_4-P_1. The small voltage unbalances the bridge, and is amplified 100 times by opamp IC_{2a} which is wired as a non-inverting amplifier. In practice, input voltages as small as +2.5 mV or –2.5 mV are detected reliably by the balanced bridge.

The output voltage of IC_{2a} controls a window comparator built around IC_{2b} and IC_{2c}. The LEDs at the opamp outputs indicate whether the battery is charged (D_4 lights), discharged (D_6 lights), or is in a `neutral' state (D_5 lights).

The two positive feedback networks associated with IC_{2b} and IC_{2c} are decoupled at the output of IC_{2a} by R_6 and R_7 to ensure that the hysteresis of the window comparator does not affect the reference voltages supplied by R_8-R_9-R_{10}. If necessary, R_9 may be made smaller to make the 'idle' (neutral) range smaller.

The fixed resistors in the bridge must be close tolerance types mounted such that they are in thermal contact with each other.

The circuit has an internal power supply based on the ubiquitous 7805 three-terminal voltage regulator. The regulator requires a heat-sink only when the circuit is installed in a vehicle with a battery voltage higher than 12 V.

Adjusting the circuit is straightforward. Start the engine and let it idle. Set preset P_2 to mid-travel. Next, adjust P_1 until the 'idle' LED, D_5, lights. Carefully readjust P_2

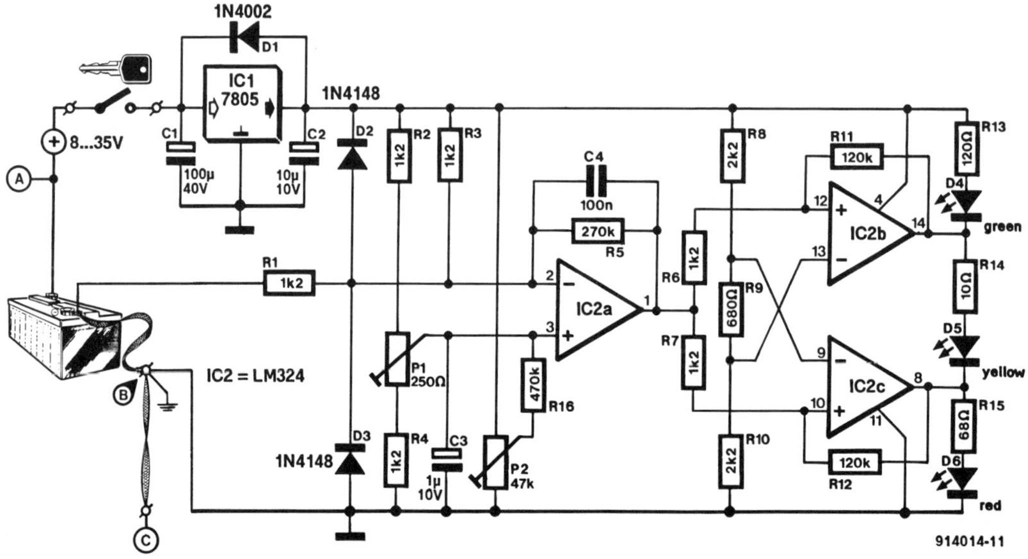

until IC_{2a} supplies an output voltage of 2.5 V. Rev up and check that the 'charge' LED, D_4, lights.

With the component values shown here, the circuit will indicate a charge or discharge current greater than about 1.5 A, which corresponds to 18 W at a battery voltage of 12 V. Suggested colours for the LEDs are green for D_4 (charge), red for D_6 (discharge) and yellow for D_5 (idle). Alternatively, the charge and discharge LEDs may be triangular types which can be fitted to point up and down respectively. The 'idle'LED is then a rectangular type fitted in between. Finally, the circuit draws about 30 mA when connected to a 12-V system.

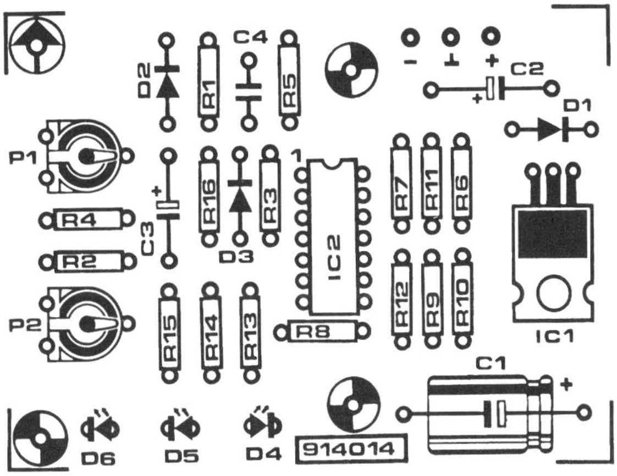

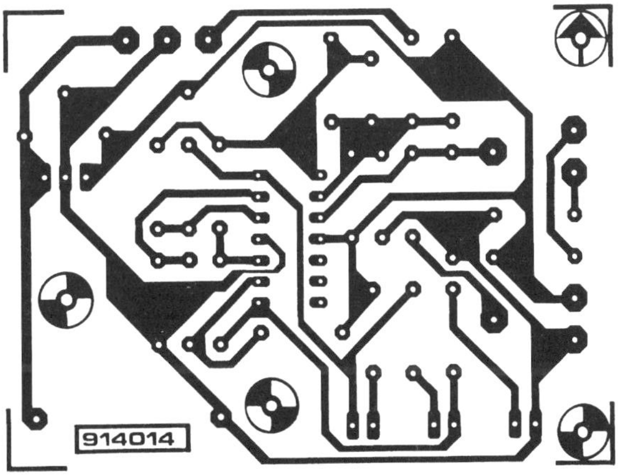

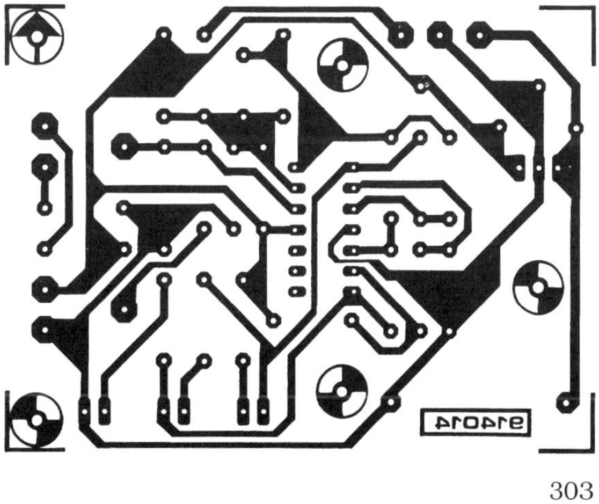

Parts list

Resistors:
R1–R4, R6, R7 = 1.2 kΩ
R5 = 270 kΩ
R8, R10 = 2.2 kΩ
R9 = 680 Ω
R11, R12 = 120 kΩ
R13 = 120 Ω
R14 = 10 Ω
R15 = 68 Ω
R16 = 470 kΩ
P1 = 250 Ω preset H
P2 = 47 kΩ preset H

Capacitors:
C1 = 100 µF, 40 V
C2 = 10 µF, 10 V
C3 = 1 µF, 10 V
C4 = 100 nF

Semiconductors:
IC1 = 7805
IC2 = LM324
D1 = 1N4002
D2, D3 = 1N4148
D4 = LED, green
D5 = LED, yellow
D6 = LED, red

Programmable LED indicator

240 The most popular current indicator in electronics is almost certainly the light-emitting diode (LED) which is available in a number of colours (red, green, blue, yellow, as well as dual colours) and variations (round, square, flashing). Unfortunately, these virtually indestructible devices are often abused: terminals cut too short, current too high, burnt by a soldering iron, and many others. Most of these abuses may be prevented by the circuit described here: it enables the LED indication to be adapted to the circuit you are working on (flashing or not flashing, red, gree, and so on).

The heart of the indicator is dual-colour diode D_2. The manner in which this will light when the input of the indicator is actuated depends on the settings of S_1-1–3. Switch S_1-4 enables the setting of a high or low level to which the input reacts. Components R_9, D_3 and D_4 protect the input from damage should the input voltage become too high.

An oscillator formed by IC_{1a} and IC_{1b} provides a flash frequency that is set by P_1. Depending on the setting of switches S_1-1–3, the oscillator signal is applied in a certain manner to the dual-colour LED via multiplexer IC_2. For convenience's sake, the positions of switches S_1-1–3 and the consequent actions of the LED are summarized in Table 1 and those of S_1-4 and the status of the LED in Table 2.

The indicator needs a supply of 8–25 V. The current drawn from a supply of 25 V in normal operation is 30 mA.

TABLE 2

S_1–4	CTRL input	LED status
off	L	disable
off	H	enable
on	L	enable
on	H	disable

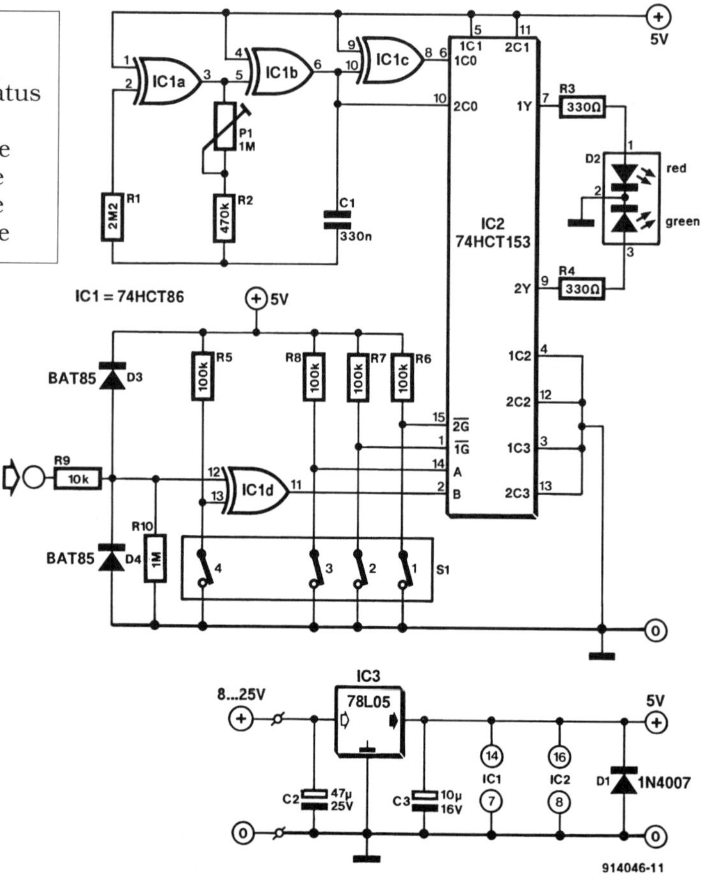

TABLE 1

S_1–1	S_1–2	S_1–3	LED action
on	on	on	flashing red/green
on	on	off	red/green
on	off	on	flashing green
on	off	off	green
off	on	on	flashing red
off	on	off	red
off	off	on	off
off	off	off	off

Slave flash trigger

241 The circuit in the diagram is intended for synchronous, wireless triggering of one or more slave flash units when the mother flash is triggered, so as to obtain better lighting of the photographic object.

Phototransistor T_1 is switched on upon the receipt of the light of the mother flash unit. The potential at the inverting input of comparator IC_{1a} rises, the comparator toggles and for a brief instant, determined by the value of C_1, the voltage level at the inverting input of IC_{1b} is lower than that at the non-inverting input. This causes IC_{1b} to toggle momentarily, which results in the thyristor being triggered and the flash contacts to be closed.

The circuit may be used in dark as well as in brightly lit places. The operating range depends on the mother flash and will normally be 5–15 m (15–50 ft). The sensitivity depends primarily on the base resistance of the phototransistor and this may be modified according to circumstances. Any tendency to instability may be cured by shunting R_2 with a 100 pF capacitor.

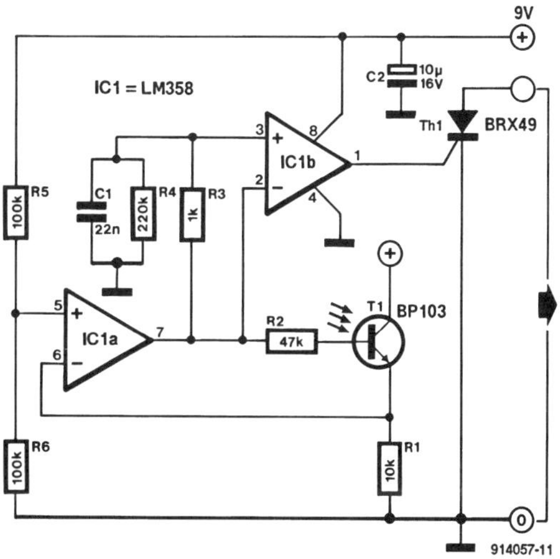

Power is supplied by a 9-V (PP3) battery.

Horse simulator

242 Here is a way of faithfully reproducing the movements of a horse: with a rotary switch that enables selection of step, trot, gallop to the right, gallop to the left and backward. The manner in which the horse puts down its hoofs is indicated clearly by LEDs. The simulator forms, therefore, a versatile demonstration model for instructors and learner riders.

Tabel 1, EPROM-Listing

00000:	00 00 00 00 00 00 00 00	85 84 92 90 A2 22 2A 09
00010:	84 00 21 00 84 00 21 00	00 00 00 00 00 00 00 00
00020:	01 45 44 54 10 00 00 00	00 00 00 00 00 00 00 00
00030:	00 00 00 00 00 00 00 00	00 00 00 00 00 00 00 00
00040:	80 A2 22 2A 08 00 00 00	00 00 00 00 00 00 00 00
00050:	00 00 00 00 00 00 00 00	00 00 00 00 00 00 00 00
00060:	00 00 00 00 00 00 00 00	00 00 00 00 00 00 00 00
00070:	00 00 00 00 00 00 00 00	00 00 00 00 00 00 00 00
00080:	5A 12 96 48 69 5A 00 00	00 00 00 00 00 00 00 00
00090:	00 00 00 00 00 00 00 00	00 00 00 00 00 00 00 00
000A0:	00 00 00 00 00 00 00 00	00 00 00 00 00 00 00 00
000B0:	00 00 00 00 00 00 00 00	00 00 00 00 00 00 00 00
000C0:	00 00 00 00 00 00 00 00	00 00 00 00 00 00 00 00
000D0:	00 00 00 00 00 00 00 00	00 00 00 00 00 00 00 00
000E0:	00 00 00 00 00 00 00 00	00 00 00 00 00 00 00 00
000F0:	00 00 00 00 00 00 00 00	00 00 00 00 00 00 00 00

The patterns for driving the LEDs are provided by an EPROM, IC_4, whose hex dump in the table shows the relevant addresses and associated data. Addressing is carried out with S_{3a} and counter IC_1. Oscillator IC_{3c} ensures that successive addresses are generated automatically. The speed with which the horse moves can be set with P_1. The oscillator can also be switched off by closing S_1. If that is done, S_2 must be pressed briefly before the next pattern is supplied to the LEDs.
When correct addresses are generated, a reset pulse is passed to the counter by IC_{3a}. The power-up reset is provided by C_2 and one of the resistors in array R_{10}.
The circuit is operated by a 6 V battery. A separate on/off switch is not required because that function is already provided by S_{3b}. In operation, the circuit draws a current of only 35 mA.
The simulator can be built on a piece of prototyping or similar board. The only important thing during the construction is to ensure that the LEDs are configured as shown, that is, two rows of four LEDs each with D_1 at the top left and D_5 at the bottom left. The head of the horse is then at the right and its tail at the left.

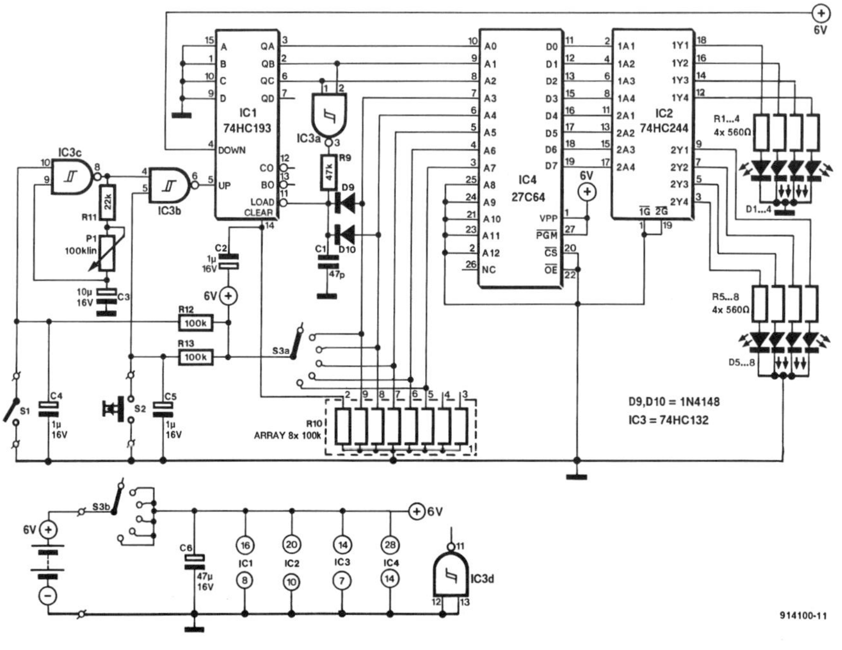

Super starter for cars

243 The super starter makes it possible to start cars with ageing batteries and obsolescent (coil-based) ignition systems, particularly during cold or damp weather. During starting, the voltage of an ageing (and possibly cold) battery will be insufficient for the coil to generate a tension high enough to create a strong spark across the spark plugs. The circuit presented ensures that the coil is powered by a battery of NiCd cells; even in these arduous conditions, such cells will last up to ten minutes. After the engine has started and the dynamo voltage has risen, the coil is powered by the car battery again.

The circuit uses the D+ terminal of the charging current indicator lamp (La_1) to check whether the engine is running, since that terminal is connected directly to the dynamo. As log as the engine does not run, and the dynamo, therefore, does not generate a voltage, relay Re_1 is energized via the ignition key (+ connection) and the low-resistance dynamo (connection to earth). The NiCd battery then provides plenty of power to the coil, irrespective of the state of the car battery. Once the engine has started, a voltage will be generated by the dynamo. There is then no potential difference across the relay coil and its contact changes over, whereupon the coil is supplied by the car battery (or the dynamo).

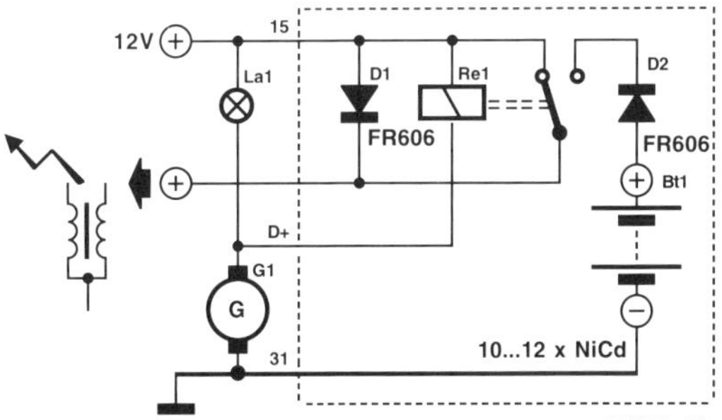

924003 - 11

Diode D_1 prevents two possible troubles. First, it prevents the relay interrupting the current to the coil when its contact changes over, which would result in a spark at a moment that the engine does not need one. Therefore, the coil is powered via the diode during the change-over period. This reduces the voltage across the coil by about 2 V, but that does not matter. Secondly, it ensures that the car can be started when the car battery is fully charged and the NiCd cells are flat, or have been removed for charging.

The relay must have a contact rated at not less than 8 A. Car relays with change-over contacts are not easily obtainable as a spare part, but can be often be found in car scrap yards (particularly in Citroen CX models).

The Type FR606 diode may be replaced by a Type BYW29-100; both can handle currents up to 6 A and their reverse voltage is high enough to withstand the inductive peaks generated by the coil.

The circuit is best built in behind the dashboard, although it is advisable to place the NiCd cells in a removable holder to enable them to be charged externally. It is, of course, possible to charge them from the car battery via a suitable resistor.

The D+ connection at the charging indicator lamp is that which is at earth potential when the engine is not running, but the ignition is switched on. At the ignition switch is a cable that goes to the coil: that cable must be connected to the output, 15, of the circuit. The ⊕ terminal of the circuit must be connected to the freed contact at the ignition switch. The –ve line of the circuit must be connected to the car chassis.

4-digit counter module

244 This compact four-digit digital counter is suitable for many applications, such as clocks, frequency meters, digital voltmeters, tachometers, stop-watches, score boards, etc. The counter module provides a reset switch and a carry output signal.

The maximum count of the module depends on which version of the IC used:

- the MM74C926 pulls its carry output high when counter state 6000 is reached. The IC counts to 9999.
- the MM74C927 is like the MM74C926, but the second most significant digit divides by 6 rather than by 10. Thus, if the clock input frequency is 10 Hz, the display would read minutes, seconds and tenths of seconds. e.g, 9:59.9.
- the MM74C928 is like the MM74C926, but the most significant digit divides by 2 rather than by 10. Also, the carry-out is an overflow indicator that goes high at counter state 2000, and goes low only when the counter is reset. Thus, this is a $3^1/_2$ digit counter.

All three MM74C92x ICs are CMOS devices. Each has an on-board display multiplexer, which requires only four external switching transistors at the common cathode connections of the 7-segment LED displays. The multiplex frequency is about 1 kHz.

The carry-out terminal may be used to cascade counter modules.

Current consumption is of the order of 100 mA.

Finally, it should be noted that clock input signals exceeding the module supply voltage are not clamped, and should not be allowed to exceed 15 V. The supply voltage range of all three MM74C92x ICs that may be used here is 3–6 V (data taken from National Semiconductor data sheets).

Parts list

Resistors:
R_1-R_4;R_{12} = 1.5 kΩ
R_5-R_{11} = 56 Ω
R_{13} = 10 kΩ

Capacitors:
C_1;C_2 = 100 nF

Semiconductors:
T_1-T_4 = BC547B
LD_1-LD_4 = HD1107

Integrated circuits:
IC_1 = 74C926/927/928 (see text)

Miscellaneous:
S_1 = PCB-mount push-button
K_1 = 4-way PCB header

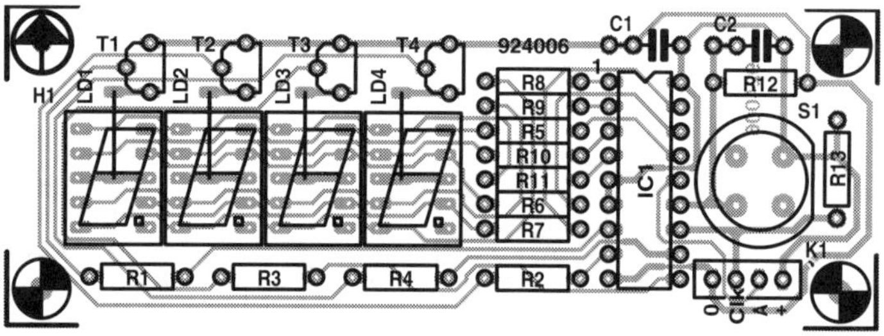

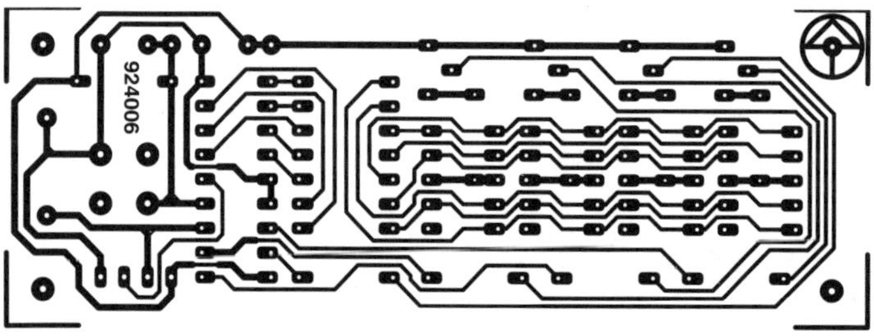

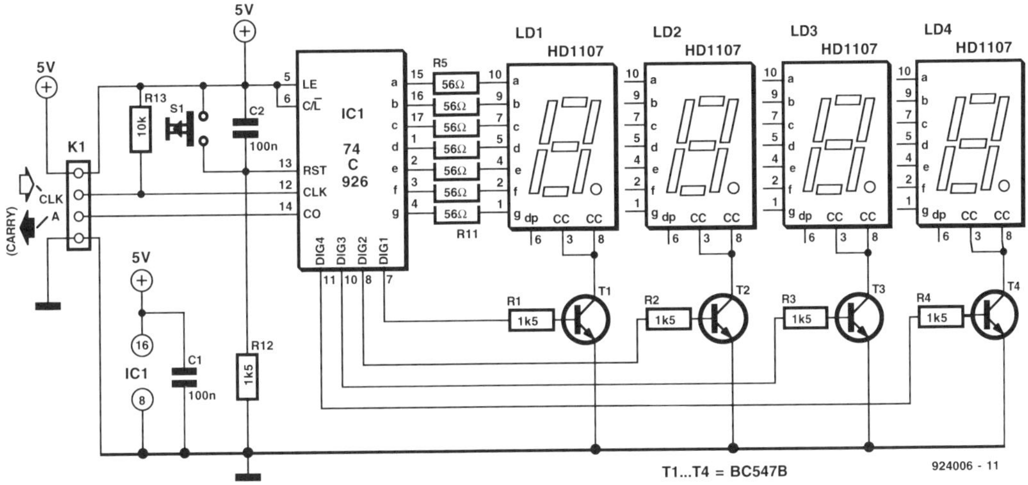

Water pump control for solar system

245 In most small solar power systems using a boiler it is required that the water circulation pump is not switched on until the temperature of the collector (the solar panel) exceeds that of the water in the vessel. Here, a two-sensor monitor is presented that enables this condition to be met. One sensor is fitted on the collector, the other on the water vessel. The control shown here has two adjustments: one for the temperature difference at which the pump starts to operate, and one for the temperature difference at which it is switched off. Although these settings are independent, the switch-off level must be lower than the switch-on level. Calibration in degrees Celcius is simple because the gradient of the voltage at the wiper of the potentiometers (or presets) that set the on/off temperatures is exactly 0.1 V °C^{-1}.

The two temperature sensors Type LM334 are adjusted to supply a temperature gradient of 1μA/°C. Unequal sensor temperatures therefore produce a current flow at their junction. The voltage across R_1 is directly proportional to the temperature difference measured. This enables the switching thresholds of the on/off control to be set with the aid of two presets: the 'on' preset (P_4) is adjusted to, say, 3 °C, and the 'off' preset (P_3) to 1 °C. The range of the two presets is about 5 °C.

The sensors used here supply a current rather than a voltage. This eliminates thermocouple effects caused by temperature changes on the connecting cables between the sensors and the circuit. If voltage-type sensors such as PTCs or NTCs were used, the circuit would have become more complex because of the required compensation. The AD590 may be used instead of the LM334. Note, however, that the AD590 does not require an adjustment preset or resistor.

Relay Re_2 switches the pump on and off. A second relay, Re_1, comes on after Re_2. It is optional, and may be used to switch the pump briefly to a higher speed, which is required in some solar heating systems to increase the intial water flow, or to fill the system.

The circuit is calibrated by setting equal sensor currents at equal sensor temperatures. The sensor current equals

$$[273 + T_a)]\ \mu A,$$

where T_a is the ambient temperature in degrees Celsius. Thus, at a room temperature of 20 °C, presets P_1 and P_2 are adjusted until the current flow through each

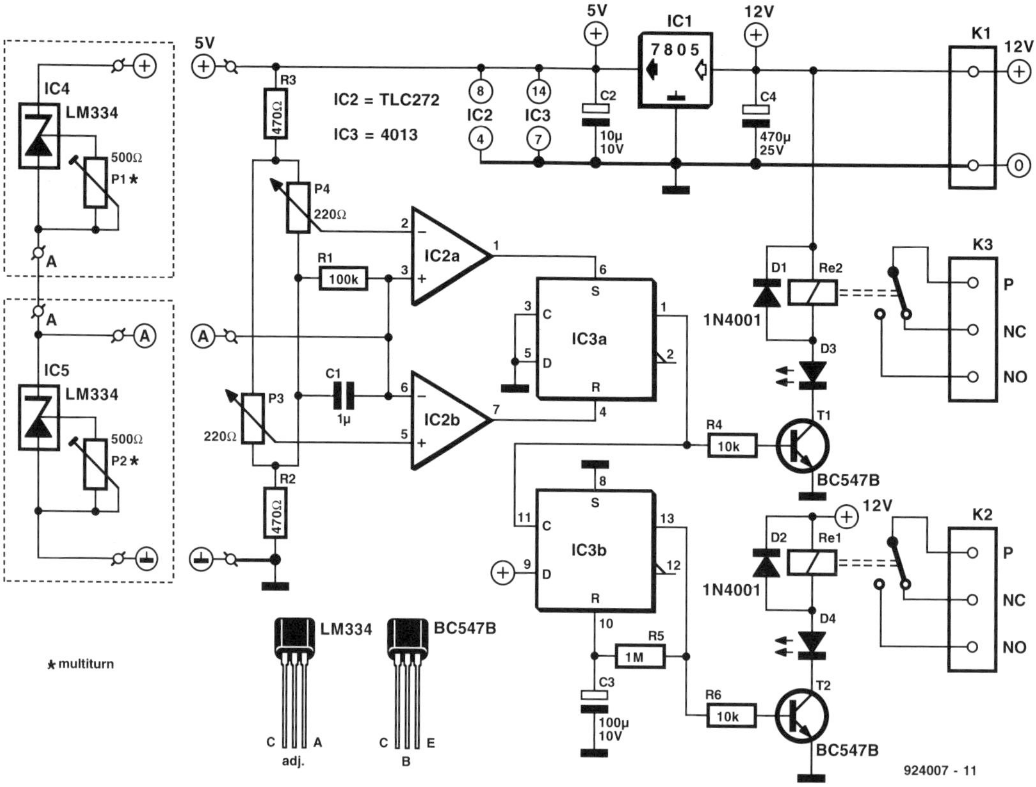

sensor is 293 µA. A few microamps more or less will not make much difference, as long as the sensor currents are equal. It is best to first adjust one sensor only. Start by connecting a microammeter between 'A' and ground, and adjust P_1. Next, adjust the other preset until the voltage across R_1 is nought. It will be clear that these initial adjustments require that the two LM334 are at the same temperature.

Current consumption of the on/off control is about 11 mA plus about 35 mA for each relay.

The dimensions of the printed circuit board are geared to the size of the box mentioned in the parts list. The potentiometers are fitted with the spindles at the track side of the board.

Parts list

Resistors:

R_1 = 100 kΩ
R_2;R_3 = 470 Ω
P_3;P_4 = 220 Ω linear potentiometer
R_4;R_6 = 10 kΩ
R_5 = 1 MΩ
P_1;P_2 = 500 Ω multiturn preset

Capacitors:

C_1 = 1 µF MKT
C_2 = 10 µF 10V radial
C_3 = 100 µF 10V radial
C_4 = 470 µF 25 V radial

Semiconductors:

D_1;D_2 = 1N4001
D_3;D_4 = LED
D_5;D_6 = LM334

Integrated circuits:

IC_1 = 7805
IC_2 = TLC272

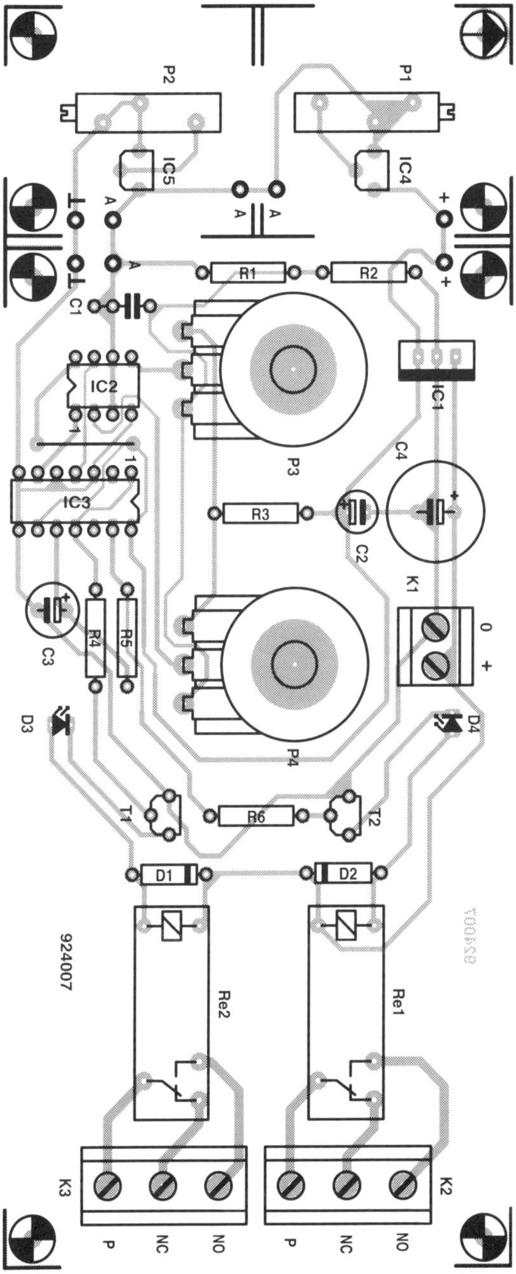

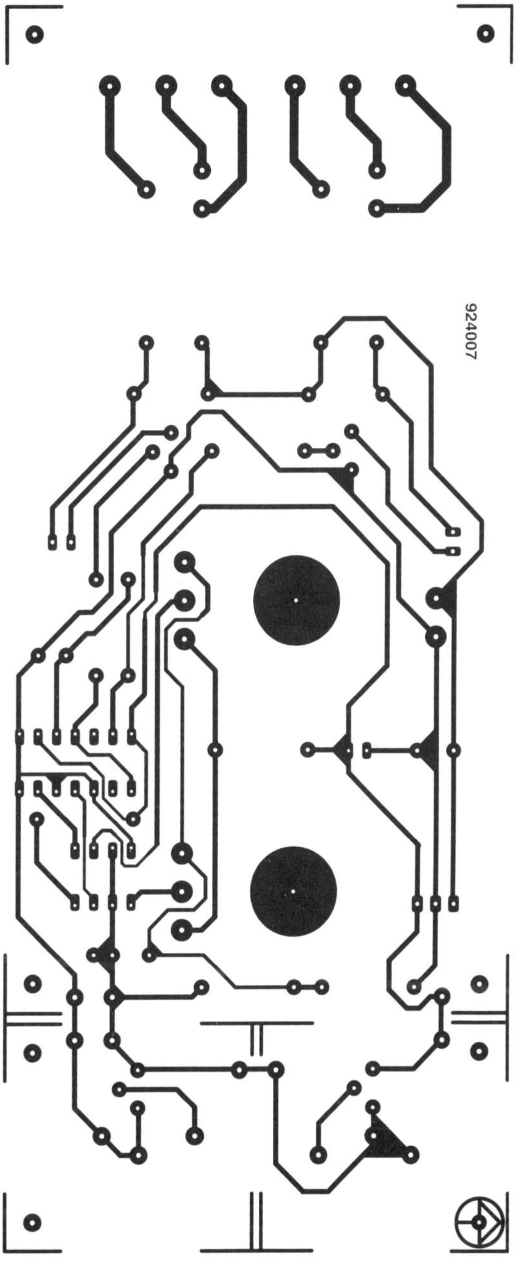

IC_3 = 4013
T_1;T_2 = BC547B

Miscellaneous:
K_1 = 2-way PCB terminal block, pitch 5 mm.
K_2;K_3 = 3-way PCB terminal block, pitch 7.5 mm.

Re_1;Re_2 = GBR 10.2-11.12 (12V coil, contact 250V @8A).
Enclosure: approx. size 155×61×90 mm; e.g., Retex Gibox Type RG3.
PCB Ref. 924007.

Starter for model aircraft

246 Model aircraft tend to be realistic reproductions of the real thing. That means, among others, that starting the petrol engine must be done by hand or with an external electric starter. Manual starting may be realized in two ways: with the circuit in Fig. 1 or with that in Fig. 2.
The circuit in Fig. 1 places a 3-V relay between the starter and the motor. When the starter switch is closed, nothing happens. But if then the propeller is turned by hand, the motor acts as a generator and, once it is turned fast enough, the generated voltage is high enough to energize the relay, whereupon the motor starts. An advantage of this is that it is a pure mechanical operation, which is readily incorporated. A disadvantage is that once the relay is energized, it is operated by a 6-V supply: a slight waste of energy.
The circuit in Fig. 2 is rather more economical, since it uses a 6-V relay. Again, the motor is used as a generator operated by turning the propeller. The generated voltage is applied to the base of T_1 via R_2. When the base is supplied with sufficient current, the transistor is switched on, the relay is energized and the motor starts. The starting point can be preset with P_1.

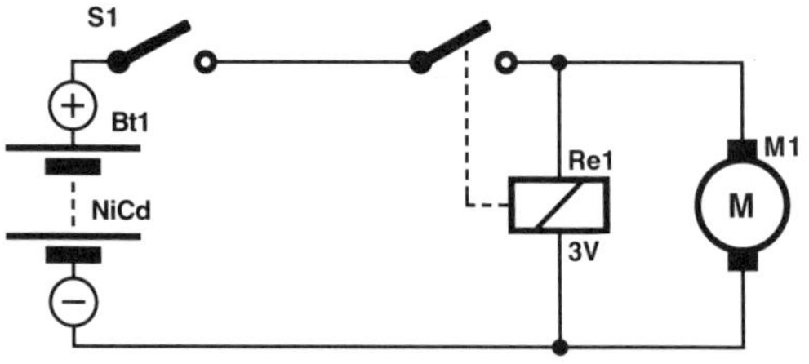

924008 - 11

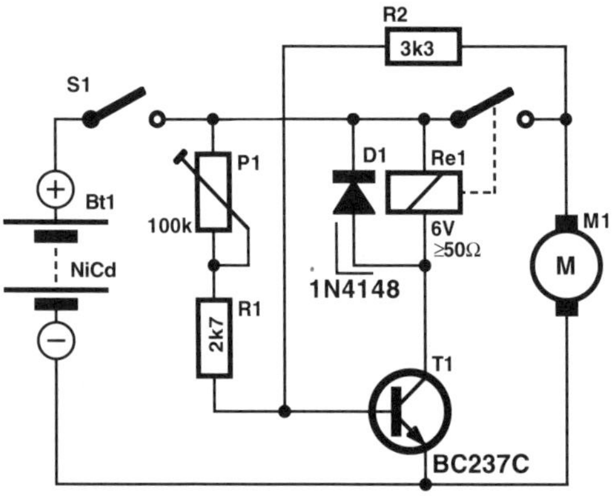

924008 - 12

Bounce-free change-over switch

247 With the aid of two D-type bistables contained in a 74HCT74 and some external components, any push-button switch can be used as a bounce-free change-over switch.
In the diagram, IC_{1b} provides the change-over function. Pin 8 of the bistable is linked with pin 12, which results in the logic levels at pins 8 and 9 alternately changing state when a leading edge appears at pin 11.
Circuit IC_{1a} serves as a pulse generator and debouncing element. The push-button switch, S_1, is connected between its reset input (pin 1) and ground. Normally, because of R_2, there is a high level at pin 1. When S_1 is pressed, IC_{1a} is reset.
The clock input (pin 3) is also connected to S_1 via R_1-C_1. When the switch is operated, C_1 is discharged rapidly via D_1; when the switch is released, it takes a little while before C_1 is recharged to a logic high level.
When S_1 is open, pin 9 is low, whereas pin 8 and pin 5 are high. When the switch is closed, IC_{1a} is reset immediately, resulting in pins 5 and 3 going low.
When S_1 is released, the reset is removed, but it takes a little while before C_1 is charged to a logic high level. Only when that level is reached, and a leading edge appears at pin 3, does pin 5 go high again. This results in IC_{1b} being clocked, whereupon its Q outputs (pins 8 and 9) change state.

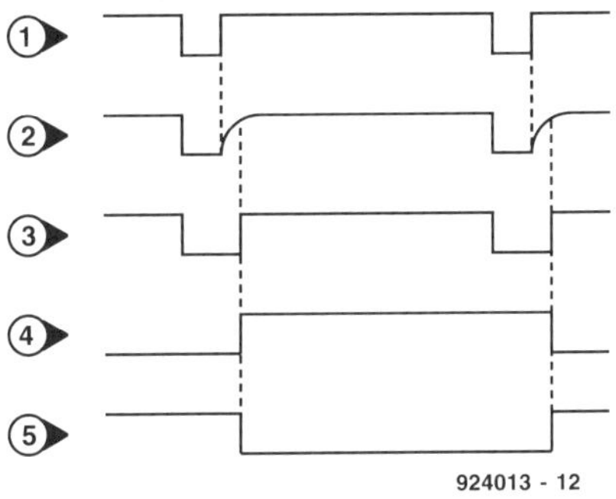

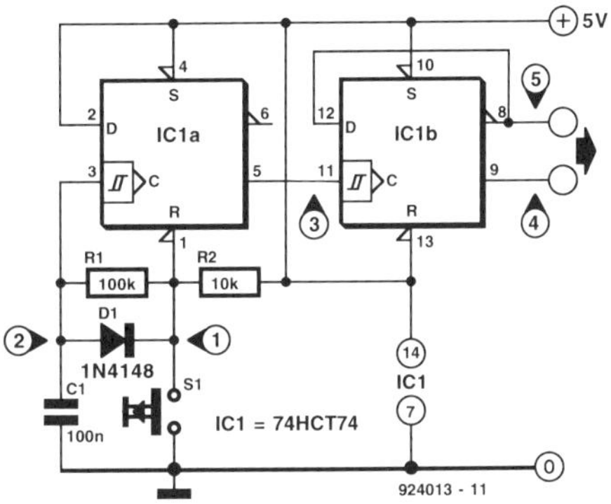

Real-time switch debounce

248 Most debounce circuits shorten or lengthen the time a switch is pressed to obviate interference pulses. The present circuit does **not** affect that time: the pulse at the output is exactly in accordance with the time elapsed between the depressing and the releasing of the switch.

When switch S_1 is pressed, monostable IC_{1a} is triggered via gates IC_{3d}, IC_{3a}, and IC_{3b}. The feedback from IC_{3a} to IC_{3d} ensures that the monostable cannot be triggered again during during the pulse duration.

When the mono time has elapsed, bistable IC_{2a} is clocked. If at that moment the output of IC_{3d} is still high, the Q output of IC_{2a} goes high and the $\overline{Q}$ output becomes low. The bistable keeps IC_{3a} and IC_{3b} off, while IC_{3c} passes the output signal of IC_{3d} to the reset input of the monostable.

As long as S_1 remains depressed, the monostable is reset via IC_{3c}. When the switch is released and its contact opens, the monostable is retriggered. If the key stays open for the length of the mono time, bistable IC_{2a} is reset after that time. If the switch contact closes during the mono time, IC_{2a} remains set.

The mono time needs to be longer than the longest open time of the switch during bounce.

Time constant R_2-C_2 must be considerably longer than the mono time, so that the feedback from IC_{3a} to IC_{3d} is available during the entire mono time.

The circuit draws a current of about 1 mA.

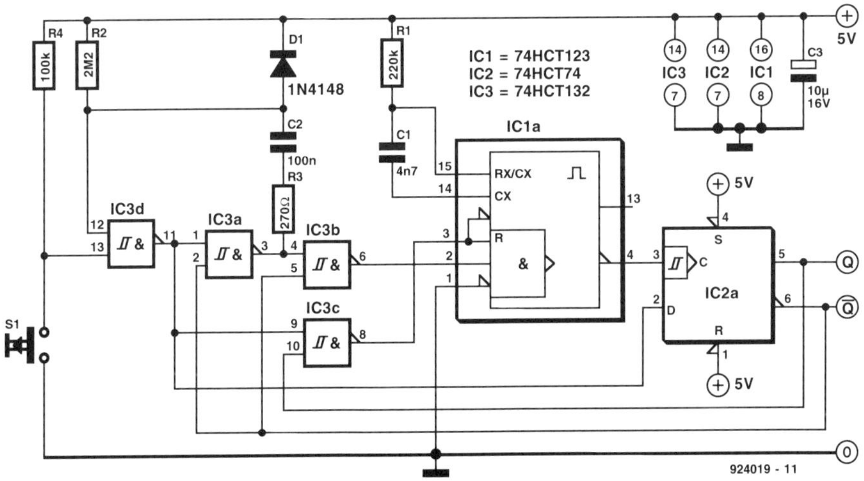

Car lights alert

249 Virtually all modern cars have a built-in alarm to warn that you have left the lights on when you are getting out, There are, however, millions of older cars that have no such useful device and their owners might find the present circuit just what they've been waiting for.

The anode of diode D_1 is connected between the lights switch and the lights. The cathode is connected to the 12 V buzzer. The other terminal of the buzzer is connected to one of the door-switch contacts or to terminal 31B of the interior lights cable. When one of the front doors (fitted with a switch, which, in older cars is not always the case with rear doors) is opened while the external car lights are still on, the buzzer begins to hum. Note that the circuit allows for only two door switches: S_2 is the driver's door contact and S_3 the front passenger door's contact.

If you don't want the buzzer to be operated by the front passenger door, fit diode D_2 as shown and disconnect the wire between the contacts on the two doors.

Note that the circuit is intended only for cars whose door contacts switch the interior lights to earth (which is the case in most cars built in the past 10 years).

Diode D_1 may be a light-duty type such as the 1N4001, but D_2 must be a somewhat heavier-duty type, such as a 1N5401, since it carries the current through the interior lights.

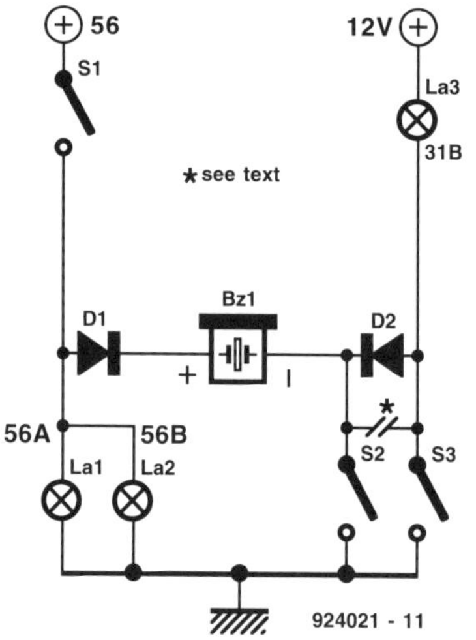

Front-to-rear wiper coupling

250 On some cars, it is convenient to couple the rear window wiper to the windscreen wiper. However, since the rear window does not get nearly as wet as the windscreen, particularly when the car is moving, the rear wiper should operate only once for every umpteen wipes of the windscreen. Note that your car may be one of the fortunate ones of which the rear window, when the car is moving in the wet, does not get wet at all because of the car's design.

The coupling shown in Fig. 2 ensures that the rear wiper operates once for every four or 16 wipes of the windscreen, depending on the setting of switch S_1 (as shown, once every four).

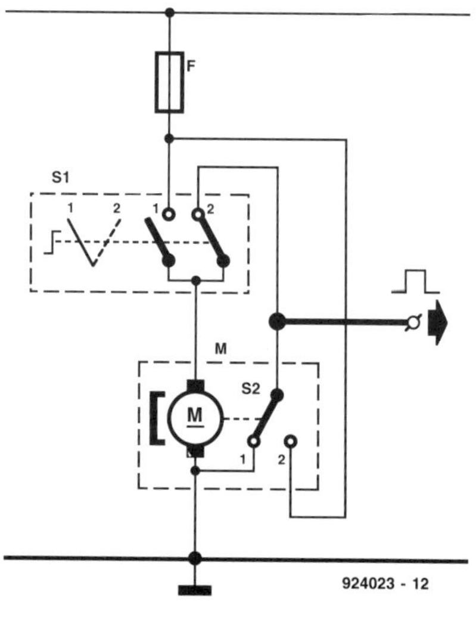

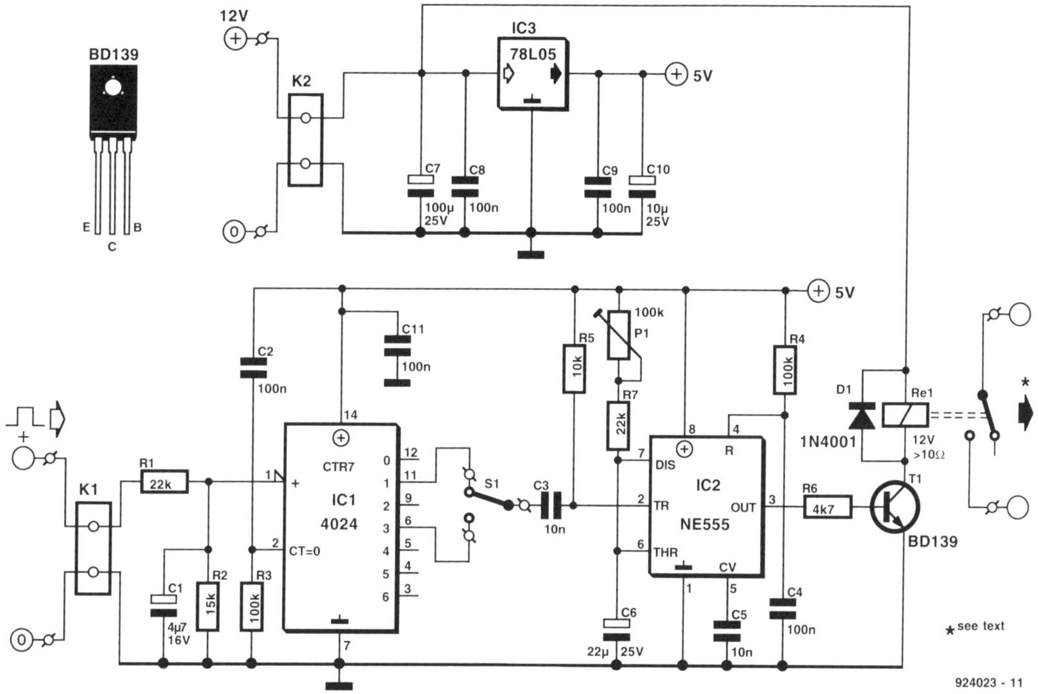

The clock for the circuit is taken from the return (terminal 53e – green/black wire on most cars) of the windscreen wiper motor—see Fig. 2. This signal, which is a square-wave, is applied to IC_1 via K_1. Its level is lowered to not more than 5 V by potential divider R_1-R_2 to prevent any damage to IC_1. Any noise from the car's electrical system is bypassed by C_1.

The Q_2 output (pin 1) of IC_1 goes high every fourth clock input, and the Q_4 output (pin 3) once every eighth clock input.

The trailing edge of the signal at S_1 is transformed into a trigger pulse by R_5-C_3. This pulse, whose length is determined by R_7-P_1-C_6, is applied to monostable IC_2. The length of the pulse should be set to about one second to give the rear wiper time to get going.

The monostable drives transistor T_1, which in turn controls relay Re_1 This relay is a motorcar type that can handle the large switch-on current of a wiper motor: it may be rated at 25 A or 35 A.

The supply for the circuit is provided by IC_3, which brings down the car voltage of 12 V to 5 V. This IC also prevents large peaks on the battery voltage from reaching the circuit.

Telephone monitor

251 Modern telephones can easily be connected in parallel, enabling a household to have one in the bedroom, kitchen and hall or study. It is, however, not always easy to see (or hear) at one position whether the phone at one of the other positions is in use or has taken up the call.

The d.c. level on most telephone lines drops from 48 V to about 8 V when one appliance is in use and to around 5 V when two telephones are in use. Also, its polarity changes over when a telephone is in use.

Since the polarity changes over, the circuit is of necessity a bridge type as shown in Fig. 1. To keep the current drain small and

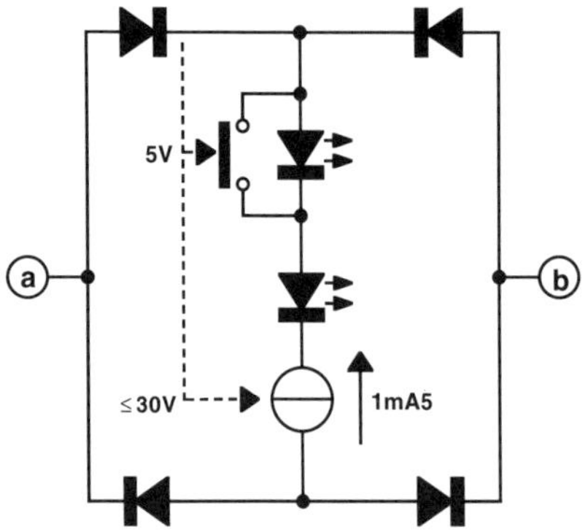

924030 - 11

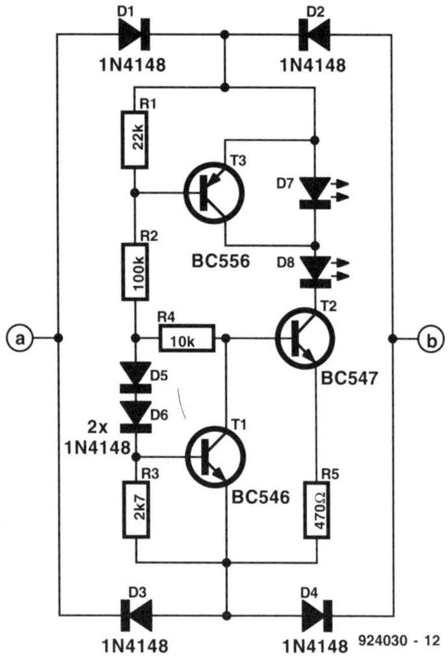

924030 - 12

relatively constant at varying voltage levels, a current source is needed for the high-efficiency indicator LEDs. At 50 V across *a* and *b*, the circuit is inactive; when that potential drops to about 8 V, the current source is on; when the level has dropped to about 5 V, the 'switch' across the upper LED opens so that both diodes light.

The diagram of the practical circuit is shown in Fig. 2. The bridge rectifier is formed by diodes D_1–D_4, while T_3 serves as the LED switch.

When the applied voltage is greater than 30 V, T_1 is on and switches off current source T_2. In this situation, the circuit draws not more than 0.5 mA. When the applied voltage drops to 8 V, T_1 is off and T_2 is on. The current source is completed by D_5, D_6, R_3 and R_5: with values as shown, the current supply is about 1.5 mA.

In this situation, because of the values of R_1 and R_2, T_3 is on, and D_7 is short-circuited. As soon as a second receiver is lifted, the line voltage drops to some 5 V, resulting in T_3 being switched off, so that D_7 lights.

The current through the LEDs is necessarily small and should in any circumstances not rise above 5 mA. With two telephones that means that R_5 must be 270–330 Ω, and for three telephones, 390–470 Ω. As already stated, this means that high-efficiency LEDs are essential.

Metal detector

252 This detector will help you find fairly large objects that consist of materials with a relatively high permeability. Also, it indicates whether the magnetic object inside the detection coil has good or bad conductive properties. Examples of materials that couple good magnetic properties to a fairly high electrical isolation are ferrites pressed from metal oxides. The detector is not suitable for 'coin digging', for which it is not sensitive enough. The more fanciful stuff like bombs and treasures left by pirates, are, however, reliably located.

The metal detector is powered symmetrically by two 9-V batteries, each of which is loaded with about 15 mA. The detection coil, L_1, forms part of a sine wave oscillator built around transistor T_1. Normally, the central frequency of the VCO (voltage-controlled oscillator) in the PLL (phase-locked loop) contained in IC_1 equals the oscillator frequency of T_1. That changes when a metal object (ferro-metallic or nonferrometallic) enters the field induced by L_1. When that happens, the sine-wave oscillator is detuned, and the voltage difference between pins 6 and 7 of IC_1 indicates the difference between the sine oscillator frequency and the VCO frequency. This difference causes moving-coil meter M_1 to deflect. The needle deflection itself is a measure of the frequency change, while the direction of the needle depends on the type of material detected by the coil. The meter used here is a centre-zero ±50 µA type.

The coil consists of 40 turns of enamelled copper wire on a plastic former with a diameter of about 10 cm (4 in). The inductance so obtained ensures that the sine-wave oscillator works at a frequency which is roughly equal to that of the VCO in the PLL.

Use an oscilloscope to check that pin 2 of IC_1 supplies a sine-wave signal with a frequency of about 75 kHz. Next, adjust P_1 such that the edges of the rectangular signal at pin 4 coincide with the positive peaks of the sine wave at pin 2. Next, null the meter by turning potentiometer P_2. Since the null adjustment will drift a little as the battery volatge drops, it will be necessary to redo the balance adjustment every now and then during use.

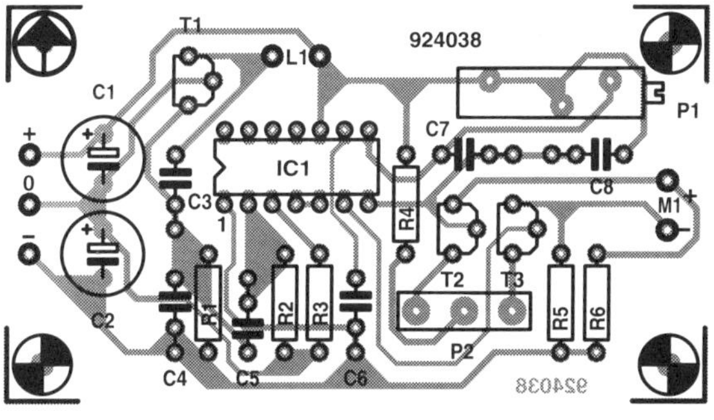

Parts list

Resistors:
R_1 = 6.8 kΩ
R_2, R_3 = 4.7 kΩ
R_4 = 680 Ω
R_5, R_6 = 5.6 kΩ
P_1 = 10 kΩ multiturn preset
P_2 = 470 Ω linear potentiometer

Capacitors:
C_1, C_2 = 100 µF 16 V radial
C_3 = 68 nF
C_4 = 15 nF
C_5;C_8 = 10 nF
C_6;C_7 = 1 nF

Inductor:
L_1 = details in text

Semiconductors:
T_1 = BC547B
T_2, T_3 = BC557B

Integrated circuits:
IC_1 = NE565N

Miscellaneous:
S_1 = miniature double-pole change-over switch.
Bt_1;Bt_2 = 9-V battery with connecting clip.
M_1 = centre-zero ±50µA moving coil meter.

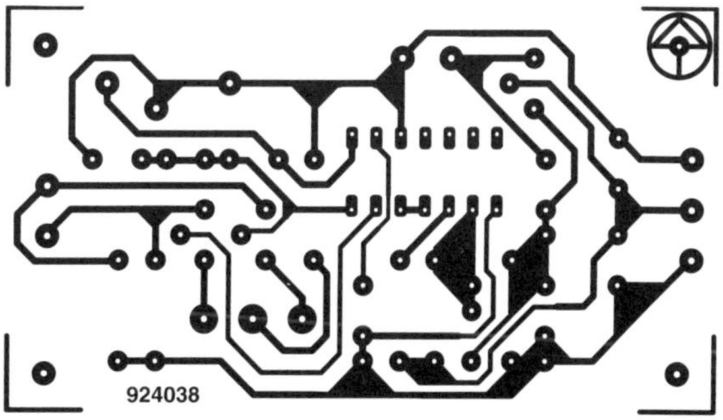

Digital tape counter

253 The tape counter described gives a ten-step indication with the aid of two ICs and a seven-segment display. Its great advantage is that it can be connected to a recorder in a straightforward manner, since it records the length of time the motor runs.

The motor of the cassette player is connected to the reset input of IC_2 and to the circuit earth. It is necessary for the motor to run only when the tape runs: in two- and three-motor decks the capstan motor usually runs continuously.

At the instant the reset input of IC_2 goes high, the timer begins to deliver very-low-frequency pulses: one pulse every three minutes with S_2 in position C60 as shown in the diagram, and one every $4^1/_2$ minutes in position C90.

The pulses are applied to decade counter/display driver IC_1. This stage raises the display position by 1 for every pulse, so that at the end of a C60 or C90 cassette the display has gone through positions 0–9. The display is reset by pressing S_1.

Owing to leakage currents and the tolerances of the electrolytic capacitors in the timing network, R_3-R_2-R_3-C_2 (or R_3-R_2-R_1-C_3), it may be necessary for the value of R_2 to be changed (empirically).

Because the first cycle of a Type 555 timer is always longer than the next, repeated stopping and starting of the tape may result in an appreciable error in the displayed time.

It is essential that the input voltage is free of noise pulses to prevent the 555 being reset erroneously.

The counter draws a current of not more than 50 mA.

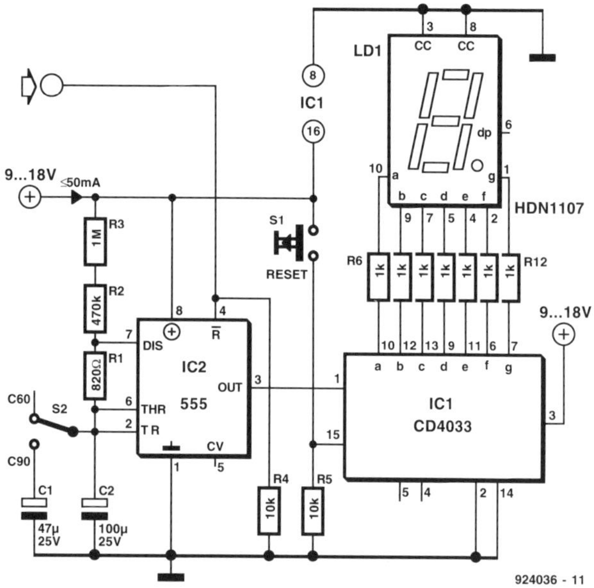

924036 - 11

Pulse-operated relay

254 With the aid of a few components, a standard relay can be made to change state every time a pulse is applied (or a switch is pressed).

When the relay is not actuated, transistor T_1 is off because its base and collector are at the same potential via R_2. Capacitor C_1 is charged via R_4, R3 and the relay coil. When a short pulse (logic zero, shorter than 0.5 s) is applied to the input, or S_1 is pressed, T_1 conducts and the relay is energized. The relay contact changes state, whereupon T_2 also delivers current to the relay coil. Capacitor C_1 is then discharged until its potential is equal to the base-emitter voltage of T_2.

When the pulse ceases, or the switch is released, T_1 is off again, but the relay remains energized by T_2.

A slightly longer input pulse is needed to deactuate the relay: it needs to be long enough for C_1 to be discharged completely (or very nearly so) via R_3. When the switch is released, or the input becomes high

again, T_1 stops conducting. Simultaneously, C_1 is charged again, so that within a short time the base current of T_2 drops to nearly zero.; T_2 then switches off and the relay is deenergized.
Short pulses at the input have no effect when the relay is actuated.
If fairly long pulses are input with the relay inoperative, the circuit functions as a standard relay that is actuated almost immediately by the pulse and is disabled as soon as the pulse ceases.
The hold voltage of the relay should be not more than half the supply voltage. The lower the hold voltage, the shorter the pulse required to actuate the relay can be.

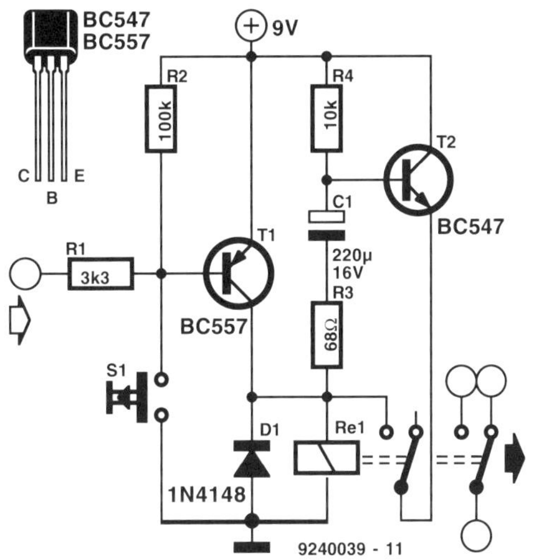

Dimmer for neon tubes

255 A neon tube cannot be dimmed as easily as an incandescent lamp because the tube can start only at a voltage much higher than the mains, after which it will remain lit at the mains voltage. The level of both the starting voltage and the working voltage depends on the temperature of the tube.
Normally, the high starting voltage is obtained by interrupting the current through a choke. This is usually done by the starter, which also ensures that a fairly large current flows through the filaments of the tube. This heats the ends of the tube, which makes starting easier.
These tasks of the starter are taken over by the circuit show in the diagram, which also enables the tube to be dimmed.
During the zero voltage crossings of the applied mains voltage, the triac will instantaneously switch off. At those instants, capacitor C_3 will be charged rapidly, which results in the instantaneous voltage, whose phase has shifted relative to that of the current, being applied across the tube. Capacitor C_3 and the choke form a resonant circuit that raises the sudden voltage across the tube to a very high value, whereupon the tube starts.
The larger the angle of the mains voltage during which the triac conducts, the larger the current through the tube filaments, which results in a lower starting voltage. At the same time, since a larger part of the current flows through the triac, that through the tube will be reduced, so that the tube will light more faintly.
When the tube is first switched on, the dimmer control, P_1, should be set for maximum brightness of the tube to facilitate starting.
The triac used should have a high du/dt value, otherwise the steep voltage transitions occurring across the tube, and thus across the triac, during the zero voltage crossings would cause the triac to remain on.

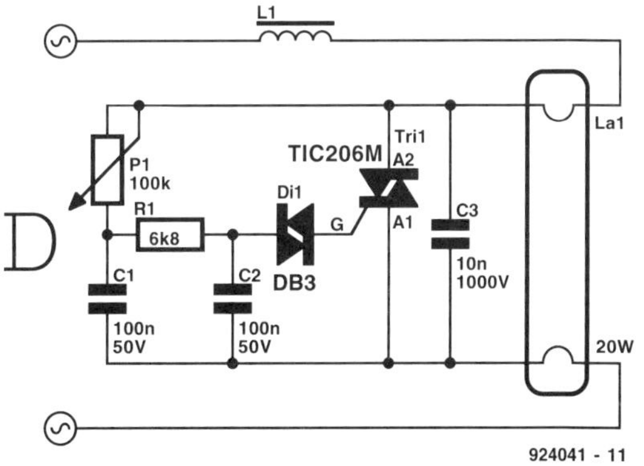

Revolution counter for diesel engines

256 Although most petrol-engined cars and lorries have a rev counter as standard, that is by no means the case in diesel-engined motor vehicles. The reason for this is that, since a diesel engine has no contact breaker, it is not so easy to derive pulses to drive a rev counter. There are, none the less, several possibilities of adding a rev counter if so desired.

Firstly, it would be possible to take the pulses from terminal W of the alternator. Unfortunately, that machine does not run at the same speed as the engine, so that some arithmetic unit would have to be added. Moreover, and more seriously, terminal W of modern alternators is no longer accessible externally.

A second way might be to attach a small magnet to each of the cranks on the crankshaft and so induce magnetic pulses in a fixed coil. The problem here is to attach these magnets securely.

A third method is an optical one proposed in this article. In this, the cranks are divided into sectors that are painted alternately white and black. A home-made light barrier is then used to evaluate the speed with which the sectors are rotating. If the cranks are divided into four sectors, pulses are generated that are suitable for driving commercially available rev counters, irrespective of whether the diesel engine has four, six or eight cylinders.

The circuit in Fig. 1 may, therefore, be considered as an adaptor for the rev counter.

A small 12-V bulb lights the crankshaft, whereupon the light reflected by the white sectors falls on to phototransistor T_1. This transistor is connected in a darlington configuration with T_2. The type of phototransistor is not important: it is thus not necessary to use the BP103 shown in the diagram. The output of T_2 is applied via C_2 to IC_1, where it is chopped and amplified by about ×50 to give a rectangular output signal of about 10 V_{pp}. That signal is perfect for driving a rev counter.

The circuit is best built on a small piece of prototyping (vero) board and then fitted in a tube, whose front is closed watertight by a circular piece of perspex. It may be necessary to separate the lamp and the phototransistor by a dark screen.

The circuit is connected to the rev counter by a three-core cable. The cable connections (+12 V, earth and pulse signal) to the circuit must be waterproof.

If the lamp is found to be too bright, it may be connected in series with a small resistor.

The circuit can be tested by measuring the voltage at the collector of T_2, which should be 1–5 V when a white sector is being illuminated.

If the sensitivity is too high, replace R_1 by the circuit in Fig. 2, which keeps the voltage at the collector of T_2 at around 5 V.

The circuit draws a current of about 10 mA plus the current through the lamp.

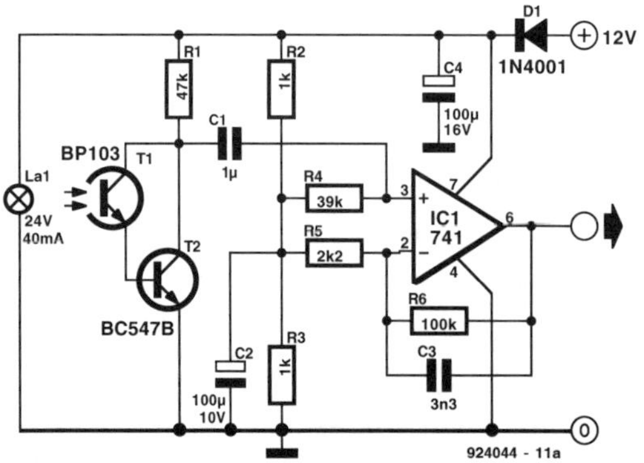

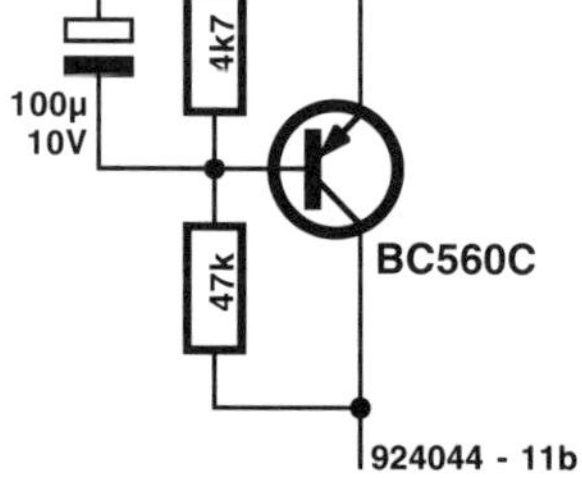

IDC-to-box header adaptor

257 Those of you who have ever worked with flatcables will know that IDCs (insulation displacement connectors) are simple to use, and give reliable connections. IDCs are available as sockets and plugs, and used extensively to connect flatcables to double-row box headers or pin-headers on computer cards that sport just about any type of interface to the outside world (a good example is the multi-purpose Z80 card described in Ref. 1). The present adaptor cards (there are six of them contained on the PCB – Ref. 924049 – shown here) are, for instance, perfect for 'changing' from IDC to eject-header style connectors (as illustrated by the photographs) when a flatcable runs from a board to a connector on the rear panel of the enclosure. Also, in many cases, an adaptor board fitted with an eject-style header will be cheaper and more flexible (when it comes to connecting and disconnecting flatcables) than a press-on (IDC style) sub-D socket or plug.

The spots on the component overlay indicate the position of the holes that have to be drilled when smaller types of eject header are used. Straight headers with 10, 14, 16, 18 or 20 pins may be fitted.

If you fit two box headers on an adaptor board, you can use it to couple flatcables terminated into IDC sockets. In that way, you can make IDC extension cables, which are particularly useful when a PCB with lots of flatcable connections is removed from an enclosure for repair or inspection.

Reference:

1. "Multi-purpose Z80 card", *Elektor Electronics*, May and June 1992.

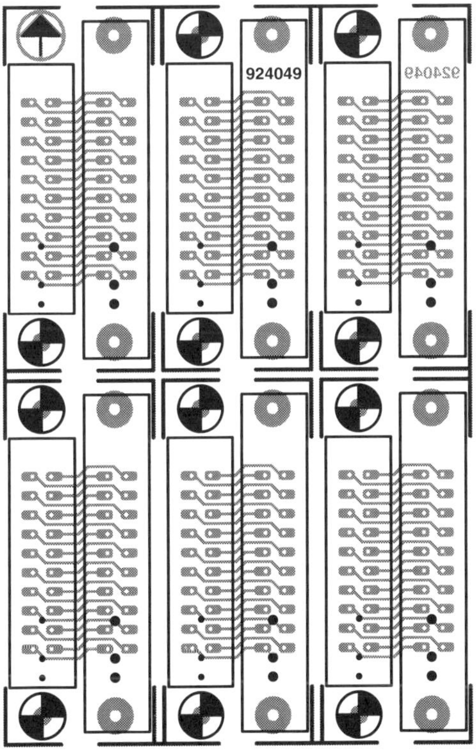

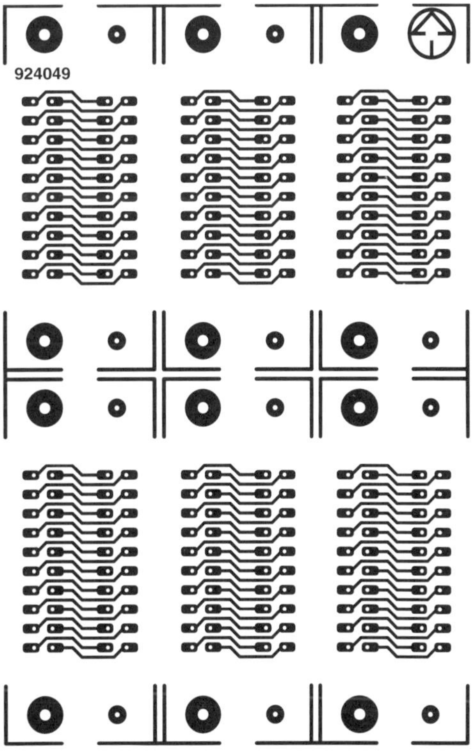

Telephone gong

258 Those of you who like the telephone bell louder may find this circuit of interest. It uses the telephone bell signal to actuate an oscillator, which in turn drives a relay that operates a standard door-gong.

The oscillator is an *RC* type based on a 4093 Schmitt trigger. Both the charging and discharge times of C_4 are set with presets. The oscillator is followed by a kind of buffer that energizes the relay via T_1.

The telephone bell signal is applied to terminals *a* and *b*. Capacitors C_1 and C_2 in the two lines isolated the circuit from the telephone network, at least as far as d.c. is concerned.To prevent the circuit responding to speech signals, level thresholds are provided by zener diodes D_1 and D_2. Network R_1-D_3 limits the 120–150 V_{pp} bell signal to about 12 V. That signal is rectified by D_4 and smoothed by C_3, after which it is used to switch IC_{1a}. Capacitor C_3 is discharged rapidly via R_3 when the bell signal ceases. Resistor R_2 prevents too high a drive voltage when there is speech on the input lines.

The relay contact is connected across the terminals via which the gong is operated. The circuit may be fed by the same transformer used for the gong. When the relay is energized, the total current drawn is only 35 mA.

The setting of presets P_1 and P_2 depends on individual taste how long the on and off periods of the gong are desired to be.

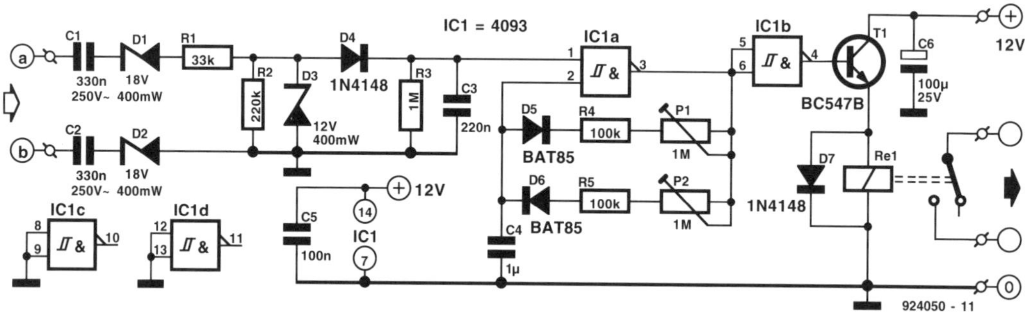

Extra brake light

259 There are many cars on the road that could do with an extra brake light, particularly one high up that can be seen by the third, fourth or fifth car behind it. That proposed here consists of a running-light bar of LEDs that starts again and again when all LEDs light. In some countries, this may not be allowed—check with your local highway or police department, but it can be used, anyway, in model cars.

Power is taken from terminals of the present brake lights. Because of bridge rectifier D_1–D_4, the polarity of that voltage is immaterial. The voltage is kept steady at 5 V by IC4.

An oscillator, based on IC_1, provides clock pulses to shift registers IC_2 and IC_3 as long as the brake pedal is pressed. The p.p.r. (pulse repetition rate) is set with P_1.

LEDs are connected to the outputs of the shift registers via series resistors. Because input pins 1 and 2 of IC_2 are connected to the positive supply line, each clock pulse generates a logic high at successive output pins, so that more and more LEDs will light. Since the last output of IC_2 is linked to input pins 1 and 2 of IC_3, the outputs of

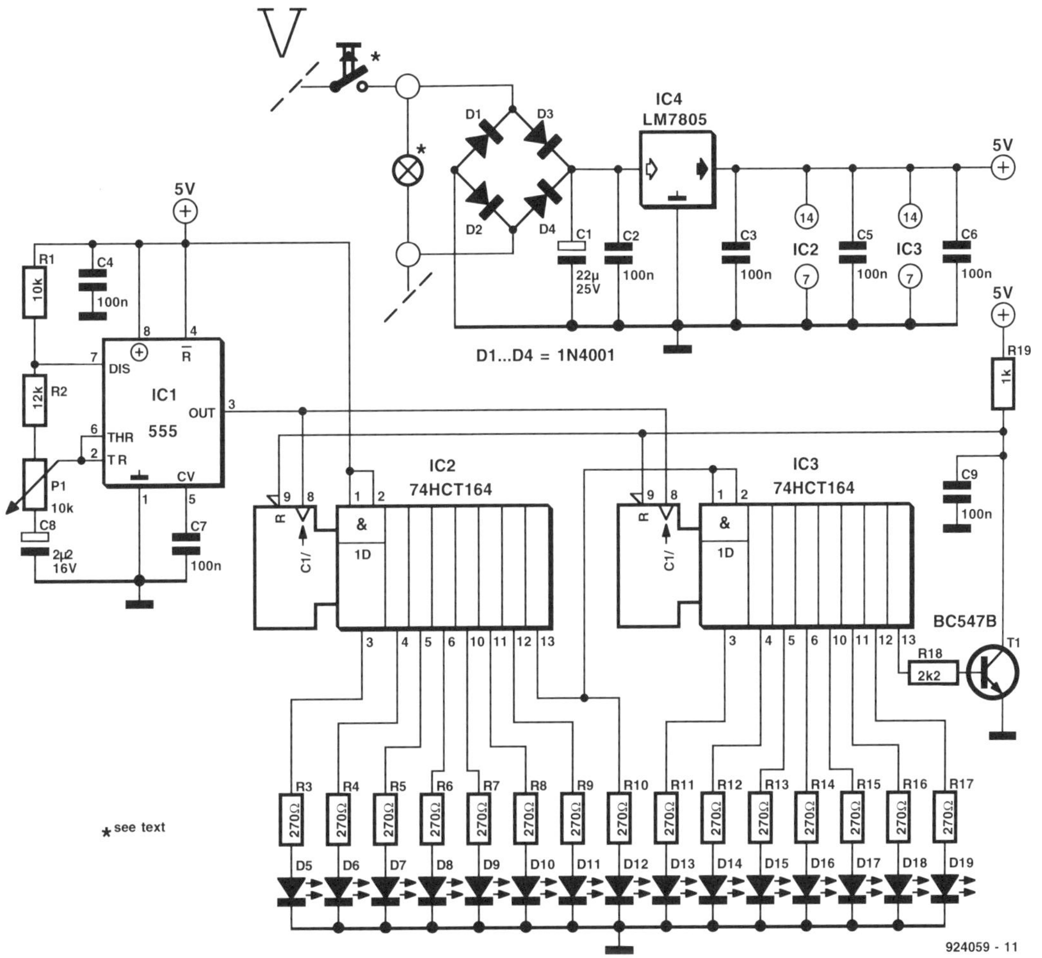

that register will also be supplied successively with logic 1s when all LEDs connected to IC_2 light. When, finally, the last output of IC_3 goes high, T_1 is switched on, which results in a reset of the shift registers and all LEDs going out. If the brake pedal is still pressed, the LEDs will light one by one again.

Network R_{19}-C_9 ensures that the shift registers are reset and thus the LEDs go out the instant T_1 is switched on.

Halogen lamp protector

260 Halogen lamps, particularly high-wattage ones, tend to draw very high currents when they are cold, because they then have a very low resistance: of the order of 0.1 Ω or lower. If such a light is operated from a 24 V battery, a switch-on peak current of over 200 A may flow. This is highly detrimental to the life span of the light, which after only a few switch-ons may give up the ghost. That costly situation may be prevented by gradually building up the

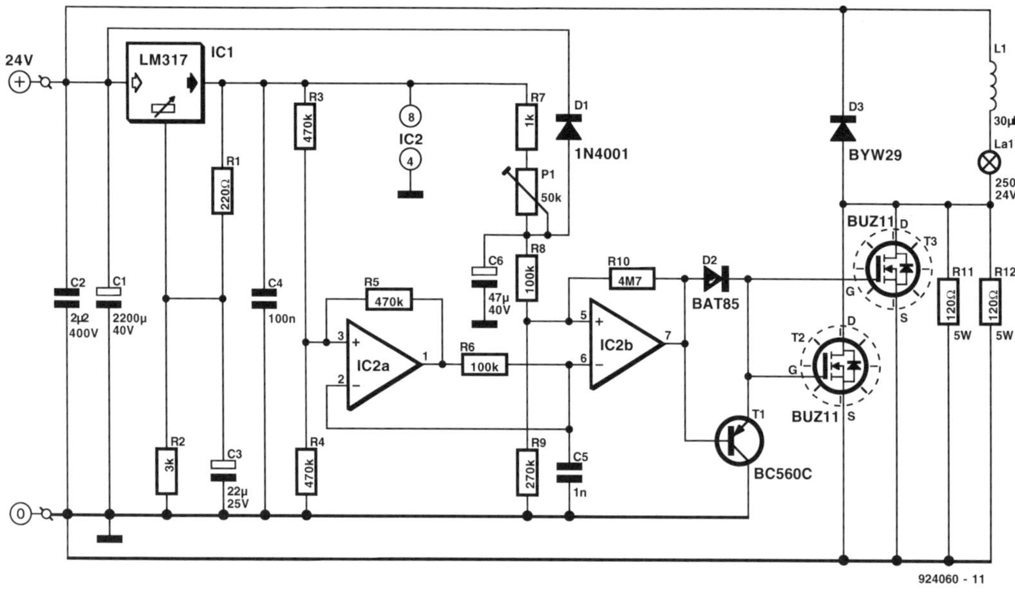

power supplied to the lamp. Since operation is from a d.c. source, the only practical way of so doing is by pulse-width modulation. With that technique, the voltage to the lamp is switched from zero with increasing pulse width, while the current is smoothed by a coil, so that its average level increases gradually. Switching is carried out with two MOSFETs Type BUZ11. This type is characterized by an extremely low channel resistance, which is typically 0.03 Ω for a gate-source voltage of 15 V; moreover, the device can handle continuous currents of up to 30 A and pulsed ones of up to 120 A. By connecting the two in parallel, the current is split two ways: not exactly 50/50, of course, but near enough to ensure that on switch-on the devices remain within their limits.

The control circuit for the MOSFETs, T_2 and T_3, provides nothing new. A regulator, IC_1, ensures that the supply to the circuit cannot rise too high: it is set for about 18.5 V. The battery voltage cannot be used directly, because, among others, the gate-source voltage of the MOSFETs must not be higher than 20 V.

A sort of triangular waveform is generated by Schmitt trigger IC_{2a}; R_2 and R_4 ensure that the operating point of the opamp is half the supply voltage. Because of the feedback via R_6-C_5, the output waveform is rectangular. However, the voltage across C_5 is an exponential waveform, that is more or less triangular. That voltage is compared by IC_{2b} with the terminal potential of C_6, which, after switch-on, rises gradually.

As long as the voltage across C_5 is lower than at the non-inverting input of IC_{2b}, the output of that opamp will be high. As soon as the inverting input reaches a potential higher than that at the +ve input, the output changes state, which is accelerated by the positive feedback via R_{10}.

The higher the terminal voltage of C_6, the longer the output of IC_{2b} will remain high. Eventually, it will reach a value higher than the maximum voltage across C_5: pin 7 of IC_{2b} is then permanently high. The lamp is then no longer switched on and off, but remains on.

The Type CA3240 dual opamp in the IC_2 position has the advantage that its output can become almost zero, but the disadvantage that it cannot sink relatively large currents. Because of that, T_1 ensures that the gate capacitances of T_2 and T_3, together 2–10 nF, are discharged rapidly. Those gate capacitances are charged again (that is, T_2 and T_3 are driven into conduction) by IC_{2b} via D_2. An additional driver is not needed, because the CA3240

can source enough current at high levels. When the FETs are on, their gate potential is about 16 V.
When the lamp is switched off, C_6 is discharged immediately, so that the circuit is ready at once to switch it on again. To ensure that on switch-off the induced potential across L_1 does not rise above the maximum level of the drain-source voltage (50 V), the coil is shunted by D_3. This needs to be a fast type (25 ns or better) that can handle currents of up to 30 A.
Resistors R_{11} and R_{12} provide the lamp with some voltage before the MOSFETs switch on.
The speed at which the circuit switches the lamp on is preset with P_1; normally, it will be sufficient to set this to the centre of its travel.
It is advisable to mount the FETs on a heat sink, although their dissipation is of the order of 1.6 W only.
Capacitors C_1 and C_2 must be able to handle high-frequency pulse currents of up to 30 A.
Inductor L_1 must ensure that the lamp current does not exceed a predetermined value: the larger the inductance, the lower the maximum level of the current. However, the physical dimensions of the coil must be acceptable. In the prototype, the maximum lamp current was set at 30 A. At a switching frequency of 7 kHz, an inductance of 30 μH is sufficient. Moreover, to avoid saturation problems, the coil is an air-cored one.
It is made by winding 45 turns of 1.5 mm ($^1/_{16}$ in) dia. enamelled copper wire in three layers on a 24 mm ($^{15}/_{16}$ in) dia. round former. During the winding, apply some glue from time to time to the turns.
The current drawn by the circuit is primarily that through the lamp: with a 250 W lamp and a 24 V battery, the current is some 10 A.

Bias compensation for bipolar opamps

261 One advantage bipolar opamps have over FET types is that their input noise level is appreciably lower. However, when a bipolar opamp is used in a high-impedance circuit, its bias current often presents a problem. For instance, the bias current of the well-known NE5534 opamp is typically 0.5 μA with a variation of about 5 nA ∞C^{-1}. Temperature variations of ±10 ∞C would thus cause current changes of almost 10%, which would make it impossible to hold the output voltage at zero volts d.c.
The circuit presented here compensates the bias current with a current source. Its design ensures a high input impedance. In practice the current source is not entirely stable with temperature variations, but that is largely nullified by placing it in a control loop.
The current source, T_2, is connected to the non-inverting input of IC_1 via a 10 MΩ resistor. The reference voltage for T_2 is established by a 2 mA ancillary current source, T_1, and D_1. This arrangement ensures that noise on the supply lines hardly affects the circuit.
To keep the noise of T_2 to a minimum, the 'reference voltage' of D_1 is divided by R_2-P_1.
Opamp IC_2 measures the direct voltage at the output of IC_1 and regulates the current source via R_3 in such a way that there is no direct voltage at the output of IC_1. This means, of course, that IC_1 cannot be used as a d.c. amplifier.
To ensure that the bias compensation is not nullified or affected by the output resistance or output offset of the preceding stage, a capacitor, C_2, is necessary at the input. It might be thought that, in view of the high input impedance, this should have a small value. For a 20 Hz cut-off point and an input impedance of 25 MΩ, a value of 330 pF would be sufficient. But this is wrong, because the high reactance of such a capacitor would cause a noise level of about 640 nV Hz^{-2} at 20 Hz. Beause of this, the reactance of the capacitor must match

the output impedance of the preceding signal source.

Preset P_1 must be set so that the output voltage of IC_2 is zero, or nearly so, at an ambient temperature at the centre of the wanted temperature range.

The circuit draws a current of about 11 mA.

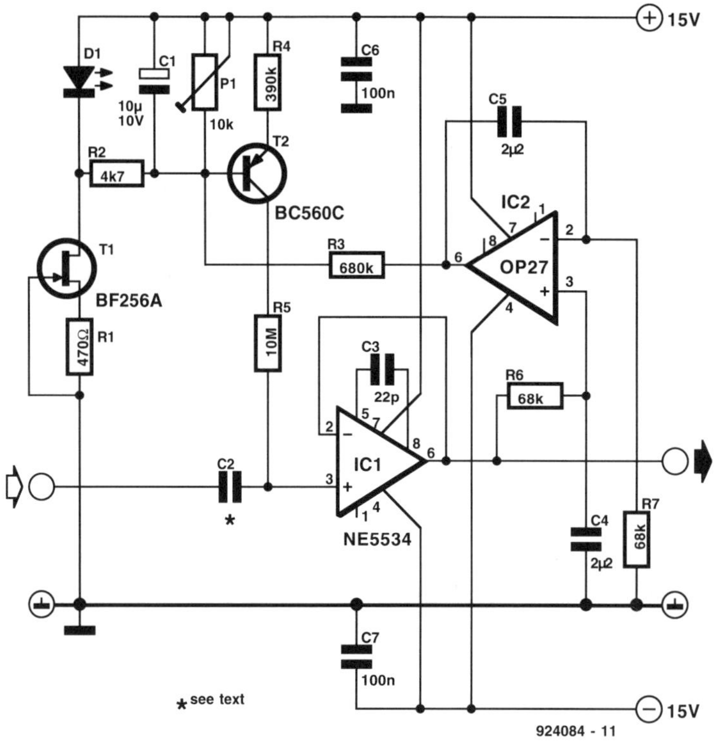

PWM governor

262 The governor described in this article is intended primarily for small 12 V motors that draw a current of not more than 2 A. Many governors limit the current through the motor, which also reduces the torque. Since it is controlled by pulse-width modulation (PWM), the present governor preserves most of the torque. The control is based on a sextuple CMOS inverter, IC_1 (Type 40106). Three of the inverters are not used here. The motor is driven via a discrete darlington, T_1-T_2.

Inverter IC_{1a} functions as an oscillator, whose on time (when T_2 conducts so that the motor is provided with power) is determined by R_2-C_2-D_3. The off time of the oscillator (when T_2 is also off) is determined by R_3-P_1-C_2-D_4, and may thus be varied with P_1.

Parallel-connected inverters IC_{1b} and IC_{1c} form a buffer between the oscillator and the power darlington.

The motor is connected across the + and 0 terminals in the diagram.

Power for the governor and the motor is provided by a mains transformer, whose secondary voltage is single-phase rectified by D_1 and D_2.

An indicator LED with current limiting resistor is connected across the motor terminals.
Network R_1-C_1 smooths the supply voltage to IC_1.

Finally, T_2 must be fitted on a suitable heat sink.

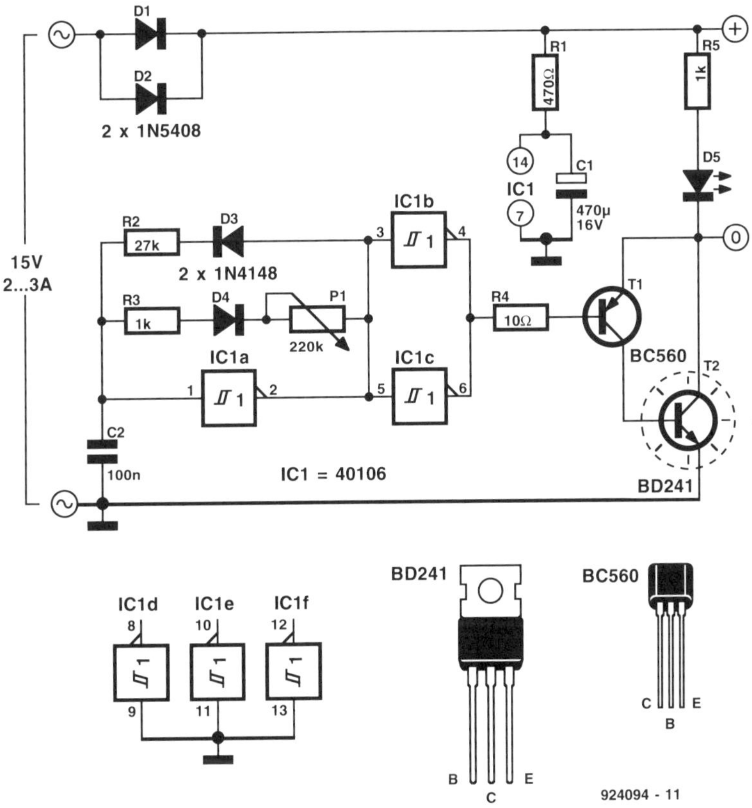

Low-drop diode

263 At high currents, many silicon diodes have a forward voltage of 1 V or more. There are types whose drop at currents of up to 2–3 A is limited to 0.5–0.6 V, but even that may cause unacceptably high losses. The circuit described here offers a possible remedy.

The cathode, C, of T_1, a SIPMOS FET, is connected to a sinousoidal-voltage source. The anode, A, thus functions as a reference point. Capacitor C_1 is charged to the peak value of the sinusoidal voltage, U_s, viaD_1. This ensures that the opamp is provided with power even during the negative half-periods of U_s.

The non-inverting input of IC_1 is set to half the peak value of U_s via potential divider R_1-P_1-R_3. Because of voltage divider R_2-R_4, the potential at the inverting input of the opamp will be higher than that at the non-inverting input only during the positive half-periods of U_s. This means that the drain-source channel of T_1 is switched on by the opamp when the voltage at the cathode tends to become lower than that

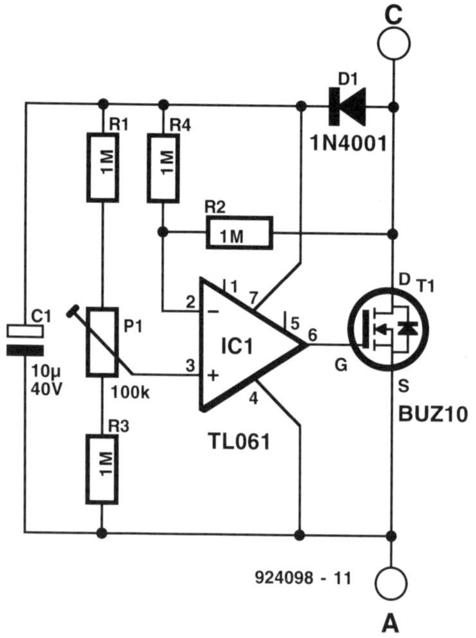

at the anode. In that case, the current through the FET flows from source to drain, parallel to the internal protection diode. In other words, the FET is used the wrong way round. The forward voltage of the FET diode so created is the product of the current through it and the on resistance (0.07 Ω).

The setting of preset P_1 determines the anode-cathode potential at which the output voltage of the (mainly) linearly operating opamp begins to rise and thus drive T_1 into conduction. The preset can be adjusted accurately only with the aid of an oscilloscope connected to the drain and source of the FET. It is set to that position where for the nominal forward current the voltage across T_1 is as small as possible during the half-periods when the FET is on. In the prototype, the forward voltage so measured was 0.5 V with an alternating current of 10 A at a frequency of 50 Hz. At 3.3 A, the drop was only 0.2 V and at 300 mA just 0.1 V. Note that the forward voltage remains constant with currents below the level at which P_1 was set.

The circuit draws a current that is not much higher than the supply current to IC_1. Although the maximum supply voltage of the opamp is 36 V, the cathode-anode voltage, that is, the 'reverse' voltage of the 'diode' must not exceed 20 V, which is the maximum permissible gate-source voltage of the BUZ10.

Sidac neon tube starter

264 The sidac from Motorola is best compared with a triac of which the gate connection is missing. It switches on whenever the voltage across it exceeds a certain level. The polarity of that potential is immaterial: like a triac, the sidac works equally well with direct and alternating voltages. Furthermore, when the sidac is on, it resembles a short-circuit and remains in that state until the level of the current drops below a certain value (the holding current), whereupon it switches off.

A series network of a sidac and a load connected to the mains results in a kind of dimmer whose, non-variable, phase angle depends on the starting voltage of the sidac. Sidacs are available for starting voltages between 104 V and 280 V.

A neon tube does not switch on as easily as an incandescent lamp because the tube can start only at a voltage much higher than the mains, after which it will remain lit at the mains voltage. The level of both the starting voltage and the working voltage depends on the temperature of the tube.

Normally, the high starting voltage is obtained by interrupting the current through a choke. This is usually done by the starter, which also ensures that a fairly large current flows through the filaments of the tube. This heats the ends of the tube, which makes starting easier.

These tasks of the starter are taken over by two 135 V sidacs (or a single 270 V one). The starting voltage is thus 270 V, which is below the peak value of the mains (about

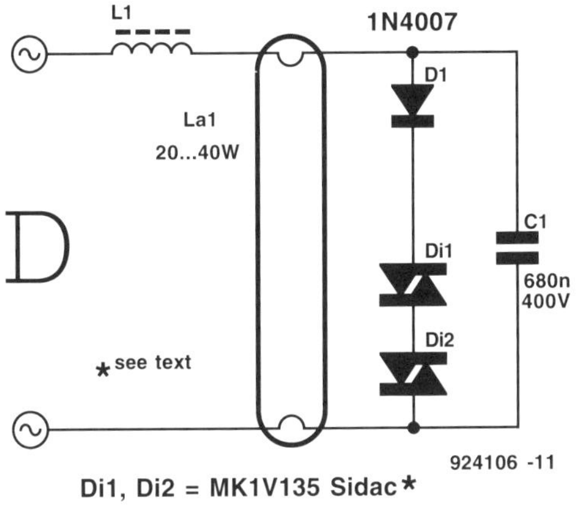

340 V), but higher than the working voltage of a 20–40 W neon tube.
As long as the tube has not started, almost the whole of the mains voltage is dropped across the starter. Assume for a moment that the polarity of the mains causes D_1 to be forward biased. When the instantaneous value of the mains voltage reaches the level of the starting voltage of the sidacs, these will short-circuit the starter, whereupon a fairly heavy current will flow through the filaments and the coil. This gives rise to a magnetic field around L_1. When the polarity of the mains voltage reverses, the positive current through L_1 will decrease gradually. When the level of the current approaches zero, the sidacs switch off, whereupon the instantaneous negative mains voltage is applied across the tube immediately, because of C_1 being charged rapidly. This capacitor and the starter form a series resonant circuit that magnifies the sudden drop across the tube to way above the level of the mains voltage.
During the next positive period of the mains voltage, the sidacs switch on again, and the sequence repeats itself until after a few cycles the tube has warmed up enough to remain lit. The drop across the lit tube does not exceed the starting voltage of the sidacs, so that the electronic starter is switched off.
Capacitor C_1 not only suppresses any r.f. interference generated by the tube, but also makes the load on the mains supply less inductive (so-called cosϕ improvement).
The capacitor and diodes can probably be fitted into the man-made fibre enclosure of the original starter.

Flash EPROM convertor

265 Flash-EPROMs are becoming increasingly popular, in spite of their being harder to reprogram than EEPROMs. That difficulty is, however, countered by their lower price, greater density and higher programming speed.
Maxim, a specialist manufacturer of all sorts of small convertor, produces a special IC for generating the necessary programming voltage of 12 V at 50 mA: the MAX732.
The MAX732 contains virtually everything to make a mini switch-mode power supply. Its input voltage requirement is 5 V, from which it produces an output of 12 V. Since that output is needed only during programming, it may be disabled via the shutdown input (pin 1).
A problem encountered in producing switch-mode supplies is the availability of certain passive components, particularly the inductor. That used here is the Sumida Type CD54-470KC, which is available from a number of retailers and also from Maxim dealers as Type MAX-L001. If neither of these can be obtained, however, a triac choke may be used, but that will lower the efficiency of the convertor to some extent. It is always possible to add or remove turns if the inductance is incorrect. Bear in mind that the inductance is proportional to the square of the number of turns.
Diode D_1 must be a Type 1N5187 as indicated or equivalent; note that a 1N4001 is not fast enough.

The prototype delivered 12 V and an output current, I_0, of up to 200 mA, more than enough for a flash-EPROM. The current drawn from the 5 V supply was about $2.4I_0$. Use a single earthing point and decouple the IC directly at its pins.

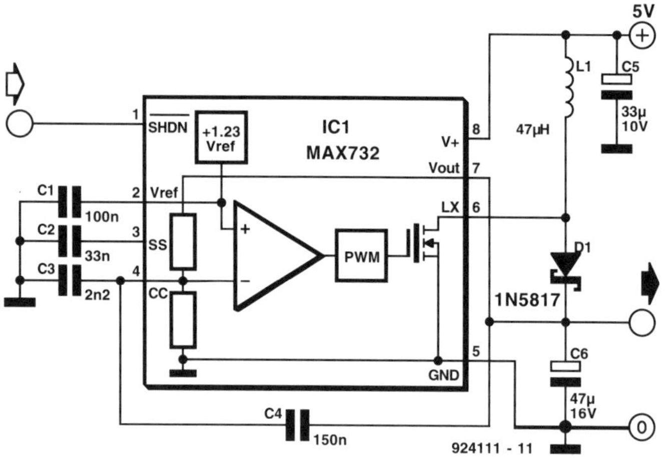

Thyristor starter for fluorescent tubes

266 The Type Y1112 fluoractor™ from Texas Instruments is a specially designed thyristor for starting fluorescent tubes. Its main advantages over a standard thyristor are its much greater hold current, its ability to cope with rapidly rising anode-cathode voltages (high du/dt value), and its overvoltage protection. These properties make it very suitable to replace a traditional electro-mechanical starter for reliable flicker-free starting and a long life.

A fluorescent tube does not switch on as easily as an incandescent lamp because it can only start at a voltage much higher than the mains, after which it will remain lit at the mains voltage. The level of both the starting voltage and the working voltage depends on the temperature of the tube.

Normally, the high starting voltage is obtained by interrupting the current through a choke. This is usually done by the starter, which also ensures that a fairly large current flows through the filaments of the tube. This heats the ends of the tube, which makes starting easier.

These tasks of the starter are taken over by the fluoractor™, Th_2. When the mains is switched on, Th_2 will start via R_1 and R_2. A fairly large current will then flow through the filaments of the tube so that this will be preheated.

Because of the potential drop across diodes D_7 and D_8, capacitor C_1 will be charged via resistors R_4 and R_5. When, after a few milliseconds, C_1 is charged to a certain value (during which time the current through Th_2 is appreciably larger than the hold current), thyristor Th_1 is switched on, whereupon the gate current of Th_2 is interrupted*. The sudden discontinuation of this fairly large current causes the choke to produce a back-e.m.f. that is high enough to start the fluorescent tube.

*This special thyristor remains on, however, until the current through it has dropped to the hold value of about 200 mA.

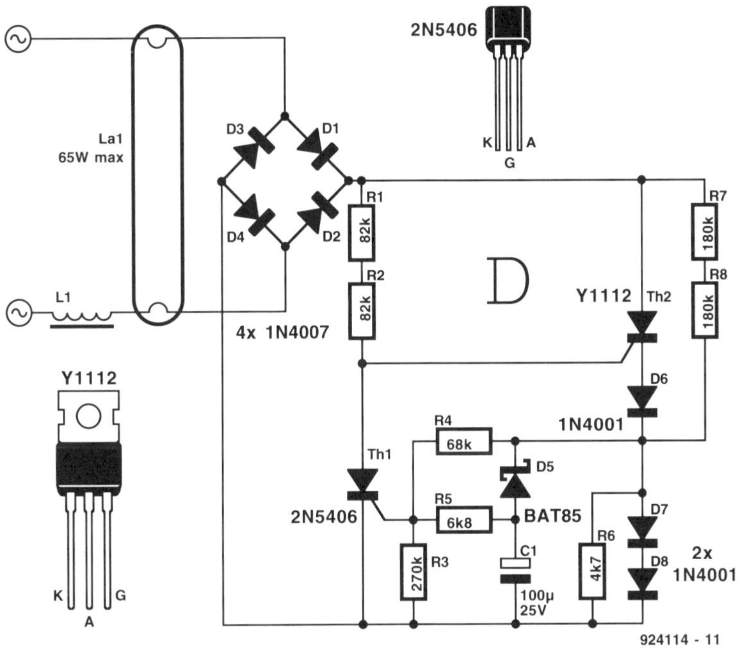

Single-IC timer

267 Timer chip ICM7250 from Maxim enables a very simple timer to be designed. The IC has eight programmable inputs for connecting to, say, two thumb wheel switches. Units are set in BCD code on the four lowest inputs, pins 1–4, and tens on the highest, pins 5–8. Thus, the user can select any time between 00 and 99 units.

The length of each time unit is determined by P_1-R_2-R_3-C_1. In this particular application time units of one second have been assumed. Other units are easily selected by a few small changes.

According to the relevant data sheet, the maximum overall resistance value of the timing network must not exceed 22 MΩ, while the value of the capacitor must not be smaller than 10 pF. Note that if longer units are set, capacitors with small leakage currents must be used.

When switch S_3 is pressed, an internal bistable is reset and the counter is started. The output of IC_1 (pin 10) then changes state (becomes low). After the elapsing of

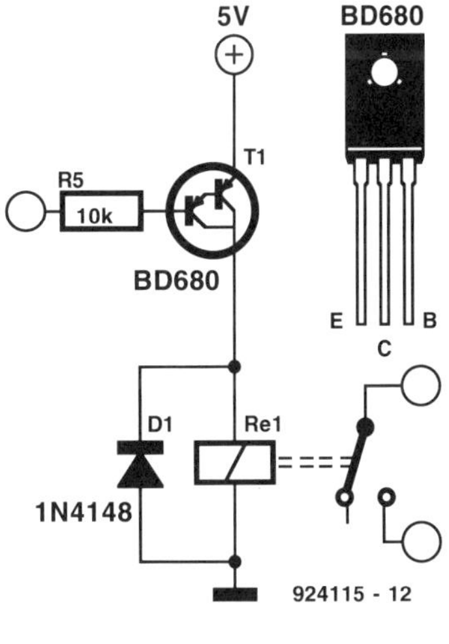

the set number of time units, the output becomes high again.

The supply voltage for the timer may be 2–9 V, in spite of the data sheet stating that up to 16 V is permitted. In the prototype, the output was permanently high with supply voltages of 9.5–16 V.

The relay connected to the output is switched by a power transistor, T_1. This darlington can switch currents of up to 4 A.

The circuit itself draws a current of only 1 mA.

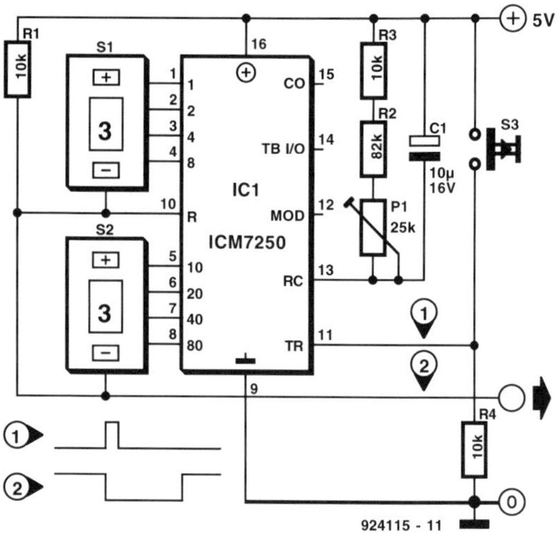

General transformer PCB

268 The board may be used either as shown in Fig. 1 or as in Fig. 2 with transformers rated from 1.5 VA to 12 VA. When used with a very small transformer, it is possible to cut off the part marked K2–K4. Place the terminal strips in the same position along the new edge as they were originally.

When non-short-circuit-proof transformers are used, a fuse in the mains entry plug is essential.

Parts list

K_1 = two-way terminal block, pitch 7.5 mm

K_2, K_3, K_4 = two-way terminal block, pitch 5 mm

Tr = transformer for PCB mounting

PCB Ref. 934004

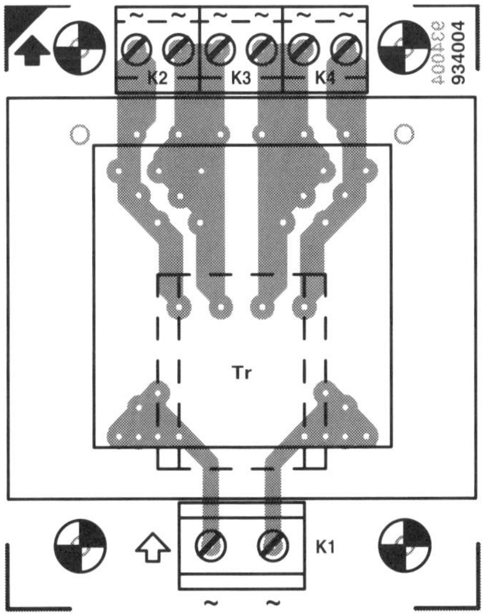

1

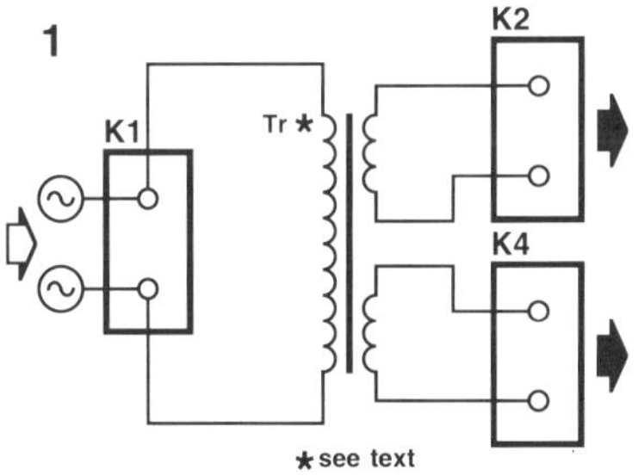

2

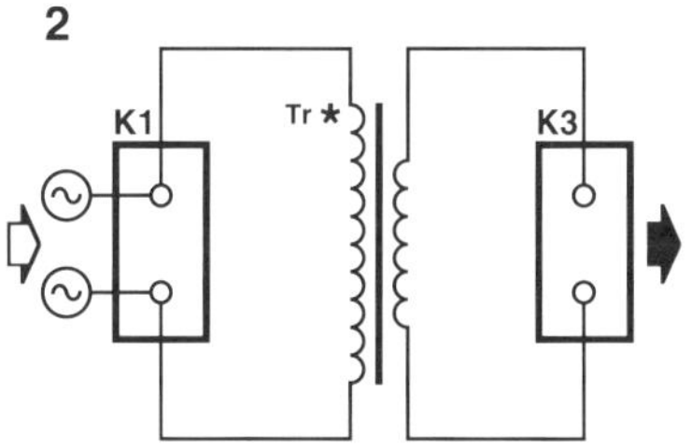

934004-11

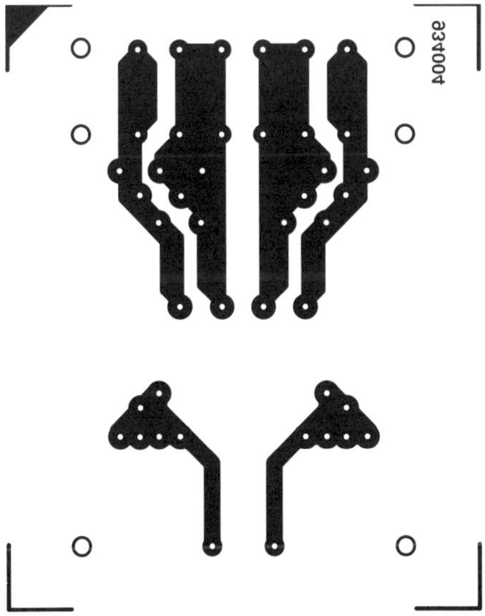

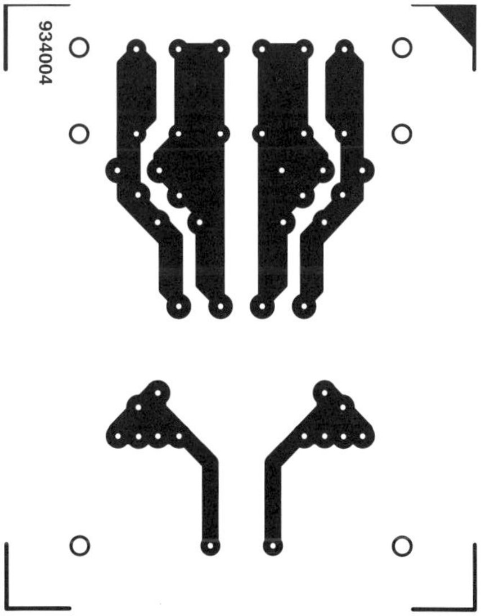

Virtual ground

269 Often only a single power supply is available for an opamp circuit. Since such circuits normally require a double supply, this means that in some way the available supply has to be converted. This is normally done by a potential divider and capacitor across the single supply, sometimes followed by an ancillary opamp.

Nowadays, a more sophisticated way is the use of the TLE2426 from Texas Instruments. This IC is not only small and easy to use, but, owing to the built-in opamp, also gives a better performance. This is particularly noticeable at low frequencies, where the impedance of the usual electrolytic capacitor across the potential divider may cause problems. In the TLE2426, a voltage follower takes the place of the capacitor, so that the impedance, even at low frequencies, remains low—typically 7.5 mΩ.

The quiescent current is only 170 μA, while the peak current that can be provided is 20 mA.

The device is available in two different enclosures: transistor-like, indicated by the suffix 'LP' in the type number, and 8-pin DIL, suffixed D, P or JG. The LP model is eminently suitable as a replacement of a conventional potential divider and capacitor (and possibly, opamp). The DIL model has the advantage that a capacitor may be connected to its pin 8, which in some cases is desirable since each of the two output voltages is half the supply voltage, which may contain much noise.

There is also the Type TLE2425, which provides a fixed voltage of 2.5 V at a peak current of 20 mA. Pin 8 of this model is not connected.

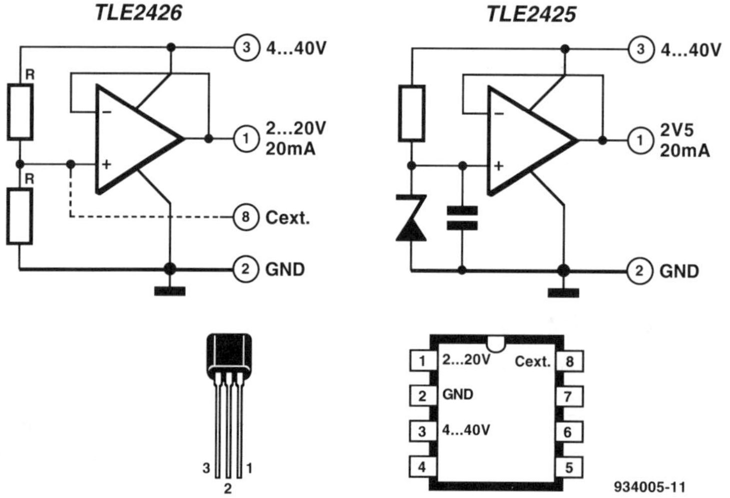

Two-way indicator light

270 The audio signal is applied to the indicator circuit via a 1:1–1:4 microphone transformer. It determines which of the two lamps, L_1 and L_2, connected to K_1 and K_2 respectively, is switched on by the relevant silicon-controlled rectifier (SCR).

In the absence of an input signal, both SCRs are switched off. During the positive half period of the mains voltage, drive voltage is applied to the gate of Th_3 via R_2 and diac Di_1. The SCR is then on and switches on L_2. When the audio input reaches a certain level, Th_1 and Th_2 also conduct and switch on L_1. At the same time, Th_2 removes the drive to Th_3, so that this SCR switches off at the next zero crossing of the mains and lamp L_2 goes out. Great care must be taken in the construction of this circuit, because potentially fatal mains voltages are present in it during operation. It must, therefore, be built into a well-insulated (main-made fibre) enclosure. The transformer must be a type whose insulation is approved for operation in mains-carrying circuits. Resistor R_2 must be high-power type, and C_1 a 400-V type.

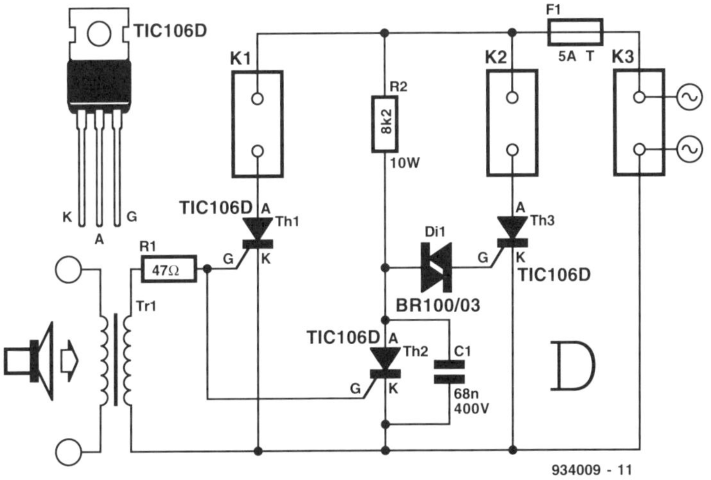

Light switch for motorcycles

271 Many motorcycles have the facility that as soon as the ignition is switched on, their dipped headlights (or day running lights) come on. Although on most modern machines these lights go off briefly while the engine is started (so as to prevent a heavy load on the battery), there are some on which this is not so. This shortcoming may be remedied by the circuit described here.

The circuit is based on optoisolator IC_1, which detects whether the 'N(eutral)' light, La_1, on the dashboard is on. As long as this light is on, the machine is not in gear. When the engine is started (with the gear box still in 'N'), La_1 remains on, so that the LED in the optoisolator lights. This switches on the transistor in IC_1. The gate of the thyristor is then at earth potential and the relay remains

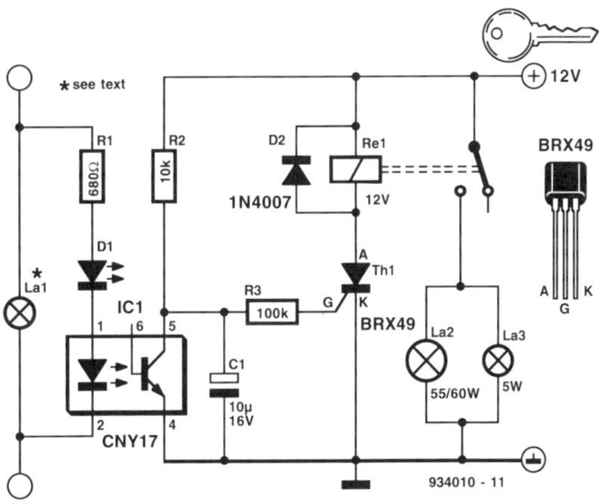

deenergized. Headlights LA_2 and La_3 are off and the engine can be started without these lights coming on. When a gear is selected, La_1 goes out and the transistor in

IC_1 switches off. The thyristor then gets a gate current via R_2 and R_3, whereupon Re_1 is energized. The headlights then come on and the machine can be driven away safely. The thyristor remains on as long as the battery is being charged, so that when 'N' is selected, for instance, at traffic lights, the lights remain on.

Sequential touch switch

272 The touch switch is based on a Schmitt trigger with hold contact, which is built from IC_{1a}, IC_{1b} and R_3. The output of IC_{1b} is fed back to IC_{1a} via R_3, resulting in both gates maintaining their stable status. As long as the circuit is quiescent, that is, the anode of D_1 is low and the cathode of D_2 is high, the gates maintain their status.
Consider that output A is low. When the touch contact is shorted with a finger, the potential across R_8 rises. This rise is translated into a short pulse by R_1-C_1. This pulse renders the input of IC_{1a} high via D_1. The output of IC_{1b}, and output A, will then also go high and, as explained before, this status is maintained.
At the same time, the potential across C_2 will rise gradually. When this capacitor is fully (or nearly so) charged, the circuit is ready to change state at the next pulse across R_1. This pulse will make the output of IC_{1a} low, whereupon the output of IC_{1b} becomes low and the potential across C_2 drops. Output A is then low again and the circuit is back to where it started.

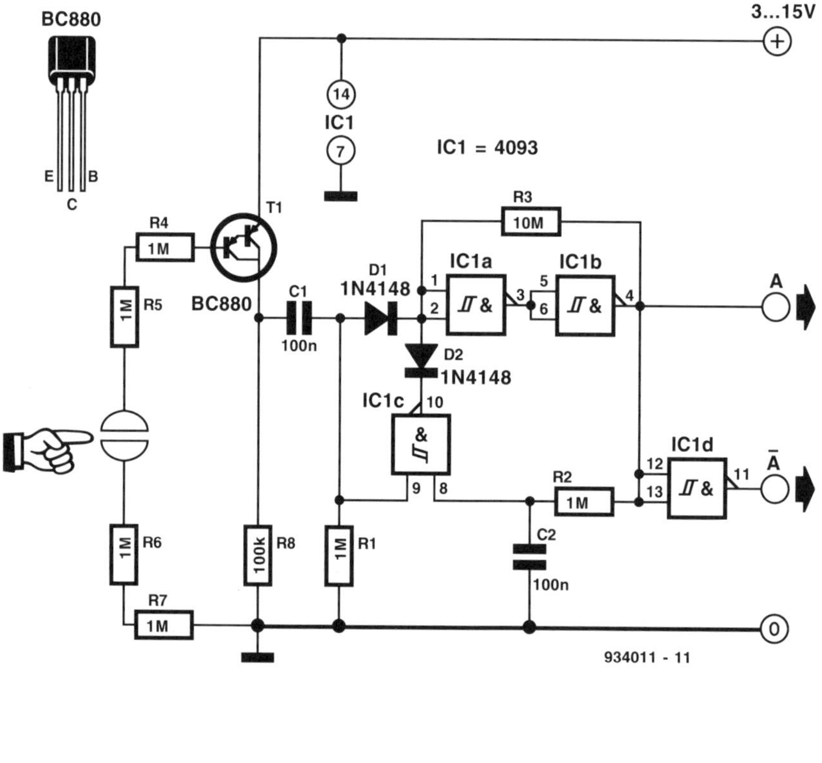

Telephone-operated night light

273 Since in some countries it may be illegal to connect this circuit to the public telephone network, check with your local telephone manager before building this circuit.

In most telephone networks, a direct voltage of 48–60 V exists at terminals *a* and *b*. When the call signal comes in, a 50–60 V alternating current at a frequency of 25–27 Hz is superimposed on the direct voltage. This open-circuit voltage is more than halved when the receiver is taken from the hook (which closes the current loop). The current in the loop is 20–100 mA.

The present circuit starts a timer when it detects the call signal.The timer energizes relay Re_1 and this switches on lamp La_1. After the last ring, the timer remains on for a short time, but then returns to its quiescent state, whereupon the lamp goes out. When, however, the receiver is taken from the hook, the timer is triggered permanently, so that the lamp remains on during the telephone conversation and for some time afterwards. Furthermore, the lamp may be switched on and off manually by S_1 being pressed.

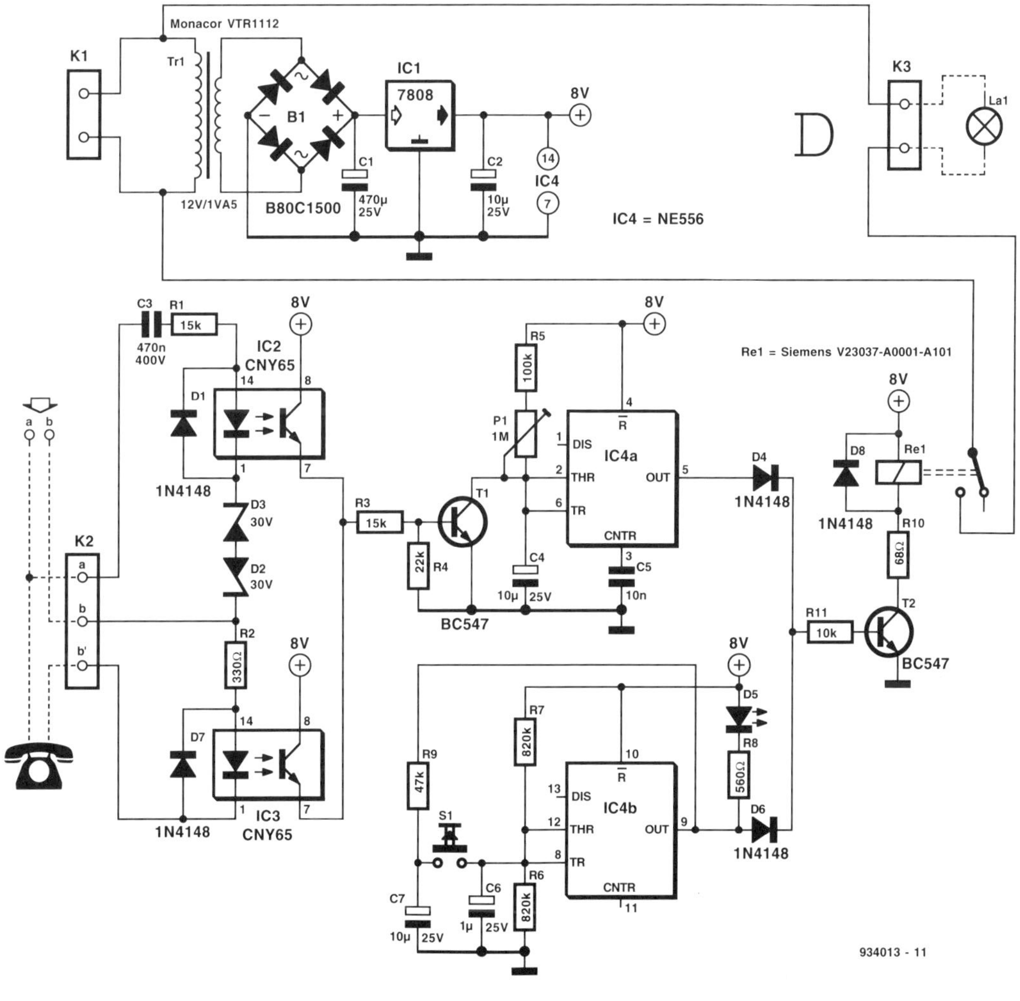

One part of the circuit is connected in parallel, and the other in series, with telephone terminals *a* and *b*. The parallel-connected part, which detects the call signal, is isolated from the direct voltage by C_3. Because of this, no current flows through the LED of optoisolator IC_2, so that the phototransistor is off. The second optoisolator, which is in series with the *a* and *b* lines, also carries no current since the circuit is broken by the cradle switch.
The call signal, however, is applied to IC_2 via C_3. Consequently, a current flows through the LED and the transistor is on. The current is limited by R_1, while diodes D_2 and D_3 chop the peaks off both the half waves to prevent any damage to the optoisolator.
The phototransistor provides sufficient base voltage to T_1 to switch on this transistor. Capacitor C_4 discharges, whereupon the potential at pins 2 and 6 of timer IC_{4a} drops to earth, which triggers the timer. Its output changes state (goes high), whereupon T_2 is switched on so that the relay is energized. The relay contact connects one of the mains lines (in the UK, the live) to lamp La_1. The time constant of the timer has been arranged so that even the longest intervals in the call signal can be bridged. If the receiver is not lifted, and the last call signal has passed, the lamp remains on for the duration of the time constant, and then the circuit returns to the quiescent state.
When the receiver is lifted off the hook, a current flows through the LED in IC_3, whereupon the phototransistor in the optoisolator, and consequently T_1, is switched on. The timer is triggered as long as the telephone conversation lasts. When the receiver is replaced into the cradle, the lamp stays on for a short while determined by the setting of P_1.
The second timer in the NE556, IC_{4b}, is used for manual operation of the lamp. Before it is triggered, half the supply voltage exists at pins 8 and 12. The output (pin 9) is low, so that C_7 cannot be charged. When S_1 is pressed, the trigger voltage briefly drops below the threshold, whereupon the output goes high. Since, owing to the time constant R_9C_7, the potential across C_7 rises only gradually, the level at the trigger input (pin 8) remains below the upper threshold. When the switch is operated again, the full supply voltage is across C_7, so that the level at the trigger input exceeds the upper threshold. The timer output then goes low again and the lamp goes out.
The power supply is a standard design, but could be replaced by a suitable 12-V mains adaptor. The lamp could then be a 12-V car type. From a safety point of view, that is not a bad alternative.

Bicycle light with afterglow

274 The design ensures that bicycle lights remain on for a little while after the bicycle (and thus the dynamo) has come to a standstill. It has the disadvantage that the frame of the bike cannot be used as the common return: a separate wire has to be run for this. This is because use is made of a bridge rectifier to provide full-wave rectification of the dynamo voltage. Consequently, the alternating and direct voltages must float with respect to one another.
Actually, bridge rectifiers are used that have one half in common. Diodes D_1–D_4 constitute the bridge via which (and R_5) the NiCd battery is charged when the bicycle is moving (and the dynamo is connected). Diodes D_1, D_3, D_6, and D_7 form the bridge via which the bicycle lights are powered when the dynamo voltage is sufficiently high.
When the dynamo is driven, C_1 is charged via D_5, which causes T_1 to conduct. If T_1 is on and the dynamo output drops below a certain level, T_2 is switched on, whereupon the bicycle lights are powered by the NiCd battery.
When the bicycle is at a standstill and the dynamo is no longer driven, C_1 is no longer

charged. Because of this, T_1 will gradually stop conducting and this will also switch off T_2. The bicycle lights will then go out.
Switch S_1 serves to prevent the NiCd battery slowly discharging via the circuit when the bicycle is not used. The switch should then be opened.
If the afterglow time is found too short, it may be lengthened by increasing the value of C_1.

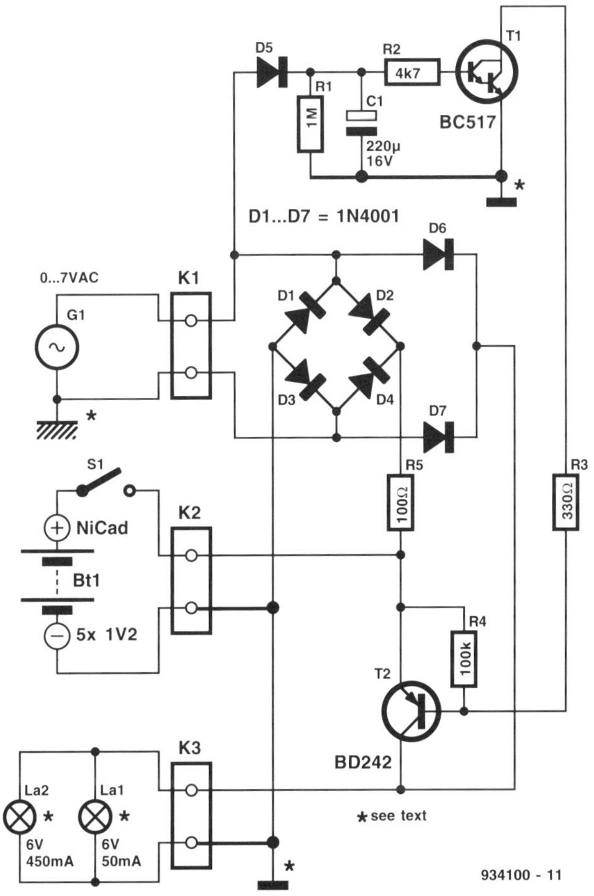

Wind direction indicator

275 The indicator shows eight wind directions by means of LEDs. The sensors are reed contacts powered by a permanent magnet. The mechanical construction must be such that it is impossible for more than one reed contact to be closed at any one time.
At the instant a reed contact closes, a leading edge is applied to the clock inputs of D-bistables (flip-flops) IC_2 and IC_3 via an OR gate in IC_1. This results in the outputs of the bistables assuming the status of the D inputs, so that only the LED associated with the closed reed contact lights.
If the wind changes direction slightly, and the reed contact opens, then, owing to the bistable construction, the LED will continue to light. Only when the wind has changed direction so much that another reed contact closes will a different LED light.

When the supply voltage is switched on (S_1), only D_9 will light owing to the power-on reset via R_{10} and C_3 . Only when one of the reed contacts has caused a leading transition (edge) at the clock inputs will one of the direction LEDs (D_1–D_8) light.

The indicator needs to be supplied by a regulated 5 V source. It draws a current of not more than 10 mA.

The LEDs are low-current Siemens types.

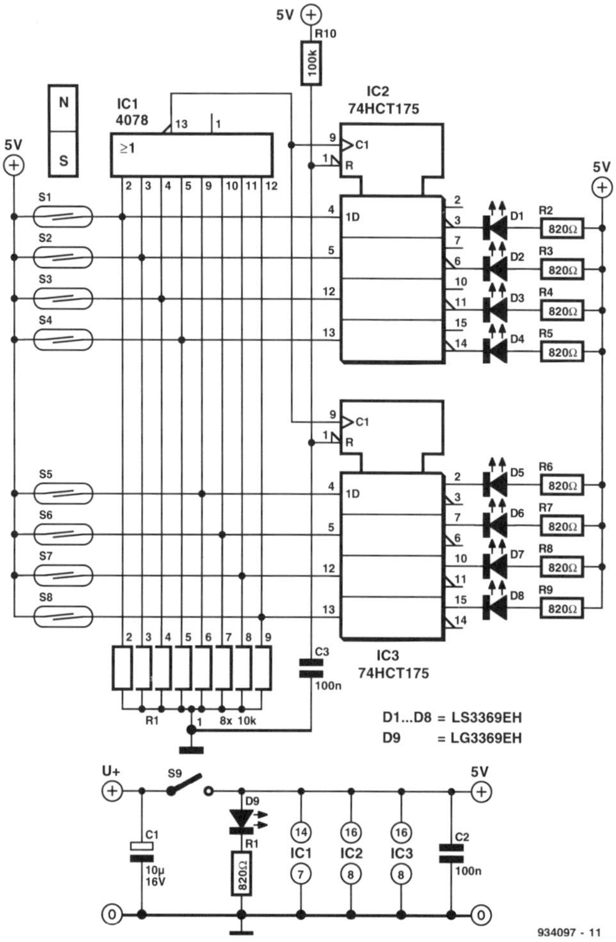

Single-chip temperature control

276 The UAA2016 from Motorola lends itself admirably to making a proportional mains-operated temperature control. A resistor with negative temperature control (NTC),R_3, which must have a value of about 100 kΩ at 25 °C, serves as sensor. In the diagram, the heating element is a 100 W light bulb, which generates enough heat for a greenhouse or a climatic chamber (for testing electronic equipment).

The triac can switch up to 3 A if it is fitted on a heat sink.

The properties of the regulator are determined by the resistors at pins 1–4. Resistors R_1 and R_2 establish over which range the temperature may be set with P_1. With values shown, the range is 0–70 °C.

Closing S_1, causes the temperature to drop by about 2 °C. The degree to which the temperature drops is determined by R_4: a value of 100 kΩ causes a drop of about 1.5 °C, while 10 kΩ results in a reduction of about 5 °C.

Resistor R_5 determines the hysteresis of the regulator: here, it is about 150 mV. How much that is in °C depends on the type of NTC and the temperature: an NTC has a non-linear temperature characteristic.

Do bear in mind that the entire circuit is connected to the mains and that, therefore, the regulator must be constructed with due care to safety regulations. It should be installed in a well-insulated, earthed (if metal) case.

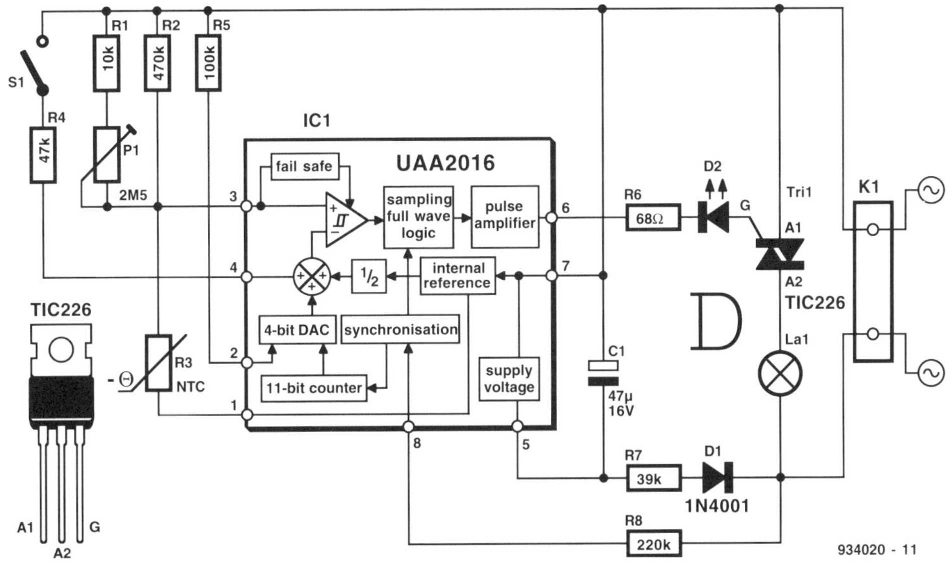

General purpose Schmitt trigger

277 The thresholds voltages, and thus the hysteresis, of the Schmitt trigger described can be set independently.

The circuit is based on opamp IC_1, which has been connected as an inverting Schmitt trigger. The two voltages at which the output of the circuit changes level are preset with P_1 (high level = U_H) and P_2 (low level= U_L). Provided that U_H is greater than U_L;

$$U_i < U_r \rightarrow U_o = U_H \rightarrow U_r = U_H;$$

$$U_i > U_r \rightarrow U_o = U_L \rightarrow U_r = U_L.$$

These two expressions are shown as a graph in Fig. 2. From this it is clear that when the output is high, it will change to low only if the input has risen above U_H, and will then change back to high again only if the input voltage has dropped below U_L.

The changing state of the output results in the changing of the reference voltage at the non-inverting input of IC_1. In the first instance, this is caused by the weak, but fast, positive feedback via R_6 and R_7. The hysteresis caused by this network amounts to only 20 or 30 mV, but that is more than enough to give the opamp good switching performance.

Once the opamp has changed state, U_r changes for the second time, because IC_{2a}–IC_{2d} alter the junction R_6–C_7 from U_H to U_L or vice versa.

Capacitor C_7, in conjunction with the on resistance of the analogue switches, slows the transfer process: the reference voltage changes smoothly from one value to the other in about 250 ns. This assumes that the speed at which the input signal changes is not too high. To enable signals with steep edges to be processed, there is a low-pass filter at the input (R_5–C_8).
The threshold voltages are set with the aid of a millivoltmeter connected successively to U_H and U_L. Presets P_1 and P_2 should preferably be multiturn types to enable the levels to be set accurately to within a millivolt.
The trigger may be used with unipolar and bipolar input signals. If only positive inputs need to be processed, the power supply for the trigger may be asymmetrical. To that effect, the negative supply line is linked to '0' and the supply (max. 16 V) connected to + and 0. With a supply voltage of 16 V, the trigger draws a current of ≤30 mA.
The switching performance is tested with a triangular voltage (0–5 V) at the input. With that input and the thresholds set to 2 V and 4 V, the trigger switched accurately at frequencies up to 5 kHz.

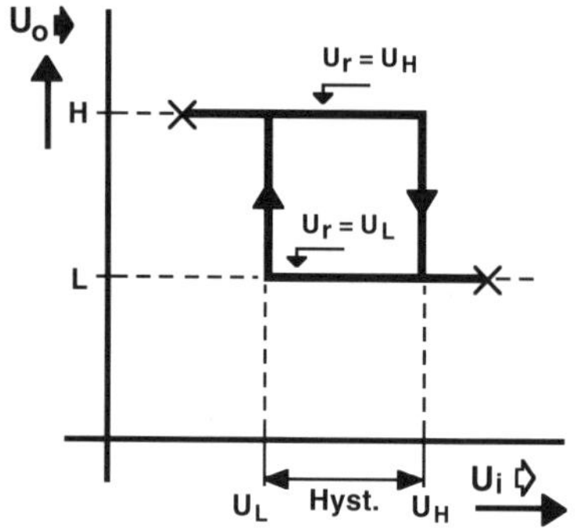

934021 - 12

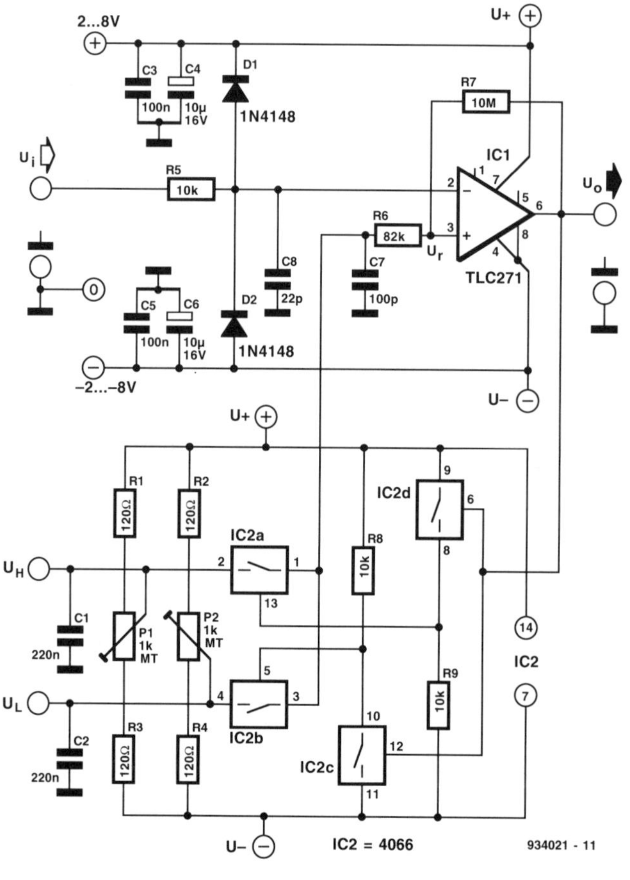

934021 - 11

Speed chess clock

278 In speed chess a player has to make his move in 5–10 seconds. If he/she does not, he/she loses the game (depending on the agreed conventions). The 'clock' described here indicates whose turn it is and who has defaulted. It may, of course, also be used for other speed games.

When player A presses S_2, D_3 lights to indicate that it is player B's turn. This player returns the game to A by pressing S_1, whereupon D_2 lights. If a player fails to return the game to his/her opponent in the time set with P_1, the buzzer will sound.

The circuit is built around a 556, the dual version of the well-known 555. One half, IC_{1b}, functions as a bistable (flip-flop) that 'remembers' which switch was pressed last: this is indicated by the associated LED lighting. The other half, IC_{1a}, operates as a monostable. It is kept in its triggered state (buzzer inactive) as long as it retriggered regularly, and timely, by switches S_1 and S_2.

The trigger input is made low by S_1 with the aid of T_1, and by S_2 with the aid of D_1. Every time IC_{1a} is retriggered, C_5 is discharged via pin 1 (DIS) of IC_{1a}. The time set with P_1 then starts afresh. If either of the switches is not pressed in time, the voltage across C_5 becomes so high that IC_{1a} is reset. Its output is then low and the buzzer sounds.

The circuit draws a current of about 20 mA (with one of the LEDs lighting). If the buzzer sounds, the current rises to around 40 mA.

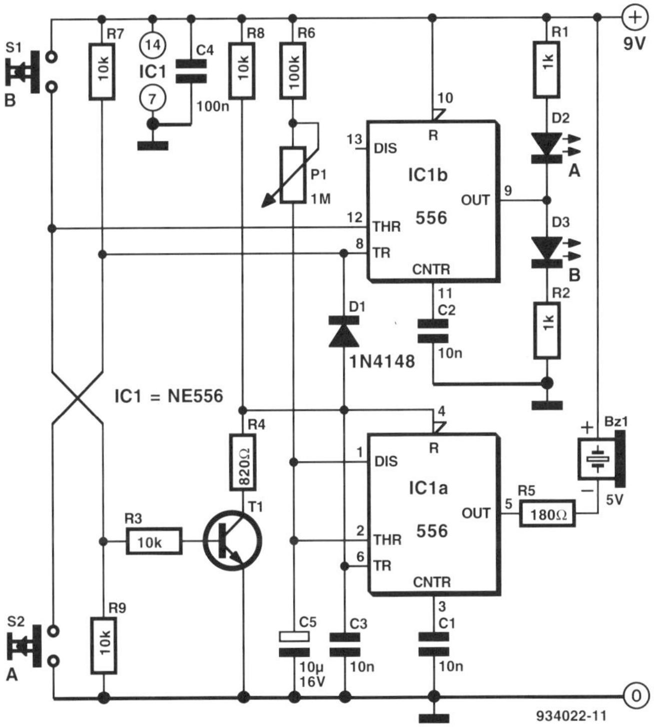

DC-motor regulator

279 The regulator shown in the diagram enables the speed of a d.c. motor to be set over a wide range. Although the most frequently used method for achieving this is pulse-width modulation, in the present design pulse position modulation (also called pulse time modulation) is used. That is, rectangular modulator IC_1 always switches the motor on for 0.5 ms (= R_1C_4). For most motors that is sufficient to start. The speed at which it will finally rotate is influenced by the pulse rate. The pulse spacing, which is the discharge time of C_4, may be set between 1 μs and 14 ms with R_2 and P_1.

T_1 can switch up to 6 A, provided it is fitted on a heat sink of not less than 20 K W^{-1}.

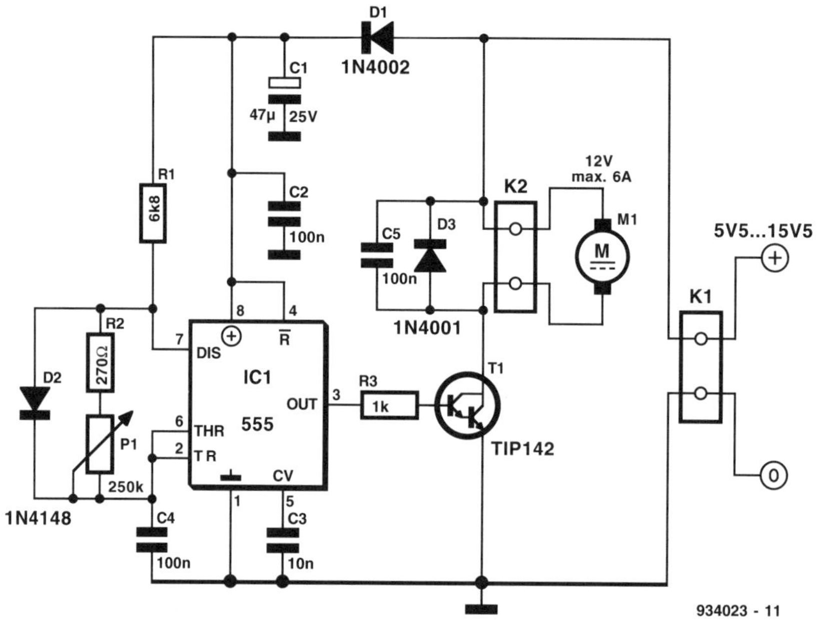

Fog light switching

280 Many cars are not factory-fitted with fog lights and, consequently, there is often no provision for switching optionally added ones. However, since in most European and North American countries it is now illegal to drive on sidelights and many cars still have a switch for these, the fog lights may be operated with this. The only trouble is that the fog lights are then on also when the headlights are on. This difficulty* can, however, be remedied with two relays and two diodes.

When the sidelights are switched on, relay Re_2 is energized and the fog lights are on. When the headlights are switched on, relay Re_1 is energized via one of the diodes. Relay Re_2 is then no longer actuated and the fog lights go out.

Relay Re_1 may be any 12-V type with a break or change-over contact, but for Re_2 a car type must be used owing to the heavy current through the fog lights. Car type relays can normally easily handle 10 A direct current; they are not normally

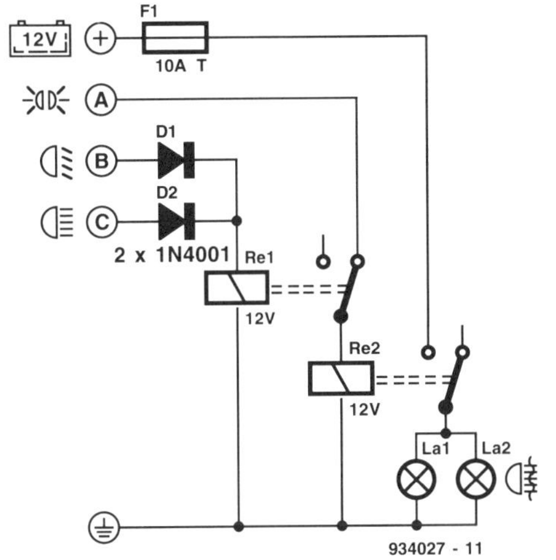

available from electronics retailers, but rather for car accessory shops.

* There is a fundamental difference of opinion between the designer and most car manufacturers, who make it impossible— rightly— for the fog lights to be on when the headlights are not. Note that it is illegal in some countries to drive with fog lights on only; even during daylight fog, headlights (dipped) must be on also. Editor.

Halogen light switch

281 Halogen lamps are rightly popular because they give a good light and have an excellent efficiency. Unfortunately, they tend to be costly. Moreover, they often give up the ghost when they are switched on, because they then draw a very high current (some ten times normal, equivalent to roughly a hundred times their normal power ating). A simple passive circuit can remedy this, but only for DC operated lamps. If they are AC operated, a rectifier could be added, but that gives relative high losses at 6 V or 12 V.

The present circuit takes a different tack: it is based on the fact that a power FET passes a current that is dependent on its gate voltage. If the gate voltage rises gradually, the current will increase gradually also. Here, the gate voltage is determined by the potential across C_1, which is charged slowly via R_1. Slowly means some tens of milliseconds, which is long enough for the filament to warm up.

Note that the FETs specified need at least 6 V gate voltage to switch on. The maximum gate voltage is 12 V, so that the circuit is suitable for 12 V lamps also. The value of R_1 for 6 V lamps should be about 100 kΩ, and for 12 V lamps, about 470 kΩ. Figure 2 shows the effect of the circuit. The lower curve represents the current without limiting: its peak is about 4.5 times the nominal current through the lamp. With the limiting circuit, the lamp current no longer assumes such high values, as shown by the upper curve.

The MOSFET may be any suitable type. The BUZ10 can handle some 20 A, so that it can switch 12 V, 20 W lamps without any problems. In practice, even 50 W lamps may be switched, because the large currents last only a very short time. A Type BUZ11 can handle up to 30 A. Losses are small: a BUZ10 has an on resistance of 0.08 Ω, which at 1.67 A gives a loss of 230 mW. This heats the transistor in free space by about 17 °C. A heat sink is, therefore, not needed.

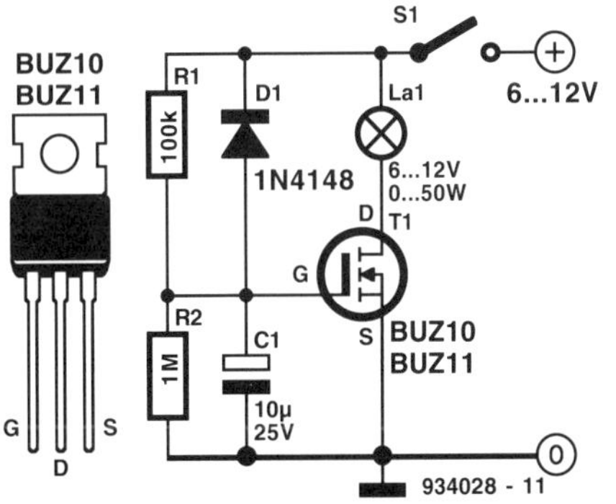

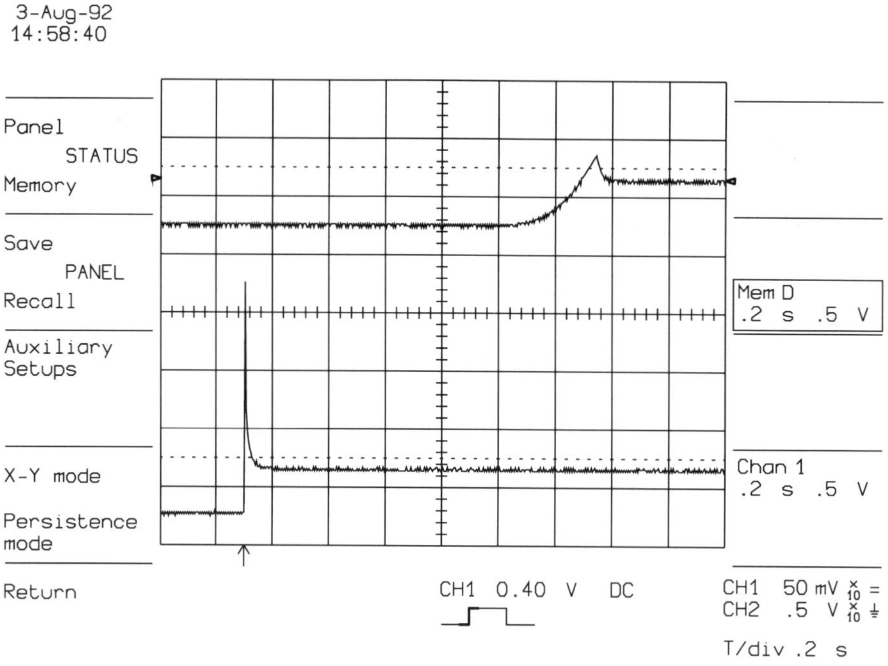

General purpose display decoder

282 Designing a suitable LED display for a particular application can be time consuming. In particular, the decoder that sits between the driver circuit and the display often needs to be made to measure. The main advantage of the circuit presented here is that the same hardware can be used to implement many different types of display decoding. This is achieved by using an EPROM, which most readers will be able to program or have programmed.

The circuit translates a 12-bit input code into suitable drive signals for a 4-digit readout built from 7-segment common-anode LED displays. Four digits (displays) require four bytes in the EPROM. If these bytes are stored at four successive addresses, they can be read by applying binary code 00 through 11 to the two least significant address lines, A0 and A1. Since the four displays are multiplexed, driving the A0 and A1 lines in this way enables codes to be fetched from the EPROM and indicated on the readout. This leaves the remaining address bits of the EPROM free to apply the code to be converted.

As an example, an EPROM is programmed to function as a code translator for an RC5 infrared remote control transmitter. The RC5 receiver described in Ref. 1 may be connected directly to the decoder input connector, K_1, via a short flatcable. An RC5 code consists of five address bits and a 6-bit key code. In this example, the two most significant digits are used to show the address, and the other two to show the key code. The six data bits, the five address bits and the toggle bit are connected to the EPROM address lines. The toggle bit is used to drive the decimal point on the digit at the extreme right, which provides an indication that a key has been pressed on the remote control unit.

The contents of the EPROM are generated with the aid of the Pascal program shown. When analysing the program, bear in mind that the outputs of the IR receiver are active low. This means that counting has to be done

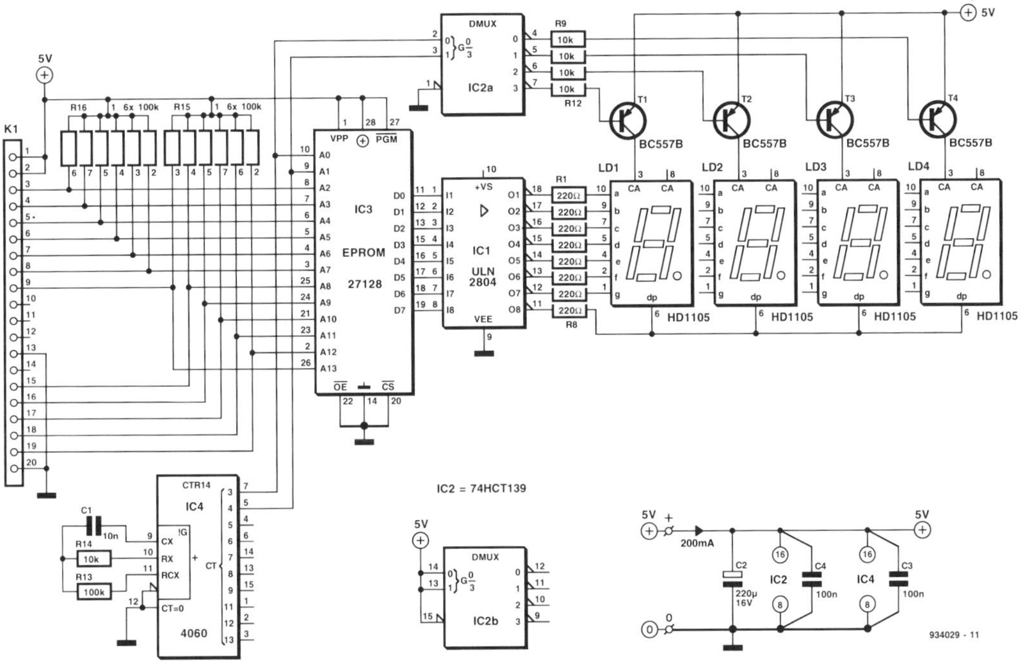

backwards, that is, from 31 to 0 for the addresses, and 63 to 0 for the data, rather than in rising order, 0 to 31 and 0 to 63, respectively. In this way, many decoder variants may be implemented simply by adapting the EPROM contents as required. The EPROM to implement the RC5 code readout is available ready-programmed from Elektor Electronics (Publishing).

Reference:
1. Universal RC5 code infra-red receiver, *Elektor Electronics* January 1992.

Parts list

Resistors:
R_1–R_8 = 220 Ω
R_9–R_{12}, R_{14} = 10 kΩ
R_{13} = 100 kΩ
R_{15}, R_{16} = 6-stage 100 kΩ SIL

Capacitors:
C_1 = 10 nF
C_2 = 220 µF, 16 V
C_3, C_4 = 100 nF

Semiconductors:
T_1–T_4 = BC557B

Integrated circuits:
IC_1 = ULN2804
IC_2 = 74HCT139
IC_3 = EPROM Type 27128 Ref. 6261*
IC_4 = 4060

Miscellaneous:
K_1 = 20-way straight box header
LD_1–LD_4 = HD11050 (Siemens)
Enclosure, e.g. Heddic Type 222
PCB Ref. 934029*

* available from Elektor Electronics (Publishing)

Representative in UK: ElectroValue, 3 Central Trading Estate, Staines TW18 4UX. Telephone (0784) 442253; Fax (0784) 460320.

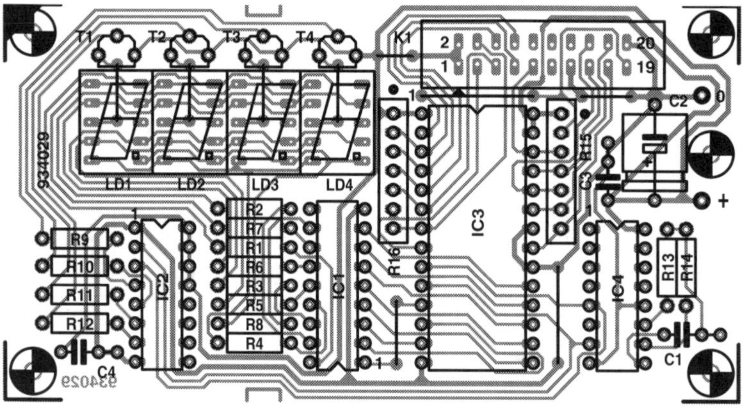
T1
T2
T3
T4
K1
2
20
1
19
LD1
LD2
LD3
LD4
934029
R2
R7
R1
R6
R3
R5
R8
R4
R9
R10
R11
R12
IC1
IC2
IC3
IC4
R13
R14
R15
R16
C1
C2
C3
C4
0
+

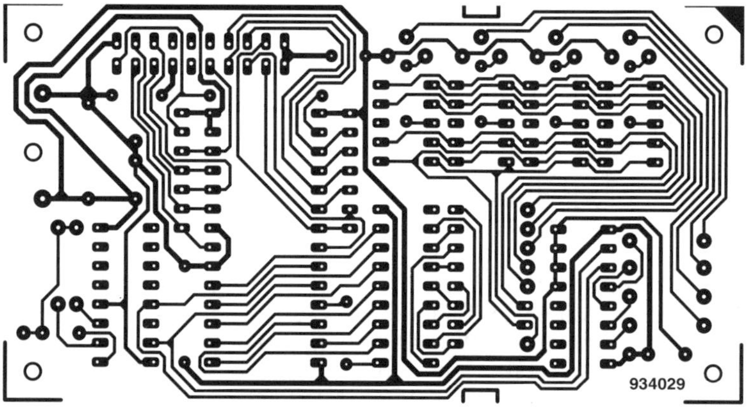
934029

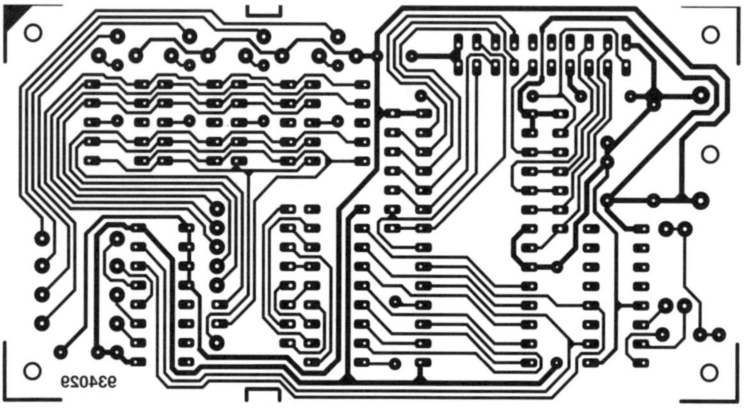

Plant humidity monitor

283 This little circuit monitors the humidity of plant soil, and should be a welcome help for forgetful readers who occasionally find their plants on the window sill suffering from acute drought.

The plant humidity sensor consists of a handful of inexpensive parts, and gives a reliable indication when the soil is too dry. Circuit IC_1 is arranged as an oscillator that supplies two complementary switching signals, Q and Q\, at a frequency of about 58 Hz. Theoreti-cally, the Q and Q\ signals are never active at the same time and thus form a perfect alternating current source that prevents electrolysis on electrodes A and B.

The potential at the wiper of P_1 is a function of the electrode resistance ($R_{A\text{-}B}$), and thus depends on the soil humidity. By comparing this potential with a reference, IC_2 indicates whether the soil humidity falls below a certain level, when the plant needs watering. The reference is derived from the symmetrical electrode current, and is taken from junction R_2-R_3, which is at a constant potential of 2.5 V with respect to ground.

The operation of the comparator is straightforward in spite of the alternating electrode current. A simple d.c. voltage comparison is, therefore, out of the question. Assuming that Q=0 V and Q\=+5 V, the wiper of P_1 is at a potential U_1 with respect to ground. Consequently, the wiper will be at 5–U_1 when the electrode current is reversed (Q=+5 V; Q\=0 V). If, in the first situation (Q=0 V and Q\=+5 V), the wiper potential is, say, higher than the reference, it follows that it is lower than the reference in the other situation. To make sure that the comparator output is actuated in both cases, the input signals are swapped by the Q and Q\ outputs of IC_1. This is achieved with the aid of the four electronic switches contained in IC_3.

The red LED, D_2, lights when the soil is too dry. This happens when U_1 is greater than 2.5 V. Depending on the setting of preset P_1, this corresponds to a soil (inter-electrode) resistance of 0–1.82 kΩ. The further P_1 is turned towards the electrode connection, the higher the soil resistance (i.e., the drier the soil) required for the green LED to go out and the red LED to light. Interestingly, the capacitance of the soil may cause the red and the green LED to light at the same time, which gives a useful indication of 'between wet and dry'.

The electrodes are best made from the carbon rods salvaged from used batteries. This is an inexpensive solution, which also prevents corrosion. The electrodes are pushed into the plant soil about 4 cm apart. The exact adjustment of P_1 will depend on the type of plant to be monitored, and has to be established em-

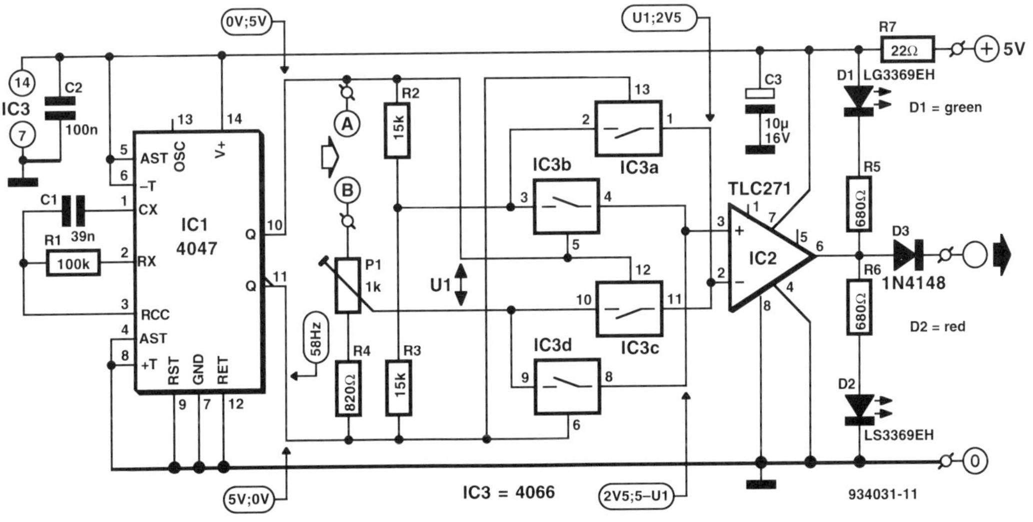

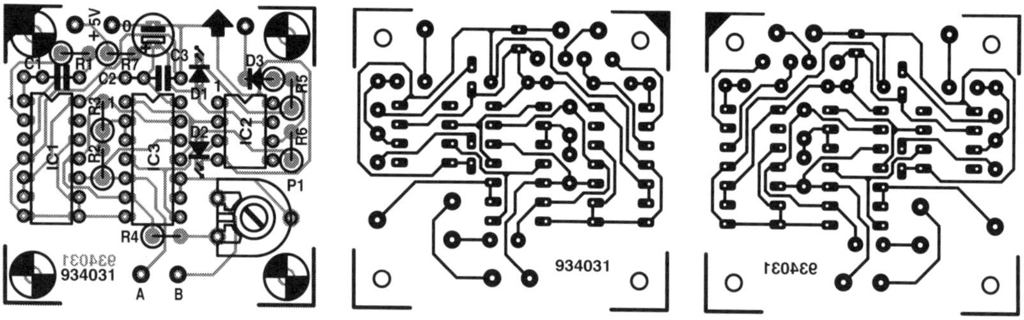

pirically. In most cases, however, good results will be obtained with P_1 set to mid-travel. Since the sensor requires a stable 5 V supply, power is derived from an inexpensive mains adaptor whose direct output voltage is 'cleaned' and stabilized with the aid of an 7805 regulator. As the sensor draws a current of only about 5 mA, the supply shown here may be used to power a number of sensors.

In addition to providing the supply voltage to the sensor units, the power supply also functions as a central (remote) indicator. LED D_1 lights when one of the sensor units connected reports 'dry soil'. If none of the sensors reports 'dry soil', but at least one of them 'between dry and wet', LED D_1 lights at reduced intensity, since the voltage applied to the LED control input is then between 2 V and 3 V. Clearly, this means that all sensor outputs must be connected to the LED control input on the supply (wired-OR configuration).

Parts list (Monitor circuit)

Resistors:
R_1 = 100 kΩ
R_2, R_3 = 15 kΩ
R_4 = 820 Ω
R_5, R_6 = 680 Ω
R_7 = 22 Ω
P_1 = 1 kΩ preset

Capacitors:
C_1 = 39 nF
C_2 = 100 nF
C_3 = 10 µF, 16 V

Semiconductors:
D_1 = LG3369EH * (3 mm, low current, green)
D_2 = LS3369EH * (3 mm, low current, red)
D_3 = 1N4148

Integrated circuits:
IC_1 = 4047
IC_2 = TLC271
IC_3 = 4066

Miscellaneous:
Enclosure 65×50×30 mm, e.g., Bopla EG406†
PCB Ref. 934031‡

Parts list: (Power supply circuit)

Resistors:

R_1, R_2 = 10 kΩ
R_3 = 220 Ω

Capacitors:
C_1 = 100 µF, 25 V, radial
C_2, C_3 = 100 nF
C_4 = 10 µF, 16 V, radial

Semiconductors:
D_1 = 1N4001
D_2 = LED, 5 mm, red
T_1 = BC547B

Integrated circuits:
IC_1 = 7805

Miscellaneous:
Enclosure 65×50×30 mm, e.g., Bopla EG406†
PCB Ref. 934032‡

* Siemens (in UK: ElectroValue at 0784 442253
† Phoenix Mecano at (0296) 398853
‡ Available from Elektor Electronics (Publishing)

Electronic bell-push

284 A solid-state bell-push is far more robust than the usual mechanical type. The touch contact may be made from an audio socket, which has a high insulation resistance and is practically indestructible.
The inverting (–) input of IC_1 is at half the supply voltage via R_4-R_5, while the non-inverting input is at earth potential via R_3. When the contact is touched (less than 10 MΩ, a light touch is fine), the output of the opamp goes high and the relay (a 9-V or 12-V type) is energized. The relay contact then actuates the (existing) bell. Resistor R_1 and capacitor C_1 ensure that the bell cannot be actuated accidentally.
The circuit is powered by the rectified bell transformer voltage (or a second bell transformer – for safety's sake, do not use a different type of transformer). A bell transformer can normally provide a current of 1 A, so the first solution is almost always possible.
The circuit draws a (quiescent) current of only 5 mA if a 741 is used and only 0.5 mA if a TLC271 is used. When the relay is energized, this rises by 30 mA.

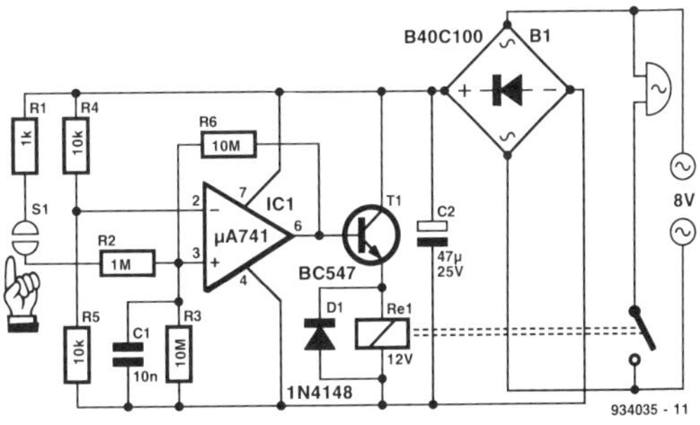

Interior light switch for cars

285 Most cars are provided nowadays with automatic interior lighting that comes on when one of the doors (or the tailgate of a hatchback) is opened. This can have a drawback if one of the doors of the car is not closed properly and the car is left for a few days in the garage.* This may effectively flatten the battery. There is, however, a remedy for this.†
For the proposed circuit, the usual lights switch takes over the control of the interior lights from the door switches. The circuit is based on a 555 (7555, 555C) timer and acts as a monostable that switches on the lights for four minutes as soon as it is switched on. This time may be altered by changing the values of R_2 and C_2. Network R_1-C_1 ensures that the IC is enabled as soon as voltage is applied.
The output of IC_1 controls relay Re_1 via T_1. The only requirement of the relay is that it should be actuated by a voltage of 12 V and a current of less than 200 mA. As soon as the timer switches the lights off, the current drain drops to 6 mA if a 555 is

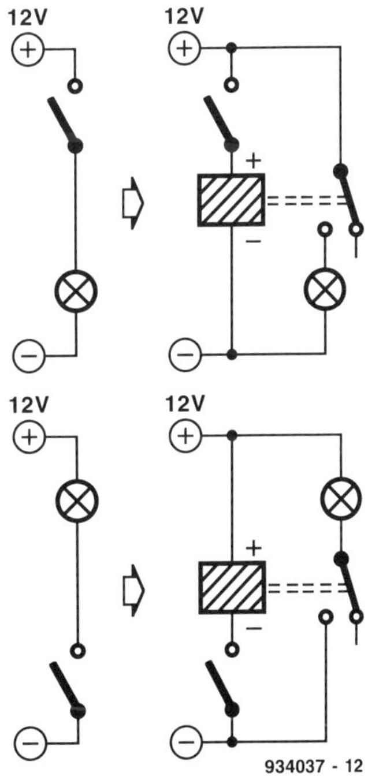

used and to 0.5 mA if a 555C is used.
Before the circuit is built into the car, check how the door switches are connected. If they are in the +ve battery line, use variant *a*; if in the –ve (earth) line, use variant *b*.
Finally, connect the circuit to the junction of the wires from the door switches, otherwise it will react to only one door.

* These days, even in the garage, it is best for security's sake to properly shut and lock all doors and, if fitted, set the alarm. Editor).

† In quite a few modern cars, the circuit described (or one like it) is a standard fitting. Editor.

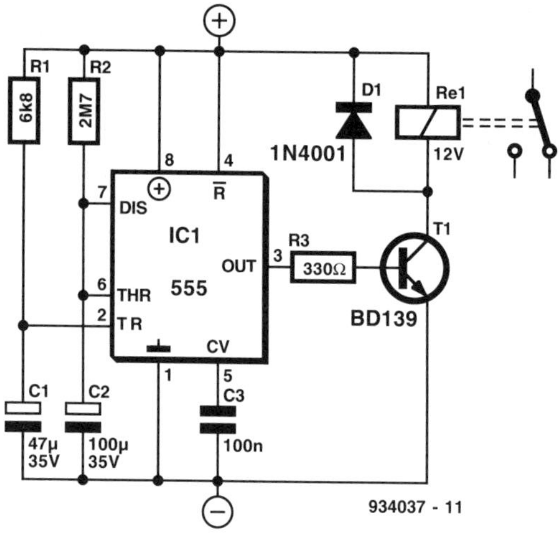

Lock code unit with auto reset

286 The code unit, which operates with a 10-switch (non-matrix) keyboard, was designed to control an electromagnetic doorlock with a four-digit code. It resets automatically after a short period.
The keys that set the code are designated S_1–S_4 in the diagram. Each of them is linked to a bistable (flip-flop) in IC_1. The connections are arranged so that when key S_1 is pressed the set input of the first bistable (pin 4) goes high. The associated Q-output at pin 2 also becomes high. When this has taken place, pressing key S_2 results in the input of the second bistable (pin 6) and the associated Q-output (pin 9) going high. Thereafter, pressing keys S_3 and S_4 has similar results. This means that if keys S_1–S_4 have been pressed in the correct sequence, a logic high level exists at the Q-output (pin 1) of the fourth bistable. If during this process one of the other keys (S_5–S_{10}) is pressed, the set bistables are reset via R_9 to make the code even more difficult to break. When pin 1 of IC_1 is high, transistor T_1 is switched on, whereupon the relay that operates the doorlock is energized.
Before key S_1 is operated, the Q-output of the first bistable is low, so that its reset input is held low via R_1, D_1 and R_5. The potential across C_1 is then almost equal to the supply voltage. When S_1 is pressed, pin 2 goes high. Because of D_1, the capacitor will discharge slowly via R_5 and R_8. After about seven seconds, the level at the reset inputs of all four bistables, for the present purpose, becomes logic high. If at that instant the relay is actuated, it will be de-energized. If the complete code has not been keyed in, a fresh start must be made. The reset time may be lengthened by increasing the value of capacitor C_1.
A keyboard with more or fewer switches may be used. Fewer than six makes breaking the code easy, however. If more than 10 switches are available, simply wire the additional ones in parallel with S_5–S_{10}.
The code may be enlarged to eight digits by connecting a second 4043 in series with IC_1. Pin 1 is then linked to key S_5 and the following three keys are connected to the second 4043 in the same way as S_2–S_4 to IC_1.
There is a slight drawback that has to be borne in mind: when switches S_1–S_4 are pressed simultaneously, the lock will open. This may be prevented by (a) making one of the keys S_5–S_{10} a reset switch, and (b) not choosing adjacent keys for S_1–S_4.
In quiescent operation, the circuit draws a current of about 10 mA.
The relay should be a 12 V type whose coil draws a current not exceeding 100 mA.

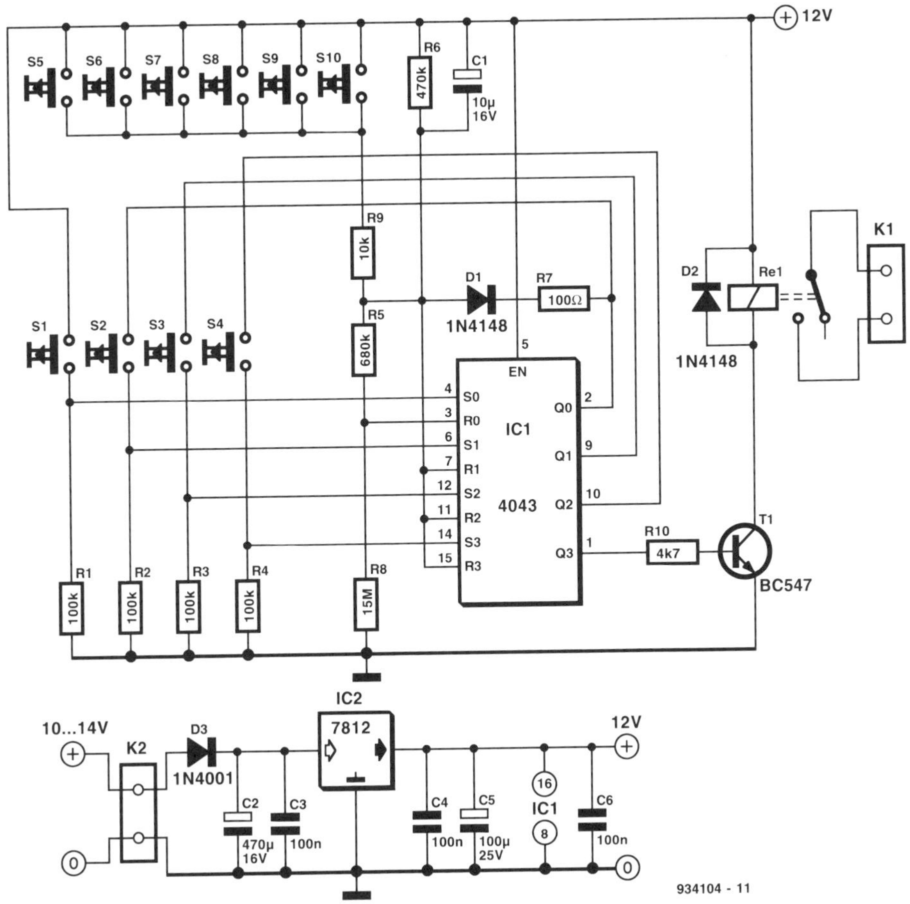

PCB drill control

287 The design presented enables the speed of an electric PCB drill to be controlled with only one preset. As an aside, the rotary direction of the drill may be reversed. When the preset is at the centre of its travel, the drill stops.

The power supply delivers (immediately after the bridge rectifier) a voltage of about 12 V for the drill, and, by means of resistors R_{19} and R_{20},and zener diodes D_6 and D_7, a symmetrical voltage of ±5 V for the electronic circuits. The ±5 V voltage is smoothed by C_6 and C_7. Diodes D_4 and D_5 prevent the motor of the drill being connected to the ±5 V section.

Speed control is effected with P_1, while rotary direction is determined by comparator IC_{1a}. This opamp ascertains whether the preset is to the left or to the right of its centre. On the basis of this, relay Re_1 is energized via T_1 or not, and this decides whether the drill turns clockwise or anti-clockwise.

The speed control operates with pulse width modulation. The control (direct) voltage set with P_1 is applied to D_2 and IC_{1b}-D_3. This latter combination functions as a rectifier. Whether the voltage at the wiper of P_1 is negative or positive, it will appear as a positive potential (minus the 0.6 V forward bias of the diode) at the junction of D_2 and D_3.

The modulator is formed by IC_{1c}. This stage is designed as a rectangular-wave

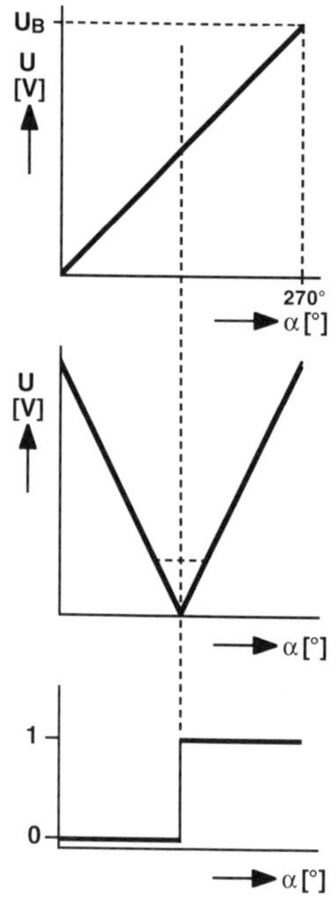

934101 - 12

generator, which causes C_4 to be charged and discharged continuously via R_{11}. The setting of P_1 determines (via R_9) in conjunction with R_{10} how much additional direct voltage is applied to C_4, and thus the pulse/spacing ratio at the output of IC_1. Inverter IC_{1d} enhances the edges of the signal, which are thereupon used to switch the power FET (T_4) via T_1.

When the drop across R_4, which is in series with the drill motor, rises to slightly more than 0.6 V (that is, when the current through the motor is about 5 A), T_3 will switch on, whereupon T_5 will reduce the pulse width slightly.

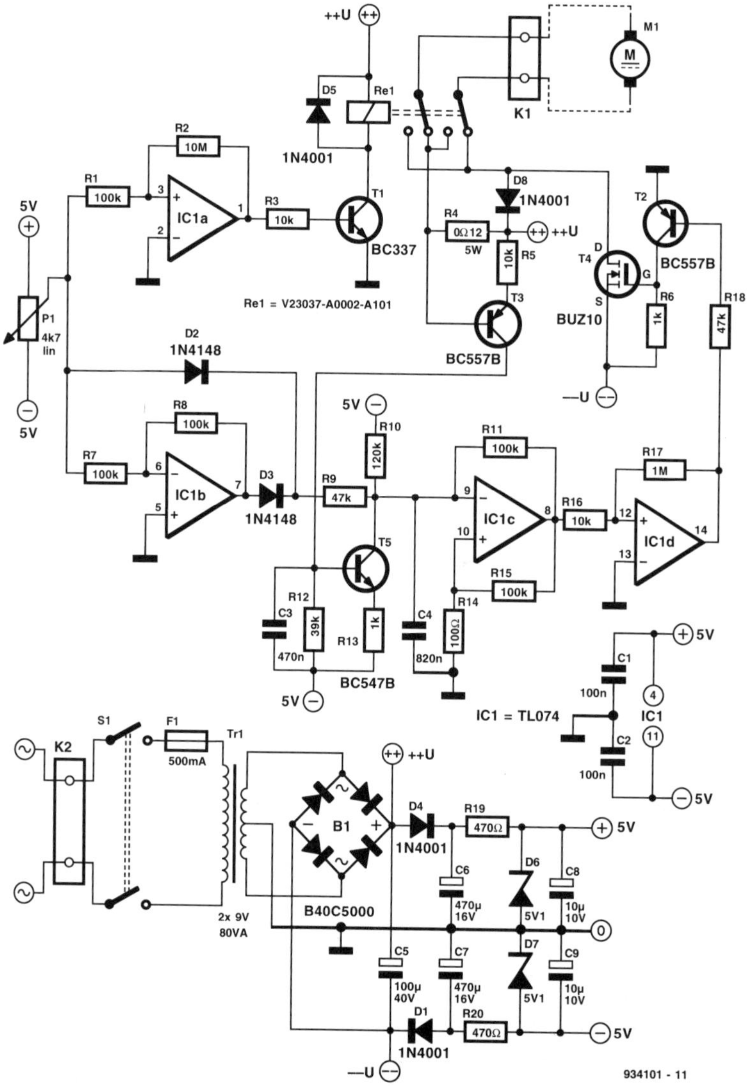

934101 - 11

Automatic twilight switch

288 This is one of the simplest twilight switches ever published in this magazine. When it gets dark, the value of light-sensitive resistor R_1 increases, whereupon T_1 switches off. Transistor T_2 then switches on and this energizes relay Re_1. At the same time a voltage drop of about 1 V develops across R_4: this is the hysteresis of the switch.

Capacitor C_1 serves to make the switch insensitive to brief changes in ambient darkness, such as caused by a passing car with its headlights blazing.

The only requirement on the transistors is high current amplification, which means the use of C types.

The varistor is a new type (Piher) that is environment-friendly (since it contains no cadmium) and comparable in size to the head of a match. If another type is used, its daylight resistance should be of the order of a few hundred ohms; this should increase to about 10 kΩ at twilight. In any case, the value of P_1 may be increased (within reason). During calibration, unsolder C_1 from earth: the circuit then reacts faster.

The relay should be a 12 V type that needs an energizing current ≤50 mA.; its contact should be able to switch 8 A. The load current, however, should not exceed 4 A. When they are switched on, most lamps, and certainly halogen types, draw a very large current. Keeping the load current down ensures a long life of the relay contacts.

The circuit draws a current of not more than 5 mA plus the relay current.

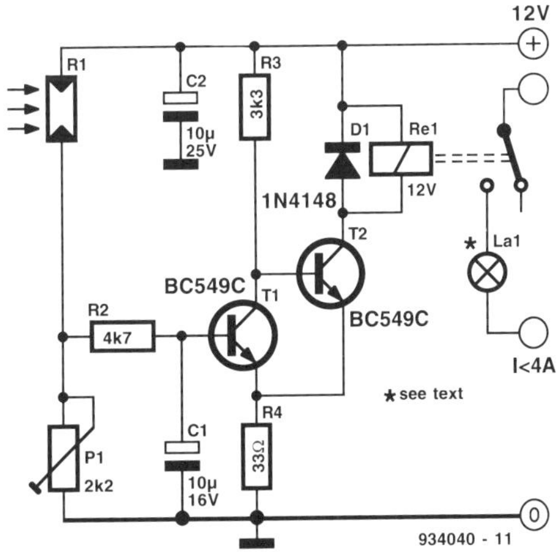

Long-duration timer

289 The timer, based on EXAR's Type XR2240, is intended for use in photographic and printed-circuit departments, as a standby switch for TV and radio equipment, as a battery charging monitor, and similar applications. The timing can be set between 15 minutes and 24 hours 45 minutes with two BCD (binary coded decimal) switches.

The timing is programmed in steps, *T*, which are determined by the time constant $R_1(C_5 + C_6) = 15$ minutes. The BCD switches allow up to 255 such periods to be selected. The timer functions are controlled by an internal bistable (flip-flop) and switches S_1 and S_2. To keep the period accurate, C_5 and C_6 must be carefully selected.

Diode D_1 lights when the timer is switched on. The voltage at junction D_1-R_2-R_3-R_8 is then so low that T_1 remains off. When the set time has elapsed, pin 10 of IC_2 becomes logic high, whereupon the supply voltage exists at the junction. The LED then goes out, T_1 conducts and relay Re_1 is energized, so that its contact changes over.

The supply voltage to the timer is stabilized by regulator IC_1. Capacitors C_1–C_4 smooth the supply voltage to prevent the timer being triggered by pulses on this voltage.

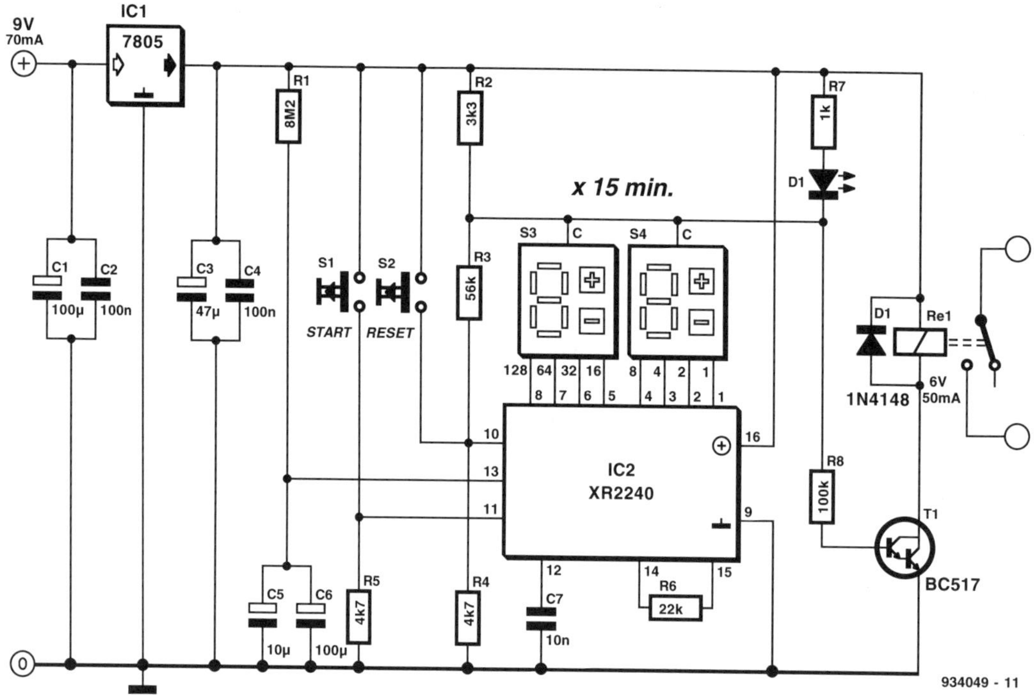

Power booster for slide projectors

290 The lamp in slide projectors is normally controlled with the aid of a triac in series with it. This method reduces the brightness of the lamp appreciably, since the triac drops about 1.5 V, which is more than 6% of the nominal lamp voltage. To compensate this loss, the mains voltage to the projector should be increased by that percentage. The transformer and ventilator can cope easily with that kind of increase, and so can any electronic circuits, since these are invariably powered via a regulator.

The a.c. voltage to the projector may be increased by connecting the secondary winding of a mains transformer in the live wire of the mains. The voltage developed across this winding is then added to the mains voltage (provided that the transformer is connected correctly as regards phase).

The voltage the secondary winding must deliver is calculated as follows. The current through the winding depends on the number of projectors and their rating. Assuming there are four, each rated at 250 W (power consumption: 300 W), the total power consumption during normal operation is 1200 W. With a mains voltage of 240 V, the current will be 5 A. Since the projectors do not all use full power simultaneously, the secondary winding may be rated at 6 A. The secondary voltage, U_s, should be the ratio of the loss across the triac and the nominal lamp voltage times the mains voltage, that is,

$$U_s = 1.5/24 \cdot 240 = 15\ \text{V}.$$

Fuse F_1 should be rated at 1.25×the maximum current of all projectors. Assuming the earlier stated values, that is,

$$I_{fuse1} = 1.25 \cdot 1200/240 = 6.25\ \text{A},$$

or, rounded off to the next standard value, 6.3 A (delayed action). The rating of fuse F_2 is calculated from

$$I_{fuse2} = 1.25 \cdot U_s \cdot I_s / 240 = 487 \text{ mA},$$

or, rounded off upwards, 500 mA.
The circuit should preferably be fitted in a man-made fibre enclosure.
If the output voltage is lower, rather than higher, than the mains voltage, the connections to either the primary or the secondary of the additional transformer must be interchanged.

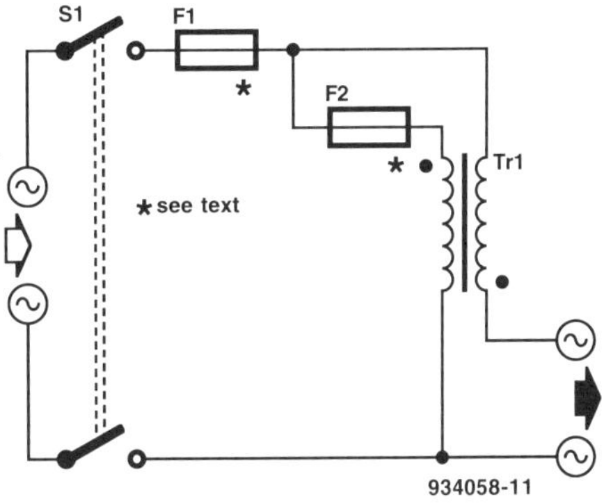

4-bit random generator

291 The generator is based on a Type 4006 quadruple shift register, IC_1. Two of the shift registers are four bits long, the other two, five bits: a total of 18 bits. Only the clock is common to the registers: all inputs and outputs are separate. Of the four-bit registers only the fourth bit is available as an output; of the five-bit ones, the fourth and fifth bit.
Feeding back the outputs of the four registers to the inputs results in a (pseudo) random generator. The outputs of the four XNOR gates serve as the outputs of the generator. The advantage of this design is that the four bits do not only form a random sequence of numbers, but also that each of them has a different pattern. This is in contrast to the shift register itself, where the levels are shifted from one output to another. This generator may, therefore, be used to produce random four-bit numbers or four random digital signals that are not related.
The BASIC program in Fig. 2 enables the operation of the shift register to be simulated. Line 140 states how long the register is (MAX) and how many output bits there are (BMAX). On line 150 all bits of the 'shift register' (array A) are set to zero. If, as is usual in practice, a random starting state is required, this line will have to be rewritten. On lines 190–220 the XNOR functions may be seen. Any changes in the way these gates are connected should be indicated here. In that case, it may also be necessary to change lines 290 and 300.
The program contains only 390 steps (the monitor screen is then full) of the 2^{18} = 262 144 possible.
In the diagram, a clock generator in the form of a Schmitt trigger (IC_{3a}) has been added. It is, however, possible to use any suitable clock at the clock input.

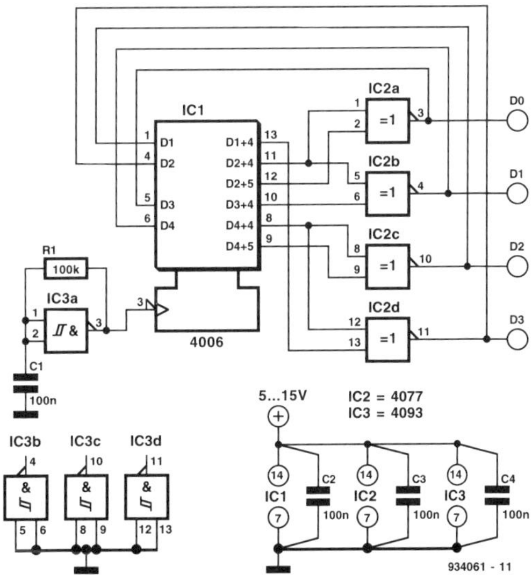

```
100 CLS : PRINT "Random Number Genera-
    tor Using 4006 Shift Registers."
140 KEY OFF: MAX = 18: BITS = 4: BMAX =
    2 ^ BITS
145 DIM A(MAX), B(BMAX): TIME = 0
150 FOR CNT = 0 TO MAX:  A(CNT) =
    0: NEXT CNT: REM CLEAR ALL
155 FOR CNT = 0 TO BMAX: B(CNT) =
    0: NEXT CNT
160 FOR CNT = 1 TO 390
170 TIME = TIME + 1: GOSUB 240: REM
    INCREASE TIME AND SHOW VARIABLES
180 FOR BIT = MAX TO 1 STEP -1: A(BIT)
    = A(BIT - 1): NEXT BIT: REM SHIFT
190 A(6) = 1 - (A(4) XOR A(5)): REM CAL-
    CULATE INPUT NEXT REGISTER (XNOR)
200 A(10) = 1 - (A(4) XOR A(9))
210 A(15) = 1 - (A(13) XOR A(14))
220 A(0) = 1 - (A(13) XOR A(18))
230 NEXT CNT
235 END
240 PRINT CHR$(11): PRINT : REM
    REPOSITION CURSOR
250 PRINT "DATA [  .  .  .  .  .  . SHIFT
    REGISTER  BITS   .  .  .  ]"
260 FOR BIT = 0 TO MAX: PRINT USING
    "###"; BIT; :    NEXT BIT: PRINT
270 FOR BIT = 0 TO MAX: PRINT USING
    "###"; A(BIT); : NEXT BIT: PRINT
280 PRINT "NUMBER OF SHIFT ACTIONS : ";
    TIME: PRINT
290 PRINT "RESULT, BINARY:   "; : PRINT
    USING "##"; A(0); A(6); A(10);
    A(15);
300 PRINT "           DECIMAL: "; : RESULT
    = 8 * A(0) + 4 * A(6) + 2 * A(10) +
    A(15): PRINT RESULT
310 B(RESULT) = B(RESULT) + 1: IF LAST
    = RESULT THEN B(BMAX) = B(BMAX) + 1
320 PRINT "RESULT   "; : FOR BIT = 0 TO
    BMAX - 1: PRINT USING "####"; BIT;
    :     NEXT BIT: PRINT
330 PRINT "QUANTITY"; : FOR BIT =
    0 TO BMAX - 1: PRINT USING "####";
    B(BIT); : NEXT BIT: PRINT
340 PRINT "NUMBER OF IDENTICAL FOLLOWING
    RESULTS", B(BMAX); : LAST = RESULT
350 FOR NUM = 0 TO 2 * TIME: PRINT
    CHR$(28); : IF NUM MOD 60 = 1 THEN
    PRINT
360 NEXT NUM: PRINT " "; CHR$(RESULT
    + 48 - 7 * (RESULT > 9));
370 RETURN
```

934061 - 12

PWM motor control

292 Pulse width modulation (PWM) is ideal for controlling small electric PCB drills that draw currents of up to 2 A. Larger currents are possible if T_2 gets addtional cooling and the value of C_2 is increased. The design also makes use of the fact that the drill works with a small d.c. motor whose rotary speed is a function of the voltage across it.

The design (Fig. 1) is based on astable multivibrator IC_{1a}, whose output is low for a period determined by R_1 and high for a period set by R_2 and P_1.

When C_1 is discharged, the level at the input of IC_{1a} is below the lower threshold, so that the output of this stage (pin 2) is high. The capacitor is then charged rapidly via D_1 and R_1, and reaches the upper threshold in about 1.5 ms. The output of IC_{1a} then goes low, whereupon C_1 isdischarged via D_2, R_2 and P_1. In the prototype, the discharge time could be set between 0.2 ms and 25 ms. This means that the duty factor of the output signal may be varied between 5% and 90%.

The signal is inverted again and then applied to the base of T_1. Transistors T_1 and T_2 are switched on, and the motor is energized, during the negative period of the output pulse of IC_{1b} (pin 4). When the resistance of P_1 is at a minimum, the rotary speed of the drill motor is at a maximum.

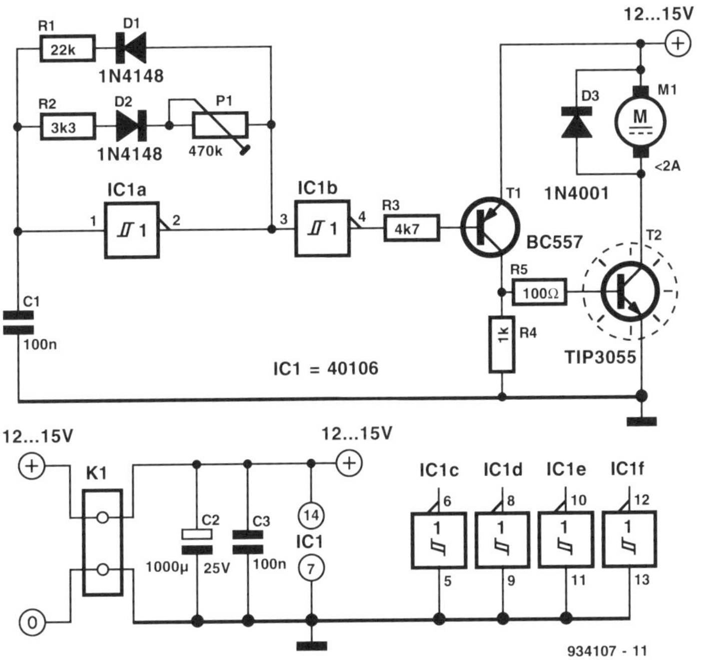

Small power converter

293 The converter enables an existing positive supply voltage to be raised, to be lowered, or to be changed into a negative potential.

The new voltage is electrically isolated from the source by a DIY transformer wound on a G2-3FT12 toroid. The primary winding consists of 30 turns. The number of secondary turns, n, is calculated from

$$n = 30U_o / U_i,$$

where U_o is the wanted voltage and U_i is the input voltage. Increase the number of turns so found by 10–20 to compensate for losses. If the output voltage is somewhat too large, it can always be reduced with P_1. Both windings may be wound with 0.3 mm dia. enamelled copper wire. Make sure that the turns of both are evenly distributed along the core.

The transformer is driven by a CMOS Schmitt trigger NAND gate that has been converted to a rectangular-wave generator by R_1 and C_1. MOSFET T_1 serves as output stage. Additional charging current for C_1 is provided via R_2 and P_1, which control the duty factor of the rectangular signal. The frequency of this signal is about 220 kHz and its duty factor must be smaller than 0.5.

When T_1 is switched on, some energy is transferred to the secondary winding and some is stored in the magnetic field. When T_1 stops conducting, the energy in the magnetic field is transferred to the secondary winding.

The idea is to make the duty factor small enough to ensure that all energy stored in the field is transferred before T_1 switches on again. If not, there is the risk that the residual magnetic field becomes stronger and stronger, which causes the core to become saturated and that in turn lowers the efficiency. Also, owing to the reduced inductance at the primary side, the current through T_1 will rise appreciably, which means the end of the transistor.

The current through T_1 will also rise dangerously when the secondary load is too heavy. The average current through the primary should not exceed 150 mA (the peak current may be several times larger). From the turns ratio it is easily calculated what the maximum secondary load current should be. With the ratio shown in the drawing (10% extra turns at the secondary side), the secondary load should be not smaller than 80 Ω,

Apart from a heavy load, no load is also to be avoided. In that case, the energy stored in the magnetic field can only move to C_2 where it is stored in the electric field. This means that the charge on C_2, and thus the voltage across it, increases to a level where it can have a serious effect on the circuit to which the converter will be connected. As a rule of thumb (just as with the maximum load current), the indicated value of 1.5 kΩ is directly proportional to the turns ratio.

Type 1N4148 rectifier diodes are fast enough to cope with the frequency of 220 kHz (1N400x types are not). These diodes can stand a constant current of 200 mA (400 mA peak).

The efficiency of the convertor with a supply voltage of 15 V is about 65%. When the load current is small, this drops to about 50%. The efficiency also drops when the supply voltage is lower than indicated.

The maximum input voltage is 15 V, since the supply to neither IC_1 nor T_1 must exceed this value. The current drawn from a 15 V source by a load of 80 Ω is about 165 mA.

The prototype worked fine with the wiper of P_1 turned completely to +; it may even be possible to lower the value of R_2 slightly.

If necessary, the duty factor of the rectangular voltage may be increased somewhat with P_1. While this is being done, the current through T_1 should be watched carefully, preferably on an oscilloscope. If the current suddenly rises too fast, the core is becoming saturated and this means that P_1 must be turned back slightly. Bear in mind that when P_1 is set to a critical position, a slight change in load can drive the core into saturation, with all the consequences of this.

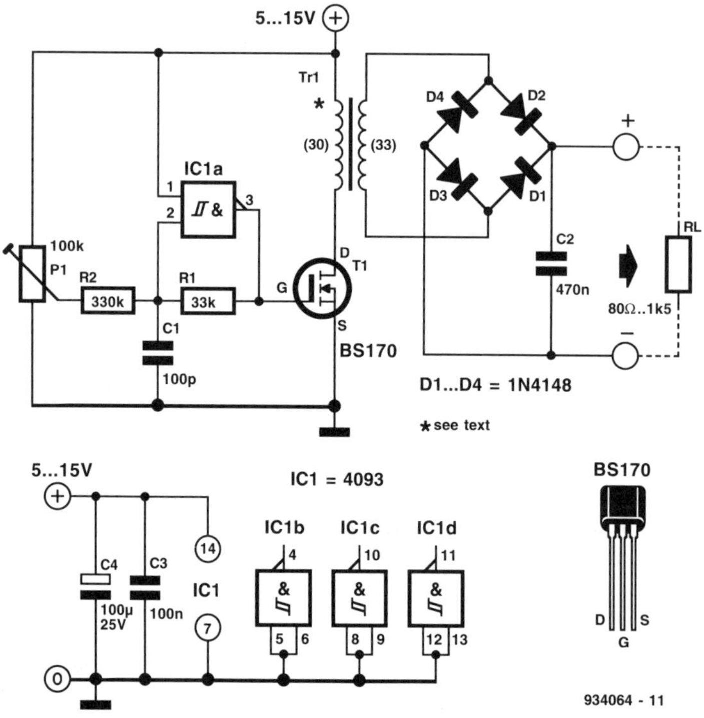

Multi-colour-led

294 It had to come: an LED that can produce all visible colours. It is the Everlight Type 339-1 VRKG-BBW. In fact, it consists of four LEDs in one case: one red, one green, and two blue.When these LEDs are driven into varying brightnesses, all visible colours and white can be produced. The circuit described makes the LEDs light in all colours in any given order.

The circuit consists of an integrator followed by a Schmitt trigger. Together these form an oscillator that produces a triangular voltage with an amplitude of about 1.5 V_{pp} at its pin 1. This signal is applied to one of the LEDs via T_1 and current limiting resistor R_5.

Three of these circuits should be built: one for the red, one for the green, and one for the two blue LEDs. Each circuit should, however, have a different value C_1, say, 470 nF, 330 nF, and 220 nF. Each of the blue LEDs, although connected to the same circuit, should have its own series resistor.

The direct voltage level of the triangular signal may be shifted with P_1. Start with the wiper at earth, and then turn the control very slowly till the LED just begins to light. The best setting is when the LED lights for two thirds of the time and is off for one third.

The value of the series resistors is low, because the sensitivity, especially of the blue LEDs, is low.

Do not turn P_1 to far, but take care that the current through the red and green LEDs remains below 30 mA; the blue ones can draw up to 40 mA. The average current drawn by the circuit is then about 70 mA.

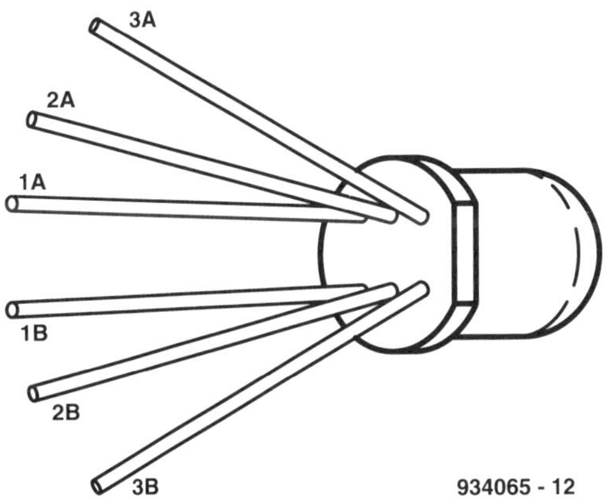

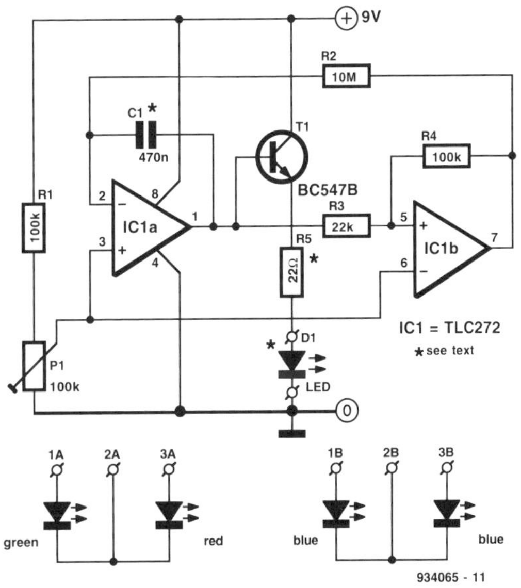

Temperature monitor

295 The monitor is intended for use in relatively small spaces (like living rooms). The sensor is a readily available LM335 from National Semiconductor. This gives an output voltage of 10 mV °C^{-1}. This voltage is compared in IC_{1a} and IC_{1b} with two reference voltages. One of these is preset with P_1 and the other with P_2. The output of the comparators is used to switch D_2, D_3 and D_5.

When the voltage produced by D_1 is smaller than either of the reference potentials, the outputs of IC_{1a} and IC_{1b} are low: D_2 will then light.

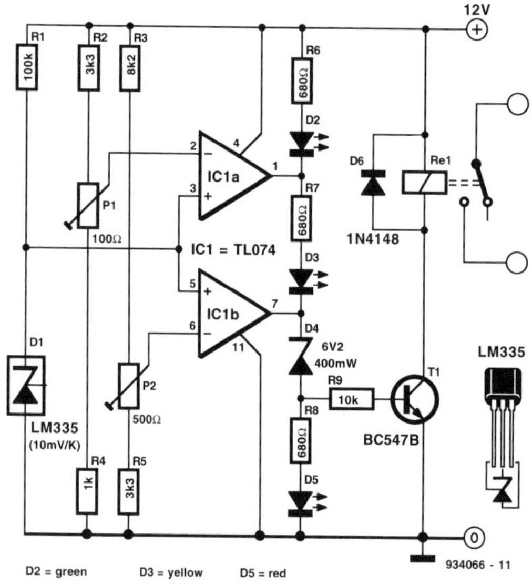

When the ambient temperature rises, the output of D_1 rises proportionally. When the level of the sensor output lies between the two reference levels, the output of IC_{1a} is high and that of IC_{1b} is low. Diode D_3 will then light, showing that the critical temperature has been reached.

At even higher temperatures, the output of IC_{1b} will also go high and D_5 lights, while the other two LEDs go out. At the same time, the relay will be energized via T_1. The relay contact may then actuate an external load (e.g., a buzzer).

Zener diode D_4 ensures that T_1 does not come on when D_3 lights (since the output voltage of IC_{1b} then rises slightly owing to the current through this IC).

The temperature at which the LEDs should light may be set with P_1 and P_2. Bear in mind, however, that the monitor is intended for 'normal' temperatures between 25 °C and 100 °C.

The monitor draws a current of about 20 mA; when the relay is energized, this rises to about 50 mA.

Zero suppressor for ICl7106

296 Many people are irritated by the superfluous zeros that precede a number on, for instance, an liquid crystal (LC) display. To them '1' is clearer than '001'. In systems where the ICL7106 display is used, this irritation may be removed by the present suppressor.

Both the backplane (BP) and the segment of an ICL7106 are driven by a rectangular signal. When the segment is actuated, the drive signals are in antiphase; when the segment is inactive, the display remains dark because the drive signals are in phase.

Suppressing superfluous zeros is, therefore the fairly simple matter of finding these zeros and inverting the signal that drives the segment.

A $3^1/_2$ digit display has three complete digits numbered 1, 2, and 3. The fourth (half) digit is driven via output AB and can only show a '1'.

Whether digit 3 has a leading 0 is easily found out, because if digit 4 is not used (AB is off) and at digit 3 segment **g** is off while segment **e** is on, there is a leading 0 on the display. In that case, the drive signal to the segments for digit 3 may be inverted with XNOR gates, whereupon the digit goes out.

In the case of digit 2, it must be checked whether digit 3 is off and whether a 0 is displayed via digit 2. The latter may be determined from the state of segments **e** and **g**. If it is the case, the drive signals for digit 2 must be inverted.

Opamp IC_{9a} buffers the level of the test output so that this may serve as the earth level for all connected logic circuits.

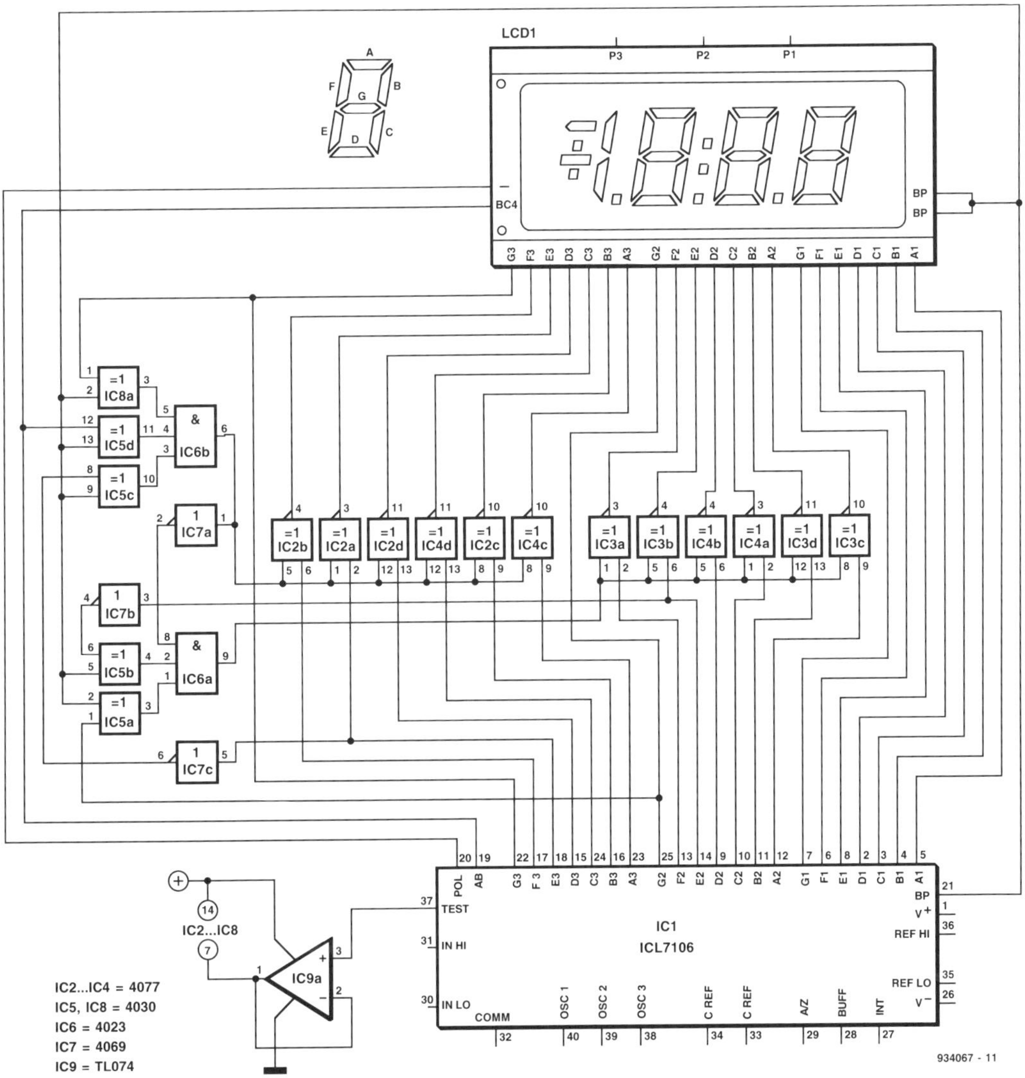

Telephone monitor

297 It is sometimes necessary to make a telephone conversation audible to more than one person at either end or to make a tape recording of it (as in many businesses). Often, this is done with the aid of a microphone fitted with a suction cup. This does not work very well, though, hence the present circuit.

Use an old, but still serviceable, telephone set connected across the telephone line in parallel with the receiver to be monitored. Connect a length of screened audio cable across the receiver inset and to an amplifier or tape recorder, as the case may be. As soon as a conversation has to be recorded or amplified, take the receiver from the hook and adjust the input level of the amplifier or recorder as required. Do not set the amplifier gain too high, because this may give rise to howling.

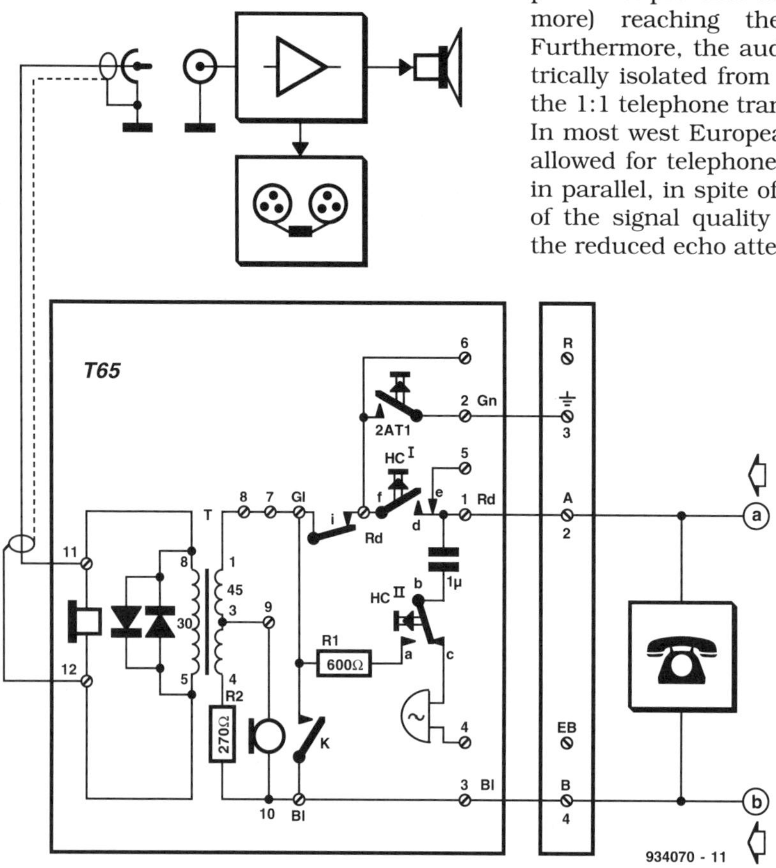

The hook contact of the additional telephone set prevents the bell voltage (50 V or more) reaching the audio equipment. Furthermore, the audio equipment is electrically isolated from the telephone line by the 1:1 telephone transformer, T.
In most west European countries it is now allowed for telephone sets to be connected in parallel, in spite of the resulting paucity of the signal quality (primarily caused by the reduced echo attenuation).

Car interior light delay

298 The delay circuit ensures that a car's interior light remains on for about five seconds after all the doors have been closed. This enables the ignition lock to be found easily in the dark (many modern cars have an illuminated ignition lock as standard).
When one of the front doors of the car is opened, the door switch (here, S_1) closes. A current will then flow through interior light L_1 via S_1 and D_2. At the same time, capacitor C_1 is charged rapidly via R_1, D_1 and S_1. When the door is closed, S_1 opens. At the same instant, T_1 is switched on by the low level at the –ve terminal of C_1. After a little while, depending on the value of C_1 and

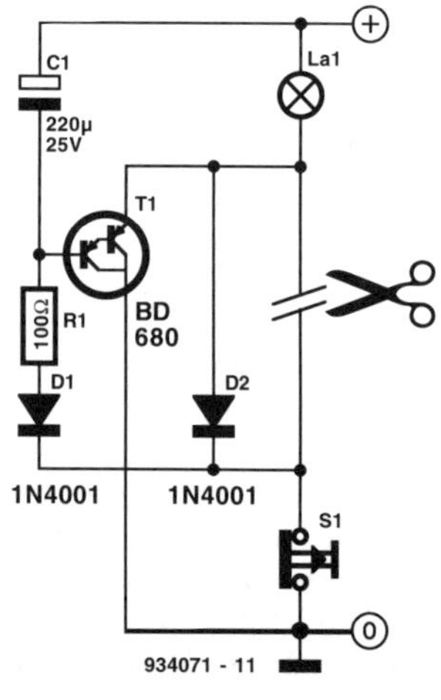

the base current of T_1, the base voltage of T_1 has risen to a level where the transistor ceases to conduct, so that L_1 goes out.

Note. In quite a few modern cars, the circuit described (or one like it) is a standard fitting. [Editor]

Angler's bait indicator

299 The indicator shows when a fish has taken the bait. Until that happy moment (at least for the angler), the bite sensor is open, so that both LEDs are out. When a fish has been hooked, the sensor closes the current loop, whereupon both LEDs light. In addition to these diodes, a buzzer may be used to give an audible signal. D_2 continues to light, even when the sensor has been reopened, for as long as the thyristor has no hold current. In this way, each tug on the hook line is indicated by D_1.

The circuit draws a current of 20–60 mA, depending on the type of LED and whether a buzzer is used.

The sensor is made from a mercury switch, cast, half a metre of stranded wire and a 3.5 mm mono jack socket. The cast is placed in the line directly under the reel. When the line is tightened, the indicator is actuated.

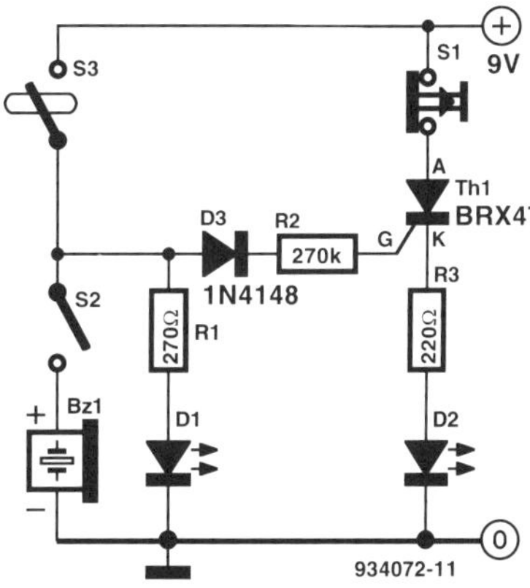

Variable differential amplifier

300 In a traditional differential amplifier with feedback to the –ve (inverting) input of an operational amplifier and a potential divider at the +ve (non-inverting) input, it is tricky to make the amplification variable. This is because the feedback network and the potential divider need to be each other's image to ensure satisfactory suppression of the common-mode signals (common-mode rejection ratio – CMRR). This means that there must be two variable resistors and these must at all times be equal.

This is not necessary with the present circuit in which one preset can arrange the amplification without affecting the CMRR. It has the following transfer function:

$$U_o = 2R_2/R_1(1+R_2/P_1)(U_2-U_1) \quad [V]$$

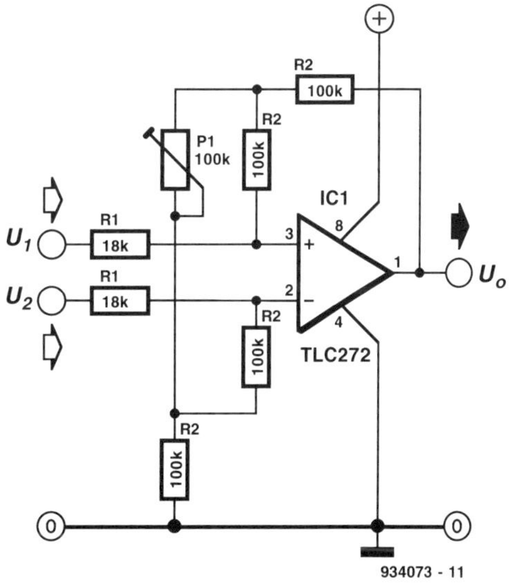

It is clear from this that only P_1 affects the difference voltage U_2–U_1. The common-mode signal does not appear in the function. This is partly because P_1 has no effect on it and partly because the resistors with the same circuit reference also have the same value. In theory, that gives total rejection of common-mode signals; in practice, tolerances of the components used will determine the CMRR. To calculate that, however, the formula will have to be expanded appreciably.

Basic timer [II]

301 The timer is based on the well-known display driver LM3915, which has a logarithmic scale. This may seem strange, but is not because a simple RC network is used for time measurements. The voltage across this network, R_6-P_1-C_1, has, while C_1 is being charged, an exponential character ($U = U_{ref}e^{-t/RC}$). When this is applied to IC_1, which takes its logarithm, the time is shown linearly on the display. The values of P_1 and R_6 have been chosen to enable the timer showing periods of 1–15 minutes. The timer is reset (C_1 is discharged) when S_2 is closed. Resistor R_1 limits the peak value of the discharge current from C_1 to an acceptable value for the switch contacts.

Resistor R_5 compensates the leakage currents of the electrolytic capacitors to obviate the risk of D_{10} going out.

If pin 9 of IC_1 is left open, the display is in the dot mode. The elapsing of time is then shown by only one LED lighting. After a reset, D_1 will light first and then, sequentially, all the other LEDs.

The total current drawn by the timer is 20 mA, so that battery supply is possible.

When pin 9 is connected to pin 3 (+ve supply line), the display is in the bar mode. After a reset, all LEDs will light, after which they will go out in turn, starting with D_1. In this mode, battery supply is not recommended.

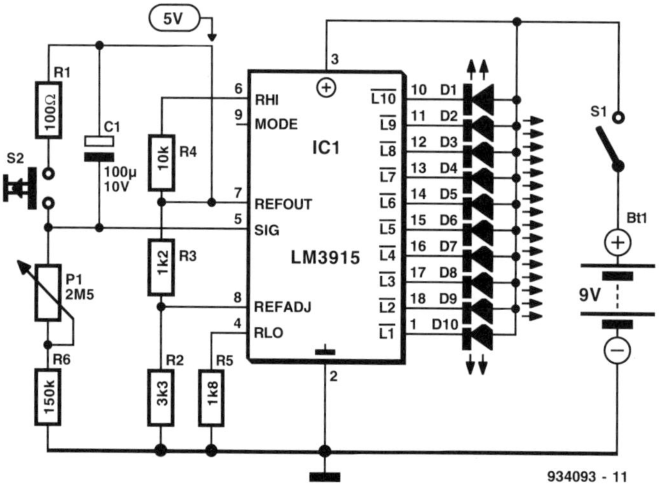

Privacy phone

302 Undesired listening-in to telephone conversations may be prevented by the present circuit, which requires only two components to protect a telephone.
In each telephone to be protected, a triac is inserted in series with the *a* line. A diac is connected between the anode and the gate of the triac. The trip voltage of the diac is about 25 V, which is appreciably lower than the line voltage of about 50 V, and considerably higher than the potential across a telephone set that is in use (5–12 V). Thus, the triac is switched only if the receivers of all the parallel telephones are on the hook. As soon as a receiver is lifted, the voltage across all sets drops from about 50 V to at most 12 V, so that no diac can be tripped. If, after the receiver has been lifted, the polarity of the line voltage reverse, the diac is not affected. When a call signal is received, all triacs conduct, so that the bell on each telephone will ring.
An added advantage of the present circuit is that it suppresses the parallel tinkling of telephones when a number is dialled on one of them (this happens only where older telephone exchanges are still in use).
The circuit is small enough to be built into the telephone line outlet box.

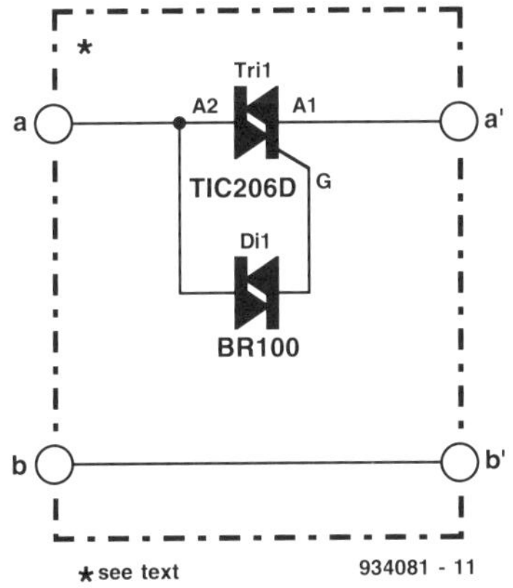

Low-drop a.c. switch for 12 V halogen lights

303 Because of their inertia, power consumption, contact wear, and often size, relays are not really suitable for switching alternating voltages. Moreover, they cannot be used in phase control circuits. Replacing them by triacs may give problems owing to the forward voltage drop across these devices.
A much better solution is the use of anti-series-connected SIPMOS transistors. Unfortunately, these need a control voltage that is isolated from the alternating voltage to be switched. This difficulty is, however, overcome by making use of the inverse diodes of the transistors as shown in the diagram.
In the off condition, when the optoisolator does not conduct, C_1 is charged during the negative half period (A positive, B negative) via D_3 and one of the inverse SIPMOS diodes. If the output is terminated, the capacitor will also be charged during the positive half period either via D_4 and the inverse diode of T_2 (load between A2 and B2), or via D_4 and the inverse diode of T_1 (load between A1 and B1).
The circuit is switched on by making the optoisolator conduct, whereupon the voltage across C_1 is connected to the gates of T_1 and T_2 via R_1 and the transistor in the optoisolator. Diode D_5 prevents the gate voltage becoming too high. When the circuit is on, C_1 continues to be charged during the negative half periods via D_3 and one of the SIPMOS transistors.
Diode D_2, in conjunction with D_1 and D_4, suppresses current peaks caused by the switching of inductive loads.

The circuit can switch alternating voltages of up to 45 V. Without a heat sink for the SIPMOS transistors, the maximum current should not exceed 3 A. For larger currents or when large inductive loads are switched regularly, a small heat sink should be used.

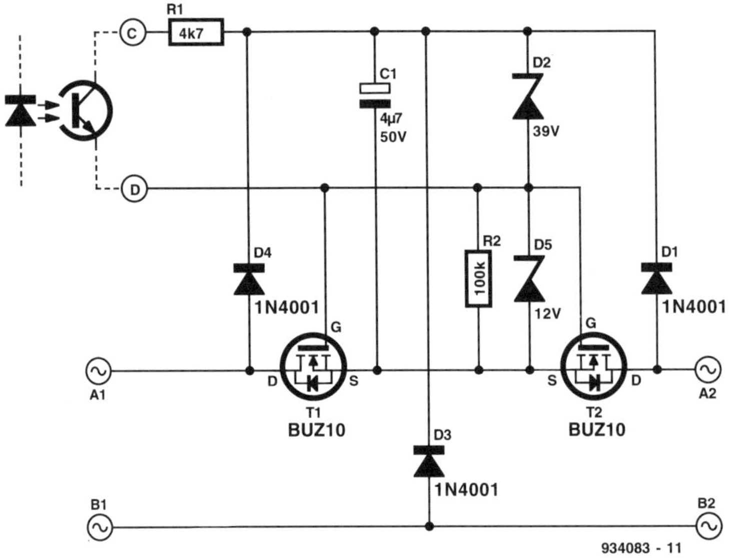

Temperature-to-current converter

304 The XTR103 is a temperature-to-current (T–I) converter that contains a linearization circuit for PT100 sensors. As usual in industrial applications, the measuring current is made available via the supply lines. The value of resistors R_2–R_4 ensures that the measuring current, I_o, rises from 4 mA to 20 mA when the temperature at the PT100 rises from 0 °C to 200 °C.

The resistance value, R_p, of a PT100 is calculated from

$$R_p = 100(1+3.90802 \cdot 10^{-3} \cdot T - 0.580195 \cdot 10^{-6} \cdot T^2).[\Omega]$$

At 200 °C, R_s = 175.8 Ω. When this value is connected to the input of the PT100 via a potentiometer, the maximum measuring current (±20 mA) is easily checked. In this way, the current can also be determined for other values of the PT100 resistance (read: temperature).

According to the manufacturer's data sheet, the supply voltage should be 9 V, but it is advisable to use a higher one, since there are always losses in the supply circuit. In the circuit diagram, R_s is the sum of the resistance of the supply lines, the value of the close-tolerance resistance used to convert the current into a voltage, and the internal resistance of an ammeter (if used). In the prototype, it was estab-

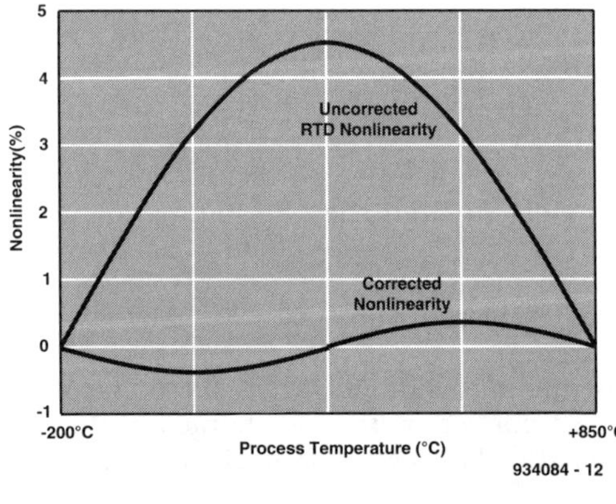

lished that R_S should not be higher than 400 Ω with a supply voltage of 15 V.
As already mentioned, the values of resistors R_2–R_4 specified are right for temperature measurements in the range 0–200 °C. These values should be altered to 100 Ω, 358 Ω, and 1200 Ω for 0–400 °C, and to 19 Ω, 184 Ω, and 838 Ω for –200 °C to 200 °C.

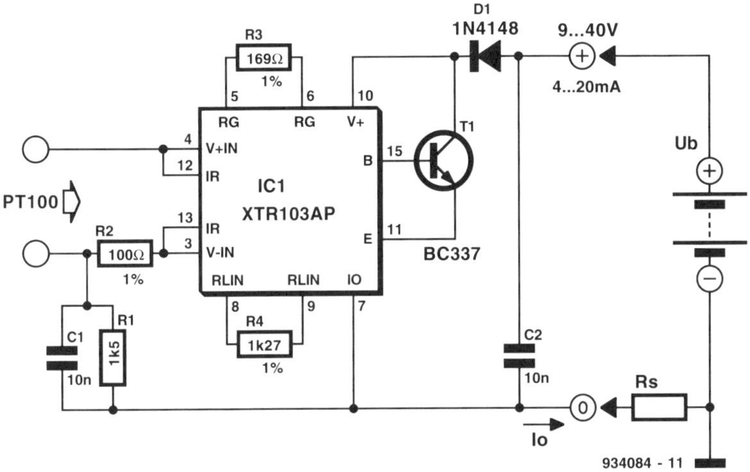

Frequency-operated switch

305 Voltage-operated switches are fairly commonplace, but frequency-operated ones are still rare. The one described here is based on a 4046, a PLL (phase-locked loop) that, internally, is largely digital.
The two frequencies that are to be compared are applied to pins 3 and 14 of IC_1. They must be rectangular with an amplitude equal to the supply voltage of 3–5 V. Their duty factor is not important, since the IC reacts only to leading transitions (edges). If f_1 is lower than f_2, the output goes low. If $f_1 = f_2$, a rectangular voltage, whose duty factor is stable and determined by the phase difference of the two signals, is present at pin 13. This voltage is converted into a direct voltage by R_1 and C_1, which is then applied to switching transistor T_1 via the source follower. If the voltage level is high enough, T_1 will switch on and energize the relay.
In theory, the error rate of the switch is 0, but in practice 0.1% must be allowed. The time constant R_1-C_1 must be about 10 times as long as the period of the input signal. Higher ratios delay the operation unnecessarily.
When the frequencies are almost identical, it may, in the worst case, take one period of the difference frequency before the circuit is enabled.
The circuit is suitable for operation from 3–15 V supplies, but make sure that the supply and relay voltages are the same. Transistor T_1 can switch up to 100 mA. The current drawn is about 0.5 mA plus the relay current.

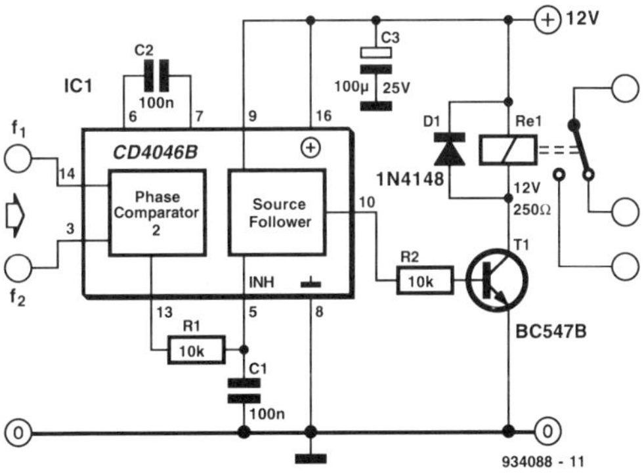